Jonathan Lytton

# ANNUAL REVIEW OF BIOPHYSICS AND BIOMOLECULAR STRUCTURE

## EDITORIAL COMMITTEE (1993)

JACQUELINE K. BARTON
CHARLES R. CANTOR
WAYNE L. HUBBELL
SUNG-HOU KIM
THOMAS D. POLLARD
ROBERT M. STROUD

Responsible for the organization of Volume 22
(Editorial Committee, 1992)

ROBERT L. BALDWIN (Guest)
CHARLES C. CANTOR
FREDERICK W. DAHLQUIST
DONALD M. ENGELMAN
WAYNE L. HUBBELL
SUNG-HOU KIM
MAURICIO MONTAL
THOMAS D. POLLARD
ROBERT M. STROUD (Guest)

Production Editor     AMANDA M. SUVER
Subject Indexer     STEVEN SORENSON

# ANNUAL REVIEW OF BIOPHYSICS AND BIOMOLECULAR STRUCTURE

VOLUME 22, 1993

DONALD M. ENGELMAN, *Editor*
Yale University

CHARLES R. CANTOR, *Associate Editor*
Center for Advanced Biotechnology

THOMAS D. POLLARD, *Associate Editor*
The Johns Hopkins University School of Medicine

ANNUAL REVIEWS INC.   4139 EL CAMINO WAY   P.O. BOX 10139   PALO ALTO, CALIFORNIA 94303-0897

**ANNUAL REVIEWS INC.**
Palo Alto, California, USA

COPYRIGHT © 1993 BY ANNUAL REVIEWS INC., PALO ALTO, CALIFORNIA, USA. ALL RIGHTS RESERVED. The appearance of the code at the bottom of the first page of an article in this serial indicates the copyright owner's consent that copies of the article may be made for personal or internal use, or for the personal or internal use of specific clients. This consent is given on the conditions, however, that the copier pay the stated per-copy fee of $2.00 per article through the Copyright Clearance Center, Inc. (21 Congress Street, Salem, MA 01970) for copying beyond that permitted by Section 107 or 108 of the US Copyright Law. The per-copy fee of $2.00 per article also applies to the copying, under the stated conditions, of articles published in any *Annual Review* serial before January 1, 1978. Individual readers, and nonprofit libraries acting for them, are permitted to make a single copy of an article without charge for use in research or teaching. This consent does not extend to other kinds of copying, such as copying for general distribution, for advertising or promotional purposes, for creating new collective works, or for resale. For such uses, written permission is required. Write to Permissions Dept., Annual Reviews Inc., 4139 El Camino Way, P.O. Box 10139, Palo Alto, CA 94303-0897 USA.

*International Standard Serial Number: 1056-8700*
*International Standard Book Number: 0-8243-1822-6*
*Library of Congress Catalog Card Number: 79-188446*

Annual Review and publication titles are registered trademarks of Annual Reviews Inc.

∞ The paper used in this publication meets the minimum requirements of American National Standard for Information Sciences—Permanence of Paper for Printed Library Materials, ANSI Z39.48-1984.

Annual Reviews Inc. and the Editors of its publications assume no responsibility for the statements expressed by the contributors to this *Review*.

TYPESET BY BPCC-AUP GLASGOW LTD., SCOTLAND
PRINTED AND BOUND IN THE UNITED STATES OF AMERICA

*Annual Review of Biophysics and Biomolecular Structure*
*Volume 22, 1993*

# CONTENTS

## STRUCTURAL PRINCIPLES

| | |
|---|---|
| Macromolecular Crowding: Biochemical, Biophysical, and Physiological Consequences, *Steven B. Zimmerman and Allen P. Minton* | 27 |
| The Control of Protein Stability and Association by Weak Interactions with Water: How Do Solvents Affect These Processes?, *Serge N. Timasheff* | 67 |
| Models of Lipid-Protein Interactions in Membranes, *Ole G. Mouritsen and Myer Bloom* | 145 |
| The Design of Metal-Binding Sites in Proteins, *Lynne Regan* | 257 |
| Hydrogen Bonding, Hydrophobicity, Packing, and Protein Folding, *George D. Rose and Richard Wolfenden* | 381 |
| Mechanisms of Membrane Fusion, *Joshua Zimmerberg, Steven S. Vogel, and Leonid V. Chernomordik* | 433 |

## STRUCTURE AND FUNCTION

| | |
|---|---|
| Prolyl Isomerase: Enzymatic Catalysis of Slow Protein-Folding Reactions, *Franz X. Schmid* | 123 |
| Functional Bases for Interpreting Amino Acid Sequences of Voltage-Dependent $K^+$ Channels, *Arthur M. Brown* | 173 |
| The Effects of Phosphorylation on the Structure and Function of Proteins, *L. N. Johnson and D. Barford* | 199 |
| The Structure of the Four-Way Junction in DNA, *David M. J. Lilley and Robert M. Clegg* | 299 |
| Structure and Function of Human Growth Hormone: Implications for the Hematopoietins, *James A. Wells and Abraham M. de Vos* | 329 |

## DYNAMICS

| | |
|---|---|
| Interaction of Proteins with Lipid Headgroups: Lessons from Protein Kinase C, *Alexandra C. Newton* | 1 |
| Realistic Simulations of Native-Protein Dynamics in Solution and Beyond, *V. Daggett and M. Levitt* | 353 |

*(continued)*

CONTENTS (*continued*)

Glycoprotein Motility and Dynamic Domains in Fluid Plasma
Membranes, *Michael P. Sheetz*   417

Fast Crystallography and Time-Resolved Structures,
*Janos Hajdu and Inger Andersson*   467

EMERGING TECHNIQUES

The Two-Dimensional Transferred Nuclear Overhauser Effect:
Theory and Practice, *A. P. Campbell and B. D. Sykes*   99

What Does Electron Cryomicroscopy Provide that X-Ray
Crystallography and NMR Spectroscopy Cannot?, *W. Chiu*   233

Artificial Neural Networks for Pattern Recognition in
Biochemical Sequences, *S. R. Presnell and F. E. Cohen*   283

INDEXES

Subject Index   499
Cumulative Index of Contributing Authors, Volumes 18–22   511
Cumulative Index of Chapter Titles, Volumes 18–22   513

# SOME RELATED ARTICLES IN OTHER *ANNUAL REVIEWS*

From the *Annual Review of Biochemistry*, Volume 62 (1993):
  *Cytoplasmic Microtubule-Associated Motors*, R. A. Walker and M. P. Sheetz
  *Analysis of Glycoprotein-Associated Oligosaccharides*, R. A. Dwek, C. J. Edge, D. J. Harvey, M. R. Wormald, and R. B. Parekh
  *Membrane Partitioning During Cell Division*, G. Warren
  *Biochemistry of Multidrug Resistance Mediated by the Multidrug Transporter*, M. M. Gottesman and I. Pastan
  *Conformational Coupling in DNA Polymerase Fidelity*, K. A. Johnson
  *Structure-Based Inhibitors of HIV-1 Protease*, A. Wlodawer and J. W. Erickson
  *Molecular Chaperone Functions of Heat-Shock Proteins*, J. P. Hendrick and F.-U. Hartl
  *Protein Tyrosine Phosphatases*, K. M. Walton and J. E. Dixon
  *New Photolabeling and Crosslinking Methods*, J. Brunner
  *The Structure and Biosynthesis of Glycosyl Phosphatidylinositol Protein Anchors*, P. T. Englund
  *Cognition, Mechanism, and Evolutionary Relationships in Aminoacyl-tRNA Synthetases*, C. W. Carter, Jr.
  *Pathways of Protein Folding*, C. R. Matthews
  *Structural and Genetic Analysis of Protein Stability*, B. W. Matthews
  *Determination of RNA Structure and Thermodynamics*, J. A. Jaeger, J. SantaLucia, Jr., and I. Tinoco, Jr.

From the *Annual Review of Cell Biology*, Volume 8 (1992):
  *Genetic Approaches to Molecular Motors*, S. A. Endow and M. A. Titus
  *The Differentiating Intestinal Epithelial Cell: Establishment and Maintenance of Functions Through Interactions Between Cellular Structures*, D. Louvard, M. Kedinger, and H. P. Hauri
  *Actin Filaments, Stereocillia, and Hair Cells: How Cells Count and Measure*, L. G. Tilney, M. S. Tilney, and D. J. DeRosier
  *Cadherins*, B. Geiger and O. Ayalon
  *Structural Framework for the Protein Kinase Family*, S. S. Taylor, D. R. Knighton, J. Zheng, L. F. Ten Eyck, and J. M. Sowadski
  *Structure and Function of the Nuclear Pore*, D. J. Forbes
  *Chromatin Structure and Transcription*, R. D. Kornberg and Y. Lorch

From the *Annual Review of Genetics*, Volume 26 (1992):
  *Communication Modules in Bacterial Signaling Proteins*, J. S. Parkinson and E. C. Kofoid
  *Genetics and Biogenesis of Bacterial Flagella*, R. Macnab
  *Genetics and Enzymology of DNA Replication in* Escherichia coli, T. A. Baker and S. H. Wickner

RELATED ARTICLES (*continued*)

From the *Annual Review of Microbiology*, Volume 47 (1993):
   *Enzymes and Proteins from Organisms That Grow Near and Above 100°C*,
      M. W. W. Adams

From the *Annual Review of Physical Chemistry*, Volume 43 (1992):
   *Protein-Water Interactions Determined by Dielectric Methods*, R. Pethig
   *Polymer Dynamics in Electrophoresis of DNA*, J. Noolandi
   *Computational Alchemy*, T. P. Straatsma and J. A. McCammon
   *Analysis of Femtosecond Dynamic Absorption Spectra of Nonstationary States*, W. T. Pollard and R. A. Mathies
   *Electron Density from X-Ray Diffraction*, P. Coppens

From the *Annual Review of Physiology*, Volume 55 (1993):
   *The Renal H-K-ATPase: Physiological Significance and Role in Potassium Homeostasis*, C. S. Wingo and B. D. Cain
   *Structural Changes Accompanying Memory Storage*, C. H. Bailey and E. R. Kandel
   *$Ca^{2+}$ Oscillations in Non-Excitable Cells*, C. Fewtrell
   *The Use of Physical Methods in Determining Gramicidin Channel Structure and Function*, D. D. Busath
   *The Intestinal $Na^+/Glucose$ Cotransporter*, E. M. Wright
   *Facilitated Glucose Transporters in Epithelial Cells*, B. Thorens
   *The Cystic Fibrosis Transmembrane Conductance Regulator*, J. R. Riorden
   *Formal Approaches to Understanding Biological Oscillators*, W. O. Friesen, G. D. Block, and C. G. Hocker
   *Molecular Approaches to Understanding Circadian Oscillations*, J. S. Takahashi, J. M. Kornhauser, C. Koumenis, and A. Eskin
   *Controlling Cell Chemistry with Caged Compounds*, S. R. Adams and R. Y. Tsien
   *Multimode Light Microscopy and the Dynamics of Molecules, Cells, and Tissues*, D. L. Farkas, G. Baxter, R. L. DeBiasio, A. Gough, M. A. Nederlof, D. Pane, J. Pane, D. R. Patek, K. W. Ryan, and D. L. Taylor

From the *Annual Review of Plant Physiology and Plant Molecular Biology*, Volume 43 (1992):
   *Spatial Organization of Enzymes in Plant Metabolic Pathways*, G. Hrazdina and R. A. Jensen
   *Superoxide Dismutase and Stress Tolerance*, C. Bowler, M. Van Montague, and D. Inzé
   *Structure and Function of Photosystem I*, J. H. Golbeck
   *Anion Channels in Plants*, S. D. Tyerman
   *Calcium-Modulated Proteins: Targets of Intracellular Calcium Signals in Higher Plants*, D. M. Roberts and A. C. Harmon
   *Regulation of Ribulose 1,5-Bisphosphate Carboxylase/Oxygenase Activity*, A. R. Portis, Jr.

ANNUAL REVIEWS INC. is a nonprofit scientific publisher established to promote the advancement of the sciences. Beginning in 1932 with the *Annual Review of Biochemistry*, the Company has pursued as its principal function the publication of high quality, reasonably priced *Annual Review* volumes. The volumes are organized by Editors and Editorial Committees who invite qualified authors to contribute critical articles reviewing significant developments within each major discipline. The Editor-in-Chief invites those interested in serving as future Editorial Committee members to communicate directly with him. Annual Reviews Inc. is administered by a Board of Directors, whose members serve without compensation.

1993 Board of Directors, Annual Reviews Inc.

J. Murray Luck, Founder and Director Emeritus of Annual Reviews Inc.
   *Professor Emeritus of Chemistry, Stanford University*
Joshua Lederberg, Chairman of Annual Reviews Inc.
   *University Professor, The Rockefeller University*
Richard N. Zare, Vice Chairman of Annual Reviews Inc.
   *Professor of Physical Chemistry, Stanford University*
Winslow R. Briggs, *Director, Carnegie Institution of Washington, Stanford*
W. Maxwell Cowan, *Howard Hughes Medical Institute, Bethesda*
Sidney D. Drell, *Deputy Director, Stanford Linear Accelerator Center*
Sandra M. Faber, *Professor of Astronomy, University of California, Santa Cruz*
Eugene Garfield, *President, Institute for Scientific Information*
William Kaufmann, *President, William Kaufmann, Inc.*
Daniel E. Koshland, Jr., *Professor of Biochemistry, University of California, Berkeley*
Donald A. B. Lindberg, *Director, National Library of Medicine*
Gardner Lindzey, *Director Emeritus, Center for Advanced Study in the Behavioral Sciences, Stanford*
Charles Yanofsky, *Professor of Biological Sciences, Stanford University*
Harriet A. Zuckerman, *Vice President, The Andrew W. Mellon Foundation*

---

Robert H. Haynes, Editor-in-Chief and President
John S. McNeil, Publisher and Secretary-Treasurer
William Kaufmann, Managing Editor

ANNUAL REVIEWS OF
Anthropology
Astronomy and Astrophysics
Biochemistry
Biophysics and Biomolecular Structure
Cell Biology
Computer Science
Earth and Planetary Sciences
Ecology and Systematics
Energy and the Environment
Entomology
Fluid Mechanics
Genetics
Immunology
Materials Science
Medicine
Microbiology
Neuroscience
Nuclear and Particle Science
Nutrition
Pharmacology and Toxicology
Physical Chemistry
Physiology
Phytopathology
Plant Physiology and
   Plant Molecular Biology
Psychology
Public Health
Sociology

SPECIAL PUBLICATIONS

Excitement and Fascination
   of Science, Vols. 1, 2,
   and 3

Intelligence and Affectivity,
   by Jean Piaget

For the convenience of readers, a detachable order form/envelope is bound into the back of this volume.

# INTERACTION OF PROTEINS WITH LIPID HEADGROUPS: Lessons from Protein Kinase C

*Alexandra C. Newton*

Department of Chemistry, Indiana University, Bloomington, Indiana 47405

KEY WORDS: phosphatidylserine, protein:lipid interactions, allostery

CONTENTS

| | |
|---|---|
| PERSPECTIVES AND OVERVIEW | 1 |
| LIPIDS AS REGULATORS OF PROTEIN FUNCTION | 2 |
| PROTEIN KINASE C | 4 |
|     *Function* | 4 |
|     *Structure* | 5 |
| INTERACTION OF PROTEIN KINASE C WITH LIPID HEADGROUPS | 7 |
|     *Measuring the Interaction of Protein Kinase C with Lipid Headgroups* | 7 |
|     *Specific Interaction with Diacylglycerol* | 9 |
|     *Specific Interaction with Phosphatidylserine* | 10 |
|     *Nonspecific Interaction with Acidic Headgroups* | 17 |
|     *Lipid-Induced Conformational Changes* | 18 |
|     *Model for Protein Kinase C:Lipid Interaction* | 19 |
| LESSONS FROM PROTEIN KINASE C | 21 |
| CONCLUSIONS | 22 |

## PERSPECTIVES AND OVERVIEW

The red cell membrane alone contains several hundred different lipid species (116). Why such diversity? One explanation is that this diversity allows considerable variation in the structure of the lipid matrix—for example, cone-shaped lipids are proposed to accumulate in areas of high surface curvature during fusion events (31). However, a major reason for the diversity in lipid structure is that it provides a remarkably sensitive, and in many cases specific, method for modulating the function of biological membranes through effects on membrane proteins. The typical membrane has a ratio of approximately 100 lipids per intrinsic protein, so that each protein is usually surrounded by several shells of lipid. This contact with the lipid matrix can have profound effects on the structure and function of proteins. Membrane proteins are sensitive to bulk properties of the lipid bilayer (e.g. fluidity) as well as to interactions with specific lipids (e.g. binding via specific headgroups). This review describes the exquisite regulation of one protein, protein kinase C, by lipids with the intention that it may serve as a paradigm for our understanding of how lipid headgroups regulate protein function.

## LIPIDS AS REGULATORS OF PROTEIN FUNCTION

Specificity in protein:lipid interactions has been described for an increasing number of diverse membrane proteins. Table 1 provides but a partial list of the numerous proteins that are preferentially activated (or inhibited) by specific phospholipid headgroups. Lipid headgroups alter the activity of transmembrane proteins such as transporters and receptors, peripheral membrane proteins, and cytoskeletal proteins. In addition to regulation by lipid headgroups, many proteins are sensitive to acyl chain composition (length and degree of saturation), bilayer fluidity, bilayer thickness, and lipid backbone structure, as well as to other lipids such as cholesterol and sphingolipids (44).

Selectivity in the interaction of membrane proteins has not been studied extensively because many membrane proteins retain biological activity in detergent (bilayer structure is not required) or are active when reconstituted into any bulk lipid matrix (a specific lipid structure is not required). However, systematic variation of lipid species usually reveals that the protein has distinct preferences for some lipids. A good example is the insulin receptor: this transmembrane protein contains a tyrosine kinase activity that is catalytically active when detergent-solubilized. However, addition of diacylglycerol to receptor:detergent micelles results in enhanced activity, whereas phosphatidic acid results in dramatic inhibition of activity (3).

Whereas most proteins examined display selectivity for certain lipid headgroups (commonly acidic), two display an absolute specificity: protein kinase C requires phosphatidylserine (PS) (68), and $\beta$-hydroxybutyrate dehydrogenase requires phosphatidylcholine, for activity (30, 104).

The lipid requirement for activity generally correlates with preferential binding of the particular lipid to the protein. Binding is monitored by spectroscopic techniques such as NMR, EPR, and fluorescence, by direct binding measurements using exclusion chromatography or centrifugation,

**Table 1** Examples of proteins whose functions are regulated by lipid headgroups

| Protein | Lipid selectivity[a] | Reference |
|---|---|---|
| **Transporters** | | |
| Glucose transporter | PS > PA > PG > PC[b] | 115 |
| Lactose carrier (*E. coli*)[c] | PE, PS | 25 |
| $Ca^{2+}$-ATPase | Acidic | 23 |
| $(Na^+, K^+)$-ATPase | Acidic | 62 |
| Putative PS flippase | PS > PE ~ PI ~ PG ~ CL > PC | 81, 121 |
| **Receptors** | | |
| Nicotinic acetylcholine receptor | Acidic | 42 |
| Rhodopsin | PE > PC | 120 |
| Vitronectin receptor | Acidic | 27 |
| Insulin receptor | PA (inhibit) | 3 |
| **Intrinsic membrane enzymes** | | |
| Adenylate cyclase | Acidic (inhibit) | 50 |
| Cytochrome *c* oxidase | CL | 117 |
| Cytochrome P-450$_{scc}$ | CL > PG > PC | 97 |
| Neutrophil respiratory burst oxidase | PS > CL > PE > PI | 114 |
| **Peripheral membrane enzymes** | | |
| β-Hydroxybutyrate dehydrogenase | PC (specific requirement) | 30 |
| Calcineurin | PG > PI > PS | 99 |
| Casein kinase 1 | $PIP_2$ > PIP > PA (inhibit) | 19 |
| Glycerol-3-phosphate acyltransferase (*E. coli*) | CL > PG > PE | 106 |
| Phosphatidylinositol-4-phosphate kinase | PA | 80 |
| Protein kinase C | PS (specific requirement) | 68 |
| SecA | Acidic | 72 |
| **Cytoskeletal proteins** | | |
| Profilin | $PIP_2$ > PIP > PI (inhibit) | 67 |
| Myosin | Acidic | 122 |

[a] Unless otherwise indicated, these lipids stimulate catalytic activity.
[b] Abbreviations: CL, cardiolipin; PA, phosphatidic acid; PC, phosphatidylcholine; PE, phosphatidylethanolamine; PG, phosphatidylglycerol; PI, phosphatidylinositol; PIP, phosphatidylinositol monophosphate; $PIP_2$, phosphatidylinositol bisphosphate; PS, phosphatidylserine.
[c] Proteins are from eukaryotic membranes unless otherwise noted.

or by the observation of lipid-induced conformational changes in the protein. The reader is referred to reviews of specificity in lipid binding to proteins (35, 73). EPR is a most effective tool for resolving selectivity in the association of lipids with membrane proteins and is also useful for determining the number of motionally restricted lipids interacting with a protein (73). For example, EPR measurements have revealed that the order of affinity of the integral membrane protein cytochrome P–450$_{scc}$ for lipid is: CL > PG > PS ~ PC > PE > PI (see Table 1, footnote b, for definitions of these abbreviations), which is similar to the order of effectiveness of these lipids in activating the enzyme (97).

For proteins whose function cannot be readily assayed, spectroscopic techniques have provided important information about selectivity for lipid headgroups. For instance, EPR and $^2$H- and $^{31}$P-NMR have established that myelin basic protein, an extrinsic membrane protein, has a higher affinity for acidic headgroups compared with the choline headgroup (15, 110). Headgroup selectivity for activity and binding supports the possibility that discrete lipid-binding sites can exist on membrane proteins.

Interaction with specific lipids provides a mechanism for fine-tuning membrane function (e.g. by modulating enzyme activity) or membrane structure (e.g. by regulating interactions with cytoskeletal proteins). If concentrations of the regulatory lipid fluctuate in response to external stimuli, short-term regulation is possible. Several lipid species are metabolized in response to changes in the cell's environment, and we are only now beginning to realize the importance of lipids and lipid turn-over in modulating the function of proteins involved in the transfer and processing of information.

## PROTEIN KINASE C

### Function

The discovery in 1977 of the $Ca^{2+}$/lipid-dependent protein kinase C provided the first indication of the importance of lipid in sensory transduction—both as second messenger and as a regulator of function (90, 91). Protein kinase C is a ubiquitous protein that plays a crucial role in the transduction of extracellular signals provoking phospholipid turnover (Figure 1). Best characterized is the receptor-mediated hydrolysis of phosphatidylinositol bisphosphate to generate two second messengers: the water-soluble headgroup, inositol trisphosphate, and the lipid backbone, diacylglycerol. Inositol trisphosphate mobilizes intracellular $Ca^{2+}$, causing cytosolic protein kinase C to translocate to the plasma membrane, where it is activated by diacylglycerol. Activity also depends on phosphatidylserine (PS) (48, 56, 87), an acidic aminophospholipid located exclusively on the inner leaflet of the plasma membrane. The activated kinase phosphorylates various proteins (see 90), in addition to catalyzing its autophosphorylation by an intrapeptide reaction (86).

Protein kinase C is also activated by phorbol esters, potent tumor promoters, which replace the diacylglycerol requirement for activity (24). The effects of many biological agonists are mimicked by phorbol esters, implicating a role for protein kinase C in the transduction of such signals.

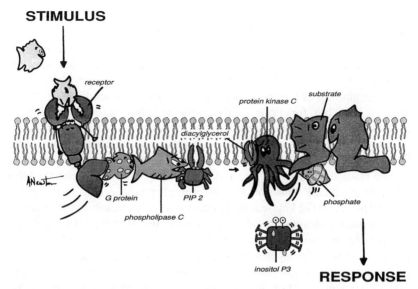

*Figure 1* Cartoon representation of the events involved in the transduction of signals that promote lipid hydrolysis. The receptor-mediated activation of phospholipases that are coupled to G proteins is illustrated; however, some phospholipases interact directly with receptors.

However, protein kinase C is not "the" phorbol ester receptor; at least two proteins with no kinase activity, n-chimaerin and the *unc-13* gene product of *Caenorhabditis elegans*, bind phorbol esters (2, 74).

A multiplicity of functions, from short-term changes such as alterations in membrane permeability to long-term changes involving regulation of cell growth, have been ascribed to protein kinase C (90). In addition, mounting evidence suggests that a major role of protein kinase C is to desensitize cells to extracellular information by affecting receptor function (4). For example, phosphorylation of the visual receptor, rhodopsin, by protein kinase C has been proposed to play a role in light adaptation (89).

## Structure

Protein kinase C is actually a family of functionally similar polypeptides: 10 different isozymes of the protein have been described to date (13). All isozymes are regulated by PS and diacylglycerol, but only some members of the family are regulated by $Ca^{2+}$ ($\alpha$, $\beta$I, $\beta$II, and $\gamma$). Each isozyme is a single polypeptide with a molecular weight of approximately 80 kDa: the kinase region is located in a 45-kDa catalytic domain, whereas the regulatory membrane-binding region is confined to the 35-kDa N-terminus of the polypeptide. The two functional domains can be separated by mild proteolysis at a hinge region comprising residues that differ amongst isozymes. The $Ca^{2+}$-regulated protein kinases C contain four conserved regions (Figure 2, open boxes) interspersed with five smaller variable sequences.

The beginning of the first conserved domain, C1, encodes a stretch of amino acids that functions as the pseudosubstrate domain of protein kinases C (49). Allosterically regulated protein kinases have been proposed to have a stretch of peptide that is structurally similar to a consensus phosphorylation-site sequence except that generally a nonphosphorylatable residue replaces the hydroxyl-containing residue—hence the term "pseudosubstrate." This domain has been proposed to bind to the active site and maintain the kinase in an inactive state by steric inhibition. Binding of activators is thought to induce a conformational change that displaces the pseudosubstrate domain and thus allows access to substrates. All protein kinase C isozymes, including yeast protein kinase C, photoreceptor-specific *Drosophila* protein kinase C, and all the known mammalian isozymes contain a pseudosubstrate sequence near their amino terminus.

Following the pseudosubstrate domain, protein kinase C has two adjacent cysteine-rich sequences that are essential for phospholipid-dependent phorbol ester-binding (21, 92). Mutants lacking both cysteine-rich regions do not bind phorbol esters, although one of the two domains is sufficient for binding (21, 92). In addition, recombinant C1 domain binds phorbol esters in a $Ca^{2+}$-independent but phospholipid-dependent manner (92).

The regulatory domain has recently been shown to bind four atoms of zinc (52, 100), and the conserved cysteines in each of the tandem repeats have been implicated in coordinating two zinc atoms per repeat. N-chima-

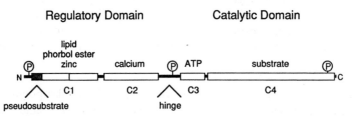

*Figure 2* Schematic of the structure of $Ca^{2+}$-regulated protein kinases C ($\alpha$, $\beta$I, $\beta$II, $\gamma$) showing regions involved in binding lipid, phorbol esters, and zinc (C1), calcium (C2), ATP (C3), and substrates (C4). Also indicated are the pseudosubstrate sequence, the proteolytically labile hinge between the regulatory and catalytic domains, and the autophosphorylation sites on the $\beta$II isozyme (41) of protein kinase C. $Ca^{2+}$-independent protein kinases C lack the calcium-binding region.

erin, which possesses a similar cysteine-rich domain, also binds zinc, and recent evidence suggests that removal of the zinc in this protein and in protein kinase C inhibits phorbol ester binding (1).

The C2 region of protein kinase C mediates calcium-binding; $Ca^{2+}$-independent isozymes do not contain this region, and deletion of most of C2 from a $Ca^{2+}$-dependent isozyme resulted in an enzyme with full activity in the absence of $Ca^{2+}$ (55). The first half of the C2 region is similar to sequences in phospholipase C, the synaptic vesicle protein p65, a GTPase activating protein, and a cytosolic phospholipase $A_2$ that translocates to acidic membranes in response to $Ca^{2+}$ (26, 98).

The catalytic domain of protein kinase C contains two conserved regions: C3, which contains the ATP-binding consensus sequence, and the much larger C4 region, which has determinants responsible for protein substrate binding (46, 59). The recent elucidation of the crystal structure of the catalytic subunit of the cAMP-dependent kinase has broad implications for understanding the structure of the protein kinase C catalytic core (64).

The lipid-binding domain of protein kinase C regulates the catalytic function of the enzyme, and proteolysis at the hinge region results in a constitutively active kinase. In addition to regulating catalytic activity, the regulatory domain may also influence the substrate specificity of protein kinase C: a mutant enzyme containing the regulatory domain of the ε isozyme and the catalytic domain of the γ isozyme displayed the same substrate specificity as the ε isozyme (96). Furthermore, recent evidence suggests that the regulatory domain is involved in lipid-mediated protein:protein recognition. The binding of protein kinase C to cytoskeletal proteins (54, 77) has been proposed to arise, in part, from a lipid-dependent interaction with the pseudosubstrate region (S. L. Hyatt, L. Liao & S. Jaken, in preparation) or the C2 domain (77a).

Protein:lipid interactions are central to the regulation and action of protein kinase C. The following section describes how lipid headgroups regulate both the structure and function of protein kinase C.

# INTERACTION OF PROTEIN KINASE C WITH LIPID HEADGROUPS

## Measuring the Interaction of Protein Kinase C with Lipid Headgroups

The interaction of protein kinase C with lipid headgroups has been studied most extensively with small or large unilamellar vesicles or Triton X-100 mixed micelles. Although the former most closely approximate a biological bilayer, Triton X-100 mixed micelles have been invaluable in studying the stoichiometry and specificity of the interaction of protein kinase C with lipids. In most instances, results obtained from both systems are qualitatively similar; the differences are discussed below.

Triton X–100 forms defined micelles containing an average of 140 molecules of detergent per micelle at 25°C. Robson & Dennis (102) established that, when detergent is in excess, addition of lipid increases the size of the micelles in direct proportion to the added lipid without changing the number of detergent molecules per micelle. Thus, the number of lipid molecules per micelle can be approximated by Equation 1:

$$\text{molecules of lipid/micelle} = \frac{140(x_{\text{lipid}})}{1-(x_{\text{lipid}})}, \qquad 1.$$

where $x$ is the mole fraction of lipid. For example, a micelle containing 10 mol% PS and 5 mol% diacylglycerol will contain approximately 16 molecules of PS and 8 molecules of diacylglycerol at 25°C. In the mid 1980s, Bell and coworkers established the utility of mixed micelles in studying protein kinase C when they discovered that one monomer of protein kinase C binds to one micelle (48) (Figure 3). Systematic variation of the lipid concentration and composition has provided information on the number and species of lipids necessary to activate one monomer of protein kinase C.

The difficulty in obtaining sufficient quantities of purified isozymes of protein kinase C has caused investigations of protein kinase C's interaction with phospholipid to focus on activity requirements. However, the pro-

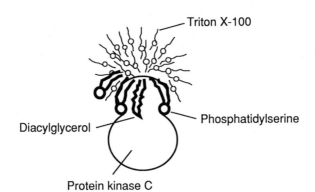

*Figure 3* Cartoon representation of a Triton X–100:lipid mixed micelle showing that one monomer of protein kinase C binds to one micelle. The illustration is roughly to scale since the diameter of a headgroup is approximately 1 nm and that of a Triton micelle is on the order of 8 nm (102), a dimension not inconsistent with that of a spherical protein having the molecular weight of the regulatory domain of protein kinase C (35 kDa).

duction of isozymes in the baculovirus-insect cell expression system (22) and the development of binding assays that require less than microgram quantities of protein kinase C (94; M. Mosior & R. M. Epand, submitted) have allowed recent progress in understanding the binding of $Ca^{2+}$-dependent protein kinases C to lipid.

ACTIVITY Protein kinase C autophosphorylates by an intrapeptide reaction, so that this activity depends only on the kinase:lipid interaction rather than the three-component substrate:kinase:lipid interaction (86). Because most protein kinase C substrates are basic, they must be targeted to acidic membranes to serve as substrates for protein kinase C (7, 88). Autophosphorylation also avoids any substrate-induced aggregation of micelles or vesicles. Nonetheless, the lipid dependences for autophosphorylation and phosphorylation of the commonly used substrate histone are similar at saturating diacylglycerol and $Ca^{2+}$ concentrations, and both serve as accurate measures of protein kinase C's interaction with PS at saturating $Ca^{2+}$ concentrations.

BINDING The binding of protein kinase C to model membranes has been investigated using resonance energy transfer from protein kinase C to dansyl phosphatidylethanolamine–containing vesicles. Although this method has provided important information on the binding of protein kinase C to model membranes (5), vesicles contain added surface charge from the fluorescent lipid (typically 10 mol%) that contributes to the binding of protein kinase C to the vesicles (M. Mosior & R. M. Epand, submitted). Binding has also been assessed by monitoring the order-of-magnitude increase in the rate of proteolysis at the hinge region of protein kinase C that accompanies membrane binding (94). Binding data acquired using the proteolytic sensitivity method or resonance energy transfer are qualitatively similar, except that slightly less cooperativity is observed using the latter method, which probably results from the contribution to binding from the negative charge of the probe (94).

## Specific Interaction with Diacylglycerol

All isozymes of protein kinase C require diacylglycerol for activity. Studies with PS vesicles suggest that one molecule of diacylglycerol binds to one molecule of protein kinase C (65), consistent with the stoichiometric requirement for diacylglycerol in activating protein kinase C in detergent:lipid mixed micelles (48). Similar stoichiometries for the binding and activation by phorbol esters, which are competitive inhibitors of diacylglycerol binding, have also been reported (47, 61). The existence of two phorbol ester–binding sites in the C1 region of protein kinase C, however, presents the possibility that the enzyme contains two diacylglycerol binding sites (21).

The activation of protein kinase C by 1,2-sn-diacylglycerol displays strict stereospecificity and tolerates little variation in the structure of the diacylglycerol backbone and ester linkages to the fatty acids (101). The importance of the hydroxyl group is illustrated by the inability of analogues with ether, halogen, methyl ester, or thiol substitutions at the hydroxyl position to activate protein kinase C (43). Similarly, inactive lipids are generated by substituting ether, thioether, or amide linkages for the ester linkages to the fatty acids or by altering the diacylglycerol backbone (43, 78). The position of the acyl chains is also critical: 1,3-diacylglycerol and 2,3-sn-diacylglycerol, the enantiomer of 1,2-sn-diacylglycerol, are inactive (17).

Variation in acyl chain length or saturation only modestly affects the ability of diacylglycerols to activate protein kinase C; the most important determinant is that the molecules are sufficiently hydrophobic to partition into membranes (33, 79). The lack of specificity in acyl chains has proven to be of great use in activating protein kinase C in cells because the relative water solubility of molecules such as 1-oleoyl-2-acetyl-sn-glycerol and 1,2-sn-dioctanoylglycerol allows efficient delivery to intact cells (33, 57).

The strict specificity in structural requirements, including stereospecificity, indicates that diacylglycerol interacts with specific structural elements of protein kinase C. Based on molecular requirements for activation by diacylglycerols and tumor promoters such as phorbol esters, three hydrophilic groups have been proposed to H-bond with protein kinase C: in the diacylglycerol structure, these three groups are the hydroxyl group and the two carbonyl oxygens (85). A hydrophobic membrane anchor is provided by the acyl chains.

What is the role of diacylglycerol? First, diacylglycerol causes a marked increase in the affinity of protein kinase C for PS. Second, diacylglycerol and PS induce a conformational change that exposes the pseudosubstrate domain of protein kinase C. Both of these functions of diacylglycerol are discussed in greater detail in following sections.

## Specific Interaction with Phosphatidylserine

Shortly after the discovery of protein kinase C, Takai et al (113) noted that a "membrane factor" was required for activity. This membrane factor is, of course, PS. Investigations with bilayers, monolayers, and detergent:lipid mixed micelles have established that protein kinase C requires PS for optimal activity (17, 48, 56, 112), with lipids sharing some of the functional properties of the serine headgroup (negative charge or amine group) able to enhance activity (56, 68, 87).

The majority of the information on the specificity and stoichiometry of protein kinase C's interaction with phospholipid has been obtained by systematically varying the number and species of phospholipids in Triton X–100 mixed micelles, work pioneered by Bell and coworkers (48). Multiplicity, specificity, and cooperativity in the interaction of protein kinase C with PS has been observed by examining the dependence of protein kinase C activity and binding on the PS content of Triton X–100 mixed micelles containing diacylglycerol, as shown in Figure 4 and discussed below.

MULTIPLICITY  Two points are worth noting regarding the PS-dependent interaction of protein kinase C with mixed micelles at saturating $Ca^{2+}$ and diacylglycerol concentrations (Figure 4). First, activity (48, 87) and binding (94, 95) plateau when ⩾12 molecules of PS are included in micelles, indicating an upper limit on the number of PS molecules interacting with protein kinase C. Second, the PS dependence of activity displays steep sigmoidal kinetics, with Hill coefficients on the order of 10 calculated for the $Ca^{2+}$-dependent isozymes (22, 87). Hill coefficients indicate the

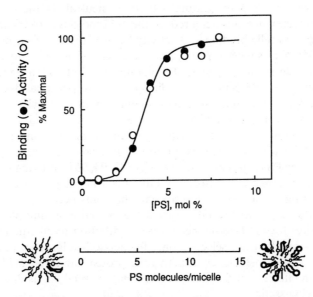

*Figure 4* The interaction of protein kinase C with PS displays sigmoidal kinetics. Binding and activity were measured as a function of the mol% of PS in Triton X–100 mixed micelles containing 5 mol% diacylglycerol. The ruler under the figure relates the mol% to the approximate number of PS molecules per micelle. Reproduced from Ref. 94 with permission.

minimum number of interacting subunits (in this case PS binding sites), so that at least this number of lipid molecules interact with protein kinase C. Thus, protein kinase C is activated by multiple PS molecules.

SPECIFICITY   The activation of protein kinase C displays remarkable specificity for the L-serine headgroup when assayed using mixed micelles. Extensive work by Lee & Bell (68) has revealed that alterations in the stereochemistry of the serine headgroup or in the distance between carboxyl and amine groups, or removal of one or more functional groups, result in phospholipids that cannot significantly activate protein kinase C. Particularly striking is the ineffectiveness of lipids with the D-serine headgroup or the L-homoserine headgroup (differing from L-serine by one additional methylene) to activate the enzyme. Thus, strict geometric and steric constraints regulate the productive binding of the lipid headgroup to specific sites on protein kinase C. Curiously, lysoPS is a poor activator of protein kinase C, suggesting that the interfacial conformation of the headgroup is important in allowing it to interact productively with protein kinase C (68).

Equally surprising is the relative inefficiency of lipids bearing some of the functional groups of the serine headgroup in reducing the number of PS molecules required for maximal activity. At neutral pH, the amino lipid phosphatidylethanolamine can reduce the $\geqslant 12$ molecules of PS necessary for maximal micelle binding and activity by only three or four molecules (70, 87, 94). This suggests that the amine group of phosphatidylethanolamine can interact with putative PS binding sites but with a significantly lower affinity than PS. Similarly, acidic lipids such as phosphatidylglycerol, phosphatidylinositol, phosphatidylpropionate, phosphatidylethanol, and the phosphoinositides are able to reduce the number of PS molecules required for full enzymatic activity, generally in proportion to their net charge (69, 70, 87). Interestingly, phosphatidic acid is relatively ineffective at altering the PS requirement for activity (87, 94, 95), although this result has not been observed consistently (70). Even in the presence of excess nonactivating acidic lipids, close to half the lipids necessary for activity must be PS (70). These results support the existence of multiple, specific binding sites for the L-serine headgroup in addition to an equal number of interactions that are selective only for acidic headgroups. In marked contrast to their inefficiency at activating protein kinase C, acidic lipids provide an effective surface for protein kinase C binding, as discussed in the following section.

At concentrations of PS approximating 15 mol% of the total lipid found on the cytoplasmic leaflet of many plasma membranes, activity studies with vesicles also reveal specificity for PS (95). However, at high relative lipid concentrations ($\geq 30$ mol%), lipids sharing some of the functional properties of PS result in partial catalytic activity of protein kinase C (6; M. Mosior & R. M. Epand, submitted). For example, protein kinase C can bind to, and be activated by, vesicles containing high concentrations of phosphatidylethanolamine (40 mol%) at a pH (7.9) where the lipid headgroups are partially negatively charged and at millimolar $Ca^{2+}$ concentrations (12). While such conditions are clearly not physiologically relevant, these results indicate that protein kinase C can be forced into an active conformation at high concentrations of lipids that have a low affinity for PS binding sites.

COOPERATIVITY  The high cooperativity in the interaction of protein kinase C with PS in mixed micelles requires diacylglycerol. Figure 5 shows that only modest sigmoidal binding curves are observed in the absence of diacylglycerol, indicating that diacylglycerol increases the affinity of protein kinase C for PS (95). Similarly, $Ca^{2+}$ increases the degree of cooperativity in the binding of protein kinase C to PS (95).

Protein kinase C also interacts cooperatively with PS in bilayers (87, 95; M. Mosior & R. M. Epand, submitted). However, the degree of cooperativity observed with bilayer PS, although still significant, is lower ($n \approx 5$) (88, 95; M. Mosior & R. M. Epand, submitted; A. C. Newton, in preparation) compared with the high cooperativity observed in the mixed-micelle system ($n \approx 10$). Differences in the accessibility of PS to protein kinase C in bilayers (where the diluent is phosphatidylcholine) compared with micelles (where the diluent is the much smaller Triton X-100) may account for the reduced cooperativity. Alternatively, structural elements of the bilayer may decrease the degree of interaction between PS binding sites. Nonetheless, the binding and activation of protein kinase C is cooperatively regulated by the mole fraction of PS in bilayers.

What is the physical basis for the steep sigmoidal binding curves observed in the presence of saturating diacylglycerol and $Ca^{2+}$ (Figure 4)? The sigmoidal binding curves are consistent with the probability of binding PS increasing as more of this lipid binds, and suggest that activity is linearly related to the amount of PS bound to protein kinase C. This is in marked contrast to the lipid-dependence for activity of the $(Na^+, K^+)$-ATPase, a protein that is active catalytically only when most of the lipid binding sites are occupied. In this case, lipid appears to bind one at a time with equal affinity, but apparent cooperativity is observed because partial occupancy of lipid sites does not result in partial activity (103). For protein kinase C, the sigmoidal curves that describe the PS-dependence for activity appear to reflect cooperativity in lipid binding.

Several models can account for the sigmoidal binding curves. Three of

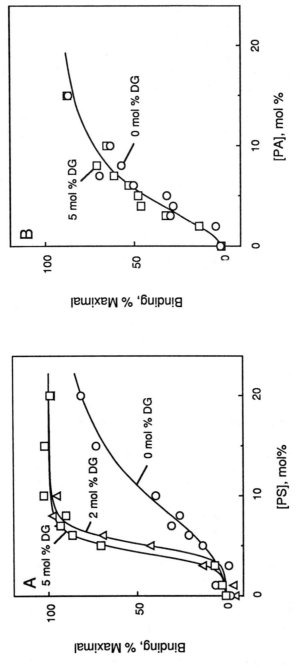

*Figure 5* Diacylglycerol (DG) increases the affinity of protein kinase C for PS but not the nonactivating acidic lipid, phosphatidic acid (PA). Reproduced from Ref. 95 with permission.

these are summarized in Figure 6 and are described below. First, sigmoidal binding kinetics could arise because the concentration of PS sensed by soluble protein kinase C is several orders of magnitude lower than that in the local environment surrounding micelle-bound protein kinase C; thus, the probability of binding PS molecules increases after the first one has bound. This reduction in dimensionality accounts for the sigmoidal curves describing the binding of positively charged peptides to negatively charged membranes (83). Two results indicate that the sigmoidal kinetics do not arise because of a reduction in dimensionality. First, protein kinase C bound to acidic micelles or vesicles (containing the nonactivating lipid, phosphatidic acid) continues to be cooperatively activated by PS (94).

1. Local concentration of PS increases after binding first PS; binding constants for each PS are the same

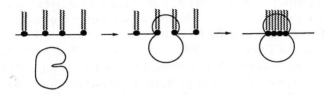

2. Protein kinase C recognizes a pre-formed domain of PS

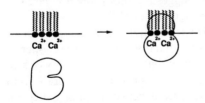

3. Binding constants for each PS increase as more lipid-binding sites become occupied

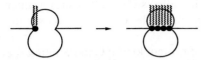

*Figure 6* Possible models accounting for the sigmoidal kinetics that describe the binding of protein kinase C to PS in the presence of diacylglycerol. *Model 1*: After the first PS molecule is bound, and protein kinase C is recruited to the membrane, apparent cooperativity in binding subsequent PS molecules is observed because the local concentration of PS sensed by protein kinase C has increased by several orders of magnitude. In this reduction-in-dimensionality model, the binding constants for individual PS molecules are the same and sites are noninteracting. *Model 2*: Protein kinase C binds to a preformed domain of PS. *Model 3*: Binding of the first PS molecule increases the affinity of other lipid-binding sites for PS, consistent with true cooperativity in lipid binding.

A slight decrease in cooperativity is observed under these conditions, suggesting that a reduction in dimensionality provides a minor contribution to the sigmoidal kinetics and/or that the acidic phosphatidic acid decreases the strength of PS:protein or PS:PS interactions (87). Second, the highly cooperative interaction with PS requires diacylglycerol, indicating that surface charge and reduction in dimensionality are not simply responsible for the apparent cooperativity. However, the modest cooperativity observed in the absence of diacylglycerol may arise because of these factors.

A second explanation for the apparent cooperativity is that the first PS molecules may be inaccessible to protein kinase C. This could arise if the enzyme recognizes a preformed domain of this lipid, perhaps in the form of a $Ca^{2+}$:$(PS)_n$ complex. However, although the aggregation of PS-containing bilayers by $Ca^{2+}$ is well documented (36, 40, 76), there is no evidence for aggregation within the same bilayer at the $Ca^{2+}$ concentrations and mole fractions of PS employed in protein kinase C experiments (17, 53). Furthermore, $Ca^{2+}$-independent isozymes of protein kinase C are cooperatively activated by PS in the absence of $Ca^{2+}$ (105), and different isozymes of protein kinase C display varying PS dependencies (22, 87), indicating that the PS requirement for activity is an intrinsic property of the enzyme rather than a property of the membrane or micelle structure. Lastly, PSs with different affinities for $Ca^{2+}$ bind protein kinase C with the same $Ca^{2+}$ dependence, suggesting that protein kinase C does not recognize a preformed $Ca^{2+}$:$(PS)_n$ complex (60).

The model most consistent with experimental data is that protein kinase C has multiple, specific PS binding sites that interact with true cooperativity (see 66) in the presence of diacylglycerol: that is, the probability of binding increasing numbers of PS molecules increases because of alterations in the affinity of binding sites for PS. Interactions between binding sites appear to be weak in the absence of diacylglycerol, suggesting that diacylglycerol induces a conformational change that allows PS binding sites to interact. Evidence for this conformational change is discussed in a later section.

OTHER PARAMETERS  The binding of protein kinase C to PS in micelles or vesicles is relatively insensitive to ionic strength in the presence (7, 95), but not the absence (A. C. Newton, in preparation), of diacylglycerol. This suggests that diacylglycerol promotes a specific interaction between protein kinase C and PS that is driven by forces other than electrostatic ones. Consistent with this, the diacylglycerol-dependent activation by PS is relatively insensitive to surface charge contributed by nonactivating lipids such as phosphatidic acid or lysoPS (70, 94, 95).

The role of $Ca^{2+}$ in regulating the interaction of $Ca^{2+}$-dependent protein kinases C with PS remains to be established. Micelle-binding studies have revealed that $Ca^{2+}$ increases the affinity of protein kinase C for PS (95). Recent reports indicate that the $\alpha$ and $\beta$ isozymes bind multiple $Ca^{2+}$ ions, with the former isozyme containing an additional high-affinity $Ca^{2+}$-binding site (75). Binding of multiple $Ca^{2+}$ ions depends on membrane binding (10). Sando and coworkers have proposed that the multiple sites mediate $Ca^{2+}$-dependent membrane association, whereas the high-affinity site in the $\alpha$ isozyme is located at or near the active site (75).

Protein kinase C activity was shown recently to depend on the degree of unsaturation of phospholipid bilayers composed of PS and phosphatidylcholine, with little effect on whether the unsaturation is of fatty acids of PS or phosphatidylcholine (16). Based on these data, Bolen & Sando (16) proposed that headgroup spacing may play an important role in allowing protein kinase C to interact effectively with the membrane, consistent with suggestions that membrane structure affects protein kinase C activity (37).

In summary, binding data indicate that diacylglycerol induces specificity and cooperativity in the interaction of protein kinase C with the L-serine headgroup. Multiple PS molecules are cooperatively sequestered around protein kinase C in the presence of saturating diacylglycerol, with lipids that share either the amine functionality or negative charge able to substitute for only some of the multiple PS molecules necessary for optimal activity in detergent:lipid mixed micelles.

## Nonspecific Interaction with Acidic Headgroups

In marked contrast to the extraordinary specificity for the L-serine headgroup for enzymatic activity, protein kinase C lipid binding displays selectivity only towards the negative charge of the headgroup (5, 68, 95). For example, in the absence of diacylglycerol, approximately twice as much of the monovalent acidic PS is required to mediate micelle binding compared with the divalent acidic phosphatidic acid (see Figure 5).

The interaction with nonactivating acidic lipids is similar to the nonspecific interaction with PS observed in the absence of diacylglycerol but differs from the specific PS interaction induced by diacylglycerol in three important aspects. First, the binding to nonactivating acidic lipids (including PS if diacylglycerol is absent) is sensitive to ionic strength whereas the specific binding to PS (in the presence of diacylglycerol) is unaffected by up to 300 mM NaCl (7, 95; A. C. Newton, in preparation). Second, neither diacylglycerol nor $Ca^{2+}$ affect the kinetics of binding of nonactivating acidic lipids to protein kinase C, whereas both second messengers cause a marked increase in the sigmoidal kinetics describing the diacylglycerol-mediated specific interaction of protein kinase C with PS (95). Third, binding to nonactivating acidic lipids does not induce the conformational change necessary for enzymatic activity (following section).

The sensitivity to ionic strength and surface charge suggest that electrostatic interactions promote the binding of protein kinase C to acidic membranes. In this regard, binding studies with a peptide derived from the pseudosubstrate sequence of protein kinase C have led Mosior & McLaughlin (82) to propose that the interaction of basic residues in the pseudosubstrate domain with acidic membranes provides 6 kcal mol$^{-1}$ of stabilization energy. Thus, basic sequences on protein kinase C likely play a role in the initial targeting of protein kinase C to acidic membranes.

## Lipid-Induced Conformational Changes

Strong support for the role of PS as a specific allosteric activator of protein kinase C was obtained recently. Using proteases as a conformational probe, we have shown that PS, but not nonactivating acidic lipids, induces a conformational change that exposes the autoinhibitory pseudosubstrate domain of protein kinase C (93). Exposure of this domain is enhanced by diacylglycerol. In contrast, the domain is masked in the absence of lipids, when protein kinase C is bound to nonactivating acidic lipids or after dilution of the activating lipids. Furthermore, autophosphorylation on the amino terminus protects the pseudosubstrate domain from proteolysis. Figure 7 presents a model illustrating the PS- and diacylglycerol-dependent displacement of the pseudosubstrate domain from the active site, presumably allowing access of substrates because exposure of the domain correlates with activity (93).

How can the above results account for diacylglycerol's ability to increase the affinity of protein kinase C for PS but not for nonactivating acidic lipids? One possibility is that diacylglycerol stabilizes the interaction of protein kinase C with PS by promoting an interaction between basic residues in the pseudosubstrate sequence with acidic headgroups. The pseudosubstrate domain is masked in the absence of PS, which would explain why diacylglycerol has no effect on the affinity of protein kinase C for nonactivating acidic lipids. Thus, specific binding sites for PS regulate the diacylglycerol-mediated exposure of the pseudosubstrate domain, which is then stabilized by the binding to any acidic lipid. The ability of nonactivating lipids to reduce the PS requirement for activity is sensitive to ionic strength (95), supporting the possibility that lipids that can substitute for PS (70) associate by electrostatic interactions with the exposed pseudosubstrate region.

Protein kinase C also undergoes a gross conformational change upon membrane binding. Specifically, the exposure of the hinge region of the enzyme increases significantly, as evidenced by an over 10-fold increase in

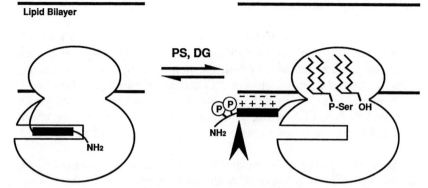

*Figure 7* Schematic showing the reversible exposure of the pseudosubstrate domain of protein kinase C by PS and diacylglycerol. The autoregulatory pseudosubstrate domain is depicted as a black box near the amino terminus; the amino-terminal autophosphorylation sites are indicated by the circles. The arrow indicates the site of proteolysis that is masked when protein kinase C is inactive but that becomes exposed upon binding of PS and diacylglycerol; binding of these two lipids is proposed to release the pseudosubstrate domain from the active site, thereby allowing access of substrates. A stabilizing electrostatic interaction between basic residues of the exposed pseudosubstrate domain and acidic lipids is indicated. Although protein kinase C binds multiple PS molecules, only one is indicated for clarity. Reproduced from Ref. 93 with permission.

proteolytic sensitivity upon membrane binding (51, 63, 87). This increased proteolytic sensitivity results from membrane binding and is not regulated by specific headgroups (94).

Conformational changes upon membrane binding have also been detected by monitoring intrinsic tryptophan fluorescence and by circular dichroism of protein kinase C (20, 107). Using the latter technique, Shah & Shipley (107) reported that a decrease in $\alpha$ helical content from 37 to 22% and an increase in $\beta$ sheet from 55 to 64% occurs upon binding to PS multilayers.

In summary, at least two lipid-mediated conformational changes exist for protein kinase C: a global conformational change involving exposure of the hinge region of protein kinase C that attends membrane binding, and a localized conformational change in the regulatory domain that is mediated by PS and diacylglycerol and accompanies activation.

## Model for Protein Kinase C:Lipid Interaction

Understanding the mechanism of protein kinase C's regulation by lipid headgroups is complicated by several seemingly anomalous facts. Most intriguing is the ability of short-chained phosphatidylcholines to activate protein kinase C only at the critical micelle concentration of the lipid (119) and the ability of unsaturated fatty acids to activate protein kinase C (84). Activation by the former requires diacylglycerol while activation by fatty acids is enhanced by diacylglycerol (109). Isozymes of protein kinase C are differentially regulated by both molecules (109, 118), suggesting some specificity in their interaction with the non-PS activators. We recently showed that activation by short-chained phosphatidylcholines is accompanied by exposure of the pseudosubstrate domain of protein kinase C (J. W. Orr & A. C. Newton, in preparation), indicating common effects on the structure of protein kinase C by PS and the PS-independent activators.

The model in Figure 8 summarizes some of the findings regarding the interaction of protein kinase C with acidic membranes in the absence (middle panel) and presence (right panel) of diacylglycerol. In the absence of diacylglycerol, the interaction of protein kinase C with acidic membranes is driven primarily by electrostatic interactions regardless of whether PS is present. This interaction is sensitive to ionic strength and surface charge and is accompanied by a global conformational change that exposes the hinge region of protein kinase C. Diacylglycerol induces remarkable specificity and cooperativity in the interaction of protein kinase C with lipids containing the L-serine headgroup. The specific binding to phosphatidylserine is likely driven by multiple, specific, interacting lipid-binding sites on the enzyme that recognize the serine headgroup. This specific interaction is insensitive to increasing ionic strength and is

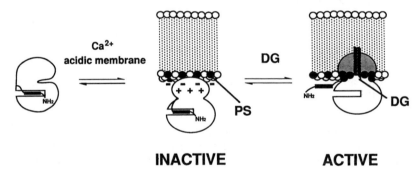

*Figure 8* Model for the interaction of protein kinase C with lipid headgroups. In the absence of diacylglycerol, electrostatic interactions drive the binding of protein kinase C to acidic lipids (*middle*); this binding results in a conformational change that exposes the hinge region of protein kinase C. Diacylglycerol promotes cooperativity and specificity in the protein:lipid interaction (causing a marked increase in the affinity of the kinase for PS) and activates the membrane-bound kinase by causing a PS-dependent release of the pseudosubstrate domain from the active site, a conformational change that may be accompanied by insertion of the protein into the hydrophobic core of the membrane (*right*).

accompanied by a conformational change that exposes the pseudosubstrate domain of the enzyme, an event that promotes catalytic activity. In the absence of PS, diacylglycerol has no effect on the binding of protein kinase C to acidic membranes or on the conformation of the enzyme.

Binding data indicate that multiple PS molecules are cooperatively sequestered around protein kinase C, which implies that a domain enriched in this lipid should form around the enzyme. In support of this, a fluorescence quenching assay has been used to show that protein kinase C clusters acidic lipids upon membrane binding (11). Further evidence that PS is enriched around protein kinase C was obtained when a photoactivatable phorbol ester exclusively labeled PS and phosphatidylethanolamine in brain membranes (34).

The role of diacylglycerol appears to be one of stabilizing the protein kinase C:membrane interaction by increasing the affinity of the enzyme for PS, an event that is insensitive to ionic strength (95) and that may be accompanied by insertion of the regulatory domain into the hydrophobic core of the membrane (8, 20, 111). In this regard, Bazzi & Nelsestuen (8, 9) have proposed that protein kinase C binds to membranes by a reversible and irreversible mechanism, with the latter promoted at high PS concentrations and by phorbol esters.

Although protein kinase C has full catalytic activity in the absence of lipid bilayers, alterations in membrane structure modulate the enzyme's function. Epand and colleagues have suggested that protein kinase C activity is sensitive to molecules that nonspecifically destabilize bilayers, pointing out that diacylglycerol promotes the formation of an inverted hexagonal ($H_{II}$) phase (37). The ability of bilayer intercalating amphipaths to affect protein kinase C activity (38) and the recent finding that acyl-chain unsaturation is essential for optimal activity (16) indicate that, in addition to specific headgroup interactions, protein kinase C is sensitive to the physical state of membranes.

## LESSONS FROM PROTEIN KINASE C

Protein kinase C illustrates that lipids can be allosteric regulators: binding to specific sites on a protein, inducing a conformational change, and thus altering the activity of the enzyme. Is protein kinase C an extreme example of regulation by lipid or will regulation through binding sites for specific lipid headgroups be a common theme as we learn more about membrane proteins? The specific interaction of phosphatidylcholine with $\beta$-hydroxybutyrate dehydrogenase shares some of the features that typify the interaction of protein kinase C with PS: more than one specific headgroup-binding site, high specificity for the headgroup, and cooperative interaction between the binding sites (30).

One mechanistic feature that many peripheral proteins share with protein kinase C is that electrostatic interactions target the proteins to acidic membranes, and then specific headgroup or hydrophobic interactions stabilize the membrane association. For example, the binding of the coagulation proteins to anionic membranes has been proposed to be initiated by an electrostatic interaction followed by association of lipid with specific sites on the protein (32). A notable difference between this and protein kinase C's interaction with PS is that diacylglycerol induces cooperativity and specificity in the headgroup-binding sites. In a manner similar to the specific protein kinase C:PS interaction (i.e. in the presence of diacylglycerol), the association of prothrombin with PS, but not with phosphatidylglycerol, is relatively insensitive to ionic strength (45) and results in a specific change in the conformation of the protein (71). As occurs for protein kinase C, forces other than electrostatic ones contribute to the interaction of various membrane proteins with acidic lipids. EPR studies as a function of ionic strength and pH indicate that only a small fraction of the selectivity displayed by the $(Na^+, K^+)$-ATPase for acidic lipids arises from electrostatic interactions (39).

In contrast to the headgroup specificity observed for protein kinase C, the interaction of CTP:phosphocholine cytidylyltransferase with anionic lipids is driven primarily by electrostatic interactions and displays little selectivity among acidic headgroups (28). Rather, Cornell and coworkers (29, 58) have proposed that electrostatic interactions target the enzyme to acidic membranes; this action is followed by insertion of a hydrophobic domain of the protein into the lipid bilayer. Such a mechanism likely accounts for the binding of synapsins and SecA to acidic membranes (14, 18). Lipid binding induces a conformational change in the latter, providing another example of lipids serving as allosteric regulators (108).

A question raised from the work with protein kinase C is whether PS has a unique feature that allows it to interact more effectively than other acidic lipids with proteins. The amine group and acidic charge may provide more possibilities for multiple interactions with amino acid side chains. Nonetheless, many examples exist where PS is less effective than other acidic lipids at activating enzymes (Table 1).

## CONCLUSIONS

The ability of lipid headgroups to alter the structure and function of membrane proteins provides a sensitive mechanism for fine-tuning membrane function. When the protein:lipid interaction is cooperative, as it is for protein kinase C, membrane proteins can sense and respond to subtle changes in the intrabilayer concentration of a particular lipid. Both lipid turn-over and lipid sequestration by proteins to form domains depleted or enriched in a particular lipid could thus significantly affect the function of a lipid-regulated membrane protein.

ACKNOWLEGMENTS

This work was supported by NIH grant GM43154 and by a Searle Scholars Award from the Chicago Community Trust. I am grateful to David Daleke, Stuart McLaughlin, and members of my research group for comments on the manuscript. I thank Daniel E. Koshland, Jr. for his inspiring insight into cooperativity and for tolerating my interest in the "grease" around protein kinase C.

*Literature Cited*

1. Ahmed, S., Kozma, R., Lee, J., Monfries, C., Harden, N., Lim, L. 1991. *Biochem. J.* 281: 233–41
2. Ahmed, S., Kozma, R., Monfries, C., Hall, C., Lim, H. H., et al. 1990. *Biochem. J.* 272: 767–73
3. Arnold, R. S., Newton, A. C. 1992. *Biophys. J.* 61: A89
4. Ashendel, C. L. 1989. In *Receptor Phosphorylation*, ed. V. K. Moudgil, pp. 163–75. Boca Raton: CRC Press
5. Bazzi, M. D., Nelsestuen, G. L. 1987. *Biochemistry* 26: 115–22
6. Bazzi, M. D., Nelsestuen, G. L. 1987. *Biochemistry* 26: 5002–8
7. Bazzi, M. D., Nelsestuen, G. L. 1987. *Biochemistry* 26: 1974–82
8. Bazzi, M. D., Nelsestuen, G. L. 1988. *Biochemistry* 27: 7589–93
9. Bazzi, M. D., Nelsestuen, G. L. 1989. *Biochemistry* 28: 3577–85
10. Bazzi, M. D., Nelsestuen, G. L. 1990. *Biochemistry* 29: 7624–30
11. Bazzi, M. D., Nelsestuen, G. L. 1991. *Biochemistry* 30: 7961–69
12. Bazzi, M. D., Youakim, M. A., Nelsestuen, G. L. 1992. *Biochemistry* 31: 1125–34
13. Bell, R. M., Burns, D. J. 1991. *J. Biol. Chem.* 266: 4661–64
14. Benfenati, F., Greengard, P., Brunner, J., Bähler, M. 1989. *J. Cell Biol.* 108: 1851–62
15. Boggs, J. M., Moscarello, M. A., Papahadjopoulos, D. 1982. In *Lipid-Protein Interactions*, ed. P. C. Jost, O. H. Griffith, 2: 1–51. New York: Wiley-Interscience
16. Bolen, E. J., Sando, J. J. 1992. *Biochemistry* 31: 5945–51
17. Boni, L. T., Rando, R. R. 1985. *J. Biol. Chem.* 260: 10819–25
18. Breukink, E., Demel, R. A., De Korte-Kool, G., De Kruijff, B. 1992. *Biochemistry* 31: 1119–24
19. Brockman, J. L., Anderson, R. A. 1991. *J. Biol. Chem.* 266: 2508–12
20. Brumfeld, V., Lester, D. S. 1990. *Arch. Biochem. Biophys.* 277: 318–23
21. Burns, D. J., Bell, R. M. 1991. *J. Biol. Chem.* 266: 18330–38
22. Burns, D. J., Bloomenthal, J., Lee, M.-H., Bell, R. M. 1990. *J. Biol. Chem.* 265: 12044–51
23. Carafoli, E. 1991. *Physiol. Rev.* 71: 129–53
24. Castagna, M., Takai, Y., Kaibuchi, K., Sano, K., Kikkawa, U., Nishizuka, Y. 1982. *J. Biol. Chem.* 257: 7847–51
25. Chen, C.-C., Wilson, T. H. 1984. *J. Biol. Chem.* 259: 10150–58
26. Clark, J. D., Lin, L.-L., Kriz, R. W., Ramesha, C. S., Sultzman, L. A., et al. 1991. *Cell* 65: 1043–51
27. Conforti, G., Zanetti, A., Pasquali-Ronchetti, I., Quaglino, D. Jr., Neyroz, P., Dejana, E. 1990. *J. Biol. Chem.* 265: 4011–19
28. Cornell, R. B. 1991. *Biochemistry* 30: 5873–80
29. Cornell, R. B. 1991. *Biochemistry* 30: 5881–88
30. Cortese, J. D., McIntyre, J. O., Duncan, T. M., Fleischer, S. 1989. *Biochemistry* 28: 3000–8
31. Cullis, P. R., de Kruijff, B., Hope, M. J., Verkleij, A. J., Nayar, R., et al. 1983. In *Membrane Fluidity in Biology*, ed. R. C. Aloia, 1: 39–81. New York: Academic

32. Cutsforth, G. A., Whitaker, R. N., Hermans, J., Lentz, B. R. 1989. *Biochemistry* 28: 7453–61
33. Davis, R. J., Ganong, B. R., Bell, R. M., Czech, M. P. 1985. *J. Biol. Chem.* 260: 5315–22
34. Delclos, K. B., Yeh, E., Blumberg, P. M. 1983. *Proc. Natl. Acad. Sci. USA* 80: 3054–58
35. Devaux, P. F., Seigneuret, M. 1985. *Biochim. Biophys. Acta* 822: 63–125
36. Ekerdt, R., Papahadjopoulos, D. 1982. *Proc. Natl. Acad. Sci. USA* 79: 2273–77
37. Epand, R. M., Lester, D. S. 1990. *Trends Pharm. Sci.* 11: 317–20
38. Epand, R. M., Stafford, A. R., Cheetham, J. J., Bottega, R., Ball, E. H. 1988. *Biosci. Rep.* 8: 49–54
39. Esmann, M., Marsh, D. 1985. *Biochemistry* 24: 3572–78
40. Feigenson, G. W. 1986. *Biochemistry* 25: 5819–25
41. Flint, A. J., Paladini, R. D., Koshland, D. E. Jr. 1990. *Science* 249: 408–11
42. Fong, T. M., McNamee, M. G. 1986. *Biochemistry* 25: 830–40
43. Ganong, B. R., Loomis, C. R., Hannun, Y. A., Bell, R. M. 1986. *Proc. Natl. Acad. Sci. USA* 83: 1184–88
44. Gennis, R. B. 1989. *Biomembranes: Molecular Structure and Function.* New York: Springer-Verlag
45. Gerads, I., Govers-Riemslag, J. W. P., Tans, G., Zwaal, R. F. A., Rosing, J. 1990. *Biochemistry* 29: 7967–74
46. Hanks, S. K., Quinn, A. M., Hunter, T. 1988. *Science* 241: 42–52
47. Hannun, Y. A., Bell, R. M. 1986. *J. Biol. Chem.* 261: 9341–47
48. Hannun, Y. A., Loomis, C. R., Bell, R. M. 1985. *J. Biol. Chem.* 260: 10039–43
49. House, C., Kemp, B. E. 1987. *Science* 238: 1726–28
50. Houslay, M. D., Needham, L., Dodd, N. J. F., Grey, A.-M. 1986. *Biochem. J.* 235: 237–43
51. Huang, F. L., Yoshida, Y., Cunha-Melo, J. R., Beaven, M. A., Huang, K.-P. 1989. *J. Biol. Chem.* 264: 4238–43
52. Hubbard, S. R., Bishop, W. R., Kirschmeier, P., George, S. J., Cramer, S. P., Hendrickson, W. A. 1991. *Science* 254: 1776–79
53. Hui, S. W., Boni, L. T., Stewart, T. P., Isac, T. 1983. *Biochemistry* 22: 3511–16
54. Hyatt, S. L., Klauck, T., Jaken, S. 1990. *Mol. Carcinog.* 3: 45–53
55. Kaibuchi, K., Fukumoto, Y., Oku, N., Takai, Y., Arai, K.-i., Muramatsu, M.-a. 1989. *J. Biol. Chem.* 264: 13489–96
56. Kaibuchi, K., Takai, Y., Nishizuka, Y. 1981. *J. Biol. Chem.* 256: 7146–49
57. Kaibuchi, K., Takai, Y., Sawamura, M., Hoshijima, M., Fujikura, T., Nishizuka, Y. 1983. *J. Biol. Chem.* 258: 6701–4
58. Kalmar, G. B., Kay, R. J., Lachance, A., Aebersold, R., Cornell, R. B. 1990. *Proc. Natl. Acad. Sci. USA* 87: 6029–33
59. Kemp, B. E., Pearson, R. B. 1990. *Trends Biochem. Sci.* 15: 342–46
60. Keranen, L. M., Orr, J. W., Newton, A. C. 1992. *Biophys. J.* 61: A89
61. Kikkawa, U., Takai, Y., Tanaka, Y., Miyake, R., Nishizuka, Y. 1983. *J. Biol. Chem.* 258: 11442–45
62. Kimelberg, H. K., Papahadjopoulos, D. 1972. *Biochim. Biophys. Acta* 282: 277–92
63. Kishimoto, A., Kajikawa, N., Shiota, M., Nishizuka, Y. 1983. *J. Biol. Chem.* 258: 1156–64
64. Knighton, D. R., Zheng, J., Ten Eyck, L. F., Ashford, V. A., Xuong, N. H., et al. 1991. *Science* 253: 407–14
65. König, B., DiNitto, P. A., Blumberg, P. M. 1985. *J. Cell. Biochem.* 29: 37–44
66. Koshland, D. E. Jr. 1970. *The Enzymes* 1: 342–96
67. Lassing, I., Lindberg, U. 1985. *Nature* 314: 472–74
68. Lee, M.-H., Bell, R. M. 1989. *J. Biol. Chem.* 264: 14797–14805
69. Lee, M.-H., Bell, R. M. 1991. *Biochemistry* 30: 1041–49
70. Lee, M.-H., Bell, R. M. 1992. *Biochemistry* 31: 5176–82
71. Lentz, B. R., Wu, J. R., Sorrentino, A. M., Carleton, J. N. 1991. *Biophys. J.* 60: 942–51
72. Lill, R., Dowhan, W., Wickner, W. 1990. *Cell* 60: 271–80
73. Marsh, D. 1983. *Trends Biochem. Sci.* 8: 330–33
74. Maruyama, I. N., Brenner, S. 1991. *Proc. Natl. Acad. Sci. USA* 88: 5729–33
75. Maurer, M. C., Sando, J. J., Grisham, C. M. 1992. *Biochemistry* 31: 7714–21
76. McLaughlin, S., Mulrine, N., Gresalfi, T., Vaio, G., McLaughlin, A. 1981. *J. Gen. Physiol.* 77: 445–73
77. Mochly-Rosen, D., Khaner, H., Lopez, J. 1991. *Proc. Natl. Acad. Sci. USA* 88: 3997–4000
77a. Mochly-Rosen, D., Miller, K. G., Scheller, R. H., Khaner, H., Lopez, J., Smith, B. L. 1992. *Biochemistry* 31: 8120–24
78. Molleyres, L. P., Rando, R. R. 1988. *J. Biol. Chem.* 263: 14832–38
79. Mori, T., Takai, Y., Yu, B., Takahashi,

J., Nishizuka, Y., Fujikura, T. 1982. *J. Biochem.* 91: 427–31
80. Moritz, A., De Graan, P. N. E., Gispen, W. H., Wirtz, K. W. A. 1992. *J. Biol. Chem.* 267: 7207–10
81. Morrot, G., Zachowski, A., Devaux, P. F. 1990. *FEBS Lett.* 266: 29–32
82. Mosior, M., McLaughlin, S. 1991. *Biophys. J.* 60: 149–59
83. Mosior, M., McLaughlin, S. 1992. *Biochim. Biophys. Acta* 1105: 185–87
84. Murakami, K., Chan, S. Y., Routtenberg, A. 1986. *J. Biol. Chem.* 261: 15424–29
85. Nakamura, H., Kishi, Y., Pajares, M. A., Rando, R. R. 1989. *Proc. Natl. Acad. Sci. USA* 86: 9672–76
86. Newton, A. C., Koshland, D. E. Jr. 1987. *J. Biol. Chem.* 262: 10185–88
87. Newton, A. C., Koshland, D. E. Jr. 1989. *J. Biol. Chem.* 264: 14909–15
88. Newton, A. C., Koshland, D. E. Jr. 1990. *Biochemistry* 29: 6656–61
89. Newton, A. C., Williams, D. S. 1991. *J. Biol. Chem.* 266: 17725–28
90. Nishizuka, Y. 1986. *Science* 233: 305–12
91. Nishizuka, Y. 1988. *Nature* 334: 661–65
92. Ono, Y., Fujii, T., Igarashi, K., Kuno, T., Tanaka, C., et al. 1989. *Proc. Natl. Acad. Sci. USA* 86: 4868–71
93. Orr, J. W., Keranen, L. M., Newton, A. C. 1992. *J. Biol. Chem.* 267: 15263–66
94. Orr, J. W., Newton, A. C. 1992. *Biochemistry* 31: 4661–67
95. Orr, J. W., Newton, A. C. 1992. *Biochemistry* 31: 4667–73
96. Pears, C., Schaap, D., Parker, P. J. 1991. *Biochem. J.* 276: 257–60
97. Pember, S. O., Powell, G. L., Lambeth, J. D. 1983. *J. Biol. Chem.* 258: 3198–3206
98. Perin, M. S., Fried, V. A., Mignery, G. A., Jhan, R., Südhof, T. C. 1990. *Nature* 345: 260–63
99. Politino, M., King, M. M. 1987. *J. Biol. Chem.* 262: 10109–13
100. Quest, A. F. G., Bloomenthal, J., Bardes, E. S. G., Bell, R. M. 1992. *J. Biol. Chem.* 267: 10193–97
101. Rando, R. R., Kishi, Y. 1992. *Biochemistry* 31: 2211–18
102. Robson, R. J., Dennis, E. A. 1978. *Biochim. Biophys. Acta* 508: 513–24
103. Sandermann, H. Jr. 1983. *Trends Biochem. Sci.* 8: 408–11
104. Sandermann, H. Jr., McIntyre, J. O., Fleischer, S. 1986. *J. Biol. Chem.* 261: 6201–8
105. Schaap, D., Parker, P. J. 1990. *J. Biol. Chem.* 265: 7301–7
106. Scheideler, M. A., Bell, R. M. 1989. *J. Biol. Chem.* 264: 12455–61
107. Shah, J., Shipley, G. G. 1992. *Biochim. Biophys. Acta* 1119: 19–26
108. Shinkai, A., Mei, L. H., Tokuda, H., Mizushima, S. 1991. *J. Biol. Chem.* 266: 5827–33
109. Shinomura, T., Asaoka, Y., Oka, M., Yoshida, K., Nishizuka, Y. 1991. *Proc. Natl. Acad. Sci. USA* 88: 5149–53
110. Sixl, F., Brophy, P. J., Watts, A. 1984. *Biochemistry* 23: 2032–39
111. Snoek, G. T., Feijen, A., Hage, W. J., Van Rotterdam, W., De Laat, S. W. 1988. *Biochem. J.* 255: 629–37
112. Souvignet, C., Pelosin, J.-M., Daniel, S., Chambaz, E., Ransac, S., Verger, R. 1991. *J. Biol. Chem.* 266: 40–44
113. Takai, Y., Kishimoto, A., Iwasa, Y., Kawahara, Y., Mori, T., Nishizuka, Y. 1979. *J. Biol. Chem.* 254: 3692–95
114. Tamura, M., Tamura, T., Tyagi, S. R., Lambeth, J. D. 1988. *J. Biol. Chem.* 263: 17621–26
115. Tefft, R. E., Carruthers, A., Melchior, D. L. 1986. *Biochemistry* 25: 3709–18
116. van Deenen, L. L. M., de Gier, J. 1974. In *The Red Blood Cell*, ed. D. M. Surgenor, 1: 147–211. New York: Academic
117. Vik, S. B., Georgevich, G., Capaldi, R. A. 1981. *Proc. Natl. Acad. Sci. USA* 78: 1456–60
118. Walker, J. M., Homan, E. C., Sando, J. J. 1990. *J. Biol. Chem.* 265: 8016–21
119. Walker, J. M., Sando, J. J. 1988. *J. Biol. Chem.* 263: 4537–40
120. Wiedmann, T. S., Pates, R. D., Beach, J. M., Salmon, A., Brown, M. F. 1988. *Biochemistry* 27: 6469–74
121. Zimmerman, M., Daleke, D. L. 1992. *Biophys. J.* 61: A38
122. Zot, H. G., Doberstein, S. K., Pollard, T. D. 1992. *J. Cell Biol.* 116: 367–76

# MACROMOLECULAR CROWDING: Biochemical, Biophysical, and Physiological Consequences[1]

## Steven B. Zimmerman

Section on Physical Chemistry, Laboratory of Molecular Biology, National Institute of Diabetes and Digestive and Kidney Diseases, National Institutes of Health, Building 5, Room 330, Bethesda, Maryland 20892

## Allen P. Minton

Section on Physical Biochemistry, Laboratory of Biochemical Pharmacology, National Institute of Diabetes and Digestive and Kidney Diseases, National Institutes of Health, Building 8, Room 226, Bethesda, Maryland 20892

KEY WORDS: excluded volume, cytoplasm, cell volume, diffusion, intermolecular interactions, proteins, DNA, nucleic acids, enzymes

---

CONTENTS

| | |
|---|---|
| PERSPECTIVES AND OVERVIEW | 28 |
|    Nomenclature | 30 |
| THEORETICAL CONCEPTS | 31 |
|    The Effect of Macromolecular Crowding on Solution Equilibria | 31 |
|    The Effect of Macromolecular Crowding on Reaction Rates | 36 |
|    The Effect of Macromolecular Crowding on Diffusive Transport of Solutes | 37 |
|    The Effective Specific Volume of a Macromolecule—What Is It? | 40 |

[1] The US government has the right to retain a nonexclusive, royalty-free license in and to any copyright covering this paper.

EXPERIMENTAL MEASUREMENT OF THE EFFECT OF VOLUME OCCUPANCY ON THE
  THERMODYNAMIC ACTIVITY OF A TRACER PROTEIN .......................................... 43
    *Solutions Containing a Single Macrosolute* ................................................. 43
    *Solutions Containing a Tracer Protein and a Single Background Species* ................. 43
EXPERIMENTALLY OBSERVED EFFECTS OF CROWDING ON EQUILIBRIA AND REACTION
  RATES IN MODEL SYSTEMS ...................................................................... 46
    *Extended Ranges of Enzymatic Activities Under Crowded Conditions* .................... 46
    *Alteration of Reactants or Products Under Crowded Conditions* .......................... 47
EXPERIMENTALLY OBSERVED EFFECTS OF CROWDING UPON TRANSPORT PROPERTIES IN
  MODEL SYSTEMS ................................................................................. 51
    *Viscosity of a Protein Solution* ............................................................. 51
    *Self-Diffusion of Proteins in Membranes* ................................................... 51
    *Self-Diffusion of Proteins in Aqueous Solution* ............................................ 54
    *Tracer Diffusion of a Protein in Solutions of a Second Protein* ........................... 54
    *Tracer Diffusion of Probe Molecules in Polymers* ......................................... 55
CROWDING IN CYTOPLASM .......................................................................... 56
    *Experimental Approaches to the Calculation of Volume Exclusion in Cytoplasm* ...... 56
    *Rationalization of Discrepancies Between Phenomena Observed in Vitro and in Vivo* 56
    *Diffusion of Probe Molecules in Cytoplasm* ............................................... 57
RAMIFICATIONS, SPECULATIONS, CONCLUSIONS ................................................ 58
    *Possible Role of Crowding in Regulation of Cellular Volume* ............................ 58
    *Efficacy of Macromolecular Drugs* ......................................................... 58
    *Synthesis by Degradative Enzymes* ......................................................... 59
    *Macromolecular Crowding as a Laboratory Tool* .......................................... 59
    *Osmotic Remedial Mutants* .................................................................. 59
    *Crowding as a Factor in Cellular Evolution* ................................................ 59
    *Crowding Versus Confinement* ............................................................... 60
    *Homeostasis or Metabolic Buffering* ........................................................ 60
    *Crowding and the Functional Importance of Condensed DNA* .......................... 61
    *Crowding and Cytoplasmic Structure* ....................................................... 61

# PERSPECTIVES AND OVERVIEW

Measurements of biochemical rates and equilibria are conventionally carried out under relatively idealized conditions selected to minimize effects of nonspecific interactions between the constituents of the reaction unit (i.e. the reactants, products, and transition-state complexes) and other species that do not directly participate in the reaction under study. Concentration-dependent apparent rate or equilibrium constants are then extrapolated to the limit of infinite dilution in order to extract intrinsic quantities reflecting the properties of isolated reactant and product molecules in an idealized bath of solvent.

Biological media differ from the idealized solvent bath in two important ways. First, a biological medium is likely to contain a high total concentration of nominally soluble macromolecules. Sometimes a single species at a high concentration predominates, such as hemoglobin within red cells ($\sim$350 g/liter) (see 3). More commonly, a medium may contain a variety of macromolecular species, none of which taken individually may be present at high concentration, but which taken collectively occupy a substantial fraction of the total volume of the medium [e.g. $\sim$340 g/liter of total RNA+protein in the cytoplasm of *Escherichia coli* (172)] (see also 19, 39). We refer to such a medium as crowded. Second, particularly in eukaryotes, a biological medium is also likely to be structured at the molecular level by the presence of a network of extended structures such as F-actin, microtubules, intermediate filaments, and membranous boundaries (40). We refer to those soluble or structural macromolecules that do not directly participate in a particular reaction as background species.

The local environment will influence reactions taking place in a biological medium when reactants, transition-state complexes, and products interact unequally with background species, as illustrated in Figure 1, and/or when interactions with background species alter the mobilities of constituents of the reaction unit. This applies to interactions of all types, including steric-repulsion, electrostatic, hydrophobic, and van der Waals interactions. In the present review, we focus on the steric-repulsive interactions deriving from the fundamentally impenetrable nature of molecules, as these interactions are always present in addition to any other interactions that may or may not be present.

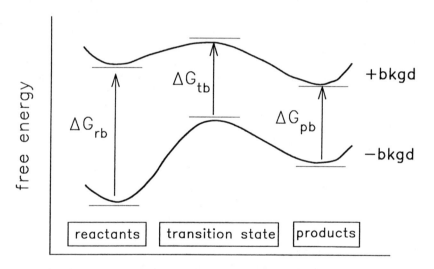

*Figure 1* Schematic free-energy profile of a simple chemical reaction in the absence (*lower curve*) and presence (*upper curve*) of background molecules. $\Delta G_{rb}$, $\Delta G_{tb}$, and $\Delta G_{pb}$, respectively, denote the free energies of nonspecific interaction between reactants and background molecules, between transition-state and background molecules, and between products and background molecules. To the extent that these three nonspecific interaction energies are unequal, the rate and/or equilibrium constants characterizing the reaction will differ in the absence and presence of background molecules.

Obviously, a living organism bears scant resemblance to a bath of solvent, yet it is not generally appreciated that, from a quantitative point of view, biochemical rates and equilibria in a living organism may likewise bear scant resemblance to those measured in a bath of solvent. Only relatively recently have investigators recognized that steric repulsion between macromolecules in a crowded medium can strongly influence both the rate and extent of a variety of macromolecular reactions, and that crowding effects must be taken into account when attempting to relate biochemical and biophysical observations made in vitro to physiological processes observed in vivo (93–95).

Thermodynamic evidence for the presence of substantial repulsive intermolecular interactions in solutions of proteins and mixtures of proteins and polymers comes from studies, dating back to the early part of this century, of colligative and partitioning properties of such solutions (2). Not until much later did researchers show that the thermodynamic data were accounted for semiquantitatively (or quantitatively in favorable cases) by simple geometric models presupposing only the existence of short-range steric repulsion between macromolecules (36, 132). These models have subsequently proven to be useful for interpreting the results of a wide variety of measurements of steady-state and time-dependent behavior in crowded media, as well for predicting the dependence of various measurable reaction rates and equilibria upon the fraction of volume occupied by macromolecules within a given medium (see below). Most recently, a growing body of experimental evidence testifies to the variety of biochemical processes that can be profoundly influenced by macromolecular crowding.

This review places particular emphasis upon research on the thermodynamic aspects of crowding done since 1983, when the last comprehensive review of this subject was presented (95), and upon topics of current interest to the authors. The reader is also referred to other discussions of crowding and crowding-related phenomena (5, 24, 29, 39, 43, 82, 124, 140).

## Nomenclature

We refer frequently to the behavior of individual *tracer* or *probe* particles (molecules) in an environment comprised of many other particles (molecules) of various species, which are termed *background* particles (molecules). *Excluded volume* refers to the volume of a solution that is excluded to the center of mass of a probe particle by the presence of one or more background particles in the medium. *Fractional volume-occupancy* ($\phi$) denotes the fraction of the total volume occupied by macromolecules.

## THEORETICAL CONCEPTS

### The Effect of Macromolecular Crowding on Solution Equilibria

Consider the general reaction

$$n_1X_1 + n_2X_2 + \cdots + n_jX_j \rightleftharpoons n_{j+1}X_{j+1} + n_{j+2}X_{j+2} + \cdots + n_qX_q,$$

where $n_i$ is the number of molecules of $X_i$ participating in the reaction. The condition of equilibrium at constant temperature and pressure requires that

$$K^\circ(T,P) \equiv \prod_{i=j+1}^{q} a_i \bigg/ \prod_{i=1}^{j} a_i,$$

where $K^\circ$ is the thermodynamic association constant; $a_i$ is the thermodynamic activity of solute species $i$, which may be written as a product of the concentration of species $i$, $c_i$, and an activity coefficient, $\gamma_i$.[2] It follows that the conventional equilibrium constant, expressing an equilibrium relation between concentrations rather than activities, is itself a function of solution composition (denoted by $\{c\}$):

$$K(T,P,\{c\}) \equiv \prod_{i=j+1}^{q} c_i \bigg/ \prod_{i=1}^{j} c_i = K^\circ(T,P) \times \Gamma(T,P,\{c\}),$$

where

$$\Gamma \equiv \prod_{i=1}^{j} \gamma_i(T,P,\{c\})^{n_i} \bigg/ \prod_{i=j+1}^{q} \gamma_i(T,P,\{c\})^{n_i}. \qquad 1.$$

The activity coefficient $\gamma_i$ is a measure of nonideal behavior arising from interactions between solute molecules:

$$\gamma_i(T,P,\{c\}) = \exp[G_i^{NI}(T,P,\{c\})/RT], \qquad 2.$$

where $G_i^{NI}(T,P,\{c\})$ denotes the average free energy of interaction between a molecule of species $i$ and all of the other solute molecules present in a real solution whose composition is denoted by $\{c\}$.[3] For the case of simple volume exclusion, $G_i^{NI}$ is greater than 0, and $\gamma_i$ is greater than 1 (95). The magnitude of $\gamma_i$ strongly depends upon the relative sizes and shapes of the

---

[2] The concentration may be expressed in any units. In the present review, we express concentrations in units proportional to the number density of molecules, i.e. molar or weight/volume, depending on the specific application.

[3] As the solution approaches the limit of infinite dilution, intersolute interactions of all types become negligible, $\gamma_i \to 1$ for all $i$, $\Gamma \to 1$, and $K \to K^\circ$.

probe and background species as well as the concentrations of background species.

It follows from Equations 1 and 2 that

$$RT \ln \Gamma = \sum_{i=j+1}^{q} n_i G_i^{NI} - \sum_{i=1}^{j} n_i G_i^{NI}.$$

Thus, the apparent equilibrium constant $K$ is increased or decreased relative to the ideal equilibrium constant $K°$ by a factor that reflects the difference between the free energies of nonspecific interaction of molecules of products with all solute species and molecules of reactants with all solute species (Figure 1) (see also 95).

The logarithm of the activity coefficient may be expanded in powers of the concentrations of solute(s):

$$\ln \gamma_i = \sum_j B_{ij} c_j + \sum_j \sum_k B_{ijk} c_j c_k + \ldots, \qquad 3.$$

where the indices $j, k, \ldots$ can have any value from 1 to $q$. The terms $B_{ij}, B_{ijk}, \ldots$ are referred to as two-body, three-body, ... interaction (or virial) coefficients, respectively. Exact formal expressions for the virial coefficients as functions of the effective potential of interaction between two or more molecules of solute in a bath of solvent have been derived via statistical thermodynamics (55, 89). In practice, evaluation of the values of the virial coefficients is only possible for small numbers of particles interacting via very simple interaction potentials. The simplest of these is the hard-particle potential, which is equal to zero for all interparticle distances (defined with respect to the particle centers) above a certain contact distance,[4] and infinite for all interparticle distances equal to or less than the contact distance. Thus it is a common (but not necessarily realistic) practice to describe macromolecular solutes as effective rigid hard particles with simple shapes for the purpose of calculating virial coefficients and estimating the value of the activity coefficient (107, 108, 110, 113, 161). Fortunately, this approximation seems to work reasonably well for solutions of a single globular protein in solutions of moderate ionic strength, i.e. under conditions such that long-range electrostatic interactions are largely damped out (95, 132). Approximate corrections for the presence of nonnegligible electrostatic interactions have been presented (18, 100, 161).

Even when macrosolutes may be reasonably approximated by rigid hard

---

[4] If at least one of the interacting particles is nonspherical, then the contact distance will be a function of the relative orientations of the particles as well as the distance between particle centers.

particles of simple shape, calculation of virial coefficients above the two-body coefficient is difficult to impossible for particles other than uniformly sized hard spheres. Thus, evaluation of activity coefficients via Equation 3 is limited to solutions in which all solute species are sufficiently dilute so that three-body and higher-order contributions would be negligible, or in which the size of background particles is sufficiently small relative to the size of the probe particle (see below).

An alternate approach to the calculation of activity coefficients of solutes in the hard-particle approximation is based upon the relation

$$\gamma_i(\{c\}) = \frac{1}{P_i(\{c\})},$$

where $P_i(\{c\})$ is the probability that a randomly selected point in the solution of composition $\{c\}$ can accommodate a probe molecule of species $i$, i.e. the point lies at the center of a solute-free region (or hole) that is at least as large as the probe molecule, as illustrated in Figure 2.

Although the value of $P_i(\{c\})$ cannot be calculated exactly, two approximate methods have proven useful. The first method, introduced by Ogston (110), which we refer to as available volume theory, or AVT (see also 23, 32, 158, 171), is based upon the assumption that centers of background particles are distributed randomly, so that the number of centers of background particles within a given element of area or volume is governed by a Poisson distribution. This approximation (which is equivalent to neglecting the interaction between molecules of background species) leads to expressions of the form of Equation 3 truncated after two-body interaction terms, and thus predicts that the logarithm of activity of each species varies linearly with the concentrations of all species; that is:

$$\ln \gamma_i = \sum_j B_{ij} c_j.$$

The second approximate method utilizes an equation of state derived from scaled particle theory, or SPT, to calculate $P_i(\{c\})$ and $\gamma_i$ (41; note corrections to 41 in 21; 76, 125). According to SPT, the logarithm of the activity coefficient of species $i$ may be expressed as a finite series in powers of $R_i$, the radius (or characteristic dimension) of the probe particle

$$\ln \gamma_i = \sum_{k=0}^{d} A_k R_i^k,$$

where $d$ is the dimensionality of the solution and the $A_k$ are positively valued functions of the number densities, characteristic dimensions, and shapes of all macrosolute species in solution. SPT has been applied to the calculation of protein activities and crowding effects in isotropic solutions

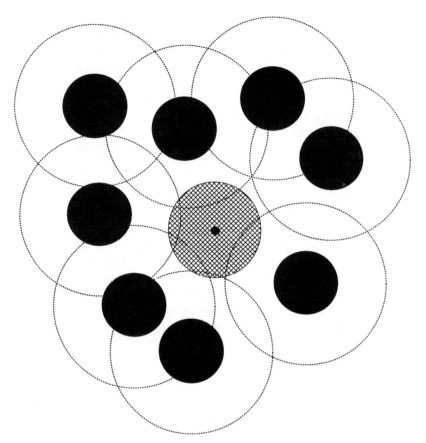

*Figure 2* Illustration of volume available to a probe particle (*crosshatched sphere*) in a solution containing background particles (*solid spheres*). Each background particle lies at the center of a spherical region, indicated by a dashed surface, the radius of which is equal to the sum of the radii of background and probe particles. This region, representing the covolume of the background and probe particles, is inaccessible to the center of mass of the probe particle, indicated by a black dot. The volume available to the probe, defined as equal to the volume accessible to the center of mass of the probe, is thus the volume exterior to all such regions (*dashed surfaces*). If the probe is of species $i$, then $P_i$ is equal to the ratio of the available to the total volume.

(94, 132, 133) and in solutions of proteins undergoing polymerization to highly elongated aggregates that spontaneously form nematic and other liquid-crystalline phases (83, 173). In both AVT and SPT, the solvent (water) is treated as a featureless continuum, but such a treatment will become increasingly less realistic as the size of background molecules approaches the size of solvent molecules (15). A modified SPT calculation of crowding effects was recently presented in which water molecules are explicitly represented by an additional species of effective hard spherical particles (15).

Statistical-mechanical models for enumerating the possible configurations of a polymer chain in the vicinity of a compact impenetrable protein molecule (54) provide an approximate means for calculating the negentropic work associated with the insertion of a protein molecule into a polymer solution. A recent model of this type allows for the presence of short-range attractive interactions between segments of the polymer chain and the surface of the protein (12).

Model calculations of the effect of crowding upon association and conformational equilibria have been carried out for membranes modeled as two-dimensional fluids (45) and for aqueous solutions (69, 93–95, 108, 133). The major qualitative predictions of equilibrium excluded-volume analysis may be summarized as follows, where $\phi$ denotes the fraction of the total volume occupied by macromolecules:

1. For self- and heteroassociation reactions, the nonideal correction factor $\Gamma$ increases monotonically with increasing $\phi$, and the value of $d\Gamma/d\phi$ increases monotonically with increasing $\phi$.
2. For fixed $\phi$, the value of $\Gamma$ increases with the degree of association. For example, $\Gamma$ for a hypothetical monomer-tetramer self-association reaction would be expected to exceed that for a monomer-dimer self-association of the same species subunit.
3. Increasing volume occupancy tends to favor compact or globular conformations (or assemblies) relative to highly anisometric conformations (or assemblies).
4. For fixed $\phi$, the value of $\Gamma$ for association reactions increases with decreasing molecular weight (i.e. increasing number density) of background species.[5]
5. Volume occupancy by macrosolutes has a greater influence on association equilibria in three dimensions than the same degree of volume (area) occupancy has upon the corresponding reaction equilibria in two dimensions.
6. For fixed $\phi$, the effect of crowding upon associations between macromolecules is expected to be much greater than upon associations between a macromolecule and a small molecule.
7. Finally, and most importantly, under conditions of macromolecular

---

[5] As pointed out above, this result would only be valid so long as background species are large relative to solvent, i.e. are also macromolecules.

volume occupancy comparable to those found in vivo ($0.1 < \phi < 0.5$), macromolecular association constants are predicted to exceed those in dilute solution by as much as several orders of magnitude. In fact, the contribution of volume exclusion in a crowded medium to the overall standard free energy change of a reaction can, under some circumstances, equal or exceed the ideal contribution that is intrinsic to the reaction unit (132).

Excluded volume theory has also been used recently to analyze the effects of small-molecule solutes (or cosolvents) upon the equilibrium properties of macrosolutes (141, 163). This analysis provides a microscopic or mechanistic alternative to traditional analyses of these effects in the context of macroscopic models of the preferential interaction of macrosolute with either water or cosolvent (see e.g. 6).

## The Effect of Macromolecular Crowding on Reaction Rates

Macromolecular crowding affects the rates of different types of biochemical reactions in distinctly different ways.

FORMATION OF MACROMOLECULAR COMPLEXES Consider the association of two macromolecules forming a homo- or heterodimer. If the rate-determining step for dimer formation is the conversion of activated complex to fully formed dimer, then separated reactants and activated complex may be treated to a first approximation as if they are in equilibrium. Under these conditions, crowding is predicted to lower the free energy of activation and increase the rate of formation of complexes roughly to the same extent that crowding increases the equilibrium constant for dimer formation (94, 95). However, if the rate-determining step for dimer formation is the formation of activated complex (i.e. the rate with which the two reactant molecules encounter each other), then the reaction is said to be diffusion-limited, and the overall rate depends upon the sum of the diffusion coefficients of the reactants (16). Because crowding lowers the diffusion coefficient (see below), it is predicted to lower rates of association in diffusion-limiting circumstances (94, 97). Since the encounter rate represents an absolute upper limit for the rate of any bimolecular reaction, ultimately the overall bimolecular reaction rates must decrease with increasing fractional volume-occupancy (independent of the effect of crowding at low fractional volume-occupancy) when crowding becomes sufficiently great (97). Figure 3 schematically illustrates the overall effect of increasing volume occupancy on bimolecular association rate.

ENZYME-CATALYZED REACTIONS The effect of macromolecular crowding upon the rate of a prototypical enzyme-catalyzed reaction (substrate →

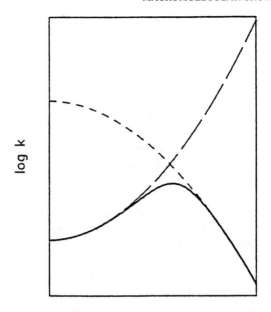

**fractional volume occupancy**

*Figure 3* Schematic dependence upon fractional volume occupancy of transition state–limited rate constant (*long-dashed curve*), diffusion-limited rate constant (*short-dashed curve*), and overall rate constant (*solid curve*). The overall reaction rate is transition-state limited at low volume occupancy and diffusion limited at high volume occupancy. Reproduced from Ref. 97.

product) has been analyzed for the cases in which each of the following elementary steps is rate-limiting: (*a*) association of substrate and enzyme (94), (*b*) conversion of enzyme-bound substrate to enzyme-bound product (69, 94), and (*c*) dissociation of enzyme-bound product (94). In addition, if the enzyme exists in several states of association that have different intrinsic catalytic activities (see e.g. 116), crowding can affect the equilibrium-average state of association and hence the equilibrium-average catalytic activity (94, 101).

## The Effect of Macromolecular Crowding on Diffusive Transport of Solutes

In the absence of an externally applied potential field, the motion of individual compact tracer macromolecules in dilute solution can be described by a random walk, so that

$$\langle r^2 \rangle = 2 \cdot d \cdot D^\circ \cdot t,$$

where $\langle r^2 \rangle$ is the squared displacement of position observed after elapsed time $t$, averaged over many observations; $d$ is the dimensionality; and $D^\circ$, the intrinsic diffusion coefficient of species $i$, is given by the Einstein relation

$$D_i^\circ = kT/f_i^\circ,$$

where $f_i^\circ$ is the frictional coefficient characteristic of an isolated tracer particle of species $i$ in a solvent of given viscosity (148). In the absence of a macroscopic concentration gradient of background molecules, the coefficient of tracer or probe diffusion is defined by

$$D^{tr} = \langle r^2 \rangle_{tr}/(2 \cdot d \cdot t),$$

where the subscript tr indicates that the average of $r^2$ is taken over all of the tracer molecules in the solution.[6] The coefficient of tracer diffusion is equal to the coefficient of self-diffusion $D^{self}$ when tracer and background species are identical.

The influence of background molecules on the diffusive motion of a probe molecule is a phenomenon that is extremely complex to analyze theoretically, and is far less tractable than the parallel phenomenon of the influence of background particles on the thermodynamic activity of a probe molecule. In principle, $D_i^R$, the reduced tracer diffusion coefficient of species $i$, may be expanded in a power series of concentrations, similar to Equation 3:

$$D_i^R \equiv \frac{D_i^{tr}}{D_i^\circ} = 1 + \sum_j Q_{ij} c_j + \sum_j \sum_k Q_{ijk} c_j c_k + \ldots,$$

where $D_i^{tr}$ and $D_i^\circ$ respectively denote the tracer and intrinsic diffusion coefficients of species $i$, and the coefficients $Q_{ij}, Q_{ijk}, \ldots$ are measures of two-body, three-body, and higher-order interactions. Each interaction coefficient reflects contributions from direct interactions between solute particles (short-range steric repulsions in the present instance) and from hydrodynamic interactions.[7] Attempts to incorporate both direct and hydrodynamic interactions into statistical-mechanical theories of the liquid state have led to complex formal relations that permit numerical com-

---

[6] In principle, the tracer diffusion coefficient may be a function of time and distance (138, 140). In the context of the present discussion, we will take this quantity to be equal to the limiting value at long time and large distance, as that is the quantity determined in most experiments.

[7] Hydrodynamic interaction is the effect on the motion of the tracer particle of transient flows of solvent at the position of the tracer caused by motions of the background particles. These effects may be compared to the effect of the wake of a ship on the motion of a second nearby ship.

putation only in the limiting case of self-diffusion of hard spherical particles at low fractional volume-occupancy (87, 122). Three general approaches to the calculation of the effect of arbitrary concentrations of a compact background species upon tracer diffusion in the absence of hydrodynamic interaction have been developed.

A statistical-mechanical theory has been developed that permits one to calculate the coefficient of tracer diffusion of species $i$ in the presence of a given concentration of background species $j$, over a broad range of background concentrations, provided that the equilibrium pair-correlation function $g_{ij}(r)$ is known (or can be calculated) for each concentration of species $j$ (115). This theory has been adapted for two-dimensional self-diffusion of membrane proteins (1). The two-dimensional pair-distribution function $g_{ii}(r)$ was computed for hard disks at fractional volume- (area-) occupancies of 0.25 and 0.50 via Monte Carlo simulation. Using these distribution functions, the reduced two-dimensional self-diffusion coefficient of hard disks was computed at the two fractional volume-occupancies (1) and found to be in good agreement with the results of simulation of walks of hexagonal particles on a hexagonal lattice for different levels of fractional lattice occupancy (120, 137). A great deal of numeric calculation is required to compute the self-diffusion coefficient for even a single value of the fractional volume-occupancy, and at the present time, rapid computation of the reduced tracer-diffusion coefficient as a function of the size and volume occupancy of background species via fundamental theories of the liquid state does not appear to be feasible in two dimensions, much less three dimensions.

The second method for obtaining information about the effect of volume occupancy on tracer diffusion is simulation of random walks of particles on lattices of varying volume occupancy. Simulations have been carried out for two-dimensional particles on planar lattices, in which tracers and background particles are characterized by different intrinsic jump rates (diffusion coefficients) and different sizes (120, 137). These simulations have yielded insight into the effect of mobile obstacles on the motion of random walkers and serve as standards against which approximate theories should be compared. However, each simulation carried out for a single set of parameters (lattice type, particle sizes, intrinsic jump rates, lattice occupancy, etc) is computationally intensive, and we are unaware of comparable calculations carried out for a three-dimensional model at this time.

A third approach to calculating the effect of crowding on tracer diffusion employs a semiempirical free-volume model, according to which an isolated tracer molecule in solution diffuses by undergoing a Brownian displacement of average distance $\Delta r$ on an average of once every $\Delta t$ seconds. When background molecules are added to the solution, the probability of undergoing a displacement is assumed to be proportional to the probability that the target volume (the element of volume into which the tracer would be relocated in the event of a successful jump) is free of any part of a background molecule. It follows that

$$D_i^R = P_i^V(\{c\}) = \exp[-\Delta G_i^V(\{c\})/RT],$$

where $P_i^V$ is the probability that the target volume is vacant, and $\Delta G_i^V$ is the (negentropic) free energy associated with the creation of a vacancy of the required size and shape. For the case of a globular protein undergoing diffusion in a solution of random-coil polymers, the value of $P_i^V$ has been estimated via AVT (112), and for globular proteins diffusing in solutions of other proteins, $\Delta G_i^V$ was estimated using SPT (106). Free-volume models provide a rapid means of estimating crowding effects in both two and three dimensions via back-of-the-envelope calculations (see also 96, 114), but because they are semiempirical and contain one or more adjustable parameters, they must be calibrated against experimental data.

All of the above-described theories and models for diffusion of a globular protein in a protein solution predict that $D_i^R$ decreases more rapidly with increasing volume fraction of background protein $j(\phi_j)$ than expected on the basis of a simple decaying exponential in $\phi_j$, and reaches zero at a value of $\phi_j < 1$ that is determined by the relative volumes of species $i$ and $j$. The free-volume theory also predicts that smaller proteins will have a larger effect per unit weight on the diffusion of larger proteins than vice versa. In contrast, free-volume theory for diffusion of a globular protein in a polymer solution (112) predicts that $D_i^R$ decreases as a decaying exponential in the square root of $\phi_j$.

## The Effective Specific Volume of a Macromolecule—What Is It?

Although the representation of macromolecules by equivalent rigid particles of simple shape for the purpose of calculating macromolecular activities in crowded solutions is an appealing and useful concept, the volume of the equivalent particle is an inherently ambiguous quantity that is more often than not evaluated by fitting approximate theoretical relations (of possibly dubious applicability) to experimental data and letting the results fall where they may. In principle, one would like to perform calculations that have predictive—as opposed to purely hand-waving—value. Thus, a discussion of the factors affecting the volume of an equivalent particle, and methods for estimating the "best" volume for a particular calculation is in order.

Hard-particle models are generally based upon several assumptions: (a) the solvent may be treated as a continuum (but see 15 for an exception); (b) a macrosolute may be treated—at low resolution—as a rigid convex particle of simple shape or as an assembly of such particles; and (c) the potential of average force acting between the actual molecules in solution is short range and effectively zero (i.e. $\ll kT$) beyond some contact distance defined as a function of the geometry and relative orientations of the interacting particles. To the extent that these assumptions are valid, the size and shape (at low resolution) of actual macromolecules do in fact closely resemble those of the equivalent particles best accounting for concentration-dependent activity coefficients (106, 130, 132). Conversely, if significant attractive or repulsive interactions exist between molecules of macrosolute in a solution containing only a single macrosolute, the equivalent hard particle may be significantly smaller or larger than the actual size of the molecule (95, 100).

Even more confounding is the fact that, in the case of multiple solute species interacting via electrostatic or other nonadditive potentials, one cannot in principle define a unique equivalent hard particle corresponding to each solute species. For instance, consider a solution containing two quasispherical protein species, A and B, of approximately the same mass and density. In the absence of any long-range interaction, we could validly represent both of these proteins by equivalent hard spheres of radius $r$. However, if one of the proteins, say A, bears a significant net charge, then the interaction between two molecules of A will be more repulsive than the interaction between two molecules of B, which will be equal to the interaction between a molecule of A and one of B. The reader may readily verify that no values of $r_A$ and $r_B$, effective or otherwise, will simultaneously satisfy these three relations in the context of a pure hard-particle model.

In the absence of a theoretical guide to the appropriate choice of effective particle size, it is commonly assumed that the effective molar volume is proportional to the molar mass of a macrosolute; the constant of proportionality is referred to as the effective specific volume or $v$ (19, 101, 172). Uncertainty in $v$ can easily be the largest source of error in crowding calculations. Past studies have variously assumed $v$ values based upon partial specific volumes ($\sim 0.6$–$0.8$ ml/g) or hydrodynamic volumes ($\sim 1.3$ ml/g), although neither of these assumptions are correct in principle for excluded-volume interactions (92). We have estimated effective specific volume values for several proteins, for cell extracts, and for cytoplasmic macromolecules based upon a variety of experimental data (Table 1). The values range from 0.8 to 1.7 ml/g, in part a reflection of differences in experimental conditions. Values of $v$ used in recent volume-occupancy calculations are 0.77 ml/g (19), and 1.0–1.3 ml/g (172).

**Table 1** Effective specific volumes derived from experiment

| Experimental material | Analysis | Effective specific volume, $v$ (ml/g)[a] | Reference |
|---|---|---|---|
| Protein crystals | Unit cell volume per molecular mass | 1.0[b] | 86 |
| | Bound water[c] | 1.0 | 86 |
| PEG-induced precipitates of proteins and cell extracts | Compositional analysis | 1.0–1.4 | 172 |
| Protein solutions | Fitting solution parameters to a hard-sphere model | | |
| Hemoglobin | | | |
| pH 7.8[d] | Scaled particle theory | 0.77 | 19 |
| pH 7.0 | Virial expansion | 0.79 | A. P. Minton[e] |
| Serum albumin | | | |
| pH 5.1 | Virial expansion | 0.8 | 100 |
| pH 6.1 | Virial expansion | 1.3 | 100 |
| pH 7.6 | Virial expansion | 1.7 | 100 |
| Cells of *E. coli* | Cytoplasmic compositions as function of osmotic conditions | 0.9 | 103 |
| | Growth rate as function of osmotic conditions | 1.1,[f] 1.4[g] | 19 |
| | | 1.2[h] | 19 |

[a] To provide a uniform albeit approximate way of converting analytical information of various types to estimates of $v$, we assume a hard-sphere model, an average partial specific volume for proteins of 0.73 ml/g. and an average partial specific volume for *E. coli* cytoplasmic macromolecules of 0.7 ml/g (corresponding to assumed protein and nucleic weight fractions in cytoplasm of 0.70 and 0.30 with partial specific volumes of 0.73 ml/g and 0.58 ml/g, respectively). For example, for 0.5 g of water of hydration per gram dry weight macromolecule, $v \approx (0.5 \text{ ml} + 0.7 \text{ ml})/1 \text{ g} = 1.2$ ml/g.
[b] Value of $v$ for most common value of unit cell volume/molecular mass ($V_M$); range is approximately threefold in $V_M$ values.
[c] Based on estimated average value for protein crystals of ~0.25 g bound water/g protein; definitions of bound water are discussed in Ref. 67.
[d] M. T. Record, Jr, personal communication.
[e] A. P. Minton, personal communication (cited in 35).
[f] Value of $v$ based upon extrapolation to 0.4 ml cytoplasmic water/g dry weight at infinite external osmolarity.
[g] Based upon volume fraction in cytoplasm inaccessible to water, additional cell composition data, and assumptions for the magnitude of the noncytoplasmic volume fraction and the constancy of this fraction over a range of external NaCl concentrations.
[h] For an estimated value of ~$0.5 \pm 0.2$ ml cytoplasmic water/g cell dry weight at limiting conditions for growth.

# EXPERIMENTAL MEASUREMENT OF THE EFFECT OF VOLUME OCCUPANCY ON THE THERMODYNAMIC ACTIVITY OF A TRACER PROTEIN

## Solutions Containing a Single Macrosolute

A solution containing a single macrosolute is a special case of the solution containing a tracer plus a single background species; in this instance the tracer species is identical to the background species. The thermodynamic activity or the activity coefficient of a macromolecular solute may be determined as a function of concentration by measurements of membrane osmotic pressure, Rayleigh light scattering, or sedimentation equilibrium (148). The measured dependence of the logarithm of the thermodynamic activity of globular proteins upon protein concentration in solutions containing a single protein solute is well described over a wide concentration range by an expansion of the form of Equation 3, in which interaction coefficients are calculated according to a hard-particle model. Higher-order interaction terms become increasingly important as protein concentration increases, and at the highest protein concentrations for which activities have been measured, six or seven terms may be required (130, 132). Equivalent expressions derived from SPT also provide satisfactory descriptions (132). Under conditions of moderate ionic strength and at pH values such that the protein is not highly charged, the size of the effective hard particle is similar to that of the actual protein molecule (95, 132). Under conditions such that the protein bears substantial net charge, the size of the effective hard particle significantly exceeds that of the actual molecule. An approximate theory for the dependence of equivalent hard-particle size upon pH (or net charge) has been presented (100).

## Solutions Containing a Tracer Protein and a Single Background Species

It was observed as early as the 1940s that viral particles and proteins could be precipitated out of solution by addition of water-soluble polymers such as heparin, hyaluronic acid, dextran, and polyethylene glycol (25, 61, 68). A series of studies carried out in the 1960s by Ogston, Laurent, and coworkers (28, 36, 68, 71, 111) led to the recognition that these and related effects resulted primarily from exclusion of the protein from part of the volume of the polymer solution.[8] More recently, the volume-excluding

---

[8] These early observations are the basis for a relatively direct assay for certain excluded volume parameters based upon the distribution coefficients of proteins, nucleic acid–related materials, and cell extracts in a two-phase system (167, 171, 172).

properties of proteins as well as polymers have become recognized (93–95).

PROTEIN IN THE PRESENCE OF A BACKGROUND POLYMER   The interaction between proteins and polymers has been characterized experimentally by means of four methods. The first is the measurement of the equilibrium distribution, or partitioning, of a protein between a polymer-free medium (medium I) and a polymer-containing medium (medium II). If the protein is sufficiently dilute, then at equilibrium

$$\gamma_{\text{protein}}^{\text{II}} = c_{\text{protein}}^{\text{I}} / c_{\text{protein}}^{\text{II}}$$

to a good approximation. Experiments have been performed in which the two media were separated by a dialysis membrane (36), by phase immiscibility (in which case phase I may contain a relatively low amount of polymer) (64, 171), and by crosslinking the polymer to create an insoluble gel (13). By and large, the results of these experiments accord qualitatively with predictions of AVT for effective hard particles (171). Some polymers (e.g. dextran) are better modeled as rod-like particles (68, 95), while others [e.g. polyethylene glycol (PEG)] are satisfactorily modeled as effective spherical particles (171). It has been suggested on the basis of analysis using a lattice-statistical model for protein-polymer interaction that a small attractive interaction exists between at least some proteins and polymers (13).

A gel chromatographic method has been used to characterize the nonideal interaction between a small polymer (PEG 4000) and each of two proteins (156). This method has two advantages over dialysis equilibrium, to which it is thermodynamically equivalent: measurements are much more rapid, and the technique may be utilized even when the two solutes are not completely separable, so long as they partition sufficiently distinctly on a given size exclusion gel matrix.

A third method for the characterization of the interaction between proteins and polymers is the measurement of protein solubility as a function of the concentration of added polymer (9, 47, 57, 90). If it is assumed that the precipitate contains only protein and that the composition of the precipitate is independent of the concentration of polymer, then at equilibrium

$$d \ln \gamma_{\text{protein}} / d c_{\text{polymer}} = -d \ln s_{\text{protein}} / d c_{\text{polymer}},$$

where $s_{\text{protein}}$ denotes the solubility. Measurements of the effect of dextran and PEG upon the solubility of a variety of proteins indicate that, when the solubility of protein is small enough so that protein-protein interactions in solution are negligible, $\ln \gamma_{\text{protein}}$ increases linearly with polymer concentration over the entire experimentally measurable range of protein concentration, i.e. over as much as three orders of magnitude in solubility (9). For a fixed weight/volume concentration of dextran, the solubility of protein is independent of the molecular weight of polymer (68), as would be predicted by a rigid rod representation of polymer in AVT. However, for a fixed weight/volume concentration of PEG, the solubility decreases (and $\ln\gamma$ increases) with increasing molecular weight of PEG (9), in contradiction to the predictions of AVT, and in contradiction to the results of partition experiments (9). Anomalously small extents of exclusion calculated from the protein solubility data in the presence of PEG were attributed to the presence of some attractive interaction between proteins and PEG (95).

Knoll & Hermans (65) measured the excess light scattering of bovine serum albumin (BSA) in solutions of polyethylene glycol and from their data calculated the limiting value of $d \ln \gamma_{BSA}/dc_{PEG}$ as $c_{PEG} \to 0$ for PEG 1000, PEG 6000, PEG 20,000, and PEG 100,000. They found that $d \ln \gamma_{BSA}/dc_{PEG}$ decreased with increasing molecular weight of PEG, as predicted by AVT and by Hermans' polymer excluded-volume model (54).[9] Using radioactive tracer measurements, Knoll & Hermans measured the amount of PEG in the condensed phase and found substantial penetration of the condensed phase by smaller PEGs. From this finding, they argue that the assumption of constant protein activity in the condensed phase that is used to relate solubility to the protein activity coefficient (see above) is not valid for protein in the presence of PEG. Alternatively, the presence of PEG in the condensed phase may be evidence of nonspecific attractive interactions between protein and PEG.[10]

In general, three-body and higher-order interaction terms in the virial expansion (Equation 3) are not required to describe the dependence of the activity coefficient of protein on polymer concentration, even at high polymer concentration. The physical basis of this observation, which accords better with the predictions of AVT than those of SPT, is not entirely clear, as the assumption in AVT that background particles are distributed randomly is strictly valid only in the limit of massless particles, i.e. points or infinitely thin rods. The apparent ability of AVT to account for protein activity in the presence of relatively high concentrations of real polymers (but not in the presence of high concentrations of other proteins of comparable size), may possibly result from the difference between the

---

[9] The quantity $d \ln \gamma_{BSA}/dc_{PEG}$ is equal to the interaction factor $B_{12}$ in Equation 3, where BSA is component 1 and PEG is component 2.

[10] Amounts of PEG in precipitates of proteins have been found to vary greatly with the salt composition of the supernatant fluid (appendix table in 172).

cross-sectional area of a typical globular protein and the cross-sectional area of a single polymeric chain; as the size of a protein increases, the approximation that polymer excludes volume to protein as would a random array of thin rods becomes increasingly realistic. By the same reasoning, one would expect that Equation 3, truncated after two-body terms, would become an increasingly realistic description of the dependence of the activity of a globular tracer species $i$ in the presence of a large excess of a compact (i.e. not polymeric) background species $j$ as the size of species $j$ diminishes relative to that of species $i$. The results of several measurements of the effect of high concentrations of smaller solutes (such as sugars) on protein activity (141) seem to agree with this expectation.

PROTEIN IN THE PRESENCE OF A BACKGROUND PROTEIN  The solubility of deoxygenated sickle hemoglobin (HbS) is reduced in the presence of other proteins that do not coprecipitate with the hemoglobin (14, 145). This effect has been interpreted in the context of excluded-volume models (133, 145). Also, excluded volume must be taken into account in order to quantitatively account for the effects on HbS solubility of other hemoglobin variants that coprecipitate with HbS to varying degrees (35, 91, 109, 145). In principle, the excluded volume interaction between proteins of two different species could be characterized quantitatively via measurement of the osmotic pressure or sedimentation equilibrium of protein mixtures (see e.g. 22, 71). However, we are unaware of published experimental studies of this kind.

# EXPERIMENTALLY OBSERVED EFFECTS OF CROWDING ON EQUILIBRIA AND REACTION RATES IN MODEL SYSTEMS

We have summarized several observations of crowding effects on reaction equilibria in Table 2; Figure 4 shows an example (the effects of crowding on the equilibrium association of ribosomal particles). Similarly, we have summarized several reported effects of crowding upon reaction rates in Table 3; Figure 5 shows an example (the effects of crowding on the rate of enzyme-catalyzed exchange labeling of DNA termini). Certain entries in the tables are discussed below.

## Extended Ranges of Enzymatic Activities Under Crowded Conditions

Crowding has been shown to greatly extend the range of solution conditions under which several enzymes or proteins are functional. The examples cited here all involve interactions with nucleic acids, probably because of the large crowding effects on reactions between macromolecules. We anticipate that such effects are much more widespread.

DNA POLYMERASE AND DNA REPLICATION SYSTEMS   The introduction of the use of hydrophilic polymers for in vitro DNA replication systems by Fuller et al (38) was an important development in the study of those and related systems (66; see references cited in Table 3). Fuller et al (38) suggested that the excluded-volume effects of the polymers might increase effective concentrations of macromolecular reactants, a prediction borne out by subsequent studies in which polymer requirements were removed by increasing the concentrations of reactants (62, 75).

In studies with isolated enzymes, synthetic activities and associated nuclease activities of DNA polymerases increased in a range of otherwise inhibitory salt concentrations (166). The increased activity seems to be a direct result of increased binding of polymerase to DNA. Crowding also enables DNA polymerase I of *E. coli* to remain active under a variety of otherwise inhibitory conditions (e.g. unfavorable pH or temperatures, inhibitory concentrations of urea, formamide, or ethidium bromide), leading to a general proposal of crowding as a source of homeostasis (169) (see below).

DNA LIGASES   T4 DNA ligase and *E. coli* DNA ligase are both inhibited at higher salt concentrations under uncrowded conditions. In the presence of PEG, however, DNA ligation in these salt concentrations can be greatly stimulated (52, 53).

The temperature optimum of a DNA ligase from the thermophilic organism *Thermus thermophilus* is increased from 37°C in uncrowded media to a more normal temperature optimum for that organism ($\sim$ 55–65°C) in 20% PEG 6000 (147). Takahashi & Uchida (147) also note previous studies (53, 168) of DNA ligases from *E. coli* and rat liver that suggest crowding-induced shifts in temperature optima into temperature ranges characteristic for the organism.

*recA* PROTEIN   Addition of PEG or polyvinyl alcohol (PVA) extended the functional range of a *recA* protein-promoted DNA strand–exchange system (73). Crowding allowed high rates of product formation at otherwise suboptimal $Mg^{2+}$ concentrations and also approximately doubled the concentration of NaCl required for half-dissociation of *recA* protein from etheno M13 DNA (73).

## Alteration of Reactants or Products Under Crowded Conditions

Enzymes of nucleic acid metabolism have provided several examples of changes in substrates or products under crowded conditions. These are discussed below.

Table 2  Crowding effects on equilibria

| Observation | Crowding agent (background species) | Magnitude of effect | Comment | Reference |
|---|---|---|---|---|
| Reduction in solubility of deoxyhemoglobin S | Bovine serum albumin, crosslinked hemoglobin A | 10-fold reduction in 250 g/liter protein | Excluded-volume interpretation in Ref. 133 | 14 |
| Reduction in solubility of proteins | PEG, dextran | | | 9, 47, 68, 68a |
| Self-association of apomyoglobin | Lysozyme, $\beta$-lactoglobulin, ribonuclease | | Monomer in absence of added protein, mostly dimer in presence of >200 g/liter added protein | 160 |
| Stabilization of thrombin against thermal denaturation | Ribonuclease | | Attributed to crowding-induced formation of heat-stable oligomers of thrombin | 102 |
| Thickening of fibrin fibrils | Bovine serum albumin | Twofold increase in birefringence at 70 g/liter albumin | Increased regularization of aggregate | 154 |
| Bundling of actin fibers | PEG | | | 146 |

| Effect | Polymer | Result | Comment | Ref. |
|---|---|---|---|---|
| Lowering of critical concentration for actin fiber formation | PEG | Twofold in 50 g/liter PEG | | 33 |
| Self-association of pyruvate dehydrogenase | PEG | Threefold in 80 g/liter PEG | | 150 |
| | PEG | 50–90% of ~22S species converted to ~55S species in 30 g/liter PEG | Excluded volume interpretation in Ref. 108 | 17 |
| Enhancement of trypsin inhibition by bovine serum albumin | Dextran | | Attributed to enhanced formation of enzymatically inactive trypsin-albumin complex | 69 |
| Enhancement of association of ribosomal particles | PEG, Dextran, Ficoll | > 10-fold increase in equilibrium constants | Effects used to estimate particle volumes of polymers (cf 15) | 170 |
| Association of T4 gene 45 protein and gene 44/62 protein complex | PEG, Dextran | 50-fold increase in association constant at 75 g/liter of PEG 12,000 | | 59 |
| Stabilization of double-stranded nucleic acids against thermal denaturation | PEG, Dextran | $T_m$ of poly(dAT) increased ~10°C in 210 g/liter PEG 6000 | Effect increases with increasing molecular weight of polymer | 72 |
| | | $T_m$ of poly(rI)·poly(rC) increased ~3°C in 90 g/liter PEG 20,000 | | 164 |

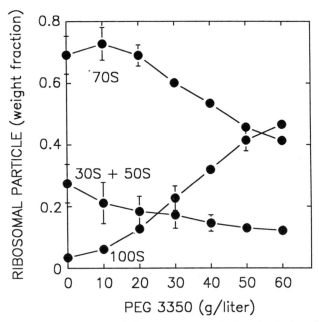

*Figure 4* Effect of concentration of PEG 3350 on the equilibrium distribution of ribosomal species isolated from *E. coli* (based upon data in Figure 2 of Ref. 170).

ALTERED SUBSTRATE SPECIFICITY  Absolute or relative substrate-specificity can be changed by crowding: (*a*) Blunt-end ligation by *E. coli* and rat-liver ligases was undetectable under normal assay conditions but became a strong reaction under crowded conditions (168), probably because of the condensation of the DNA under crowded conditions. (*b*) The relative rates of phosphorylation of blunt or recessed DNA termini by T4 polynucleotide kinase was increased under crowded conditions (49, 50).

ALTERED REACTION PRODUCTS  Mixtures of cyclic and linear ligation products from both DNA ligases (118) and the T4 RNA ligase (48) shift strongly toward linear products under crowded conditions, which may reflect the increased thermodynamic activity of DNA termini under crowded conditions (cf 34). In agreement with such a mechanism, Sobczak & Duguet (143) obtained increases in intermolecular ligation by T4 DNA ligase resulting from an increase in the DNA concentration that are comparable to those obtained by crowding with PEG (see also 11).[11] The

---

[11] Lower molecular weight PEGs tend to favor circle formation in many of these ligation systems for unknown reasons.

similarities between the DNA ligases and T4 RNA ligase occur despite the difference in substrates (duplex DNA vs single-stranded DNA or RNA, respectively).

DECREASED SOLUBILITY OF NUCLEIC ACID COMPONENTS   DNA (and perhaps RNA) can undergo intramolecular collapse in the presence of polymer and salt ("psi" DNA) (see 8, 77, 78). Such preparations are often readily sedimentable (48, 49, 118). It would be interesting to know how well the various salt and temperature effects on PEG stimulation of DNA (or RNA) ligases or *recA* protein (see above) correlate with the presence of psi DNA (or RNA), as measured, for example, by ready sedimentation of the DNA (or RNA).

# EXPERIMENTALLY OBSERVED EFFECTS OF CROWDING UPON TRANSPORT PROPERTIES IN MODEL SYSTEMS

## Viscosity of a Protein Solution

The viscosity of human hemoglobin has been measured as a function of concentration at concentrations up to 450 g/liter. The dependence of viscosity upon concentration is quantitatively accounted for by a semiempirical equation (131) that was originally formulated to describe the concentration dependence of the viscosity of suspensions of rigid colloidal particles (105). This finding supports the appealingly simple notion that the effects of volume exclusion on both hydrodynamic and thermodynamic properties of solutions of compact globular proteins may be accounted for in a reasonably realistic fashion by models in which the protein molecules are treated as rigid particles of regular shape.

## Self-Diffusion of Proteins in Membranes

Scalettar & Abney (140) recently reviewed this subject comprehensively. The concentration dependence of the two-dimensional self-diffusion of proteins in membranes, or irreversibly adsorbed onto membranes, has been measured by fluorescence recovery after photobleaching (117, 149, 153). The coefficient of self-diffusion decreases more strongly with increasing protein concentration than predicted by any model or simulation presuming only hard-particle interactions between diffusing molecules (140). However, one can model the experimental results by assuming that patches of the membrane surface are inaccessible to diffusing proteins, by assuming that there are immobile obstacles to surface diffusion, or by assuming the presence of long-range attractive or repulsive forces between diffusing proteins (136, 137, 139, 140). Apparently, the systems studied so

Table 3  Crowding effects on reaction rates

| Observation | Crowding agent (background species) | Magnitude of effect | Comment | Reference |
|---|---|---|---|---|
| Reduction in specific activity of glyceraldehyde 3-phosphate dehydrogenase | Ribonuclease, β-lactoglobulin, bovine serum albumin | 30-fold reduction in 300 g/liter protein | Attributed to crowding-induced formation of low-activity tetramer of high-activity subunits | 101 |
| Reduction in specific activity of hyaluronate lyase | PEG | Six- to eightfold reduction in 125 g/liter polymer | | 69 |
| Acceleration of actin polymerization | PEG 6000 | Threefold at 80 g/liter polymer | | 150 |
| Acceleration of actin filament growth at barbed end | Dextran, PEG, ovalbumin | Threefold at 100 g/liter dextran | | 33 |
| Acceleration of deoxyhemoglobin S polymerization | Hemoglobin F | | Hemoglobin F does not copolymerize with hemoglobin S | 145 |
| Acceleration of fibrin gel formation | Bovine serum albumin, ovalbumin, hemoglobin, γ-globulin | Five- to sevenfold at 80 g/liter protein | | 159 |
| Acceleration of DNA renaturation | Bovine serum albumin PEG, dextran sulfate | Five- to sixfold at 70 g/liter 50-fold at 175 g/liter dextran sulfate 500,000; 90-fold at 175 g/liter PEG 8000 | | 154 142, 158a |
| Acceleration of cohesion of restriction fragments of lambda DNA | Bovine serum albumin, PEG, Ficoll | 100-fold at 250 g/liter albumin; >1000-fold at 150 g/liter PEG 8000 | | 165 |
| Acceleration of cohesion of lambda DNA | Bovine serum albumin, PEG, Ficoll PEG | ~10-fold at 126 g/liter PEG 6000, 55°C | Rate approximates in vivo rate | 165 79 |

| Reaction | Crowding agent | Effect | Comment | Reference |
|---|---|---|---|---|
| Acceleration of DNA ligase reaction | PEG, bovine serum albumin, Ficoll | Orders of magnitude depending on system and polymer | | 4, 7, 48, 52, 53, 118, 123, 147, 151, 152, 168 |
| Acceleration of enzymatic catenation of DNA circles | PEG, PVA | No measurable reaction in absence of crowding agent | | 80 |
| Acceleration of enzymatic supercoiling of DNA by topoisomerase I of *Sulfolobus acidocaldarius* | PEG | | Lack of reaction intermediates suggests processive mechanism in PEG solution | 37 |
| Acceleration of DNA replication system from *E. coli* | PEG, PVA, methyl cellulose | Absolute requirement for crowding agent at lower reactant concentrations (cf 62, 75) | | 38 |
| Acceleration of in vitro replication system for λdv DNA | PEG | Eightfold increase at 60 g/liter PEG 20,000 | | 155 |
| Acceleration of in vitro transposition of bacteriophage Mu | PVA | 155-fold increase at 50 g/liter PVA 49,000 | | 104 |
| Acceleration of in vitro replication of single-stranded DNA | PVA | Fourfold increase at 50 g/liter PVA 24,000 | | 74 |
| Acceleration of nuclease and polymerase activities of DNA polymerase | PEG, Ficoll, dextran, bovine serum albumin | Large stimulations of rate under variety of inhibitory conditions | Attributed to increased binding of enzyme to DNA under otherwise dissociating conditions | 166, 169 |
| Acceleration of T4 polynucleotide kinase | PEG, glycogen, Ficoll 70 | Orders of magnitude stimulation depending on DNA substrate and polymer | Stimulates both forward and back reactions; attributed to stabilization of oligomeric form of enzyme | 49, 50 |
| Acceleration of *recA*-protein promoted DNA strand exchange at low Mg$^{2+}$ concentration | PEG, PVA | Stimulation depends upon product type and reaction conditions | Reduction in Mg$^{2+}$ requirement under crowded conditions | 73 |

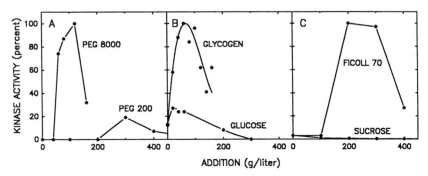

*Figure 5* Effects of macromolecules and low molecular weight derivatives on the rate of exchange labeling by T4 polynucleotide kinase of recessed DNA termini (redrawn from Ref. 50).

far do not provide a satisfactory experimental model for the study of pure crowding effects.

## Self-Diffusion of Proteins in Aqueous Solution

The concentration dependence of three-dimensional self-diffusion of proteins in aqueous solution has been measured with classical methods (reviewed in 46). Results obtained for several proteins are plotted in Figure 6, together with curves calculated using a semiempirical free-volume model (106) with best-fit values of parameters given in the figure caption. The self-diffusion coefficient typically decreases by a factor of 10 as the protein concentration increases to $\sim 300$ g/liter.

## Tracer Diffusion of a Protein in Solutions of a Second Protein

The diffusion of trace amounts of myoglobin, hemoglobin, and chromophorically labeled bovine serum albumin and aldolase has been measured as a function of the concentration of each of several different background proteins (ribonuclease, ovalbumin, serum albumin, aldolase) at concentrations up to 200 g/liter (106). For all proteins, $D^{tr}$ decreases with increasing weight/volume concentration of background protein. Plots of $D^{tr}$ vs the concentration of background protein resemble the data plotted in Figure 6. For three of the four proteins, the dependence of $D^{tr}$ upon background-protein concentration becomes stronger as the molecular weight of the background protein decreases, as predicted by the free-volume model for tracer diffusion (106). Extrapolation of the data to higher concentrations using the free-volume model suggests that the tracer diffusion coefficient of large proteins could be reduced by two or more orders of magnitude in solutions with $\phi > 0.3$.

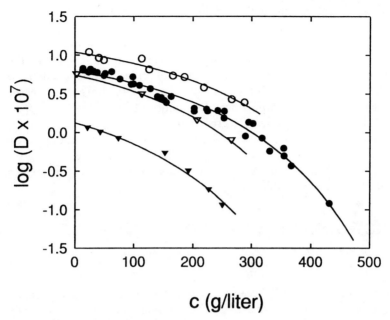

*Figure 6* Self-diffusion coefficient of globular proteins plotted as a function of protein concentration. Symbols represent the combined data of several authors, summarized by Gros (46). The curves are best-fits of Equation 6 of Muramatsu & Minton (106) to each of the data sets, with $v$ assumed equal to 0.8 ml/g for all proteins. The values of best-fit parameters for each protein are as follows: Myoglobin (*open circles*), log $(D_0 \times 10^7) = 1.04$, $\Delta r' = 4.34 \times 10^{-8}$ cm. Hemoglobin (*closed circles*), log $(D_0 \times 10^7) = 0.792$, $\Delta r' = 7.26 \times 10^{-8}$. Ovalbumin (*open triangles*), log $(D_0 \times 10^7) = 0.742$, $\Delta r' = 8.36 \times 10^{-8}$ cm. Invertebrate hemoglobin (*closed triangles*), log $(D_0 \times 10^7) = 0.126$, $\Delta r' = 4.54 \times 10^{-7}$ cm.

## Tracer Diffusion of Probe Molecules in Polymers

The diffusion of compact tracer particles (proteins and latex spheres) in random coil polymers has been studied using boundary spreading (70, 121), inelastic laser light scattering (reviewed in 119), and fluorescence recovery after photobleaching (FRAP) (44). The diffusion coefficient of a dilute probe particle in a particular polymer solution may be described to within experimental error by an empirical relation

$$D/D° = \exp(\alpha \cdot c^\nu \cdot R^\delta \cdot M^\gamma),$$

where $D°$ is the diffusion coefficient of the probe in the absence of polymer,

$c$ is the weight/volume concentration of polymer, $R$ is the radius of the probe particle, $M$ is the molecular weight of polymer, and $\alpha$, $\nu$, $\delta$, and $\gamma$ are fitting parameters. The parameter $\alpha$ increases with the net electrostatic charge of polymer (44). According to Phillies et al (119), inelastic light-scattering measurements for a broad range of probe-polymer combinations yield $\nu$ in the range 0.6–1.0, $\gamma = 0.7$–$0.9$, and $\delta \approx 0$. However, other techniques and other probe-polymer combinations have yielded $\nu$ as approximately 0.5 and $\delta$ significantly greater than zero (44, 70, 121). No single theory can account for these varied results. However, all studies reveal a large retarding effect of the polymer; the diffusion coefficient of a protein may be reduced by one or two orders of magnitude at polymer concentrations less than 100 g/liter.

## CROWDING IN CYTOPLASM

### Experimental Approaches to the Calculation of Volume Exclusion in Cytoplasm

Two groups (19, 20, 172) recently estimated the effect of volume occupancy upon the thermodynamic activity of a globular protein in the cytoplasm of *E. coli*. The two approaches are similar in outline, both starting with molecular weight distributions representing the cytoplasmic complement of macromolecules of *E. coli* and estimated or measured values for the actual cytoplasmic concentration of those macromolecules. The molecular-weight distributions are then converted to molecular-volume distributions by application of values for effective specific volumes (see above) and the molecular-volume distributions are finally expressed at cytoplasmic macromolecule concentrations to obtain a working description of the cytoplasmic background macromolecules for use in SPT calculations of activity coefficients. The two studies differ at each stage in the way parameters are chosen and the extent to which parameters are experimentally based.

### Rationalization of Discrepancies Between Phenomena Observed in Vitro and in Vivo

Binding of the *lac* repressor to the *lac* operator (10, 127, 162) and RNA polymerase binding to the $\lambda P_R$ promoter (128, 129) are both exquisitely sensitive to salt inhibition in vitro, yet Richey et al (126) found no significant salt effect on either system in vivo. Both Cayley et al (19) and Zimmerman & Trach (172) agree that crowding is one of the keys to resolving the apparent discrepancy, but their respective analyses differ in the type of data on which their respective conclusions are based and in the emphasis placed upon the contributions from various factors, as described below.

Zimmerman & Trach (172) calculate activity coefficients for test particles of arbitrary size under anticipated cytoplasmic crowding conditions over a range of values of the effective specific volume parameter. Application of these activity coefficients to parameters characterizing the dilute solution behavior of the *lac* repressor-operator system resulted in calculated trends of *lac* operator function that agree in general with the results of Richey et al (126) in vivo: crowding greatly decreases the expected salt dependence of *lac* operator function. According to this interpretation, crowding effects in vivo at any external osmolarity cause very large increases in the affinity of *lac* repressor for both specific and nonspecific binding sites in DNA. Hence under crowded conditions, an increase in internal salt concentration caused by shifts in external salt levels will not cause significant dissociation of *lac* repressor from DNA binding sites—even though such salt concentrations readily cause dissociation in dilute solution conditions in vitro. If essentially all protein is bound, then the expression of the *lac* operator is largely controlled by the (salt-independent) ratio of specific to nonspecific DNA sites.

The interpretation of Cayley et al (19) is based upon the decreased cytoplasmic volumes they observe as extracellular salt concentrations are raised; decreased cytoplasmic volumes lead to increased concentrations of cytoplasmic macromolecules and hence increased crowding effects. These increased crowding effects are suggested to balance the protein-dissociating effects of the increased salt, resulting in approximate salt independence in vivo. They estimate changes in the activity coefficient of RNA polymerase of at least two orders of magnitude in response to an increase in (RNA + protein) of 275 to 440 mg/ml cytoplasm, yielding final values consistent with the salt independence of the RNA polymerase–$\lambda P_R$ promoter interactions observed by Richey et al (126).

## *Diffusion of Probe Molecules in Cytoplasm*

Mastro, Keith, and coworkers (84, 85) measured the translational and rotational diffusion of a small spin probe in the cytoplasm of mammalian cells by measuring electron spin resonance. They found that the translational mobility was about half that in water but was greatly retarded upon cell shrinkage in hypertonic media, whereas rotational mobility was only slightly affected, suggesting that the spin probe was distributed within compartments in the cytoplasm. The mobility of the spin probe within a given compartment is thought to be similar to that in a dilute aqueous solution, but translational diffusion between compartments is hindered by structural elements of the cytomatrix. The extent of hindrance, and thus the overall long-range translational diffusion of the probe, is likely to reflect alterations in cytomatrix structure. Jacobson & Wojcieszyn (58) measured the apparent diffusion coefficients of several fluorescently labeled proteins in fibroblast cytoplasm via FRAP. No obvious dependence of the rate of diffusion upon the size of the protein was observed, leading them to suggest that associations with structural components of the cytomatrix, rather than hindrance due to volume exclusion, were the main factors regulating protein diffusion. Luby-Phelps, Lanni, and coworkers (56, 81, 82) have measured the diffusion of size-fractionated fluorophore-labeled Ficoll in the cytoplasm of mammalian cells by using FRAP. The diffusion coefficient of Ficoll in cytoplasm decreased with increasing polymer size in a fashion that could not be unambiguously interpreted in the context of any simple theoretical model, but which could be mimicked by the diffusion of Ficoll in mixtures of F-actin and concentrated unlabeled Ficoll or bovine serum albumin. On the basis of these findings, they propose that the cytoplasm may be described as a network of entangled fibers interpenetrated by a fluid phase containing a high concentration of soluble proteins.

## RAMIFICATIONS, SPECULATIONS, CONCLUSIONS

### Possible Role of Crowding in Regulation of Cellular Volume

In a variety of eukaryotic cell types, small changes in cellular volume can result in the activation of compensatory mechanisms that restore the original volume (135). Volume changes of a few percent reportedly can result in disproportionately large increases in ion fluxes (26, 60), prompting questions regarding the nature of the volume-change signal and mechanisms of response (30, 60). Recent studies demonstrated that in resealed dog red cell ghosts containing protein mixtures, activation of swelling and shrinkage-activated membrane ion transporters correlates with concentration of total cytoplasmic protein, as opposed to cell volume per se, or the concentration of any particular cytoplasmic protein constituent (26, 27). Because of macromolecular crowding in cytoplasm, small changes in volume (i.e. total macromolecular concentration) result in large changes in the thermodynamic activity of all macromolecular species, dilute as well as concentrated, and changes in the thermodynamic activity of one or more soluble regulatory proteins could trigger activation of compensatory mechanisms (26; also suggested in 166). Quantitative models for swelling-activated transporter-mediated ion flux embodying this notion were proposed recently (98a, 99).

### Efficacy of Macromolecular Drugs

The binding of macromolecular ligands to their complementary sites is affected by the volume occupancy of the surrounding medium (94, 95, 98a,

99). Apparent binding constants may be altered by one or more orders of magnitude. Thus, we believe that the screening of potential macromolecular drugs or any other macromolecular species for pharmacological activity should take place under conditions that mimic the crowding effects of the physiological medium in which ligand binding would actually occur. Surrogate crowding solutions that mimic estimated prokaryotic cytoplasmic backgrounds may be useful adjuncts (172).

## Synthesis by Degradative Enzymes

Given the enhanced tendency of macromolecules to associate to form compact complexes in crowded media, it seems probable that at least some enzymes that catalyze degradative reactions such as limited proteolysis in vitro (i.e. in dilute solution) might actually catalyze the reverse reaction in vivo, and thus function primarily as synthetic rather than degradative enzymes.

## Macromolecular Crowding as a Laboratory Tool

Inert volume-occupying macromolecules such as PEG have been added to increase the efficiency of common laboratory procedures involving nucleic acids, such as labeling DNA termini, ligation of DNA fragments, and renaturing of DNA (134, 142).

## Osmotic Remedial Mutants

Auxotrophic or temperature-sensitive mutants of prokaryotes commonly become functional in media of high osmolarity (31, 51). Cayley et al (19) showed that the steady-state cytoplasmic volume of growing cells decreased with increasing external osmolarity. They proposed that this enhances intracellular crowding effects at high osmolarity. Resulting shifts in intracellular equilibria and reaction rates could potentially compensate for a variety of mutational defects.

## Crowding as a Factor in Cellular Evolution

Living cells have high total macromolecular content—we can think of no exceptions to this statement. Does this mean that macromolecular crowding is essential to life? Life is thought to have originated in a primordial soup of prebiotic organic molecules; was that soup crowded? Reversible interactions that evolved in a crowded soup—or in a living system—would be characterized by affinities that are no stronger than they need to be in order to confer function in the crowded environment. If the cell is lysed and total macrosolute concentration significantly reduced, the resulting reduction in thermodynamic activities would lead to the dissolution of some relatively weak interactions that are present in the intact cell (88) and the loss of high-level structural information.

Berg (15) has suggested that intracellular crowding may act as an evolutionary force tending to bias conformational equilibria toward compact conformations. In this context, we note that native structures of proteins, ribosomes, and other intracellular moieties tend to be compact; even DNA, perhaps the most intrinsically anisometric of all biological materials, tends to occur in highly compact structures such as nucleoids or nucleosomes (157), and may even function in a condensed conformation (142).

## Crowding Versus Confinement

In the present review, the term crowding has been used to denote the exclusion of solution volume to macrosolute molecules by other macrosolute molecules. Solution volume may also be excluded to macrosolutes by stationary structural elements, such as membranes and fiber lattices, that restrict the translational and rotational freedom of soluble species. The cytoplasm of most eukaryotic cells contains a significant volume fraction of immobile[12] structural elements of various types, collectively referred to as the cytomatrix (40). A statistical-thermodynamic theory of the thermodynamic activity of rigid particles in confined spaces (42) was recently used to calculate the effects of confinement on various types of reaction equilibria (98). Confinement was found to substantially affect thermodynamic activity of a particular solute species when the characteristic spacing between confining boundaries became smaller than about three times the maximum dimension of the confined particle.

If confinement is uniform throughout the fluid, then confinement enhances association equilibria. Unlike crowding, the extent to which confinement affects associations depends substantially upon the relationship between the shape of the aggregate and the shape of the confining boundaries. If confinement is nonuniform, then solutes (and aggregates) will partition so that larger particles will be preferentially found in larger spaces.

## Homeostasis or Metabolic Buffering

The increased binding often caused by crowding may be a global tendency in cells that offsets a variety of perturbing influences, such as increased salt levels, drug binding, or pH changes (166, 169). This projected homeostatic tendency, termed metabolic buffering, could in principle apply to equi-

---

[12] That is, the structural elements are immobile on the time scale of translational motions of individual molecules of macrosolute.

librium and kinetic aspects of isomerization reactions, associations in solution, phase equilibria, and binding of soluble ligands to surface sites. A quantitative example has been presented (172).

## Crowding and the Functional Importance of Condensed DNA

Sikorav & Church (142) demonstrated that factors such as the addition of volume-occupying polymers that favor the formation of condensed DNA (8, 77, 78) greatly accelerate the rate of renaturation of single-stranded DNA and can also support a protein-free DNA strand–exchange reaction. They suggest that the process of matching complementary base-pair sequences is more efficient in the condensed phase than in solution, and that condensed rather than extended DNA may be the physiologically relevant structure.

## Crowding and Cytoplasmic Structure

The two major qualitative manifestations of crowding are enhancement of macromolecular associations and hindrance of macromolecular diffusion. Assuming that most of the cytoplasm is characterized by a fractional volume occupancy that is at least as great as the whole-cell average value, we speculate that a typical mature, functioning protein molecule within the cytoplasm is likely to exist as part of a complex rather than as a free-floating moiety, and consequently would have a greatly reduced rate of long-range translational diffusion (cf 24, 63). Because diffusion of small molecules is affected only slightly by macromolecular crowding, metabolism would therefore be expected to proceed predominantly via the diffusion of small-molecule metabolites between the various active sites of relatively immobile enzymes catalyzing sequential reactions. While these considerations do not necessarily mandate the formation of ordered complexes of enzymes catalyzing successive metabolic reactions, such as the metabolons proposed by Srere (144), they do render such complexes teleologically attractive: if proteins are to be complexed (as favored by crowding), would not evolution select for the formation of ordered complexes that could speed metabolic throughput and enhance control mechanisms (144) over the formation of random, functionally neutral complexes?

ACKNOWLEDGMENTS

We thank Philip Ross, NIH, and M. Thomas Record, Jr., University of Wisconsin, for helpful comments on a preliminary draft of this review, and Lizabeth Murphy, NIH, for assistance in its preparation.

## Literature Cited

1. Abney, J. R., Scalettar, B. A., Owicki, J. C. 1989. *Biophys. J.* 55: 817–33
2. Adair, G. S. 1928. *Proc. R. Soc. London Ser. A* 120: 573–603
3. Altman, P. L. 1961. *Blood and Other Body Fluids.* Washington, DC: FASEB
4. Aoufouchi, S., Hardy, S., Prigent, C., Philippe, M., Thiebaud, P. 1991. *Nucleic Acids Res.* 19: 4395–98
5. Aragón, J. J., Sols, A. 1991 *FASEB J.* 5: 2945–50
6. Arakawa, T., Timasheff, S. N. 1985. *Biochemistry* 24: 6756–62
7. Arrand, J. E., Willis, A. E., Goldsmith, I., Lindahl, T. 1986. *J. Biol. Chem.* 261: 9079–82
8. Auer, C. 1978. *The influence of the ionic environment on the structure of DNA.* PhD thesis. Vanderbilt Univ., Nashville, Tenn. 323 pp.
9. Atha, D. H., Ingham, K. C. 1981. *J. Biol. Chem.* 256: 12108–17
10. Barkley, M. D. 1981. *Biochemistry* 20: 3833–42
11. Barringer, K. J., Orgel, L., Wahl, G., Gingeras, T. R. 1990. *Gene* 89: 117–22
12. Baskir, J. N., Hatton, T. A., Suter, U. W. 1987. *Macromolecules* 20: 1300–11
13. Baskir, J. N., Hatton, T. A., Suter, U. W. 1989. *J. Phys. Chem.* 93: 2111–22
14. Behe, M. J., Englander, S. W. 1978. *Biophys. J.* 23: 129–45
15. Berg, O. G. 1990. *Biopolymers* 30: 1027–37
16. Berg, O. G., von Hippel, P. H. 1985. *Annu. Rev. Biophys. Biophys. Chem.* 14: 131–60
17. Bosma, H. J., Voordouw, G., De Kok, A., Veeger, C. 1980. *FEBS Lett.* 120: 179–82
18. Brenner, S. L. 1976. *J. Phys. Chem.* 80: 1473–77
19. Cayley, S., Lewis, B. A., Guttman, H. J., Record, M. T. Jr. 1991. *J. Mol. Biol.* 222: 281–300
20. Cayley, S., Lewis, B. A., Record, M. T. Jr. 1992. *J. Bacteriol.* 174: 1586–95
21. Chatelier, R. C., Minton, A. P. 1987. *Biopolymers* 26: 507–24
22. Chatelier, R. C., Minton, A. P. 1987. *Biopolymers* 26: 1097–1113
23. Chun, P. W., Thornby, J. I., Saw, J. G. 1969. *Biophys. J.* 9: 163–72
24. Clegg, J. S. 1984. *Am. J. Physiol.* 246: R133–51
25. Cohen, S. S. 1942. *J. Biol. Chem.* 144: 353–62
26. Colclasure, G. C., Parker, J. C. 1991. *J. Gen. Physiol.* 98: 881–92
27. Colclasure, G. C., Parker, J. C. 1992. *J. Gen. Physiol.* 100: 1–11
28. Comper, W. D., Laurent, T. C. 1978. *Biochem. J.* 175: 703–8
29. Comper, W. D., Laurent, T. C. 1978. *Physiol. Rev.* 58: 255–315
30. Cossins, A. R. 1991. *Nature* 352: 667–68
31. Csonka, L. N. 1989. *Microbiol. Rev.* 53: 121–47
32. Doi, M. 1975. *J. Chem. Soc. Faraday Trans. 2* 71: 1720–29
33. Drenckhahn, D., Pollard, T. D. 1986. *J. Biol. Chem.* 261: 12754–58
34. Dugaiczyk, A., Boyer, H. W., Goodman, H. M. 1975. *J. Mol. Biol.* 96: 171–84
35. Eaton, W. A., Hofrichter, J. 1990. *Adv. Protein Chem.* 40: 63–279
36. Edmond, E., Ogston, A. G. 1968. *Biochem. J.* 109: 569–76
37. Forterre, P., Mirambeau, G., Jaxel, C., Nadal, M., Duguet, M. 1985. *EMBO J.* 4: 2123–28
38. Fuller, R. S., Kaguni, J. M., Kornberg, A. 1981. *Proc. Natl. Acad. Sci. USA* 78: 7370–74
39. Fulton, A. B. 1982. *Cell* 30: 345–47
40. Gershon, N. D., Porter, K. R., Trus, B. L. 1985. *Proc. Natl. Acad. Sci. USA* 82: 5030–34
41. Gibbons, R. M. 1969. *Mol. Phys.* 17: 81–86
42. Giddings, J. C., Kucera, E., Russell, C. P., Myers, M. N. 1968. *J. Phys. Chem.* 72: 4397–4408
43. Goodsell, D. S. 1991. *Trends Biochem. Sci.* 16: 203–6
44. Gorti, S., Ware, B. R. 1985. *J. Chem. Phys.* 83: 6449–56
45. Grasberger, B., Minton, A. P., DeLisi, C., Metzger, H. 1986. *Proc. Natl. Acad. Sci. USA* 83: 6258–62
46. Gros, G. 1978. *Biophys. J.* 22: 453–68
47. Haire, R. N., Tisel, W. A., White, J. G., Rosenberg, A. 1984. *Biopolymers* 23: 2761–79
48. Harrison, B., Zimmerman, S. B. 1984. *Nucleic Acids Res.* 12: 8235–51
49. Harrison, B., Zimmerman, S. B. 1986. *Anal. Biochem.* 158: 307–15
50. Harrison, B., Zimmerman, S. B. 1986. *Nucleic Acids Res.* 14: 1863–70
51. Hawthorne, D. C., Friis, J. 1964. *Genetics* 50: 829–39
52. Hayashi, K., Nakazawa, M., Ishizaki, Y., Hiraoka, N., Obayashi, A. 1985. *Nucleic Acids Res.* 13: 7979–92
53. Hayashi, K., Nakazawa, M., Ishizaki, Y., Obayashi, A. 1985. *Nucleic Acids Res.* 13: 3261–71
54. Hermans, J. 1982. *J. Chem. Phys.* 77: 2193–2203
55. Hill, T. L. 1960. In *An Introduction to*

*Statistical Thermodynamics*, pp. 340–70. Reading, MA: Addison-Wesley
56. Hou L., Lanni, F., Luby-Phelps, K. 1990. *Biophys. J.* 58: 31–43
57. Ingham, K. C. 1978. *Arch. Biochem. Biophys.* 186: 106–13
58. Jacobson, K., Wojcieszyn, J. 1984. *Proc. Natl. Acad. Sci. USA* 81: 6747–51
59. Jarvis, T. C., Ring, D. M., Daube, S. S., von Hippel, P. H. 1990. *J. Biol. Chem.* 265: 15160–67
60. Jennings, M. L., Schulz, R. K. 1990. *Am. J. Physiol.* 259: C960–67
61. Juckes, I. R. M. 1971. *Biochim. Biophys. Acta* 229: 535–46
62. Kaguni, J. M., Kornberg, A. 1984. *Cell* 38: 183–90
63. Kempner, E. S., Miller, J. H. 1968. *Exp. Cell Res.* 51: 150–56
64. Kim, C. W. 1986. *Interfacial condensation of biologicals in aqueous two-phase systems.* PhD thesis. MIT, Cambridge, Mass. 291 pp.
65. Knoll, D., Hermans, J. 1983. *J. Biol. Chem.* 258: 5710–15
66. Kornberg, A., Baker, T. A. 1992. In *DNA Replication*, pp. 314, 485, 525. New York: Freeman. 2nd ed.
67. Kuntz, I. D. Jr., Kauzmann, W. 1974. *Adv. Protein Chem.* 28: 239–345
68. Laurent, T. C. 1963. *Biochem. J.* 89: 253–57
68a. Laurent, T. C. 1963. *Acta Chem. Scand.* 17: 2664–68
69. Laurent, T. C. 1971. *Eur. J. Biochem.* 21: 498–506
70. Laurent, T. C., Björk, I., Pietruszkiewicz, A., Persson, H. 1963. *Biochim. Biophys. Acta* 78: 351–59
71. Laurent, T. C., Ogston, A. G. 1963. *Biochem. J.* 89: 249–53
72. Laurent, T. C., Preston, B. N., Carlsson, B. 1974. *Eur. J. Biochem.* 43: 231–35
73. Lavery, P. E., Kowalczykowski, S. C. 1992. *J. Biol. Chem.* 267: 9307–14
74. LeBowitz, J. H., McMacken, R. 1984. *Nucleic Acids Res.* 12: 3069–88
75. LeBowitz, J. H., Zylicz, M., Georgopoulos, C., McMacken, R. 1985. *Proc. Natl. Acad. Sci. USA* 82: 3988–92
76. Lebowitz, J. L., Helfand, E., Praestgaard, E. 1965. *J. Chem. Phys.* 43: 774–79
77. Lerman, L. S. 1971. *Proc. Natl. Acad. Sci. USA* 68: 1886–90
78. Lerman, L. S. 1973. In *Physico-Chemical Properties of Nucleic Acids*, ed. J. Duchesne, 3: 59–76. New York: Academic
79. Louie, D., Serwer, P. 1991. *Nucleic Acids Res.* 19: 3047–54
80. Low, R. L., Kaguni, J. M., Kornberg, A. 1984. *J. Biol. Chem.* 259: 4576–81
81. Luby-Phelps, K., Castle, P. E., Taylor, D. L., Lanni, F. 1987. *Proc. Natl. Acad. Sci. USA* 84: 4910–13
82. Luby-Phelps, K., Lanni, F., Taylor, D. L. 1988. *Annu. Rev. Biophys. Biophys. Chem.* 17: 369–96
83. Madden, T. L., Herzfeld, J. 1992. *Mater. Res. Soc. Symp. Proc.* 248: 95–100
84. Mastro, A. M., Babich, M. A., Taylor, W. D., Keith, A. D. 1984. *Proc. Natl. Acad. Sci. USA* 81: 3414–18
85. Mastro, A. M., Keith, A. D. 1984. *J. Cell Biol.* 99: 180s–87s
86. Matthews, B. W. 1977. In *The Proteins*, ed. H. Neurath, R. L. Hill, 3: 403–590. New York: Academic. 3rd ed.
87. Mazur, P. 1987. *Faraday Discuss. Chem. Soc.* 83: 33–46
88. McConkey, E. H. 1982. *Proc. Natl. Acad. Sci USA* 79: 3236–40
89. McMillan, W. G. Jr., Mayer, J. E. 1945. *J. Chem. Phys.* 13: 276–305
90. Middaugh, C. R., Tisel, W. A., Haire, R. N., Rosenberg, A. 1979. *J. Biol. Chem.* 254: 367–70
91. Minton, A. P. 1977. *J. Mol. Biol.* 110: 89–103
92. Minton, A. P. 1980. *Biophys. Chem.* 12: 271–77
93. Minton, A. P. 1980. *Biophys. J.* 32: 77–79
94. Minton, A. P. 1981. *Biopolymers* 20: 2093–2120
95. Minton, A. P. 1983. *Mol. Cell. Biochem.* 55: 119–40
96. Minton, A. P. 1989. *Biophys. J.* 55: 805–8
97. Minton, A. P. 1990. *Int. J. Biochem.* 22: 1063–67
98. Minton, A. P. 1992. *Biophys. J.* 63: 1090–1100
98a. Minton, A. P. 1993. In *Cellular and Molecular Physiology of Cell Volume Regulation*, ed. K. Strange. Boca Raton, FL: CRC. In press
99. Minton, A. P., Colclasure, G. C., Parker, J. C. 1992. *Proc. Natl. Acad. Sci. USA* In press
100. Minton, A. P., Edelhoch, H. 1982. *Biopolymers* 21: 451–58
101. Minton, A. P., Wilf, J. 1981. *Biochemistry* 20: 4821–26
102. Minton, K. W., Karmin, P., Hahn, G. M., Minton, A. P. 1982. *Proc. Natl. Acad. Sci. USA* 79: 7107–11
103. Mitchell, P., Moyle, J. 1956. In *Bacterial Anatomy*, *6th Symp. Soc. General Microbiology*, pp. 150–80. Cambridge, UK: Cambridge Univ. Press
104. Mizouchi, K. 1983. *Cell* 35: 785–94
105. Mooney, M. 1951. *J. Colloid Sci.* 6: 162–70

106. Muramatsu, N., Minton, A. P. 1988. *Proc. Natl. Acad. Sci. USA* 85: 2984–88
107. Nichol, L. W., Jeffrey, P. D., Winzor, D. J. 1976. *J. Phys. Chem.* 80: 648–49
108. Nichol, L. W., Ogston, A. G., Wills, P. R. 1981. *FEBS Lett.* 126: 18–20
109. Noguchi, C. T., Schechter, A. N. 1985. *Annu. Rev. Biophys. Biophys. Chem.* 14: 239–63
110. Ogston, A. G. 1958. *Trans. Faraday Soc.* 54: 1754–57
111. Ogston, A. G., Phelps, C. F. 1961. *Biochem. J.* 78: 827–33
112. Ogston, A. G., Preston, B. N., Wells, J. D. 1973. *Proc. R. Soc. London Ser. A* 333: 297–316
113. Ogston, A. G., Winzor, D. J. 1975. *J. Phys. Chem.* 79: 2496–500
114. O'Leary, T. J. 1987. *Biophys. J.* 52: 137–39
115. Ohtsuki, K. 1982. *Physica A* 110: 606–16
116. Ovadi, J., Batke, J., Bartha, F., Keleti, T. 1979. *Arch. Biochem. Biophys.* 193: 28–33
117. Peters, R., Cherry, R. J. 1982. *Proc. Natl. Acad. Sci. USA* 79: 4317–21
118. Pheiffer, B. H., Zimmerman, S. B. 1983. *Nucleic Acids Res.* 11: 7853–71
119. Phillies, G. D. J., Ullman, G. S., Ullman, K., Lin, T.-H. 1985. *J. Chem. Phys.* 82: 5242–46
120. Pink, D. A. 1985. *Biochim. Biophys. Acta* 818: 200–4
121. Preston, B. N., Snowden, J. M. 1973. *Proc. R. Soc. London Ser. A* 333: 311–13
122. Pusey, P. N., Tough, R. J. A. 1983. *Faraday Discuss. Chem. Soc.* 76: 123–36
123. Rabin, B. A., Chase, J. W. 1987. *J. Biol. Chem.* 262: 14105–11
124. Ralston, G. B. 1990. *J. Chem. Educ.* 67: 857–60
125. Reiss, H. 1965. *Adv. Chem. Phys.* 9: 1–84
126. Richey, B., Cayley, D. S., Mossing, M. C., Kolka, C., Anderson, C. F., et al. 1987. *J. Biol. Chem.* 262: 7157–64
127. Riggs, A. D., Bourgeois, S., Cohn, M. 1970. *J. Mol. Biol.* 53: 401–17
128. Roe, J.-H., Burgess, R. R., Record, M. T. Jr. 1984. *J. Mol. Biol.* 176: 495–521
129. Roe, J.-H., Record, M. T. Jr. 1985. *Biochemistry* 24: 4721–26
130. Ross, P. D., Briehl, R. W., Minton, A. P. 1978. *Biopolymers* 17: 2285–88
131. Ross, P. D., Minton, A. P. 1977. *Biochem. Biophys. Res. Commun.* 76: 971–76
132. Ross, P. D., Minton, A. P. 1977. *J. Mol. Biol.* 112: 437–52
133. Ross, P. D., Minton, A. P. 1979. *Biochem. Biophys. Res. Commun.* 88: 1308–14
134. Sambrook, J., Fritsch, E. F., Maniatis, T. 1989. In *Molecular Cloning: A Laboratory Manual*, 1: 1.70–1.71, 4.35. Cold Spring Harbor, NY: Cold Spring Harbor Lab. 2nd ed.
135. Sarkadi, B., Parker, J. C. 1991. *Biochim. Biophys. Acta* 1071: 402–27
136. Saxton, M. J. 1982. *Biophys. J.* 39: 165–73
137. Saxton, M. J. 1987. *Biophys. J.* 52: 989–97
138. Saxton, M. J. 1989. *Biophys. J.* 56: 615–22
139. Saxton, M. J. 1990. *Biophys. J.* 58: 1303–6
140. Scalettar, B. A., Abney, J. R. 1991. *Comments Mol. Cell. Biophys.* 7: 79–107
141. Shearwin, K. E., Winzor, D. J. 1988. *Arch. Biochem. Biophys.* 265: 458–65
142. Sikorav, J.-L., Church, G. M. 1991. *J. Mol. Biol.* 222: 1085–1108
143. Sobczak, J., Duguet, M. 1988. *Eur. J. Biochem.* 175: 379–85
144. Srere, P. A. 1987. *Annu. Rev. Biochem.* 56: 89–124
145. Sunshine, H. R., Hofrichter, J., Eaton, W. A. 1979. *J. Mol. Biol.* 133: 435–67
146. Suzuki, A., Yamazaki, M., Ito, T. 1989. *Biochemistry* 28: 6513–18
147. Takahashi, M., Uchida, T. 1986. *J. Biochem. (Tokyo)* 100: 123–31
148. Tanford, C. 1961. *Physical Chemistry of Macromolecules*. New York: Wiley & Sons
149. Tank, D. W., Wu, E.-S., Meers, P. R., Webb, W. W. 1982. *Biophys. J.* 40: 129–35
150. Tellam, R. L., Sculley, M. J., Nichol, L. W. 1983. *Biochem. J.* 213: 651–59
151. Teraoka, H., Tsukada, K. 1987. *J. Biochem. (Tokyo)* 101: 225–31
152. Tessier, D. C., Brousseau, R., Vernet, T. 1986. *Anal. Biochem.* 158: 171–78
153. Tilton, R. D., Gast, A. P., Robertson, C. R. 1990. *Biophys. J.* 58: 1321–26
154. Torbet, J. 1986. *Biochemistry* 25: 5309–14
155. Tsurimoto, T., Matsubara, K. 1982. *Proc. Natl. Acad. Sci. USA* 79: 7639–43
156. Van Damme, M.-P. I., Murphy, W. H., Comper, W. D., Preston, B. N., Winzor, D. J. 1989. *Biophys. Chem.* 33: 115–25
157. Watson, J. D., Hopkins, N. H., Roberts, J. W., Steitz, J. A., Weiner, A. M. 1987. *Molecular Biology of the Gene*, Vol. 1. Menlo Park, CA: Benjamin Cummings. 4th ed.

158. West, R. 1988. *Biopolymers* 27: 231–49
158a. Wieder, R., Wetmur, J. G. 1981. *Biopolymers* 20: 1537–47
159. Wilf, J., Gladner, J. A., Minton, A. P. 1985. *Thromb. Res.* 37: 681–88
160. Wilf, J., Minton, A. P. 1981. *Biochim. Biophys. Acta* 670: 316–22
161. Wills, P. R., Nichol, L. W., Siezen, R. J. 1980. *Biophys. Chem.* 11: 71–82
162. Winter, R. B., Berg, O. G., von Hippel, P. H. 1981. *Biochemistry* 20: 6961–77
163. Winzor, D. J., Wills, P. R. 1986. *Biophys. Chem.* 25: 243–51
164. Woolley, P., Wills, P. R. 1985. *Biophys. Chem.* 22: 89–94
165. Zimmerman, S. B., Harrison, B. 1985. *Nucleic Acids Res.* 13: 2241–49
166. Zimmerman, S. B., Harrison, B. 1987. *Proc. Natl. Acad. Sci. USA* 84: 1871–75
167. Zimmerman, S. B., Murphy, L. D. 1992. *Biopolymers.* 32: 1365–73
168. Zimmerman, S. B., Pheiffer, B. H. 1983. *Proc. Natl. Acad. Sci. USA* 80: 5852–56
169. Zimmerman, S. B., Trach, S. O. 1988. *Biochim. Biophys. Acta* 949: 297–304
170. Zimmerman, S. B., Trach, S. O. 1988. *Nucleic Acids Res.* 16: 6309–26; Erratum. *Nucleic Acids Res.* 16: 9892
171. Zimmerman, S. B., Trach, S. O. 1990. *Biopolymers* 30: 703–18
172. Zimmerman, S. B., Trach, S. O. 1991. *J. Mol. Biol.* 222: 599–620
173. Taylor, M. P., Herzfeld, J. 1991. *Phys. Rev. A* 43: 1892–1905

# THE CONTROL OF PROTEIN STABILITY AND ASSOCIATION BY WEAK INTERACTIONS WITH WATER: How Do Solvents Affect These Processes?

*Serge N. Timasheff*

Graduate Department of Biochemistry, Brandeis University, Waltham, Massachusetts 02254

KEY WORDS: guanidine hydrochloride, urea, preferential interactions, transfer free energy, preferential hydration

CONTENTS

| | |
|---|---|
| PERSPECTIVES AND OVERVIEW | 67 |
| THERMODYNAMIC CONTROL OF EQUILIBRIA BY COSOLVENTS | 70 |
| RELATION BETWEEN PREFERENTIAL INTERACTIONS AND MOLECULAR CONTACTS | 76 |
| EXPERIMENTAL OBSERVATIONS: APPLICATIONS | 81 |
|     *Denaturants* | 82 |
|     *Stabilization: Salting-Out* | 84 |
|     *Sources of Preferential Exclusions: Stabilization or Not?* | 85 |
| BALANCE BETWEEN COSOLVENT BINDING AND EXCLUSION | 89 |
| COMPENSATION: OSMOLYTES | 92 |
| CONCLUSIONS | 94 |

## PERSPECTIVES AND OVERVIEW

For the better part of the century, a common practice among biologists and biochemists, when they isolate a biologically active entity (e.g. an enzyme activity) or dissect an organelle out of a cell, has been to add a high concentration of sucrose ($\sim 1$ M) or glycerol ($\sim 10\%$) to the medium in order to keep the activity stabilized or the organelle functional. For just as long, when isolating enzymes and other proteins, biochemists have used high concentrations (1–4 M) of salts, most commonly ammonium sulfate, to precipitate the proteins. This process has been called salting-out (32). Similarly, investigators have known that addition of concentrated urea (8 M) or guanidine hydrochloride (GuaHCl) (6 M) leads to the denaturation of proteins (enzymes), i.e. a loss of their biochemical activity, and a drastic alteration in solution properties. Researchers interpreted these observations to represent an uncoiling of the native structure of proteins (61).

The mechanisms of the salting-out and denaturation processes have been the subject of investigations for well over 50 years, and several excellent reviews explore these areas (14, 40, 64, 65, 89, 91). Precipitation by salts has been analyzed in a series of experimental and theoretical studies that are summarized in the classic Cohn & Edsall (21) monograph. The effects of salts on proteins was defined in terms of two opposing phenomena: salting-in, i.e. increase in protein solubility due to Debye-Hückel screening, at low salt concentrations, and salting-out at higher salt concentrations ($>1.0$ ionic strength). The salting-out process was found to be linear in ionic strength. It was characterized by a parameter, $K_s$, called the salting-out constant, which is the slope of the dependence of the log of protein solubility on the ionic strength.

In the case of denaturation by urea and GuaHCl, once again, investigators realized early on that the unfolding process is accompanied by the direct interaction (binding) of the denaturant with protein molecules (82). As a result of extensive studies on model compounds, Tanford (88, 91) showed that the thermodynamic facilitation of protein unfolding by addition of urea or GuaHCl can be accounted for by the free-energy changes that accompany the transfer of the protein in the native and denatured states from water to a denaturant solution, i.e. the free energy of interaction with the denaturants of protein groups that become exposed to solvent upon unfolding.

The third process, protein stabilization by sucrose, glycerol, and related compounds, had attracted little interest, probably because in this process nothing happens to the protein. A vague, commonly held belief was that these substances form some sort of protective coating—a shell or a glue that prevents the proteins from unfolding—but there was no experimental probing until some 15 years ago, when studies of both sucrose (53) and glycerol (30) showed that the opposite is true: these compounds not only do not form a shell around the protein, they are excluded from the immediate domain of protein molecules, which in their presence are surrounded by an environment enriched in water.

Although on the surface these three phenomena may seem unrelated, closer examination reveals one common basic fact: solvent denaturation, protein structure stabilization, and precipitation (salting-out) all require a high concentration of the added agent (1–8 M). Therefore, whatever interactions take place can be neither strong nor specific. They fall, therefore, into the realm of thermodynamics of weakly interacting systems. In fact, recent systematic studies of the interactions of soluble globular proteins with agents that stabilize their structure, induce precipitation, and enhance self-assembly processes have shown that the three processes are governed by the same set of weak interactions (1, 2). They can all be described in terms of three-component thermodynamic theory (18, 43, 87), in which water is explicitly taken into account. All structure-stabilizing, precipitating, and self-assembly–inducing agents are preferentially excluded from the domain of the protein, i.e. in their presence the proteins are preferentially hydrated (1, 2, 100). Denaturants, on the other hand, are preferentially bound to proteins (33, 38, 62, 71). Clearly stabilization, denaturation, and salting-out are particular consequences of a single general thermodynamic phenomenon, the preferential binding of solvent components by proteins. In fact, stabilizers (sucrose, glycerol, etc) and denaturants (urea, guanidine hydrochloride, etc) are a single class of compounds that form a continuum from strong globular-protein–structure enhancers to strong protein unfolders. The same holds true for agents that affect solubility. The action of an agent, whether it be structure stabilization or destabilization, precipitation or solubilization, is defined only by the balance between the affinities of the protein for water and the particular agent (e.g. sucrose, urea), which varies from strong preference for water to strong preference for cosolvent. This review, therefore, is devoted to the development of the unifying principles and their application to these various processes.

The principles that govern the above cited laboratory processes turn out, in fact, to already be widespread in nature, where they evolved eons ago. Nature protects life from freezing or osmotic shock by the accumulation of selected compounds within organisms to very high ($>1$ M) concentrations. It is amazing how small the number of compounds is that are used for these purposes by organisms ranging from single cells to amphibians and higher plants (110). These compounds, known as osmolytes and cryoprotectants, belong, almost without exception, to the class of compounds that are preferentially excluded from the protein surface and that act as structure stabilizers (99). This observation allows us to bring this broad natural phenomenon under the same umbrella of preferential interactions. It also makes one wonder at the thermodynamic fine tuning that certain microorganisms, frogs, and desert plants have developed to protect themselves from extinction.

# THERMODYNAMIC CONTROL OF EQUILIBRIA BY COSOLVENTS

What are the thermodynamic principles that govern all of the above phenomena, whether in nature or in the laboratory? All the controls can be described in terms of three thermodynamically related parameters and their changes during the course of a reaction. They are:

1. The *transfer free energy*, $\Delta\mu_{2,tr}$: this is the change in the interactions of the protein with the solvent system when it is transferred from pure water to the cosolvent system:

$$\Delta\mu_{2,tr} = \mu_2 \text{(cosolvent)} - \mu_2 \text{(water)}, \qquad 1.$$

where $\mu_i$ is the chemical potential of component $i$. It is defined as $\mu_i = \mu_i^\circ(P, T) + RT\ln a_i$, where $a_i$ is the activity of component $i$, i.e. $a_i = m_i\gamma_i$, where $m_i$ is its molal concentration and $\gamma_i$ is its activity coefficient. The subscript 2 indicates protein in the Scatchard (73) notation of solution components; water and cosolvent are designated as components 1 and 3.

2. The *preferential interaction parameter* $(\partial\mu_2/\partial m_3)_{T,P,m_2} = (\partial\mu_3/\partial m_2)_{T,P,m_3}$: this is the mutual perturbations of the chemical potentials of protein and cosolvent. It is the gradient of the transfer free energy with cosolvent concentration, i.e. a measure of the thermodynamic interaction at the given salt concentration. Thus,

$$\Delta\mu_{2,tr} = \int_0^{m_3} (\partial\mu_2/\partial m_3)_{T,P,m_2}\, dm_3. \qquad 2.$$

3. The *preferential binding parameter*, $(\partial m_3/\partial m_2)_{T,P,\mu_3}$: this is the observable manifestation of the chemical potential perturbation. It is the expression of the amount of cosolvent that would have to be added to (or removed from) the solvent system to restore thermodynamic equilibrium when protein is added:

$$\left(\frac{\partial m_3}{\partial m_2}\right)_{T,P,\mu_3} = -\frac{(\partial\mu_2/\partial m_3)_{T,P,m_2}}{(\partial\mu_3/\partial m_3)_{T,P,m_2}}, \qquad 3.$$

where the denominator is the cosolvent nonideality. Within the negligibly small approximation that $(\partial m_3/\partial m_2)_{T,P,\mu_3} \approx (\partial m_3/\partial m_2)_{T,\mu_1,\mu_3}$ (86), the preferential binding parameter is equal to the binding measured experimentally at dialysis equilibrium, $\bar{v}_3$ (moles ligand/mole protein) in Scatchard (74) notation.

The effect of an agent (cosolvent) on a macromolecular equilibrium reaction is examined usually from either of two points of reference: one is the manner in which it affects the reaction relative to pure water as solvent; the other is the direction in which it displaces the equilibrium at the given solvent composition. These two approaches are related thermodynamically in an exact manner. Taking water as the reference state first, the parameter of interest is the change in the standard free-energy change of the reaction, $\delta \Delta G°$, when the system is transferred from water to the cosolvent system. For the equilibrium $R \rightleftharpoons Pr$ (Reactant $\rightleftharpoons$ Product), this change is equal to the change in the transfer free energy of the protein from water to the cosolvent system, $\delta \Delta \mu_{2,tr}$, during the course of the reaction (88). Since $\delta \Delta G°$ equals $\delta \Delta \mu_{2,tr}$,

$$\delta \Delta G° = \Delta G_{m_3}^{\circ(R-Pr)} - \Delta G_w^{\circ(R-Pr)} = \Delta \mu_{2,tr}^{Pr} - \Delta \mu_{2,tr}^{R} = \delta \Delta \mu_{2,tr}. \qquad 4.$$

For the three processes of specific interest, we may now write the appropriate relations:

1. Stability ($N \rightleftharpoons D$; Native $\rightleftharpoons$ Denatured):

$$\Delta G_{m_3}^{\circ, N-D} - \Delta G_w^{\circ, N-D} = \Delta \mu_{2,tr}^{D} - \Delta \mu_{2,tr}^{N}. \qquad 5.$$

2. Solubility (protein dispersed in solution $\rightleftharpoons$ precipitate):

$$RT(\ln S_{2,m_3} - \ln S_{2,w}) = \Delta \mu_{2,tr}^{ppt} - \Delta \mu_{2,tr}^{sol}, \qquad 6.$$

where $S_2$ is protein solubility, the subscript w is water, and the superscripts ppt and sol mean precipitate and solution, respectively. Equation 6, when combined with Equation 2 and the definition of the salting-out constant, $K_s$, given by the empirical protein solubility equation at high salt concentration, $\log S = \beta - K_s m_3$ (where $\beta$ is an empirical constant) (21), leads to the thermodynamic definition of the solubility constant (100a):

$$K_s = \frac{1}{2.303 RT} \left[ \left(\frac{\partial \mu_2}{\partial m_3}\right)_{T,P,m_2}^{sol} - \left(\frac{\partial \mu_2}{\partial m_3}\right)_{T,P,m_2}^{ppt} \right]. \qquad 6a.$$

3. Self-assembly ($P_n + P \rightleftharpoons P_{n+1}$):

$$\Delta G_{m_3}^{\circ(growth)} - \Delta G_w^{\circ(growth)} = \Delta \mu_{2,tr}^{P_{n+1}} - \Delta \mu_{2,tr}^{P}, \qquad 7.$$

where the superscripts P and $P_{n+1}$ mean protein subunits freely dispersed in solution and incorporated into the assembled structure, respectively.

The direction into which any of these processes will be shifted by a cosolvent relative to water clearly depends on the balance between the transfer free energies of the protein from water to the solvent system in the two end states of the reaction. The value of $\Delta \mu_{2,tr}$ in either end state may itself be negative or positive, depending on whether the interaction of the protein with the cosolvent system is favorable or unfavorable. The two examples shown in Figure 1 illustrate how the stability of proteins reflects the balance in transfer free energies: the denaturation of chymotrypsinogen by polyethylene glycol 1000 (PEG) (50) and the stabilization of ribonuclease (RNase) by sorbitol (G. Xie & S. N. Timasheff, in preparation). In both cases, the effect of the cosolvent on the standard free-energy change of the denaturation equilibrium is equal to the difference between the transfer free energies of denatured and native protein.

In the second reference state, solvent of the given composition $m_3$, the effect of the cosolvent is examined best through the variation of the equilibrium constant with solvent composition and its relation to the change in the preferential interaction parameter during the course of the reaction, namely through the Wyman (108) linkage relation:

$$\left(\frac{\partial \ln K}{\partial \ln a_3}\right)_{T,P,m_2} = \frac{(\partial \mu_3/\partial m_2)^R_{T,P,m_3} - (\partial \mu_3/\partial m_2)^{Pr}_{T,P,m_3}}{(\partial \mu_3/\partial m_3)_{T,P,m_2}} \qquad 8.$$

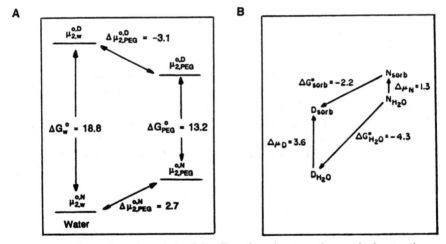

*Figure 1* Thermodynamic analysis of the effect of cosolvents on the protein denaturation equilibrium. (*A*) Unfolding of chymotrypsinogen by 20% polyethylene glycol 1000 (PEG) at 20°C; contact between the protein and PEG is unfavorable ($\Delta\mu_{2,tr}$ positive) in the native state, but favorable ($\Delta\mu_{2,tr}$ negative) in the denatured state. Hence, $\delta\Delta\mu_{2,tr} = -5.8$ kcal/mol reduces $\Delta G^{\circ,N-D}$ by an equal amount, making PEG into a denaturant [calculated from the data of Lee & Lee (50)]. (*B*) Stabilization of ribonuclease by 30% sorbitol at 48°C; the contact between the protein and sorbitol is unfavorable ($\Delta\mu_{2,tr}$ positive) for both the native and denatured protein, but more so for the latter. Hence, $\delta\Delta\mu_{2,tr} = +2.6$ kcal/mol raises $\Delta G^{\circ,N-D}$ by an equal amount, which makes sorbitol into a stabilizer (G. Xie & S. N. Timasheff, unpublished data).

Combination of Equations 8 and 3 shows that the slope of the Wyman plot (log $K$ vs log $a_3$) is determined by the change in preferential binding during the course of the reaction (90, 109). At any given cosolvent concentration,

$$\left(\frac{\partial \ln K}{\partial \ln a_3}\right)_{T,P,m_2} = v_3^{Pr} - v_3^{R} = \Delta\left(\frac{\partial m_3}{\partial m_2}\right)_{T,P,\mu_3}. \qquad 9.$$

This slope defines the direction in which a cosolvent displaces the equilibrium, because $\Delta v_3$ may be positive (enhancer), negative (inhibitor), or zero (no effect). A typical example appears in Figure 2, which shows the effect of glycerol on the self-assembly of collagen fibrils (58).

In either end state of the reaction, the preferential interaction parameter itself may be positive, negative, or zero, depending on whether the

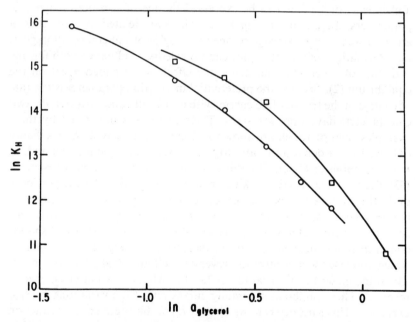

*Figure 2* Effect of glycerol on the self-assembly of collagen fibrils (○, NaBH$_4$ reduced collagen; □, native collagen). The slopes of the Wyman plots are curvilinear, with $\Delta(\partial m_3/\partial m_2)_{T,P,\mu_3}$ values varying from $-1.5$ to $-6.5$ mol glycerol per mole of native collagen [reproduced from Na (58)]. From these data, one can estimate a value of $\delta\Delta\mu_{2,tr}$ of approximately $+2.5\pm1$ kcal/mol at 1 mM glycerol for the addition of each collagen molecule to the growing fibril, which is consistent with the measured $\delta\Delta G°$ of $+3.7$ kcal/mol. The destabilization of the collagen fibrils, therefore, results from the decrease of favorable interactions of collagen with glycerol as it enters the assembled structures.

interaction between the protein and the solvent system at the given solvent composition is unfavorable, favorable, or indifferent. This means that the corresponding preferential binding measured by a thermodynamic technique, such as dialysis equilibrium, will be negative, positive, or zero (see Equation 3).

What is a measured negative binding? It means that the solvent in the domain of the protein is enriched in water with respect to the bulk solvent, i.e. the protein is preferentially hydrated (60). The corresponding preferential hydration, $(\partial m_1/\partial m_2)_{T,\mu_1,\mu_3}$, is obtained by applying the Gibbs-Duhem equation to Equation 3 (93):

$$\left(\frac{\partial m_1}{\partial m_2}\right)_{T,\mu_1,\mu_3} = -\frac{m_1}{m_3}\left(\frac{\partial m_3}{\partial m_2}\right).$$ 10.

The above presentation clearly shows that the effect of a cosolvent on a chemical equilibrium can be expressed by one of two thermodynamic parameters, $\Delta\mu_{2,\text{tr}}$ and $(\partial\mu_2/\partial m_3)_{T,P,m_2}$. These are related to each other in an exact way. Yet knowledge of one at a single-solvent composition gives no information on the other, nor can knowledge of either or both for one end state of an equilibrium define the effect of the added agent on the equilibrium (2). Because the preferential interaction parameter is the rate of change of the transfer free energy with cosolvent concentration, the two can, in fact, have opposite signs. Their relation is illustrated by three examples, two of which are shown in Figure 3, namely 2-chloroethanol (38, 94, 102), a denaturant; and $MgCl_2$, a weak salting-out agent and, at times, a stabilizer (1, 2). The third example is a good stabilizer, sucrose (53). Figure 3 displays the striking apparent complexity of the dependence of the three interaction parameters on solvent composition and their seeming unrelatedness. In fact, the parameters follow very systematic variations. In both solvent systems, the preferential binding displays a bell-shaped concentration dependence, strongly increasing binding of chloroethanol, which above 40% concentration reverses itself and finally becomes negative at the highest concentration (80%). For $MgCl_2$, the pattern is exactly reversed. These preferential binding curves reflect the variations in $(\partial\mu_2/\partial m_3)_{T,P,m_2}$. This parameter shows a highly favorable preferential interaction for 2-chloroethanol at 20% [the protein unfolds between 10 and 20% denaturant (38)], which becomes less favorable as the denaturant concentration increases; at >65% it is unfavorable. For $MgCl_2$, the pattern of $(\partial\mu_2/\partial m_3)_{T,P,m_2}$ is exactly opposite: the preferential interaction, which starts at a highly unfavorable level, decreases linearly with $MgCl_2$ concentration and becomes favorable at high salt concentration. Although the preferential-interaction parameter changes signs in both systems, the

# SOLVENT EFFECTS ON PROTEIN STABILITY 75

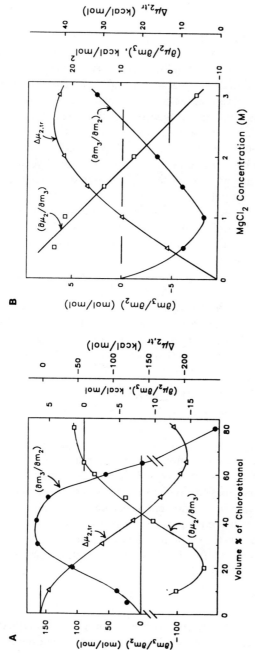

*Figure 3* Variation of the thermodynamic interaction parameters with cosolvent concentration: preferential binding $(\partial m_3/\partial m_2)_{T,\mu_1,\mu_3}$ (●), preferential interaction parameter (□), and transfer free energy, $\Delta\mu_{2,tr}$ (△). (*A*) β-Lactoglobulin in aqueous 2-chloroethanol; (*B*) β-lactoglobulin in aqueous $MgCl_2$ solution at pH 3.0 [plotted from the data of Inoue & Timasheff (38) and Arakawa et al (1)]. Note that $\Delta\mu_{2,tr}$ is at its peak (minimal for the favorably interacting 2-chloroethanol and maximal for the unfavorably interacting $MgCl_2$) at the point where the binding measured by dialysis equilibrium is zero. (The drawn points at 65% 2-chloroethanol are not experimental data; they were interpolated from the preferential binding measurements.)

same is not true for the transfer free energy in either case. The transfer free energy is highly favorable for 2-chloroethanol over the entire solvent composition (96), and highly unfavorable for the salting-out salt (1, 2). In fact, at the point at which preferential interaction is zero in both systems (dialysis equilibrium shows no binding), $\Delta\mu_{2,tr}$ is at its greatest value, minimal for 2-chloroethanol (strongest binding) and maximal for $MgCl_2$ (strongest exclusion). Therefore, the point at which the measured binding (preferential interaction) is zero, suggesting thermodynamic neutrality, occurs when the interactions are at their peak. For thermodynamic indifference, $\Delta\mu_{2,tr}$ would have to be zero at all cosolvent concentrations, i.e. no binding would be measured in dialysis equilibrium over the entire solvent-composition range. Such an example is found, in fact, in the system $\beta$-lactoglobulin-$MgCl_2$ at pH 5.5 (1).

The third example is sucrose, a good stabilizer (53). For this solvent system, the observed preferential interaction is one of monotonically increasing sugar exclusion with increase in sucrose concentration. For example, the interaction for $\alpha$-chymotrypsin attains values of $(\partial m_3/\partial m_2)_{T,\mu_1,\mu_3} = -3.4$ and $-7.6$ mol sugar/mol protein in 0.5 and 1 M sucrose, respectively. These values reflect a preferential hydration of 0.24 g water/g protein, and $(\partial\mu_2/\partial m_3)_{T,P,m_3} = +3.75$ kcal $mol^{-2}$, that are independent of sugar concentration. This means that the transfer free energy increases linearly with sucrose concentration, i.e. the interaction relative to water is increasingly unfavorable.

## RELATION BETWEEN PREFERENTIAL INTERACTIONS AND MOLECULAR CONTACTS

The rigorous thermodynamic definition of preferential interactions, given by Equations 2 and 3, clearly indicates that binding measured using thermodynamic techniques such as equilibrium dialysis may be positive, negative, or zero. What does this mean in terms of actual contacts of the protein surface with water and cosolvent molecules? By definition, in pure water, all points on the protein surface must be in contact with water. In order to occupy a site on the protein, a ligand must, therefore, displace water molecules. This is illustrated in Figure 4. Therefore, the free energy of binding at the site, $\Delta G^b$, is (103):

$$\Delta G^b = \Delta G^L - \Delta G^W, \qquad 11.$$

where $\Delta G^L$ and $\Delta G^W$ are the intrinsic free energies of interaction of the protein with the ligand and water. The corresponding binding constant, measured in aqueous medium, is then an exchange constant, $K_{ex}$ (78, 80):

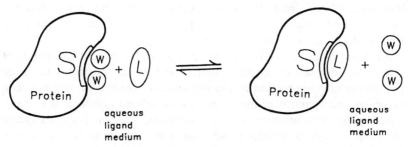

*Figure 4* Schematic representation of the competition between a ligand molecule, L, and water, W, at site S on a protein, as defined by Equations 11 and 12.

$$P \cdot H_2O_m + L \underset{}{\overset{K_{ex}}{\rightleftharpoons}} PL + mH_2O; \quad K_{ex} = \frac{[PL][H_2O]^m}{[P \cdot H_2O_m][L]}. \qquad 12.$$

Equations 11 and 12 define preferential interactions in terms of a competition equilibrium between water and the ligand for a locus on the protein surface. The sign of $\Delta G^b$ (negative, preferential binding of ligand; or positive, preferential hydration) is defined, therefore, by the relative affinities. Over the entire protein surface, $\Delta \mu_{2,\text{tr}} = \Sigma \Delta G^b$, i.e. the transfer free energy, is the sum of all the local protein-solvent interactions, and $(\partial \mu_2/\partial m_3)_{T,P,m_2} = (\partial \Sigma \Delta G^b/\partial m_3)_{T,P,m_3}$. This relation permits one to formally decompose the preferential interaction parameter into contributions from the occupancy of sites by ligand and water in equilibrium with each other (93, 103).

$$\left(\frac{\partial \mu_2}{\partial m_3}\right)_{T,P,m_2} = \left(\frac{\partial \mu_2}{\partial m_3}\right)^{(3)}_{T,P,m_2} - \left(\frac{\partial \mu_2}{\partial m_3}\right)^{(1)}_{T,P,m_2}, \qquad 13.$$

where (3) and (1) refer to occupancy of sites by ligand and by water, respectively. This heuristic analysis, in conjunction with the formalism of Equations 3 and 10, permits one to deduce the relation between the preferential binding measured by dialysis equilibrium and the total effective numbers of ligand and water molecules found at loci on the protein molecules, $B_3$ and $B_1$, respectively (39, 46, 71, 94, 102):

$$\left(\frac{\partial m_3}{\partial m_2}\right)_{T,P,\mu_3} = B_3 - \frac{m_3}{m_1} B_1. \qquad 14.$$

It must be stressed that $(\partial m_3/\partial m_2)_{T,P,\mu_3}$ is a true thermodynamic quantity; $B_1$ and $B_3$, on the other hand, are not. They are a description of the interactions in terms of a model of binding in which water and ligand molecules make contacts with the protein at discrete loci (102). Because, in the case of weak interactions such as those dealt with in this review, the two terms on the right-hand side of Equation 14 are of similar magnitude, it must always be true that $B_3 > (\partial m_3/\partial m_2)_{T,P,\mu_3}$.

As stated by Equation 14, binding can be regarded from two points of view, thermodynamic and molecular. The first one measures the total thermodynamic interaction of the protein with the solvent components, i.e. water and ligand over the entire protein-solvent interface. It is manifested by the solution composition perturbation, $(\partial m_3/\partial m_2)_{T,P,\mu_3}$, measured by using dialysis equilibrium and related techniques such as isopiestic equilibrium (33), light scattering (101), small-angle X-ray scattering (95), or sedimentation equilibrium (41). The second counts the number of protein-ligand contacts, $B_3$, manifested by e.g. the evolution or uptake of heat in a calorimetric titration (54) or the extent of perturbation of a spectral property as a function of ligand concentration. These techniques are nonthermodynamic. They cannot give a complete thermodynamic description of the interaction of a protein with a solvent system, nor, as a corollary, can they give a thermodynamic description of the effect of any given weakly interacting ligand on an equilibrium such as denaturation or self-assembly (97). Conversely, the thermodynamic equilibrium measurements cannot give a molecular description in terms of the number of protein-ligand contacts.

A combination of the information gained from the two types of measurements, however, can, within certain assumptions, yield a molecular description of the interactions of weak binding ligands with proteins in terms of numbers of effective sites that differ in their affinities for the ligand (97). When immersed in solvent, all loci on the surface of a protein molecule must be occupied either by water or cosolvent. The contacts may be classified into three general categories, as illustrated in Figure 5: (a) those that are indifferent to contact with cosolvent or water—their occupancy will be in proportion to the bulk solvent composition, i.e. $B_3/B_1 = m_3/m_1$ and $(\partial m_3/\partial m_2)_{T,P,\mu_3} = 0$; (b) those that have a significant affinity for ligand and, therefore, undergo water-ligand exchange—for them $B_3/B_1 > m_3/m_1$ and $(\partial m_3/\partial m_2)_{T,P,\mu_3} > 0$; and (c) those that have little affinity for the ligand relative to water and are occupied predominantly by water—for them $B_3/B_1 < m_3/m_1$ and $(\partial m_3/\partial m_2)_{T,P,\mu_3} < 0$. The total observed thermodynamic interaction is the sum of contributions from all sites. The first kind of site makes no contribution and can be neglected. The second kind will carry at any instant either water or ligand depending on the values of the relative affinities (exchange constants) and the solvent composition. Their occupancy will be accounted for by $B_3$ molecules of ligand and $B_1^{\text{exch}}$ molecules of water on the $n$ exchangeable sites that may be nonidentical in ligand affinity. Because the sites from which ligand is excluded

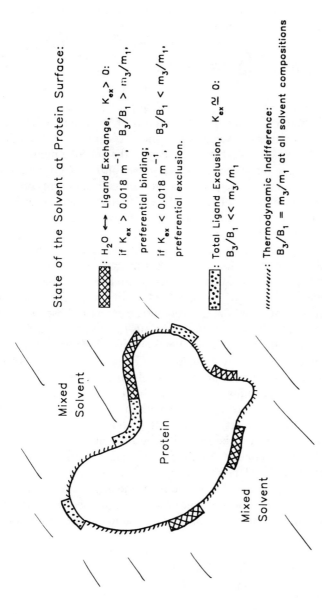

*Figure 5* Schematic representation of the thermodynamic state relative to bulk solvent of the surface of a protein dissolved in a mixed solvent. The total protein surface must make contact with water or cosolvent molecules. Depending on the free energy of interaction with the cosolvent system, three types of interactions may occur: (*a*) the cosolvent exchanges with water (these areas will be occupied by water or cosolvent depending on their relative affinities); (*b*) the cosolvent is excluded; (*c*) the ratio of cosolvent to water molecules is the same as in the bulk-solvent medium at all solvent compositions (there are no interactions with the protein other than van der Waals contacts).

make no contribution to $B_3$ (46, 71), the third kind of site will be occupied by water molecules that are not exchangeable with ligand, designated by $B_1^{\text{Nex}}$. The measured preferential binding, being the sum of contributions from all sites, is then given by (97):

$$\left(\frac{\partial m_3}{\partial m_2}\right)_{T,P,\mu_3} = \sum_1^l B_3 - \frac{m_3}{m_1}\left(\sum_1^l B_1^{\text{exch}} + B_1^{\text{Nex}}\right). \qquad 15.$$

The protein-ligand–contact counting techniques directly yield $\Sigma B_3$ as a function of $m_3$. Analysis of such results in terms of Scatchard or similar plots yields $n$, the number of exchangeable sites. Assignment of a model to the binding at the exchangeable sites (such as the effective number of water molecules interacting at each site) gives values of $B_1^{\text{exch}}$. Combination with the thermodynamic (dialysis) equilibrium results then permits one to deduce $B_1^{\text{Nex}}$, i.e. the number of water molecules that occupy sites not accessible to the cosolvent. Alternately, the effective number of water molecules interacting with the protein ($\Sigma B_1^{\text{exch}} + B_1^{\text{Nex}}$) can be estimated from NMR techniques, such as those developed by Kuntz (44, 45) and others (63). Their combination with experimental values of $(\partial m_3/\partial m_2)_{T,P,\mu_3}$ gives the value $B_3$. Such analyses of preferential binding for unfolded proteins in 8 M urea (70) and 6 M guanidine hydrochloride (51) have resulted in $B_3$ values that are in good agreement with those recently determined by careful calorimetric measurements (54).

Schellman addressed the relations between ligand-water exchange, preferential interactions, and site occupancy and their significance to protein stability in a series of important theoretical papers (76–81). He summarized much of his reasoning in the *Annual Review of Biophysics and Biophysical Chemistry* five years ago (79), so this argument is not repeated here. In his most recent papers, Schellman has shown how preferential (thermodynamic) binding can be related to equilibrium constants measured by ligand-protein–contact counting techniques in terms of specifically assigned models of the ligand-water exchange at the binding sites (78, 80). Treating explicitly the model of one independent site at which one molecule of ligand replaces one molecule of water, he (80) showed that, per site, the equilibrium exchange constant, $K_{\text{ex}}$, is related to the preferential binding by

$$\left(\frac{\partial m_3}{\partial m_2}\right)_{T,P,\mu_3}^{\text{per site}} = \frac{(K'_{\text{ex}}-1)X_3}{1+(K'_{\text{ex}}-1)X_3} = \frac{(K_{\text{ex}}-1/m_1)}{K_{\text{ex}}}\theta$$

$$\theta = \frac{K_{\text{ex}}m_3}{1+K_{\text{ex}}m_3}, \qquad 16.$$

where $K'_{ex}$ is the exchange constant in mole-fraction units, $K_{ex}$ is the constant in molal units, $X_3$ is the mole fraction of component 3, and $\theta$ is the extent of site occupancy by component 3 molecules. Summation over all the $n$ exchangeable sites of identical $K_{ex}$, and combination of Equations 14 and 16, leads to:

$$n\frac{(K_{ex}-1/m_1)m_3}{1+K_{ex}m_3} = B_3 - \frac{m_3}{m_1}B_1^{exch}. \qquad 17.$$

The experimentally measured preferential interaction over the entire protein molecule, however, is the sum over all sites (94, 96, 97). Introduction of Equation 17 into Equation 15 leads to the complete relation:

$$\left(\frac{\partial m_3}{\partial m_2}\right)_{T,P,\mu_3} = \Sigma n_i \frac{(K_{ex}^{(i)}-1/m_1)m_3}{1+K_{ex}m_3} - \frac{m_3}{m_1}B_1^{Nex}. \qquad 18.$$

Therefore, values calculated using Equation 17 must always be equal to or greater than the preferential binding measured by dialysis equilibrium.

In this analysis, Schellman examined the significance of the concept of exchange with water and, specifically, its expression in terms of Equation 16 (80). Clearly, the equilibrium-exchange model (Equation 16), in a manner similar to the molecular-occupancy model (Equations 14 and 15) and the thermodynamic statement of protein-ligand interactions (Equation 3), permits equilibrium-binding measurements to give negative, as well as positive, stoichiometries, a situation that is encountered frequently (100, 104). In Equation 16, this possibility is defined by the magnitude of $K_{ex}$ relative to $(1/m_1) = 0.018$. Thus, for $K_{ex} > 0.018$, the result is positive preferential binding; when $K_{ex} < 0.018$, the observation is preferential exclusion. As shown by Schellman (78–80), such negative values of binding cannot be accounted for by classical stoichiometric binding equations. The two approaches coalesce, however, as $K_{ex}$ increases. Thus, for $K_{ex} \geq 2$, neglect of $(1/m_1)$ introduces an error of less than 1% (80), and so, in the usual measurements of binding of biologically active ligands (such as enzyme substrates or inhibitors), neglect of water introduces no detectable error, even though conceptually this neglect is incorrect.

## EXPERIMENTAL OBSERVATIONS: APPLICATIONS

The destabilization of proteins by denaturants and the corresponding thermodynamic analysis have generated a vast literature over the years. This review does not address these reports; rather this section is devoted to the measured preferential interactions and their application to the analysis of protein stability, solubility, and self-assembly. The two types of information that are desired are: the thermodynamic contribution of the cosolvent interactions to each process and the decomposition of measured preferential binding into site occupancy by cosolvent and water, i.e. the role of bound water in these processes. As shown by Equations 4–7, a complete thermodynamic description of the effect of cosolvents on these processes requires knowledge of transfer free energies in the two end states. This means that preferential binding measurements as a function of cosolvent concentration must be carried for the proteins in both end states of the reaction (see Equations 2 and 3). To my knowledge, this has been done in only three instances: the denaturation of chymotrypsinogen by PEG 1000 (50), the stabilization of RNase A by sorbitol (G. Xie & S. N. Timasheff, in preparation) (both shown in Fig. 1), and the denaturation of collagen by glycerol (58). The reason for why analyses of this sort are rare resides in the very great technical difficulties encountered in dialysis-equilibrium measurements at high temperature (above the transition temperature for denaturation), as well as in dealing with proteins in the precipitated and self-assembled states. However, extensive literature is available on cosolvent interactions with globular proteins in the native state. Correlation of the patterns of interaction with effects on stability, solubility, and self-assembly have permitted investigators to deduce a number of rules that are discussed below. Decomposition of preferential binding according to Equation 14 that leads to the effective number of water molecules involved has been possible in only four systems for which sufficient data are available. The following sections discuss these systems.

## *Denaturants*

Preferential interactions of urea (70), guanidine hydrochloride (33, 51, 62, 71), and several organic denaturants (38, 102) with unfolded proteins have been measured. As a rule, all are preferentially bound to denatured proteins. As shown on Figure 3, 2-chloroethanol is already bound before the onset of protein unfolding (at 10% cosolvent). Measurements on a number of proteins in 8 M urea (70) and 6 M guanidine hydrochloride (33, 51), as well as on reduced bovine serum albumin (BSA) in 3 to 7 M guanidine hydrochloride (71) all show preferential binding, except for RNase A, which gives $(\partial m_3/\partial m_2)_{T,P,\mu_3}$ values of essentially zero. The absence of data for native proteins precludes calculation of the contribution of the transfer free energy to the unfolding process (Equation 4).

However, the fact that both preferential binding and protein-ligand contact numbers, measured by thermodynamic equilibrium (33, 51, 70) and calorimetric titration techniques (54, 67), respectively, are available for unfolded RNase and lysozyme in 8 M urea and 6 M GuaHCl has enabled investigators to evaluate for these systems the numbers of water molecules involved in the interactions and to carry out a thermodynamic analysis of the exchangeable sites (97) using the Schellman (80) model. Let us take RNase A in 6 M GuaHCl as an example. The pertinent experimental values are: $(\partial m_3/\partial m_2)_{T,\mu_1,\mu_3} = 0$ (51), $B_3 = 57$ molecules of GuaHCl occupying sites on denatured RNase A, and a total number of exchangeable sites, $n = 74$ (54). These values of $(\partial m_3/\partial m_2)_{T,\mu_1,\mu_3}$ and $B_3$ result, using Equation 14, in a total number of water molecules bound, $B_1^{\text{total}} = 300$, a value that is consistent with the effective hydration calculated by the NMR procedure of Kuntz (51). If all of these waters are exchangeable, they must be located on $(n - B_3) = 17$ sites, i.e. 17.6 water molecules exchanging with one guanidinium ion. This unreasonably high value leads to the conclusion that some water molecules must be interacting with sites of extremely low affinity for the denaturant, i.e. nonexchangeable sites (Equation 15). [In view of the similarity of the interactions of GuaHCl and urea with proteins, it seems pertinent to note that the interactions of urea with apolar residues are thermodynamically more unfavorable than those of water (19).] Assignment of these waters to the exchangeable and nonexchangeable categories requires the use of a model for the exchange equilibrium. If the model treated specifically by Schellman (80) is applied, the analysis results in 283 effective water molecules that cannot be replaced by GuaHCl. The corresponding exchange constant and calculated preferential binding at exchangeable sites are $K_{\text{ex}} = 0.327$ M$^{-1}$ and $(\partial m_3/\partial m_2)_{T,P,\mu_3} = 54$. This must be compared with the measured global preferential interaction, which is zero. Therefore, there must be a compensation by the contribution of the nonexchangeable sites that manifest themselves by the preferential exclusion of 54 GuaHCl molecules, i.e. a preferential hydration of 283 water molecules per protein molecule. Thermodynamic analysis using Equations 2 and 3 of the preferential binding at the exchangeable sites leads to $\Delta\mu_{2,\text{tr}}^{\text{exch}} = -26.5$ kcal/mol, i.e. a strongly favorable interaction between RNase and 6 M GuaHCl at these sites. The measured value, $\Delta\mu_{2,\text{tr}} = 0$, however, indicates that this is balanced by $+26.5$ kcal/mol of unfavorable interaction at the 283 nonexchangeable sites, i.e. $+94$ cal/mol per site, which corresponds to $K_{\text{ex}} \sim 0$. These numbers have been obtained within the exchange model treated explicitly by Schellman (80). Other exchange stoichiometries and possible cooperativity will result in different numbers. The arguments and general conclusions, however, will remain the same. The same kind of analysis for the other three systems has yielded similar results (97). This example makes it clear, therefore, that site contact–recognition measurements, even if analyzed in terms of exchange, can describe neither the thermodynamic binding nor the thermodynamic effect of the denaturant on unfolding, because they cannot detect sites of high preferential affinity for water, which must be accounted for in any such analysis.

## Stabilization: Salting-Out

Preferential interaction measurements have been carried out on many agents known to stabilize globular proteins and to reduce their solubility (salting-out). These include sugars (3, 53), glycerol (30, 60, 84) and other polyols (26–28), salts (1, 2, 4, 6, 7, 11), amino acids and methyl amines (5, 8, 9), and the two protein-crystallizing agents, polyethylene glycol (PEG) (10, 13, 48, 49) and 2-methyl-2,4-pentanediol (MPD) (69). Table 1 presents typical results. The universal observation, without any exceptions, is: all are preferentially excluded from native globular proteins; $(\partial m_3/\partial m_2)_{T,\mu_1,\mu_3}$ is negative, i.e. their interaction with the protein surface is thermodynamically unfavorable; and $(\partial \mu_3/\partial m_2)_{T,P,m_3}$ is positive. The consequences of this exclusion are that all of the above agents for which solubility measurements exist are salting-out agents (1). Yet not all are stabilizers of the native globular-protein structure (2, 98). Why is this so? The answer is found in Equations 4–6, namely that the determining factor is $\delta\Delta\mu_{2,\text{tr}}$ for the particular reaction. In the case of precipitation, by definition, the

**Table 1** Typical patterns of preferential interaction of proteins with precipitating and stabilizing cosolvents

| Cosolvent | Protein | $(\partial\mu_2/\partial m_3)_{T,P,m_2}$ kcal/mol protein/mol sucrose | $R^a$ | Mechanism[b] |
|---|---|---|---|---|
| Sucrose (0.3–1.0 M) | Chymotrypsinogen | 5.0[c] | 0.73 | I |
| Glucose (0.5–2.0 M) | Chymotrypsinogen | 5.6 decreasing to 3.7 | 0.65 | I |
| Glycine (0.7–2.0 M) | BSA | 17.0[c] | 0.90 | I |
| NaGlu (0.5–2.0 M) | BSA | 40.0 to 33.0 | 0.85–0.75 | I |
| NaGlu (1.0 M) | Lysozyme | 6.0 | 0.38 | II |
| LysHCl (0.5–2.0 M) | BSA | 14.9 to 19.7 | 0.38–0.50 | II |
| $Na_2SO_4$ (0.5–1.0 M) | BSA | 31.0 to 37.0 | 0.65–0.78 | I |
| $MgSO_4$ (1.0 M) | BSA | 20.0 | 0.55 | I |
| $MgCl_2$ (0.5–2.0 M), pH 4.5 | BSA | 13.2 to 4.5 | 0.24–0.08 | II |
| $Gua_2SO_4$ (0.5–2.0 M) | BSA | −4.6 to 15.7 | neg–0.69 | III |
| GuaHCl (1.0–3.0 M) | BSA | −13.8 to −7.3 | neg | IV |
| Glycerol (10–40%) | α-Chymotrypsin | 2.7[c] | neg | V |
| Sorbitol (5–40%) | BSA | 7.0 to 9.4 | — | V |
| PEG 600 (10–30%) | β-Lactoglobulin | 10.2[c] | neg | VI |

[a] $R$ is the ratio of the experimental value of $(\partial\mu_2/\partial m_3)_{T,P,m_2}$ to that calculated from the surface-tension increment.
[b] Mechanisms of preferential interactions: I, exclusion resulting from surface-tension increase; II, surface-tension increase compensated by binding; III, binding overcome by surface-tension increase; IV, binding; V, solvophobic effect; VI, steric exclusion.
[c] Independent of concentration.

structure of the protein molecules is identical in the two end states of the process, meaning that the chemical nature of contacts between protein and solvent remains unchanged. Coalescence reduces the total protein–solvent interface, as illustrated in Figure 6a, so that $\delta\Delta\mu_{2,\text{tr}}$ is negative and the protein is in a thermodynamically less unfavorable state when aggregated into a precipitate. The same argument holds true for the self-assembly reaction, e.g. of organelles such as microtubules (8, 23, 34, 36, 52, 60, 107). When the preferential interaction with the dispersed protein is favorable, however, the situation reverses itself, as is the case of the effect of glycerol on the self-assembly of collagen into fibrils, shown in Figure 2 (58, 59).

In the case of stabilization, the situation is more complicated. Here, the structure of the protein changes by definition during the course of the reaction. As a consequence, the nature of the interactions between solvent components and protein may be different in the two end states, and it is, a priori, impossible to state that a cosolvent that is preferentially excluded from a native protein (and is a precipitant) will be a stabilizer, because $\delta\Delta\mu_{2,\text{tr}}$ may be positive (stabilizer) or negative (denaturant). What is the difference between the preferentially excluded stabilizers and destabilizers? Examination of the preferential hydration patterns of a multitude of such compounds has permitted the identification of two distinct categories (1, 2, 100). In the first, the preferential hydration, $(\partial m_1/\partial m_2)_{T,\mu_1,\mu_3}$, is almost totally independent of cosolvent concentration and solvent pH. In Table 1, this is expressed by $(\partial \mu_2/\partial m_3)_{T,P,m_2}$. This category includes sugars, most nonhydrophobic amino acids, and good salting-out salts, such as NaCl, $MgSO_4$, $Na_2SO_4$, and $(NH_4)_2SO_4$. These always act as protein stabilizers. In the second category, $(\partial m_1/\partial m_2)_{T,\mu_1,\mu_3}$ and, hence, $(\partial \mu_2/\partial m_3)_{T,P,m_2}$ strongly depend on either concentration or pH, or both. This category includes $MgCl_2$, MPD, PEG, and guanidine sulfate (100). Their effect on protein stability cannot be predicted from their preferential interactions with proteins in the native state, although all are precipitants under some conditions. What determines the two patterns of preferential interaction?

## Sources of Preferential Exclusion: Stabilization or Not?

The two categories of salting-out cosolvents are determined by the mechanisms of preferential exclusion. The causes of preferential exclusion can be either (100): (a) those that are totally independent of the chemical nature of protein surface, i.e. the cosolvent and protein are mutually inert toward each other, and the role of the protein is only to present a surface, and (b) those in which the chemical nature of the protein surface plays a role, i.e. the cosolvent recognizes particular chemical features of the protein surface with which it interacts, whether by attraction or repulsion. The first category consists of two principal mechanisms: steric exclusion and per-

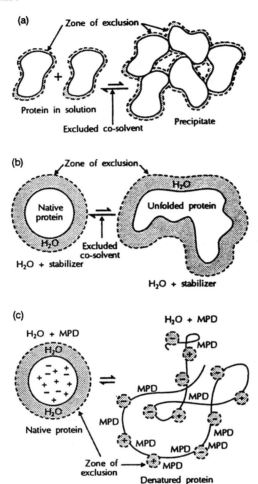

*Figure 6* Patterns of changes in protein-solvent interactions for preferentially excluded cosolvents during protein reactions. (*a*) Schematic representation of the salting-out reaction; the preferential exclusion of cosolvent per protein subunit is reduced in the precipitated state. (*b*) Schematic representation of the denaturation reaction when the predominant interaction of the cosolvent with the protein is nonspecific exclusion resulting from e.g. the cosolvent's increase of the surface tension of water. Because in the asymmetric denatured state, the exclusion is greater per protein molecule than in the compact native state, the equilibrium is displaced to the left. (*c*) Denaturation equilibrium with the mode of protein-solvent interaction different in the native and unfolded states. This is shown for the water-MPD system; strong MPD exclusion due to the high charge density on the native protein is replaced by weak MPD exclusion resulting from individual charged sites and MPD binding to newly solvent-exposed nonpolar regions in the denatured state, and the equilibrium is displaced to the right.

turbation by the cosolvent of the surface free energy of water. Steric exclusion, first proposed by Kauzmann, as cited by Schachman & Lauffer (75), is simply a reflection of the difference in size between cosolvent and water molecules. Essentially, contact between the large cosolvent molecules and the protein results in a shell that cannot be further penetrated by the cosolvent, but which is occupied by the smaller water molecules. The consequence is an excess of water at the protein surface that translates thermodynamically into preferential hydration. Steric exclusion is the source of the preferential hydration in the water-PEG systems (10, 13, 48).

The perturbation of the surface free energy of water, i.e. the surface tension, is a reflection of the fact that contact between protein and solvent constitutes an interface at which there must be an interfacial tension. Gibbs (31) showed in 1873 that a perturbation of the surface tension by an additive must lead to the perturbation of the concentration of that additive in the surface layer (in this case the protein-solvent interface). In the present notation, the Gibbs Adsorption Isotherm is:

$$\left(\frac{\partial m_3}{\partial m_2}\right)^{(\sigma)}_{T,P,\mu_3} = -\frac{S_2}{RT}\left(\frac{\partial \sigma}{\partial \ln a_3}\right)_{T,P,m_2}, \qquad 19.$$

where $S_2$ is the molar surface of the protein and $\sigma$ is surface tension. Clearly, if an agent increases the surface tension of water it will be depleted from the surface layer. Sugars (47), nonhydrophobic amino acids (16), and most salts (57) increase the surface tension of water. Hence, their concentration in the surface layer must be reduced relative to that of the bulk solvent, i.e. the observation will be one of preferential hydration and a positive value of the chemical potential perturbation. The contribution of this mechanism to preferential interactions can be evaluated from the ratio, $R$, (3, 4) of the experimental to the calculated value of $(\partial \mu_2/\partial m_3)_{T,P,m_2}$. As a rule, values of this ratio greater than 0.5 indicate a predominance of this effect. As shown in Table 1, this is the mechanism that causes the preferential exclusion of sugars, amino acids, and structure-stabilizing salts. In fact, this appears to be the predominant cause of preferential hydration and, therefore, of protein stabilization by cosolvents.

The role of surface tension in the stabilization of proteins by cosolvents has also been demonstrated in the examination of the stabilization by sucrose of three proteins, for which the surface tension was established at the transition temperature, $\sigma_{T_m}$, as a function of sucrose concentration (53). The surface tension at the transition temperature at any cosolvent concentration is given by:

$$\sigma_{T_m}^{m_3} = \sigma_{T_m}^{\circ} + \left(\frac{d\sigma}{dm_3}\right)m_3 + \left(\frac{d\sigma}{dT}\right)\Delta T_m, \qquad 20.$$

where $\sigma_{T_m}^\circ$ is the value of the surface tension at the transition temperature in water and $\Delta T_m$ is the difference between the transition temperatures in the presence of cosolvent of concentration $m_3$ and in pure water. The surface tension of water is known to decrease with an increase in temperature. It was found that at $T_m$ the increase in surface tension resulting from the presence of the sugar was compensated by its decrease resulting from the temperature increase from $T_m$ in water to $T_m$ at the given sugar concentration. As a result, $\sigma_{T_m}$ remains constant, i.e. the transition occurs at a constant value of the surface free energy when the interaction between the protein and the cosolvent is that of preferential exclusion resulting from nonspecific solvent properties. This observation could be interpreted to mean that protein expansion is possible only when the cavity free energy is below a limiting value.

In the second category of preferential exclusion, the interactions are determined by the chemical nature of the protein surface. The predominant cause of exclusion here is the solvophobic effect, which is the cause of preferential hydration in glycerol (30) and polyols (24, 25, 29). Polyols fit well in the water lattice (106), and their structure permits them to form the proper hydrogen bonds that reinforce water interactions. This makes contacts between nonpolar residues on proteins and the polyol solution even more unfavorable than contacts with water. As a consequence, polyol molecules migrate away from the protein surface. Thermodynamically, this is observed as preferential hydration. Furthermore, the interaction of polyols with peptide groups is also unfavorable (24, 25). A particular case in the second category is MPD, which is strongly repelled by charges (69) and, therefore, migrates away from the densely charged protein surface. The high value of the preferential hydration (1.03 g water/g protein for RNase A) makes MPD into a good precipitant, which explains its efficacy as a protein-crystallizing agent (42).

The above analysis of preferential hydration leads to an understanding of the reasons why some preferentially excluded cosolvents are stabilizers of the structure of globular proteins while others are denaturants, even though all are salting-out agents. This is illustrated in Figures 6$b$ and $c$. If the source of preferential exclusion is totally nonspecific, such as the increase in surface tension, then the only effect on unfolding will be an increase in the interface, and an increase in the hypothetical zone of exclusion, and therefore an increase in the unfavorable free energy of interaction. For such a system, $\delta\Delta\mu_{2,\text{tr}}$ must be positive and the equilibrium will be displaced to the left. This is true for the sucrose system discussed earlier in this review for which $\Delta\mu_{2,\text{tr}}$ (for the native protein) increases linearly with sugar concentration (53).

The stabilizing action of glycerol and polyols may be placed into the same category. Here, the interaction is one of recognition of unfavorable sites on proteins (nonpolar residues and peptide groups). Since the number of both increases upon unfolding, the preferential hydration must increase, which again drives the equilibrium toward the globular native structure.

MPD presents an opposite situation (2), depicted in Figure 6c. It interacts specifically with both end states of the unfolding equilibrium, being repelled by charges (69), but having a strong affinity for nonpolar residues on proteins (68). When the protein unfolds, its surface charges diverge, so that repulsion is weakened and MPD can penetrate to the newly exposed nonpolar residues, interact favorably with them, and, in this manner, stabilize the unfolded structure.

PEG is an intermediate case. It is excluded nonspecifically from native globular proteins (10, 48, 49), yet it also has a strong nonpolar character (35): hence the reversal of the sign of $\Delta\mu_{2,tr}$ when the protein unfolds, shown in Figure 1, as PEG interacts favorably with the newly exposed nonpolar residues and stabilizes the unfolded structure (50). Because they are strongly excluded from the globular state of the protein, both cosolvents are excellent precipitants (12, 48) and protein crystallizers (55, 56).

## BALANCE BETWEEN COSOLVENT BINDING AND EXCLUSION

The net interaction of globular proteins with salting-out and stabilizing cosolvents is that of preferential exclusion. This, however, does not mean that cosolvent molecules cannot penetrate to the surface of the protein and bind to it at some particular loci. Equation 15 describes the net interaction. In the present context, preferential exclusion caused by the various specific (e.g. solvophobic) and nonspecific (e.g. surface tension) factors corresponds to $B_1^{Nex}$. Other sites may exist, however, at which exchange occurs. Their occupancies would be expressed as $B_3$ and $B_1^{exch}$. For compounds, such as sucrose, which have no affinity for protein sites (possible specific strong interactions are excluded from this discussion), the preferential interaction is fully expressed by $(m_3/m_1)B_1^{Nex}$. In many cases, however, there are real contributions from $B_3$ and $B_1^{exch}$, as the cosolvent binds at particular sites on the protein. This is true for many salts, of which one ion, e.g. $Mg^{2+}$ or $Gua^+$, can bind weakly at protein sites (6, 7), as well as some amino acid salts such as Arg·HCl and Lys·HCl (8) (see Table 1, Lys·HCl-BSA). When the mechanism of exclusion is the increase in surface tension, binding can be recognized from the value of the ratio $R$. Thus, looking at Table 1, one can see that e.g. $MgCl_2$ binds to BSA at pH 4.5, compensating in part for the exclusion due to the surface-tension effect. Such compensation is well illustrated by the interactions of $\beta$-lactoglobulin with $MgCl_2$, and the comparison of this reaction with $MgSO_4$ (1). The two salts raise the surface tension of water to similar extents (57). $MgSO_4$ is a classical salting-out agent (32); it is also a good stabilizer of proteins (2). $MgCl_2$ displays an anomalous behavior (2). At pH 3.0, 0.5 M $MgCl_2$ decreases the solubility of $\beta$-lactoglobulin to 7% of its value in water. This trend is reversed at 1.0 M and the solubility increases until at 3.0 M it becomes too high to measure. Concomitantly, as shown in Figure 3, the preferential interaction parameter decreases until it becomes negative at 3.0 M, and the ratio $R$ decreases gradually from 0.6 at 0.5 M to negative at 3.0 M (1). At pH 5.1, the isoelectric point of the protein, $MgCl_2$ has no effect on the solubility of the protein. The preferential interaction is essentially zero at all solvent compositions. $MgSO_4$, on the other hand, is equally strongly excluded from the protein at pH 3.0 and 5.1. Such data are manifestations of the competition (6) between exclusion due to the surface-tension effect and weak binding of $Mg^{2+}$ ions to the protein. At the point of zero charge on the protein, binding balances out the exclusion due to the surface-tension effect when the cation is $Cl^-$, but not when it is $SO_4^{-2}$. When the pH is lowered, the strong positive charge on the protein repels the divalent cation, so that the surface tension effect predominates. An increase in $MgCl_2$ concentration increases the binding simply because of the Law of Mass Action, which again compensates for the exclusion caused by the surface-tension effect—hence the observed increase in solubility.

This example shows that in salts the contributions to exclusion and binding by each ion must be taken into account. Years ago, researchers observed that the effect of ions on a reaction can be additive. A classic example is the solubility of acetyltetraglycine ethyl ester (72). For this model peptide, NaCl was found to be a salting-out agent, LiBr was a salting-in agent. Yet neither LiCl nor NaBr had any effect. In the case of denaturation, GuaHCl is a destabilizer of RNase at pH 7.0 ($T_m$ was lowered from 60°C in water to 25°C in 3 M GuaHCl), yet $Gua_2SO_4$ has been found to be a stabilizer ($T_m$ was raised by 10°C in 3 M salt) (105). Such behavior reflects the balance between cations and anions in their preferential interactions with proteins. A comparison of several salts has permitted the classification of ions according to their contribution to preferential interactions, as shown in Table 2 (1, 7). The relative effectiveness in inducing the preferential hydration of native proteins increases in the orders $SCN^- < Cl^- < CH_3CO_2^- < SO_4^{2-}$ (Hoffmeister series) for anions and $Gua^+ < X^{2+} < Na^+$ for cations. Thus, $Na_2SO_4$ is much more strongly excluded than NaCl or $MgSO_4$. It should be the better precipitant

**Table 2** Effectiveness of ions in inducing preferential hydration[a]

|  | | Anion | | |
| Cation | SCN⁻ | Cl⁻ | OAc⁻ | SO₄²⁻ |
| --- | --- | --- | --- | --- |
| Na⁺ | 3[b] | −17 | −22 | −35 |
| Mg²⁺, Ca²⁺, Ba²⁺ | — | −3 to 3 | −8 to −13 | −27[c] |
| Guanidinium | — | 18 | −6 | −16 |

[a] Expressed as $(\partial m_3/\partial m_2)_{T,\mu_1,\mu_3}$ of bovine serum albumin in 1 M salt solutions [data from Arakawa & Timasheff (7) and Arakawa et al (1)].
[b] The cation was $K^+$.
[c] Value for $MgSO_4$.

and structure stabilizer. Among the guanidinium salts, this series explains why GuaHCl is a denaturant, while $Gua_2SO_4$ is a stabilizer. Very simply, the preferential exclusion of $SO_4^{2-}$ ions because of their effect on the surface tension of water overwhelms the binding of $Gua^+$ ions to proteins; $Cl^-$ ions, on the other hand, are not sufficiently excluded to do this. This is consistent with the relative effectivenesses of $Na_2SO_4$ and NaCl in inducing preferential hydration.

In the analysis of an equilibrium, such as denaturation, all of the binding and exclusion effects must be taken into account in both end states of the reaction. The slope of the Wyman plot then becomes, by Equations 9 and 15,

$$\left(\frac{\partial \ln K}{\partial \ln a_3}\right)_{T,P,m_3} = \Delta B_3 - \frac{m_3}{m_1}\Delta(B_1^{\text{exch}} + B_1^{\text{Nex}}) = \Delta v_3. \qquad 21.$$

Evidently, a variation of the equilibrium constant with ligand concentration can result either from a change in ligand binding, from a change in water binding, or from both. An interesting example can be drawn from the elegant denaturation studies of RNase T1 by Pace and coworkers (37, 66, 92). In their study of the effect of salts on the stability of this protein, they found (37) a negative slope for Equation 9. They explicitly assumed that the entire stabilization effect results from the binding of more ions to the native than the denatured protein (i.e. $\Delta B_1 = 0$) and calculated a $\Delta B_3$ value of $-2.1 \pm 0.4$. NaCl, however, is known to be preferentially excluded from proteins, essentially because it increases the surface tension of water. Consequently, if the alternate assumption is made that $\Delta v_3$ results only from an increase in the nonspecific hydration of a newly exposed surface upon unfolding it is found that $\Delta B_1 = 250$–400 additional water molecules make contact with the protein upon unfolding, with $\Delta B_3 = 0$, i.e. there is no binding of ions. This example illustrates the uncertainty of analysis in molecular terms of thermodynamic binding when the interactions are weak.

## COMPENSATION: OSMOLYTES

The compensation between binding and exclusion of a single cosolvent or of the ions of a salt, in fact, extends to four-component systems as well, i.e. mixtures of denaturants (such as urea) and stabilizers (such as a sugar). This phenomenon occurs widely in nature in the protection of organisms from osmotic shock and freezing. Most compounds used in nature for these purposes belong to the category of protein-structure stabilizers (9, 110) that are excluded by nonspecific mechanisms and have little propensity to bind to proteins (99). Compounds used in nature include sugars, polyols, amino acids and their derivatives, and methylamines. They do not interfere with biochemical processes and are known as compatible solutes (15). All are electrically neutral molecules. Compounds that can bind, such as arginine, lysine, and amino acids with large hydrophobic side chains are not used, nor are salts, except by halophilic bacteria (22, 113). These are incompatible solutes (83). One incompatible compound, urea, is nevertheless accumulated by some amphibians to build up the cellular osmotic pressure. In fact, in such cases, accumulation of urea, which is a denaturant, is accompanied by the accumulation of a structure protectant, such as trimethylamine-N-oxide (TMAO), sarcosine, or betaine (111).

Somero and coworkers (83, 110–112) have carried out detailed studies of the effect of such mixtures found in nature on the behavior of pure proteins. Some of their results are presented in Figures 7a and b, which show the effect of such mixtures on enzyme activity and protein stability. These studies revealed, for example, that the muscle system of a fish accumulates urea and a mixture of known structure stabilizers in a 2:1 molar ratio. Examination of the effect of this mixture on enzyme activity (Figure 7a) shows that it maintains $K_m$ at a constant value, whereas pure urea raises it, while pure methylamines lower it (112). A parallel study on the unfolding of ribonuclease has also shown a compensation, i.e. urea lowers $T_m$, TMAO raises it, while their 2:1 molar ratio mixture maintains it at a close to constant value (111). This compensation between a denaturant and a stabilizer demonstrates the use in nature of the principles of additivity between preferential binding and exclusion, described by Equation 15. In fact, preferential-interaction experiments in progress in our laboratory indicate that the quantity $(\Delta\mu_{2,\text{tr}}^D - \Delta\mu_{2,\text{tr}}^N)$ of Equation 5 has opposite signs for RNase T1 in urea and TMAO, respectively (T.-Y. Lin & S. N. Timasheff, in preparation). It is quite striking that, in selecting the osmolyte system, Nature did not choose just a stabilizer. This might

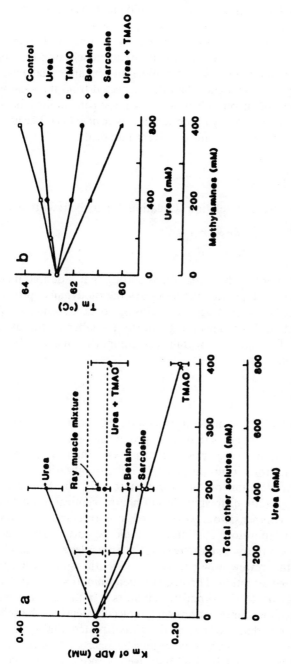

*Figure 7* Compensation between the denaturant urea and stabilizing methylamines. (*a*) Effect of cosolvents on an enzymic reaction expressed by the apparent $K_m$ of ADP of pyruvate kinase from a stingray. The ray muscle mixture contained 400 mM urea, 65 mM TMAO, 55 mM sarcosine, 50 mM β-alanine, and 30 mM betaine. This mixture maintained a constant $K_m$, while urea alone raised it and the methylamines lowered it; a 2:1 molar ratio of urea to TMAO mimicked exactly the natural system. (*b*) Compensation of the urea destabilization of RNase A by TMAO stabilization [reproduced from Yancey et al (110)].

have rendered the biochemical system too rigid and too active, but Nature chose a mixture that least perturbs the biochemical activities.

Another demonstration of the selection by Nature of the thermodynamically most favorable situation is its frequent selection of "superosmolytes," i.e. compounds, such as proline and betaine, that raise the osmotic pressure at a rate higher than that expected from the ideal van't Hoff law. The thermodynamic advantage of using superosmolytes is found in the three-component theory described in this review (99). Because the osmotic increment is related to the cosolvent nonideality by

$$\left(\frac{\partial \pi}{\partial m_3}\right)_{T,P} = \frac{m_3}{1000}\left(\frac{\partial \mu_3}{\partial m_3}\right)_{T,P}, \qquad 22.$$

the preferential interaction relation (Equation 3) may be rewritten as

$$\left(\frac{\partial \pi}{\partial m_3}\right)_{T,P,m_2} = -\frac{m_3}{1000}\frac{(\partial \mu_2/\partial m_3)_{T,P,m_2}}{(\partial m_3/\partial m_2)_{T,P,\mu_3}}. \qquad 23.$$

Nature might interpret this equation as stating that a desired increase in osmotic pressure will provide the necessary thermodynamic protection (stabilization) at a lower level of exclusion of the osmolyte from cell-compartment surfaces. This must lead to a smaller perturbation of the distribution of components within such compartments, a biologically desirable situation.

## CONCLUSIONS

The accumulation of a vast amount of data on the thermodynamic interactions of solvent components that affect protein stability, solubility, and state of aggregation has made possible, at present, the rational selection of such compounds for various purposes, be it the stabilization of proteins during storage in solution or the frozen state (17), the fine-tuning of refolding pathways (20), or formulation of pharmaceutical products (85). The principle of compensation, widely used in nature and only recently discovered under laboratory conditions, should be particularly useful in these endeavors. On the more esoteric side, the principles and observations summarized in this review now permit us to identify some very specific questions that need attention. These are the determination of transfer free energies of particular stabilizers and destabilizers at both ends of a reaction and the identification of interactions on a molecular level, both exchange and total exclusion, that give rise to the global preferential interactions that control the various equilibria undergone by macromolecules. Answering these questions should become increasingly more feasible with the continued improvement of techniques aimed specifically at the detection of water or ligand immobilization, such as NMR and calorimetry. Finally, the gained degree of understanding of the basic physico-chemical mechanisms of preferential exclusion is enabling investigators to examine the contribution of basic solvent properties, such as surface tension and osmotic pressure, to the stability of proteins in solution.

*Literature Cited*

1. Arakawa, T., Bhat, R., Timasheff, S. N. 1990. *Biochemistry* 29: 1914–23
2. Arakawa, T., Bhat, R., Timasheff, S. N. 1990. *Biochemistry* 29: 1924–31
3. Arakawa, T., Timasheff, S. N. 1982. *Biochemistry* 21: 6536–44
4. Arakawa, T., Timasheff, S. N. 1982. *Biochemistry* 21: 6545–52
5. Arakawa, T., Timasheff, S. N. 1983. *Arch. Biochem. Biophys.* 224: 169–77
6. Arakawa, T., Timasheff, S. N. 1984. *Biochemistry* 23: 5912–23
7. Arakawa, T., Timasheff, S. N. 1984. *Biochemistry* 23: 5924–29
8. Arakawa, T., Timasheff, S. N. 1984. *J. Biol. Chem.* 259: 4979–86
9. Arakawa, T., Timasheff, S. N. 1985. *Biophys. J.* 47: 411–14
10. Arakawa, T., Timasheff, S. N. 1985. *Biochemistry* 24: 6756–62
11. Arakawa T., Timasheff, S. N. 1987. *Biochemistry* 26: 5147–53
12. Atha, D. H., Ingham, K. C. 1981. *J. Biol. Chem.* 256: 12108–17
13. Bhat, R., Timasheff, S. N. 1992. *Protein Sci.* 1:1133–43
14. Brandts, J. F. 1969. In *Structure and Stability of Biological Macromolecules*, pp. 213–90. New York: Marcel Dekker. 694 pp.
15. Brown, A. D., Simpson, J. R. 1972. *J. Gen. Microbiol.* 72: 589–91
16. Bull, H. B., Breese, K. 1974. *Arch. Biochem. Biophys.* 161: 665–70
17. Carpenter, J. F., Crowe, J. H. 1988. *Cryobiology* 25: 244–55
18. Casassa, E. F., Eisenberg, H. 1964. *Adv. Protein Chem.* 19: 287–395
19. Cheek, P. J., Lilley, T. H. 1988. *J. Chem. Soc. Faraday Trans. I* 84: 1927–40
20. Cleland, J. L., Wang, D. I. C. 1990. *Biotechnology* 8: 1274–78
21. Cohn, E. J., Edsall, J. T. 1943. *Proteins, Amino Acids and Peptides.* New York: Reinhold
22. Eisenberg, H., Wachtel, E. J. 1987. *Annu. Rev. Biophys. Biophys. Chem.* 16: 69–92
23. Erickson, H. P., Voter, W. A. 1976. *Proc. Natl. Acad. Sci. USA* 73: 2813–17
24. Gekko, K. 1981. *J. Biochem. (Japan)* 90: 1633–41
25. Gekko, K. 1981. *J. Biochem. (Japan)* 90: 1643–52
26. Gekko, K. 1982. *J. Biochem. (Japan)* 91: 1197–1204
27. Gekko, K., Koga, S. 1984. *Biochim. Biophys. Acta* 786: 151–60
28. Gekko, K., Morikawa, T. 1981. *J. Biochem. (Japan)* 90: 39–50
29. Gekko, K., Morikawa, T. 1981. *J. Biochem. (Japan)* 90: 51–60
30. Gekko, K., Timasheff, S. N. 1981. *Biochemistry* 20: 4667–76
31. Gibbs, J. W. 1878. *Trans. Conn. Acad.* 3: 343–524
32. Green, A. A. 1932. *J. Biol. Chem.* 95: 47–66
33. Hade, E. P. K., Tanford, C. 1967. *J. Am. Chem. Soc.* 89: 5034–40
34. Hamel, E., del Campo, A. A., Lowe, M. C., Waxman, P. G., Lin, C. M. 1982. *Biochemistry* 21: 503–9
35. Hammes, G. G., Schimmel, P. R. 1967. *J. Am. Chem. Soc.* 89: 442–46
36. Himes, R. H., Burton, P. R., Gaito, J. M. 1977. *J. Biol. Chem.* 252: 6222–28
37. Hu, C.-Q., Sturtevant, J. M., Thomson, J. A., Erickson, R. E., Pace, C. N. 1992. *Biochemistry* 31: 4876–82
38. Inoue, H., Timasheff, S. N. 1968. *J. Am. Chem. Soc.* 90: 1890–97
39. Inoue, H., Timasheff, S. N. 1972. *Biopolymers* 11: 737–43
40. Kauzmann, W. 1959. *Adv. Protein Chem.* 14: 1–63
41. Kielley, W. W., Harrington, W. F. 1960. *Biochim. Biophys. Acta* 41: 401–21
42. King, M. V., Magdoff, B. S., Adelman, M. B., Harker, D. 1956. *Acta Crystallogr.* 9: 460–65
43. Kirkwood, J. G., Goldberg, R. J. 1950. *J. Chem. Phys.* 18: 54–57
44. Kuntz, I. D. 1971. *J. Am. Chem. Soc.* 93: 514–16

45. Kuntz, I. D. Jr., Kauzmann, W. 1974. *Adv. Protein Chem.* 28: 239–345
46. Kupke, D. W. 1973. In *Physical Principles and Techniques of Protein Chemistry*, Part C, ed. S. J. Leach, pp. 1–75. New York: Academic
47. Landt, E. 1931. *Z. Ver. Dtsch. Zucker-Ind.* 81: 119–24
48. Lee, J. C., Lee, L. L. Y. 1979. *Biochemistry* 18: 5518–26
49. Lee, J. C., Lee, L. L. Y. 1981. *J. Biol. Chem.* 256: 625–31
50. Lee, J. C., Lee, L. L. Y. 1987. *Biochemistry* 26: 7813–19
51. Lee, J. C., Timasheff, S. N. 1974. *Biochemistry* 13: 257–65
52. Lee, J. C., Timasheff, S. N. 1977. *Biochemistry* 16: 1754–64
53. Lee, J. C., Timasheff, S. N. 1981. *J. Biol. Chem.* 256: 7193–7201
54. Makhatadze, G. I., Privalov, P. L. 1992. *J. Mol. Biol.* 226: 491–505
55. McPherson, A. Jr. 1976. *J. Biol. Chem.* 251: 6300–3
56. McPherson, A. 1985. *Methods Enzymol.* 114: 120–25
57. Melander, W., Horvath, C. 1977. *Arch. Biochem. Biophys.* 183: 200–15
58. Na, G. C. 1986. *Biochemistry* 25: 967–73
59. Na, G. C., Butz, L. J., Bailey, D. G., Carroll, R. J. 1986. *Biochemistry* 25: 958–66
60. Na, G. C., Timasheff, S. N. 1981. *J. Mol. Biol.* 151: 165–78
61. Neurath, H., Greenstein, J. P., Putnam, F. W., Erickson, J. O. 1944. *Chem. Rev.* 34: 157–265
62. Noelken, M. E., Timasheff, S. N. 1967. *J. Biol. Chem.* 242: 5080–85
63. Otting, G., Liepinsh, E., Wüthrich, K. 1991. *Science* 254: 974–80
64. Pace, C. N. 1975. *C.R.C. Crit. Rev. Biochem.* 3: 1–43
65. Pace, C. N. 1986. *Methods Enzymol.* 131: 266–80
66. Pace, C. N., Grimsley, G. R. 1988. *Biochemistry* 27: 3242–46
67. Pfeil, W., Welfle, K., Bychkova, V. E. 1991. *Stud. Biophys.* 140: 5–12
68. Pittz, E. P., Bello, J. 1971. *Arch. Biochem. Biophys.* 146: 513–24
69. Pittz, E. P., Timasheff, S. N. 1978. *Biochemistry* 17: 615–23
70. Prakash, V., Loucheux, C., Scheufele, S., Gorbunoff, M. J., Timasheff, S. N. 1981. *Arch. Biochem. Biophys.* 210: 455–64
71. Reisler, E., Haik, Y., Eisenberg, H. 1977. *Biochemistry* 16: 197–203
72. Robinson, D. R., Jencks, W. P. 1965. *J. Am. Chem. Soc.* 87: 2470–79
73. Scatchard, G. 1946. *J. Am. Chem. Soc.* 68: 2315–19
74. Scatchard, G. 1949. *Ann. N.Y. Acad. Sci.* 51: 660–72
75. Schachman, H. K., Lauffer, M. A. 1949. *J. Am. Chem. Soc.* 71: 536–41
76. Schellman, J. A. 1975. *Biopolymers* 14: 999–1018
77. Schellman, J. A. 1978. *Biopolymers* 17: 1305–22
78. Schellman, J. A. 1987. *Biopolymers* 26: 549–59
79. Schellman, J. A. 1987. *Annu. Rev. Biophys. Biophys. Chem.* 16: 115–37
80. Schellman, J. A. 1990. *Biophys. Chem.* 37: 121–40
81. Schellman, J. A., Hawkes, R. B. 1980. In *Protein Folding*, ed. R. Jaenicke, pp. 331–44. New York: Elsevier
82. Simpson, R. B., Kauzmann, W. 1953. *J. Am. Chem. Soc.* 75: 5139–52
83. Somero, G. N. 1986. *Am. J. Physiol.* 251: R197–R213
84. Stauff, J., Mehrotra, K. N. 1961. *Kolloid Z.* 176: 1–8
85. Stevenson, C. L., Hageman, M. J. 1988. *Pharmaceutical Res.* 5: S30
86. Stigter, D. 1960. *J. Phys. Chem.* 64: 842–46
87. Stockmayer, W. H. 1950. *J. Chem. Phys.* 18: 58–61
88. Tanford, C. 1964. *J. Am. Chem. Soc.* 86: 2050–59
89. Tanford, C. 1968. *Adv. Protein Chem.* 23: 121–282
90. Tanford, C. 1969. *J. Mol. Biol.* 39: 539–44
91. Tanford, C. 1970. *Adv. Protein Chem.* 24: 1–95
92. Thomson, J. A., Shirley, B. A., Grimsley, G. R., Pace, C. N. 1989. *J. Biol. Chem.* 264: 11614–20
93. Timasheff, S. N. 1963. In *Electromagnetic Scattering*, ed. M. Kerker, pp. 337–55. New York: Pergamon
94. Timasheff, S. N. 1970. *Acc. Chem. Res.* 3: 62–68
95. Timasheff, S. N. 1973. *Adv. Chem.* 125: 327–42
96. Timasheff, S. N. 1992. In *Protein-Solvent Interactions*, ed. R. Gregory. New York: Marcel Dekker. In press
97. Timasheff, S. N. 1992. *Biochemistry.* 31: 9857–64
98. Timasheff, S. N. 1992. *Curr. Opin. Struct. Biol.* 2: 35–39
99. Timasheff, S. N. 1992. In *Water and Life*, ed. C. B. Osmond, C. L. Bolis, G. N. Somero, pp. 70–84. New York: Springer-Verlag. 371 pp.
100. Timasheff, S. N., Arakawa, T. 1988. In *Protein Structure & Function: A Practical Approach*, ed. T. E. Creighton, pp. 331–45. Oxford: IRL Press. 355 pp.

100a. Timasheff, S. N., Arakawa, T. 1988. *J. Cryst. Growth* 90: 39–46
101. Timasheff, S. N., Dintzis, H. M., Kirkwood, J. G., Coleman, B. D. 1957. *J. Am. Chem. Soc.* 79: 782–91
102. Timasheff, S. N., Inoue, H. 1968. *Biochemistry* 7: 2501–13
103. Timasheff, S. N., Kronman, M. J. 1959. *Arch. Biochem. Biophys.* 83: 60–75
104. von Hippel, P. H., Peticolas, V., Schack, L., Karlson, L. 1973. *Biochemistry* 12: 1256–64
105. von Hippel, P. H., Wong, K.-Y. 1965. *J. Biol. Chem.* 240: 3909–23
106. Warner, D. T. 1962. *Nature (London)* 196: 1055–58
107. Wehland, J., Herzog, W., Weber, K. 1977. *J. Mol. Biol.* 111: 329–42
108. Wyman, J. Jr. 1964. *Adv. Protein Chem.* 19: 223–86
109. Wyman, J., Gill, S. J. 1990. *Binding and Linkage. Functional Chemistry of Biological Macromolecules.* Mill Valley, CA: University Science Books. 330 pp.
110. Yancey, P. H., Clark, M. E., Hand, S. C., Bowlus, R. D., Somero, G. 1982. *Science* 217: 1214–22
111. Yancey, P. H., Somero, G. N. 1979. *Biochem. J.* 183: 317–23
112. Yancey, P. H., Somero, G. N. 1980. *J. Exp. Zool.* 212: 205–13
113. Zaccai, G., Eisenberg, H. 1990. *Trends Biochem. Sci.* 15: 333–37

# THE TWO-DIMENSIONAL TRANSFERRED NUCLEAR OVERHAUSER EFFECT: Theory and Practice

## A. P. Campbell and B. D. Sykes

Department of Biochemistry and MRC Group in Protein Structure and Function, University of Alberta, Edmonton, Alberta, T6G 2H7, Canada

KEY WORDS: ligand-macromolecule interactions, chemical exchange, 2D NMR

CONTENTS

| | |
|---|---:|
| PERSPECTIVES AND OVERVIEW | 99 |
| THEORY | 102 |
|     Complete Relaxation Matrix Analysis for a Multiple Spin System in Chemical Exchange | 102 |
|     The Relaxation Matrix in the Limits of Fast Chemical Exchange | 104 |
|     The Relaxation Matrix with Exchange Rates Included in the Calculation: Slow and Intermediate Exchange | 105 |
|     Intermolecular vs Intramolecular TRNOEs | 107 |
| EXPERIMENT | 108 |
|     Calculation of the Average Relaxation Matrix | 108 |
|     Chemically Equivalent Protons | 109 |
|     Effects of Internal Motions | 110 |
| RESULTS AND DISCUSSION | 110 |
|     Effect of Fraction Bound and Mixing Time | 111 |
|     Effects of Internal Motions | 115 |
| CONCLUSION | 118 |

## PERSPECTIVES AND OVERVIEW

One of the most important issues in biology is the interaction of ligands with macromolecules. Examples include the interaction of substrates with enzymes, antigens with antibodies, hormones with receptors, nucleotides with regulatory proteins, and peptides with phospholipids. Excepting those instances where the ligand has been crystallized in a complex with the macromolecule, nuclear magnetic resonance (NMR) spectroscopy provides the best method for determining the structure of the bound ligand. Ligand-macromolecule systems fall into two situations that require different NMR approaches. For ligands that are bound very tightly and exchange slowly, the structure of the complex must be solved directly. Here, several approaches can help, including isotope enrichment of the ligand and/or macromolecule and the use of spectral editing approaches to simplify the NMR spectrum. For ligands that are bound more weakly and exchange with free ligand at reasonable rates, the intramolecular transferred nuclear Overhauser enhancement (TRNOE) experiment, originally proposed by Balaram & Bothner-By (8, 9), provides the best technique for the determination of the structure of the bound ligand and is the subject of this review.

Concommittant with the rapid developments in two-dimensional (2D) NMR techniques and the use of NMR to determine the structure of proteins in solution (67), interest in the TRNOE has emerged recently, as evidenced by the large number of bound ligand structures that have been investigated using the TRNOE within the past 10 years. This methodology has been used to study the conformations of nucleotides bound to peptides and proteins (2, 3, 11, 26–29, 32, 33, 35–38, 46, 61–63), peptides to proteins (20, 56–59), peptides to phospholipids (42, 54, 65, 66), hormones to proteins (48), nucleic acids to proteins (28, 41), substrates and substrate analogues to enzymes (25, 38, 53), sugars to lectins (12), and antigens to antibodies (4–7, 30, 31, 39, 45, 69). Although the majority of these studies have utilized the one-dimensional, steady-state TRNOE experiment, the most recent studies have utilized the two-dimensional TRNOESY experiment (2, 3, 5–7, 20, 25, 45, 53, 56–59, 69), for which theoretical and practical evaluations of the complete relaxation matrix analysis have been published (21, 43, 49, 55).

The TRNOE is the extension of the two-dimensional nuclear Overhauser effect (NOE) to exchanging systems such as ligand-protein complexes. The intramolecular TRNOE allows the transfer of information concerning cross-relaxation between two nuclei in the bound ligand to the free ligand resonances via chemical exchange. In the unbound form, the ligand is generally characterized by short correlation times ($\tau_c \sim 10^{-10}$ s) and thus may be either in the extreme narrowing limit ($\omega\tau_c \ll 1$), where the NOEs are small and positive, or at the boundary of the extreme narrowing limit ($\omega\tau_c \approx 1$), where the NOEs approach zero. When bound to the protein, the ligand is characterized by the long correlation time of the protein ($\tau_c > 10^{-8}$ s), and thus is in the spin diffusion limit ($\omega\tau_c \gg 1$), where the

NOEs are large and negative. Figure 1 shows a typical plot of the maximum NOE between two spins as a function of rotational correlation time/molecular weight. More informative, especially in terms of the strengths and limitations of the method to be discussed later, is a plot of the relative cross-relaxation rate between two nuclei as a function of molecular weight (Figure 1). This plot shows that while the relative magnitude of the NOE increases and then levels off at high molecular weights, the cross-relaxation rate is a very strong function of molecular weight from small peptides up to very large proteins and does not level off. In the presence of chemical exchange of the ligand between its free and bound states, negative NOEs conveying conformational information of the bound ligand are transferred to the free ligand resonances where they are more easily measured because these resonances are much narrower. Equation 1 represents the exchange system in which $L_F$ and $L_B$ are the free and bound ligand, P is the macromolecule, and $\tau_F^{-1}$ and $\tau_B^{-1}$ are the exchange rates between the free and bound states.

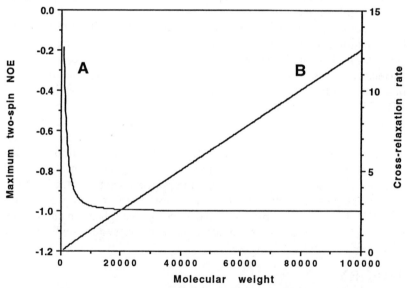

*Figure 1* (A) Maximum two-spin NOE as a function of molecular weight. Relationship between molecular weight and rotational correlation time taken from Brauer & Sykes (18). (B) Cross-relaxation rate (in $s^{-1}$) between two protons separated by 2.5 Å as a function of molecular weight. Larmor frequency is 500 MHz.

$$P + L_F \underset{\tau_B^{-1}}{\overset{\tau_F^{-1}}{\rightleftharpoons}} PL_B.\qquad\qquad 1.$$

Extreme narrowing limit       Spin diffusion limit
Cross-relaxation slow       Cross-relaxation rapid
NOEs near zero, positive, or negative      NOEs large and negative

In practical cases involving large proteins ($M_R > 20,000$), distances derived from measurement of two-dimensional TRNOESY cross-peak intensities are only approximate, because cross-relaxation becomes very rapid and direct cross-relaxation between two spins in the bound ligand is modified by indirect cross-relaxation with other spins. A rigorous approach to analyzing the TRNOE data involves the implementation of a full relaxation matrix, in which a multiple spin system in the presence of chemical exchange is analyzed. The full relaxation matrix treatment of NOE data has been developed and used to refine many protein and nucleic acid structures (10, 14–17, 40, 44). Landy & Rao (43) first developed an analogous relaxation matrix procedure for the analysis of the TRNOE. They showed the complications involved in calculating the NOE for exchanging systems and how to circumvent these complications by assuming fast chemical exchange between the free and bound ligand—the situation in which the TRNOE approach can generally be most accurately and fruitfully applied.

We recently published a theoretical evaluation of the TRNOE using the full relaxation matrix approach (21) in which we examined the effects of variables such as mixing time, fraction of bound peptide, free and bound correlation times, and other contributions to spin-lattice relaxation rates in an attempt to offer practical experimental guidelines for the design of a TRNOE experiment. We have also addressed the question of internal motions in the free and bound peptide and how these internal motions affect the development of the TRNOE (22). However, in both these sets of calculations, we considered the TRNOE only under conditions of fast chemical exchange. Two groups (49, 55) have subsequently reported methods that allow a complete relaxation matrix analysis of the TRNOE with exchange rates included in the calculation. In this review, we examine the TRNOE and present practical guidelines for its application based upon a complete relaxation matrix analysis of a model system under a wide variety of conditions.

## THEORY

### Complete Relaxation Matrix Analysis for a Multiple Spin System in Chemical Exchange

A description of homonuclear dipolar relaxation for a multiple spin system requires the complete set of coupled differential equations describing the evolution of longitudinal magnetizations of the individual spins. The differential equations are given by the Bloch equations (13) and are of the form

$$dM_{zi}/d\tau_m = -W_{ii}(M_{zi}-M_{oi}) - \sum_{j \neq i} W_{ij}(M_{zj}-M_{oj}),  \qquad 2.$$

where $W_{ii}$ is given by

$$W_{ii} = (1/10)\sum_{j \neq i} \gamma^4 \hbar^2/(r_{ij})^6 [J_0(\omega) + 3J_1(\omega) + 6J_2(\omega)], \qquad 3.$$

and $W_{ij}$, the cross-relaxation rate between spins $i$ and $j$, is given by

$$W_{ij} = (1/10)\gamma^4 \hbar^2/(r_{ij})^6 [6J_2(\omega) - J_0(\omega)]. \qquad 4.$$

The spectral density functions take the form

$$J_n(\omega) = \tau_c/[1 + (n\omega_0 \tau_c)^2], \qquad 5.$$

where $\tau_c$ is the rotational correlation time of the interproton vector between $i$ and $j$, which is identical for all interproton vectors if the spin system is tumbling isotropically in solution; $\tau_m$ is the mixing time of the TRNOESY experiment that considers the evolution of magnetization during the mixing time of a two-dimensional TRNOESY experiment, and $\omega_0$ is the NMR Larmor frequency. These coupled differential equations can be written in matrix form (1, 50, 64):

$$d\mathcal{M}/d\tau_m = -\mathcal{W} \cdot \mathcal{M}, \qquad 6.$$

where $\mathcal{M}$ is the matrix of magnetizations and $\mathcal{W}$ is the $n$-dimensional relaxation matrix with diagonal elements $W_{ii}$ and off-diagonal elements $W_{ij}$. Equation 6 can be solved as (19):

$$\begin{aligned}\mathcal{M}(\tau_m) &= \exp(-\mathcal{W}\tau_m)\mathcal{M}(0) \\ &= \chi \exp(-\lambda \tau_m)\chi^{-1}\mathcal{M}(0) \\ &= a(\tau_m)\mathcal{M}(0),\end{aligned} \qquad 7.$$

where $\chi$ is the matrix of eigenvectors of the relaxation matrix $\mathcal{W}$, $\lambda$ is the diagonal matrix of eigenvalues, and $a$ is the matrix of mixing coefficients, $a_{ij}(\tau_m)$, which are proportional to the measured TRNOESY intensities.

If the spin system undergoes chemical exchange between a free (F) and bound (B) state, the coupled differential equations must be modified to include exchange terms. These equations have been derived by McConnell (52) and have the form

$$dM_{zi}^B/d\tau_m = -W_{ii}^B(M_{zi}^B - M_{oi}^B) + \sum_{j \neq i} W_{ij}^B(M_{zj}^B - M_{oj}^B)$$
$$- M_{zi}^B/\tau_B + M_{zi}^F/\tau_F, \quad 8a.$$

$$dM_{zi}^F/d\tau_m = -W_{ii}^F(M_{zi}^F - M_{oi}^F) + \sum_{j \neq i} W_{ij}^F(M_{zj}^F - M_{oj}^F)$$
$$+ M_{zi}^B/\tau_B - M_{zi}^F/\tau_F, \quad 8b.$$

where $\tau_F$ and $\tau_B$ are the chemical exchange lifetimes of the free and bound states, respectively. When these equations are presented in matrix form, relaxation can be described by (43):

$$\frac{d}{d\tau_m}\begin{bmatrix} \mathbf{m}_B \\ \mathbf{m}_F \end{bmatrix} = -\mathcal{R}\begin{bmatrix} \mathbf{m}_B \\ \mathbf{m}_F \end{bmatrix}, \quad 9.$$

where $\mathbf{m}_B$ and $\mathbf{m}_F$ are $n$-dimensional vectors.

## The Relaxation Matrix in the Limits of Fast Chemical Exchange

To obtain the mixing coefficients $a_{ij}(\tau_m)$, $\mathcal{R}$ must be diagonalized to yield $\lambda$, the eigenvalue matrix. But because of the lack of symmetry in the matrix $\mathcal{R}$ when exchange is included, computer methods for the diagonalization are not so readily available, and the eigenvalues are not guaranteed to be real (43). These complications may be avoided in the regime of fast chemical exchange; here, the calculations result in an average relaxation matrix (43) that is real and symmetric. This can been seen as follows.

$\mathcal{R}$ may be separated into two parts representing exchange and relaxation (43) as $\mathcal{R} = \mathcal{E} + \mathcal{W}$, where $\mathcal{E}$ includes all the matrix elements involving $\tau_B$ and $\tau_F$, and $\mathcal{W}$ is the $2n$-dimensional relaxation matrix for the $n$-spin system in the free and bound forms:

$$\mathcal{R} = \begin{bmatrix} \tau_F^{-1}\mathbf{1} & -\tau_B^{-1}\mathbf{1} \\ -\tau_F^{-1}\mathbf{1} & \tau_B^{-1}\mathbf{1} \end{bmatrix} + \begin{bmatrix} \mathcal{W}_F & 0 \\ 0 & \mathcal{W}_B \end{bmatrix}. \quad 10.$$

By choice of an appropriate transformation matrix, Equation 9 takes the form (43):

$$\frac{d}{d\tau_m}\begin{bmatrix} \mathbf{m}_B + \mathbf{m}_F \\ p_B\mathbf{m}_F - p_F\mathbf{m}_B \end{bmatrix} = -\left( \begin{bmatrix} 0 & 0 \\ 0 & (\tau_B^{-1} + \tau_F^{-1})\mathbf{1} \end{bmatrix} \right.$$
$$\left. + \begin{bmatrix} p_B\mathcal{W}_B + p_F\mathcal{W}_F & \mathcal{W}_F - \mathcal{W}_B \\ p_Bp_F(\mathcal{W}_F - \mathcal{W}_B) & p_B\mathcal{W}_F + p_F\mathcal{W}_B \end{bmatrix} \right)\begin{bmatrix} \mathbf{m}_B + \mathbf{m}_F \\ p_B\mathbf{m}_F - p_F\mathbf{m}_B \end{bmatrix} \quad 11.$$

where $\mathcal{W}_B$ and $\mathcal{W}_F$ are relaxation matrices for the bound and free states respectively, and $p_B$ and $p_F$ are the fractions of bound and free ligand. In the limit of fast chemical exchange, if $\tau_B^{-1}, \tau_F^{-1}$ is much greater than $W_{ij}^B$, $W_{ij}^F$, then $m_B^i/m_F^i$ equals $p_B/p_F$, and Equation 11 then simplifies (43) to

$$\frac{d(\mathbf{m}_B+\mathbf{m}_F)}{d\tau_m} \cong (p_b \mathcal{W}_B + p_F \mathcal{W}_F)(\mathbf{m}_B+\mathbf{m}_F). \qquad 12.$$

Thus, for fast exchange at equilibrium, the longitudinal magnetization of the spin system, summed over the exchanging species, decays as a function of the population-weighted average of the individual relaxation matrices for the bound and free states $\mathcal{R} \approx p_B \mathcal{W}_B + p_F \mathcal{W}_F$. The average relaxation matrix may then be easily diagonalized to generate the desired mixing coefficients.

If one seeks an approximate solution, one can now expand the exponential term of the solution in the form of Equation 7 as a Taylor series,

$$a(\tau_m) = \exp(-\mathcal{R}\tau_m) = 1 - \mathcal{R}\tau_m + 0.5\mathcal{R}^2\tau_m^2 + \cdots \qquad 13.$$

At sufficiently short mixing times, the NOE cross-peak intensity is represented by the term in Equation 13 that is linear in $\tau_m$. One may thus arrive at a linear relationship between the measured intensities $a_{ij}(\tau_m)$ and the cross-relaxation rates

$$a_{ij}(\tau_m) \approx -\mathcal{R}_{ij}\tau_m, \qquad 14.$$

which for the TRNOESY in fast exchange becomes

$$a_{ij}(\tau_m) \approx -[p_B W_{ij}^B + p_F W_{ij}^F]\tau_m. \qquad 15.$$

This is the result presented earlier by Clore et al (23, 24, 34) and basically amounts to a two-spin approximation. However, the cross-relaxation rate between spins $i$ and $j$ in the bound state is so large for larger proteins (see Figure 1) that Equation 14 is only valid for unreasonably short values of $\tau_m$ (see below).

## *The Relaxation Matrix with Exchange Rates Included in the Calculation: Slow and Intermediate Exchange*

In some situations of interest, especially those including very large macromolecules, not all of the relevant cross-relaxation rates may fall into the fast-exchange limit, and the conclusions derived from the analysis under fast-exchange conditions will not necessarily be valid. It then becomes necessary to calculate the TRNOE including the effect of finite exchange rates. Early studies by Clore & Gronenborn (23, 24) examined the effects of chemical exchange on the build-up rate of the one-dimensional steady-state TRNOE. Their analysis distinguished three regions of chemical exchange. In the case of fast exchange between free and bound ligand, when the ligand off-rate is faster than 10 times the spin-lattice relaxation rate, and when the resonances of the bound and free ligand are averaged, they found that the magnitude of the TRNOE between two protons on the ligand was directly proportional to the bound cross-relaxation rate between these same two protons, neglecting in their calculations the effects of indirect cross-relaxation between spins. In the case of intermediate exchange between free and bound ligand, when the chemical shifts of the bound and free peptide resonances are not averaged but the off-rate is still faster than the spin-lattice relaxation rate and the bound cross-relaxation rate, the TRNOE was still found to be approximately proportional to the bound cross-relaxation rate, but only when the ligand was present in large molar excess. However, when the exchange between bound and free ligand was slow on the cross-relaxation and spin-lattice relaxation scale, then the TRNOE was found to be vanishingly small.

Other groups (49, 55) have very recently examined the effect of chemical exchange on the magnitude of the two-dimensional TRNOE using a full relaxation matrix analysis. London et al (49) have used a complete relaxation matrix analysis of the TRNOE for model spin systems of various nuclear geometries, using a range of finite exchange rates. They found that calculated build-up rates corresponding to mixing times of 10–100 ms exhibited a strong dependence on the exchange-rate constant, and thus they cautioned against the use of initial build-up rates in the interpretation of TRNOE studies without an independent determination of the rate of chemical exchange. In most cases, they detected significant discrepancies between the calculated initial TRNOE build-up rates and the parameter $p_B W_{ij}^B$, an indication that some of the limiting conditions defined by Clore & Gronenborn were not being satisfied.

Ni (55) has also reported on a method that allows a complete relaxation-matrix analysis of the TRNOE with exchange rates included in the calculation. Ni points out that despite the inherent asymmetry in the relaxation matrix, stable solutions must exist on physical grounds, and he describes an algorithum for the calculation of the TRNOE for all exchange rates. TRNOEs were calculated for a peptide derived from residues 7 to 16 of the Aα chain of human fibrinogen as a function of exchange rate. For slow exchange rates ($\tau_B^{-1} = 1 \text{ s}^{-1}$), no NOE transfers occur up to a mixing time of 500 ms, whereas for intermediate exchange rates, lag phases occurred in the TRNOE build-up, in agreement with what was observed experimentally for this system (41). Both London et al (49) and Ni (55) predict an initial lag in the TRNOE build-up curve for intermediate exchange and caution against the interpretation of this lag as characterizing indirect relaxation pathways unless the fast-exchange condition is known to be satisfied.

## *Intermolecular vs Intramolecular TRNOEs*

The majority of TRNOE studies in the literature have been performed under conditions of fast chemical exchange, and directed to the use of the intramolecular TRNOE to determine the structure of the bound ligand. However, some TRNOE studies have been performed under conditions of intermediate chemical exchange, in which the chemical shifts of the bound and free peptide resonances are not averaged but the off-rate is still fast relative to the spin-lattice relaxation time of the protein and ligand protons. These studies have been directed towards the use of the intermolecular TRNOE to determine the contacts between the bound ligand and the target macromolecule. Anglister and coworkers (4–7, 45, 69) have studied peptide-antibody complexes with TRNOE difference spectroscopy under conditions of intermediate chemical exchange. They have used the TRNOE to identify residues in the combining site of an antibody diverted against a peptide of cholera toxin (4, 5, 7, 45), as well as to generate distances in the variable-region fragment (Fv) of the antibody in order to calculate a three-dimensional model of the complex.

Under conditions of slow chemical exchange on the chemical-shift scale, two separate sets of resonances exist for free and bound ligand and macromolecule. This introduces a new level of complexity in interpreting the spectrum. Anglister and coworkers have categorized the cross-peaks appearing in their two-dimensional TRNOE spectra of peptide-antibody complexes as arising from: (*a*) chemical exchange between free and bound peptide resonances, (*b*) chemical exchange between free and bound protein resonances, (*c*) magnetization transfer within the protein (protein NOEs), (*d*) intermolecular magnetization transfer between antibody protons and free peptide protons via the bound peptide (peptide-protein TRNOEs), and (*e*) intramolecular magnetization transfer within the bound peptide via exchange with the free peptide (peptide TRNOEs). The recent paper by London et al (49) provides a theoretical analysis of the various possible TRNOE cross-peak intensities and the relationship between the slow- and fast-exchange situations. Only those cross-peaks belonging to categories *d* and *e* actually constitute TRNOE cross-peaks, whereas cross-peaks belonging to categories *a* and *b* are exchange peaks, and the cross-peaks belonging to category *c* are the standard protein NOEs. Furthermore, only the intramolecular TRNOE cross-peaks arising from magnetization transfer between free and bound ligand resonances yield information on the bound conformation of the ligand. However, the intermolecular TRNOE cross-peaks arising from magnetization transfer between the macromolecule and the bound ligand are extremely useful for determining those residues on the ligand and the macromolecule that participate at the binding interface.

Anglister and coworkers have assigned the cross-peaks in these complicated 2D TRNOE spectra by a combination of specific deuteration of the aromatic amino acids of the antibody and the peptide, and difference TRNOE spectroscopy of peptide-saturated antibody vs antibody with many-fold excess peptide. These methods have allowed editing out of all peaks other than the intramolecular and intermolecular TRNOEs in a spectrum complicated by the high protein concentrations needed to observe the intermolecular TRNOEs. Fortunately, such involved methods are not necessary for the majority of TRNOE studies used to determine the structure of ligands bound to macromolecules. This is simply because the concentration of protein required to have a large effect on the magnitude of the intramolecular TRNOE is normally so low that it is essentially unobserved in the spectrum. The protein peaks are very broad in comparison to the ligand peaks and are generally lost in the baseline of the spectrum. In addition, when fast chemical exchange prevails, the additional complications of exchange peaks in the two-dimensional spectrum need not be considered.

## EXPERIMENT

We have simulated the TRNOE build-up curves so that we might present a computer-guided theoretical evaluation of the TRNOE in which we examine the effects of variables such as mixing time, fraction of bound peptide, and free and bound correlation times, as well as internal motions in the free and bound peptide. The details of the equations used are described below; a copy of the program written for SUN computers is available from the authors upon request.

### Calculation of the Average Relaxation Matrix

Calculating the average relaxation matrix, $\mathscr{R} \approx p_B \mathscr{W}_B + p_F \mathscr{W}_F$, requires setting up individual relaxation matrices, $\mathscr{W}_B$ and $\mathscr{W}_F$. Each relaxation matrix is set up identically, in the following manner. Equation 6 may be then be represented as

$$\frac{d}{d\tau_m} \begin{bmatrix} m_i \\ m_j \\ \vdots \\ m_n \end{bmatrix} = - \begin{bmatrix} W_{ii} & W_{ji} & \cdots & W_{in} \\ W_{ji} & W_{jj} & \cdots & W_{jn} \\ & & \vdots & \\ W_{ni} & W_{nj} & \cdots & W_{nn} \end{bmatrix} \begin{bmatrix} m_i \\ m_j \\ \vdots \\ m_n \end{bmatrix} \qquad 16.$$

where the diagonal elements are given by

$$W_{ii} = 2(m_{oi}-1)(\mathscr{W}_{1ii}+\mathscr{W}_{2ii})+\sum_{k}m_{ok}(\mathscr{W}_{0ik}+2\mathscr{W}_{1ik}+\mathscr{W}_{2ik})+R_{1\text{ext}} \quad 17\text{a.}$$

and the off-diagonal elements are given by

$$W_{ij} = m_{0i}(\mathscr{W}_{2ik}-\mathscr{W}_{0ik}). \quad 17\text{b.}$$

To take into account other possible relaxation pathways, we have added an external relaxation component, $R_{1\text{ext}}$, to the diagonal element, $W_{ii}$. The transition probabilities are given by (50):

$$\mathscr{W}_{1ii} = 3/2 q_{ii} J(\omega_i) \quad 18\text{a.}$$

$$\mathscr{W}_{2ii} = 6 q_{ii} J(2\omega_i) \quad 18\text{b.}$$

$$\mathscr{W}_{1ij} = 3/2 q_{ij} J(\omega_i) \quad 18\text{c.}$$

$$\mathscr{W}_{0ij} = q_{ij} J(\omega_i - \omega_j) \quad 18\text{d.}$$

$$\mathscr{W}_{2ij} = q_{ij} J(\omega_i - \omega_j). \quad 18\text{e.}$$

Here the spectral density functions take the form

$$J(\omega_i) = \tau_c/[1+(\omega_i \tau_c)^2] \quad 19.$$

and

$$q_{ij} = 1/10 \gamma_i^2 \gamma_j^2 h^2 (r_{ij})^{-6} (\mu_o/4\pi), \quad 20.$$

where $\tau_c$ is the overall tumbling time of the peptide that modulates the interaction between spins $i$ and $j$, and $r_{ij}$ is the internuclear distance.

## Chemically Equivalent Protons

The off-diagonal elements, $W_{ij}$ and $W_{ji}$, are equivalent only if we ignore the presence of chemically equivalent groups of protons for which $m_{oi} > 1$ (such as methyl groups or aromatic rings undergoing fast internal motion). From Equation 17b, one can see that this results in an asymmetric relaxation matrix ($W_{ij} \neq W_{ji}$, because $m_{oi} \neq m_{oj}$) for which the eigenvalues are not guaranteed to be real. In addition, most computer algorithms for diagonalizing relaxation matrices are available for symmetric matrices only (16, 17). When the relaxation matrix, $\mathscr{W}$, is real and symmetric, then the transformation matrix of orthonormal eigenvectors, $\chi$, is a unitary matrix that has the mathematical property of having its inverse equal to its transpose ($\chi^{-1} = \chi^T$). Thus, Equation 7 simplifies to:

$$\mathscr{M}(\tau_m) = \chi \exp(-\lambda \tau_m) \chi^T \mathscr{M}(0). \quad 21.$$

The computer algorithm required to calculate the transpose of a matrix (as opposed to its inverse) is much simpler and computationally faster, especially as the size of the matrix increases, and thus there are computational advantages in starting with a symmetric relaxation matrix as a starting point for relaxation-matrix analysis.

Recently, Olejniczak (60) showed how nonsymmetric matrices as described above can be transformed into symmetric matrices. We considered two alternate choices for the treatment of groups of equivalent protons: the use of a single coordinate position for a pseudo proton situated geometrically in the center of the equivalent group of protons, or the explicit consideration of each proton in the group. Treating each proton as an individual distinguishable proton generates a symmetric matrix ($m_{oi} = 1$). This choice is valid as long as no second-order spin coupling is involved. Under these conditions, the TRNOE between a given proton and a group of equivalent protons is then the sum of the contributions from each equivalent proton in the group to the given proton. The same reasoning is involved when considering the scaling of experimental cross-peak intensities when groups of equivalent protons are involved (68). Cross-relaxation amongst the group of equivalent protons is calculated explicitly.

## Effects of Internal Motions

To take into account differential motion, we incorporated the order parameters $S_i^F$ and $S_i^B$ for spin $i$ in the free and bound peptide (10), using Lipari & Szabo's model-free approach (47). In this approach, internal motions in macromolecules may be specified by two independent quantities: a generalized order parameter, $S^2$, and an effective internal correlation time, $\tau_e$. The order parameter, $S^2$, can vary between 0 and 1. A value of one indicates that the correlation time of the interproton vector is the same as the overall tumbling time of the peptide ($\tau_c$); a value of zero indicates complete motional freedom.

The spectral density function of Equation 19 then becomes

$$J(\omega_i) = \frac{S^2 \tau_c}{1+(\omega_i \tau_c)^2} + \frac{(1-S^2)\tau}{1+(\omega_i \tau)^2}, \qquad 22.$$

where

$$1/\tau = 1/\tau_c + 1/\tau_e. \qquad 23.$$

## RESULTS AND DISCUSSION

In this section, we examine the behavior of the TRNOE using the average relaxation matrix treatment for a multispin system in fast chemical exchange. As a model system, we have chosen the binding of a 12-residue peptide, spanning the inhibitory region of the muscle protein troponin I (TnI), to the muscle protein troponin C (TnC). This system is of general biochemical interest, is of a size for which a quantitative treatment is appropriate, and is one for which experimental data are available (21).

The computer simulation takes as input the coordinates of the protons of the free and bound conformations of the TnI peptide, the overall rotational correlation times of the free and bound peptide ($\tau_c^F$, $\tau_c^B$), the effective internal correlation times of the free and bound peptide ($\tau_e^F$, $\tau_e^B$), the external leakage relaxation rates of the free and bound peptide protons ($R_{1ext}^F$, $R_{1ext}^B$), order parameters for individual free and bound peptide protons ($S_i^F$, $S_i^B$), the fraction of the free and bound peptide ($p_F$, $p_B$), the mixing time ($\tau_m$), and the frequency of the experiment ($\omega$).

For the first set of simulations examining the effects of fraction bound and mixing time on the development of the TRNOE, the coordinates for the free TnI peptide were generated from the standard $\phi,\psi$ angles of a $\beta$-strand ($\phi = -118°$, $\psi = +118°$), whereas the coordinates for the bound peptide were generated from the standard $\phi,\psi$ angles of an $\alpha$-helix ($\phi = -60°$, $\psi = -40°$). This is a reasonable model to adopt since the peptide is considered to be unstructured and extended in solution, but binds to TnC with helix-like structure (21). For the second set of simulations examining the effects of internal motions in the peptide on the development of the TRNOE, the coordinates for the free TnI peptide were again generated from the standard $\beta$-strand dihedral angles, whereas the coordinates for the bound peptide were generated from the actual TRNOE-derived structure of the inhibitory peptide N$\alpha$-acetyl-TnI(104–115) amide bound to Ca(II)-saturated skeletal turkey TnC (20). The proton coordinates of the TnC protein to which the TnI peptide binds were not included explicitly in the calculation, but rather as a bath of protons in which the bound TnI peptide experiences a longer rotational correlation time and a larger external leakage relaxation rate.

## Effect of Fraction Bound and Mixing Time

The TRNOE was calculated using the following standard parameters: $\omega = (2\pi) \cdot 500 \times 10^6$ s$^{-1}$; $R_{1ext}^F = 0.00$ s$^{-1}$; $R_{1ext}^B = 0.50$ s$^{-1}$. The correlation time for all protons in the free peptide was taken to be 0.4 ns, an appropriate estimate for a 12-residue peptide; at 25°C, four different correlation times were chosen for all protons in the bound peptide, i.e. $\tau_c^B = 12, 25, 45, 90$ ns, corresponding approximately to molecular weights of 25 through 200 kDa (19). The effect of fraction bound and mixing time was studied for four proton pairs with varying internuclear distances. Table 1 summarizes these proton pairs and their internuclear distances in the free and bound

**Table 1** Internuclear distances of reference proton pairs in the free and bound peptide

| Proton pair | $i$ | $j$ | $r_{ij}^F$ | $r_{ij}^B$ |
|---|---|---|---|---|
| 1 | F106$\beta$CH | F106$\beta$CH' | 1.76 Å | 1.76 Å |
| 2 | F106$\delta$CH | F106$\zeta$CH | 2.46 Å | 2.46 Å |
| 3 | R112NH | R113NH | 4.41 Å | 2.68 Å |
| 4 | R112$\alpha$CH | R113$\alpha$CH | 4.74 Å | 4.86 Å |

peptide. The geminal proton pair #1, F106$\beta$CH-F106$\beta$CH', is representative of a very short internuclear distance (1.76 Å) that does not change from the free to the bound TnI conformation. Although a stereospecific assignment is often possible with geminal protons of nondegenerate chemical shifts, it is not advisable to use their measured TRNOE as a reference TRNOE when determining relative distances, as spin diffusion becomes a major problem at short distances and all calculated distances will then appear too short. A much better reference distance is offered by the vicinal proton pair #2, F106$\delta$CH-F106$\zeta$CH, on the phenylalanine ring. This internuclear distance (2.46 Å) again does not change from free to bound peptide. It has been used as the reference TRNOE to calculate distances from the computer simulated TRNOE values for the other three proton pairs in the following section of this review. Proton pair #3, R112NH-R113NH is interesting in that it represents an internuclear distance diagnostic of an $\alpha$-helix, but not of a $\beta$-strand, and correspondingly the internuclear distance decreases from 4.41 to 2.68 Å in the free vs the bound peptide. Finally, proton pair #4, R112$\alpha$CH-R113$\alpha$CH represents a long internuclear distance, increasing only marginally from 4.74 to 4.86 Å in the free vs the bound peptide.

The first set of simulations illustrates the effect of fraction bound on the magnitude of the TRNOE, for various mixing times. The TRNOE was calculated for four values of $\tau_m$ (50, 100, 150, 200 ms) and $p_B$ varying from 0 to 1.0. Figure 2 shows TRNOE intensity as a function of fraction bound for various mixing times, interproton distances in the bound peptide, and molecular weights for the target protein. The panels from left to right represent an increase in $\tau_m$ from 50 ms to 200 ms, and from top to bottom an increase in $\tau_c^B$ from 12 to 90 ns (mol wt $\sim$ 25,000–200,000). The intensity of the TRNOE is essentially zero for $p_B$ approaching zero, indicating that the contribution to the TRNOE intensity from the peptide in the free state is negligible, because $(\omega \tau_c^F) \approx 1$, and therefore the structure of the free peptide will contribute little to the calculated distances (we return to this point later).

One can see from Figure 2 that the TRNOE does not display a linear build-up in magnetization intensity as $p_B$ is increased, for constant $\tau_m$. The linear relationship between TRNOE intensity and $p_B$ exists only at very small values of $p_B$ and very short values of $\tau_m$. For longer $\tau_m$ and larger $\tau_c^B$ values, TRNOE intensities corresponding to the three shorter internuclear distances (1.76, 2.46, and 2.68 Å) display a fast build-up followed by rapid decay. This phenomenon of rapid decay is especially prevalent for proton pair #1 (1.76 Å). For $\tau_c^B = 12$ ns (Panels $A$, $B$, $C$, $D$), the TRNOE corresponding to pair #1 begins to visually diverge from linearity at $p_B > 0.3$ for $\tau_m = 50$ ms ($A$) but has begun to diverge from linearity at $p_B > 0.05$ for $\tau_m = 200$ ms ($D$). For $\tau_m = 50$ ms (Panels $A$, $E$, $I$, $M$), the TRNOE corresponding to proton pair #1 diverges from linearity at $p_B > 0.3$ for $\tau_c^B = 12$ ns ($A$) and has begun to diverge from linearity at $p_B > 0.05$ for $\tau_c^B = 90$ ns ($M$). For longer $\tau_m$ and larger $\tau_c^B$, all magnetization begins to approach the same intensity, indicating that spin diffusion has spread magnetization uniformly over all protons in the peptide.

Figure 3 shows the TRNOE values of Figure 2 translated into distances, calculated using the proton pair #2, F106δCH-F106ξCH, as the reference TRNOE and reference distance (2.46 Å). For larger values of $\tau_m$ and $p_B$, all distances converge to an average value of approximately 2.5 Å so that it is best to work with short $\tau_m$ and small $p_B$, especially when $\tau_c^B$ is large. This can be a positive factor in the design of a TRNOE experiment, since an appreciable TRNOE can be observed for very low fraction of bound protein ($p_B < 0.05$)—a plus if the protein is difficult to obtain in any quantity.

An appropriate question to ask is how does the conformation of the free peptide contribute to the calculated conformation of the bound peptide. At very low fraction bound, one may expect the free conformation to contribute somewhat to the final conformation. Figure 4 shows the effect of changing the correlation time of the free peptide, $\tau_c^F$, on the calculation of $r$ apparent. We have taken the worst case scenario, where $\tau_c^B$ is at its lowest value ($\tau_c^B = 12$ ns). Three $\tau_c^F$ values were chosen, one of 0.2 ns, corresponding approximately to a 6-residue peptide, one of 0.4 ns, corresponding to a 12-residue peptide, and one of 0.8 ns, corresponding to a 24-residue peptide. The proton pair chosen was proton pair #3, R112NH-R113NH where $r_{ij}^F = 4.41$ Å and $r_{ij}^B = 2.68$ Å. According to Figure 4, one can see at $p_B = 0.0$ that $r$ apparent is equal to 4.4 Å, the distance in the free peptide. But even at very low $p_B$, ($p_B = 0.010$ or 1% bound), $r$ apparent is already approaching the distance in the bound peptide. The curve corresponding to $\tau_c^F = 0.8$ ns approaches $r_{ij}^B$ the most slowly, whereas the curve corresponding to $\tau_c^F = 0.2$ ns approaches $r_{ij}^B$ the most quickly. The $\tau_c^F = 0.2$ ns curve is discontinous at $p_B = 0.012$ because the TRNOE starts from a

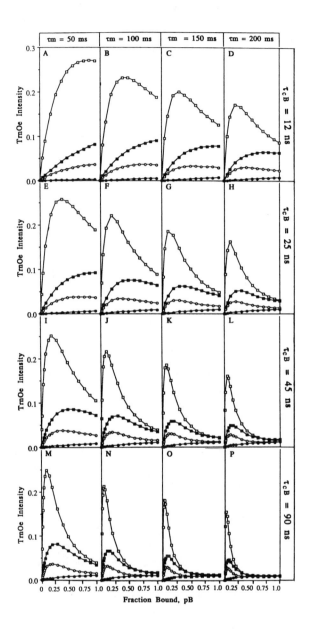

small positive value and passes through zero to a negative value ($\tau_c^F = 0.2$ ns is in the extreme narrowing limit at 500 MHz). When the TRNOE is null, $r$ apparent approaches infinity. The other curves are continuous, because the TRNOE starts at a small negative number and continues to build to a larger negative number (for $\tau_c^F = 0.4$ and 0.8 ns, NOEs are very small, but negative). Thus, one can see that for peptides as large as 24 residues, the contribution of the free peptide conformation to the final conformation is completely obviated for $p_B \geq 0.020$. For larger peptides, one need only work at slightly higher values of fraction bound, i.e. $p_B \geq 0.050$.

This set of simulations demonstrates how rapidly spin diffusion in larger proteins can influence the development of the TRNOE. Given that one has the ability to control at least two variables in the TRNOE experiment, $\tau_m$ and $p_B$, it is best to use very low $p_B$ values but work at a reasonable $\tau_m$. This method gives decent experimental signal-to-noise. This may be one of the few examples in biochemistry where less protein gives better results!

## Effects of Internal Motions

The TRNOE-derived structure of the bound inhibitory peptide used in these calculations has a central bend in the peptide that brings the residues on the hydrophobic face into closer proximity with each other, and the aromatic ring of F106 is somewhat buried by the side chains of L111 and V114. The specific TRNOEs calculated to investigate the effects of internal motions were the intraresidue interaction between the $\alpha$CH of V114 and the six chemically shift-degenerate $\gamma$CH$_3$ protons of the same residue, and the longer range interresidue interaction between these six $\gamma$CH$_3$s of V114 and three chemically shift-degenerate protons of F106 (two $\zeta$CH and one $\zeta$CH aromatic ring protons). If internal motions are important, they should be obvious in these interactions.

To estimate the values of $S^2$ to use in these simulations, we considered the fact that for a methyl group rotating around a carbon-carbon single bond, $\tau_c$ should be decreased by a factor of $S^2 = [(3\cos^2\theta - 1)/2]^2$ (51)

---

*Figure 2* TRNOE intensity vs fraction of bound peptide ($p_B$) for various bound correlation times ($\tau_c^B$) and mixing times ($\tau_m$). Calculations are presented for several pairs of nuclei with different internuclear separations: □, F106$\beta$CH-F106$\beta$CH′, 1.76 Å; ■, F106$\delta$CH-F106$\xi$CH, 2.46 Å; ○, R112NH-R113NH, 2.68 Å; ●, R112$\alpha$CH-R113$\alpha$CH, 4.86 Å. For horizontal rows of panels (i.e. A–D) the correlation time used in the calculations is indicated to the right. For vertical columns of panels (i.e. A, E, I, M), the fraction of bound peptide used in the calculations is indicated at the top. Other parameters used in the simulation were $\omega = 500 \times 10^6$ s$^{-1}$, $R_{1ext}^F = 0.00$ s$^{-1}$, $R_{1ext}^B = 0.50$ s$^{-1}$, $\tau_c^F = 0.4$ ns. ($S_i^F = S_i^B = 1.0$ for the L111 and V114 methyl protons, and $S_i^F = S_i^B = 1.0$ for the F106 ring protons.)

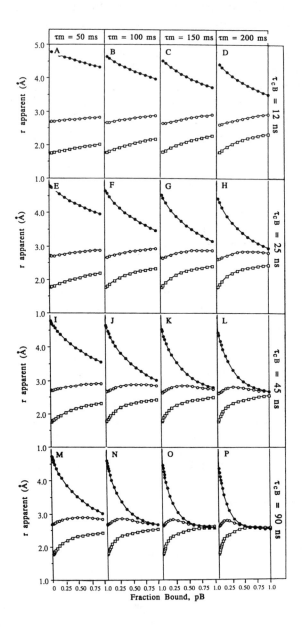

where $\theta$ is the angle between the *i-j* interproton vector and the axis of rotation (the carbon-carbon single bond). For a methyl group, $\theta = 90°$, making $S^2 = 0.25$ for rotation around the methyl axis. Thus, for the methyl protons *i* and *j* of the V114 residue in the bound and free peptide, we have chosen $S_i * S_j = 0.25$, or $S_i = S_j = 0.5$. As far as aromatic rings are concerned, the geometry leads to different estimations of $S_i * S_j$ values for the different proton pairs of the ring, so we have somewhat arbitrarily chosen $S_i = 0.5$ for the $\delta$, $\xi$, and $\zeta$ aromatic protons of the F106 residue in the free and bound peptide. Clearly this approach is not exact in detail, but allows us to introduce internal motions into regions of the peptide so that we can evaluate their effects on the TRNOE calculations. The TRNOE was calculated using the following standard parameters: $\omega = (2\pi)\cdot 500 \times 10^6$ s$^{-1}$; $R^F_{1\text{ext}} = 1.00$ s$^{-1}$; $R^B_{1\text{ext}} = 2.00$ s$^{-1}$; $\tau^F_c = 0.4$ ns; and $\tau^B_c = 25$ ns.

Figure 5 shows the effect of varying the effective internal correlation times for methyl groups' motions in the free and bound peptide on the development of the TRNOE for the intraresidue V114αCH-($\gamma$CH$_3$)$_2$ interaction. In Panel *A*, $\tau^F_e$ is kept constant at 0.01 ns, whereas $\tau^B_e$ is varied from 0.001 to 0.1 ns. In Panel *B*, $\tau^B_e$ is kept constant at 0.01 ns, whereas $\tau^F_e$ is varied from 0.001 to 0.1 ns. The order parameters for individual free and bound peptide methyl protons were set at $S^F_i = S^B_i = 0.5$. One can see that changing the effective correlation time for methyl group motions has little effect on the build-up behavior of the TRNOE (at the low fraction bound used), but a larger effect on how fast it decays. This effect is greater for internal motions in the free peptide than it is for internal motions in the bound peptide, a consideration that may be easily overlooked. Looking at Panel *B*, one can see that the less rapid the internal motion about the methyl group axis in the free peptide, the more strongly is overall relaxation affected, a function of the fact that for longer $\tau_e$ values, $\tau$ approaches the inverse of the Larmor frequency, and spin-lattice T$_1$ relaxation becomes more effective. We have chosen $\tau^F_e = \tau^B_e = 0.01$ ns for the subsequent TRNOE calculations, so that they do not depend strongly on the actual rate of the internal motion.

Figure 6 shows the TRNOE build-up curves for the intraresidue

---

*Figure 3* *r* apparent calculated from TRNOE values of Figure 2*A* vs fraction of bound peptide ($p_B$) for various bound correlation times ($\tau^B_c$) and mixing times ($\tau_m$). Calculations are presented for several pairs of nuclei with different internuclear separations: □, F106$\beta$CH-F106$\beta$CH', 1.76 Å; ○, R112NH-R113NH, 2.68 Å; ●, R112αCH-R113αCH, 4.86 Å. For horizontal rows of panels (i.e. *A–D*) the correlation time used in the calculations is indicated to the right. For vertical columns of panels (i.e. *A, E, I, M*), the fraction of bound peptide used in the calculations is indicated at the top. Distances were calculated using F106$\delta$CH-F106$\xi$CH as the reference TRNOE intensity and reference distance, 2.46 Å.

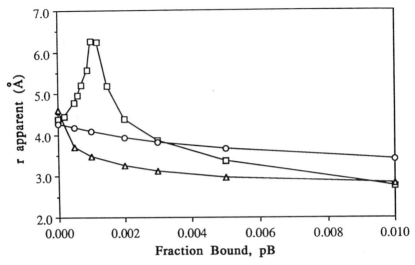

*Figure 4* r apparent for R112NH-R113NH proton pair vs low values of fraction of bound peptide ($p_B$) for various free correlation times ($\tau_c^F$). □, $\tau_c^F = 0.2$ ns; △, $\tau_c^F = 0.4$ ns; ○, $\tau_c^F = 0.8$ ns. Other parameters used in the simulation were $\omega = 500 \times 10^6$ s$^{-1}$, $R_{1ext}^F = 0.00$ s$^{-1}$, $R_{1ext}^B = 0.50$ s$^{-1}$, $\tau_c^B = 12$ ns, $\tau_m = 50$ ms. ($S_i^F = S_i^B = 1.0$ for the L111 and V114 methyl protons, and $S_i^F = S_i^B = 1.0$ for the F106 ring protons.)

V114αCH-(γCH$_3$)$_2$ and interresidue F106(ξCH$_2$ζCH)-V114(γCH$_3$)$_2$ interactions calculated in the absence of methyl and ring motions in Panels *A* and *D*, in the presence of methyl rotation in Panels *B* and *E*, and in the presence of both methyl and ring rotation in Panels *C* and *F*. One can see that the effect of internal motion is not only to dampen the build-up of the TRNOE, but to also dampen its relaxation. This effect is evident for the intraresidue V114αCH-(γCH$_3$)$_2$ interaction (Figures 6*A–C*) only for the methyl rotation, but the effects of both methyl and ring motion is evident for the interresidue F106(ξCH$_2$ζCH)-V114(γCH$_3$)$_2$ interaction.

## CONCLUSION

The calculations presented are intended as a guide to the experimentalist who is setting up a TRNOE experiment and wants to decide on experimental conditions such as fraction bound, mixing time, etc, and as an aid to the analysis of the TRNOE data. However, a necessary prerequisite to the successful application of TRNOE is that one knows the stoichiometry, affinity, and kinetics for the ligand-macromolecule interaction in advance

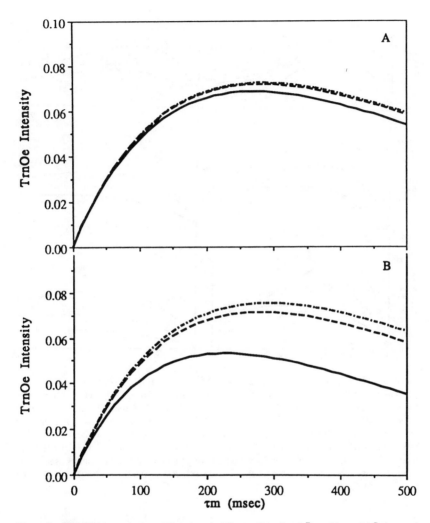

*Figure 5* TRNOE intensity vs mixing time ($\tau_m$) for various free ($\tau_e^F$) and bound ($\tau_e^B$) internal correlation times. (A) $\tau_e^F = 0.01$ ns and $S_i^F = S_i^B = 0.5$ for the V114 methyl protons: (*solid line*) $\tau_e^B = 0.1$ ns; (*dashed line*) $\tau_e^B = 0.01$ ns; (*dashed-dotted line*) $\tau_e^B = 0.001$ ns. (B) $\tau_e^B = 0.01$ ns and $S_i^F = S_i^B = 0.5$ for the V114 methyl protons: (*solid line*) $\tau_e^F = 0.1$ ns; (*dashed line*) $\tau_e^F = 0.01$ ns; (*dashed-dotted line*) $\tau_e^F = 0.001$ ns. Other parameters used in the simulation were $\omega = 500 \times 10^6$ s$^{-1}$, $R_{1\text{ext}}^F = 1.00$ s$^{-1}$, $R_{1\text{ext}}^B = 2.00$ s$^{-1}$, $\tau_c^B = 25$ ns, $\tau_c^F = 0.4$ ns, $p_B = 0.1$.

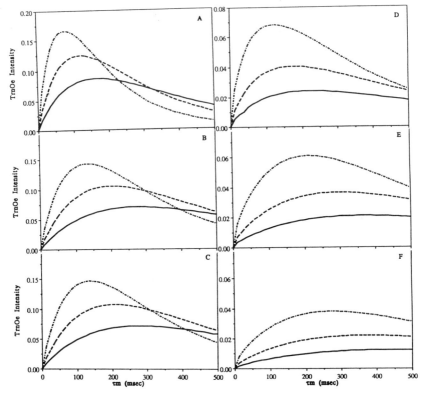

*Figure 6* TRNOE intensity vs mixing time ($\tau_m$) for various fractions of bound peptide ($p_B$): (*solid line*) $p_B = 0.1$; (*dashed line*) $p_B = 0.2$; (*dashed-dotted line*) $p_B = 0.4$. Panels *A–C* represent the intraresidue V114 $\alpha$CH-($\gamma$CH$_3$)$_2$ interaction, and Panels *D–F* represent the interresidue F106($\zeta$CH$_2$,$\zeta$CH)-V114($\gamma$CH$_3$)$_2$ interaction. (*A,D*) $S_i^F = S_i^B = 1.0$ for the V114 methyl protons and $S_i^F = S_i^B = 1.0$ for the F106 ring protons. (*B,E*) $S_i^F = S_i^B = 0.5$ for the V114 methyl protons and $S_i^F = S_i^B = 1.0$ for the F106 ring protons. (*C,F*) $S_i^F = S_i^B = 0.5$ for the V114 methyl protons and $S_i^F = S_i^B = 0.5$ for the F106 ring protons. Other parameters used in the simulation were $\omega = 500 \times 10^6$ s$^{-1}$, $R_{1ext}^F = 1.00$ s$^{-1}$, $R_{1ext}^B = 2.00$ s$^{-1}$, $\tau_c^B = 25$ ns, $\tau_c^F = 0.4$ ns, $\tau_e^B = 0.01$ ns, $\tau_e^F = 0.01$ ns.

so that one can decide whether or not the approach will work, and whether or not the theories presented herein are applicable.

ACKNOWLEDGMENTS

The financial support of the Medical Research Council of Canada (MRC Group in Protein Structure and Function) and the Alberta Heritage Medical Research Foundation (Fellowship for A. P. C) is gratefully acknowledged.

*Literature Cited*

1. Abragam, A. 1961. *The Principles of Nuclear Magnetism.* London: Oxford Univ. Press
2. Andersen, N. H., Eaton, H. L., Nguyen, K. T. 1987. *Magn. Reson. Chem.* 25: 1025
3. Andersen, N. H., Nguyen, K. T., Eaton, H. L. 1987. *J. Magn. Reson.* 63: 365
4. Anglister, J., Jacob, C., Assulin, O., Ast, G., Pinker, R., Arnon, R. 1988. *Biochemistry* 27: 717
5. Anglister, J., Levy, R. H., Scherf, T. 1989. *Biochemistry* 28: 3360
6. Anglister, J., Naider, F. 1991. *Methods Enzymol.* 203: 228
7. Anglister, J., Zilber, B. 1990. *Biochemistry* 29: 921
8. Balaram, P., Bothner-By, A. A. 1972. *J. Am. Chem. Soc.* 94: 4015
9. Balaram, P., Bothner-By, A. A. 1972. *J. Am. Chem. Soc.* 94: 4017
10. Baleja, J. D., Moult, J., Sykes, B. D. 1990. *J. Magn. Reson.* 87: 375
11. Banerjee, A., Levy, H. R., Levy, G. C., Chan, W. W. C. 1985. *Biochemistry* 24: 1593
12. Bevilacqua, V. L., Thomson, D. S., Prestegard, J. H. 1990. *Biochemistry* 29: 5529
13. Bloch, F. 1957. *Phys. Rev.* 105: 1206
14. Boelens, R., Koenig, T. M. G., Kaptein, R. 1988. *J. Mol. Struct.* 173: 299
15. Boelens, R., Koenig, T. M. G., Van Der Marel, G. A., Van Boom, J. H., Kaptein, R. 1989. *J. Magn. Reson.* 82: 290
16. Borgias, B., James, T. L. 1988. *J. Magn. Reson.* 79: 493
17. Borgias, B., James, T. L. 1989. *Methods Enzymol.* 176(A): 169
18. Brauer, M., Sykes, B. D. 1987. In *Phosphorus NMR in Biology,* ed. C. T. Burt, p. 153. Boca Raton, FL: CRC Press
19. Bull, T. E. 1987. *J. Magn. Reson.* 72: 397
20. Campbell, A. P., Sykes, B. D. 1991. *J. Mol. Biol.* 222: 405
21. Campbell, A. P., Sykes, B. D. 1991. *J. Biomol. NMR* 1: 391
22. Campbell, A. P., Sykes, B. D. 1991. *J. Magn. Reson.* 93: 77
23. Clore, G. M., Gronenborn, A. M. 1982. *J. Magn. Reson.* 48: 402
24. Clore, G. M., Gronenborn, A. M. 1983. *J. Magn. Reson.* 53: 423
25. Clore, G. M., Gronenborn, A. M., Carlson, G., Meyer, E. F. 1986. *J. Mol. Biol.* 190: 259
26. Ehrlich, R. S., Colman, R. F. 1985. *Biochemistry* 24: 5378
27. Ferrin, L. J., Mildvan, A. S. 1985. *Biochemistry* 24: 6904
28. Ferrin, L. J., Mildvan, A. S. 1986. *Biochemistry* 25: 5131
29. Fry, D. C., Kuby, S. A., Mildvan, A. S. 1985. *Biochemistry* 24: 4680
30. Glasel, J. A. 1989. *J. Mol. Biol.* 209: 747
31. Glasel, J. A., Borer, P. N. 1986. *Biochem. Biophys. Res. Commun.* 141: 1267
32. Gronenborn, A. M., Clore, G. M. 1982. *Biochemistry* 21: 4040
33. Gronenborn, A. M., Clore, G. M. 1982. *J. Mol. Biol.* 157: 155
34. Gronenborn, A. M., Clore, G. M. 1985. *Prog. Nucl. Magn. Reson. Spectrosc.* 17: 1
35. Gronenborn, A. M., Clore, G. M., Brunori, M., Giardina, B., Falcioni, G., Perutz, M. F. 1984. *J. Mol. Biol.* 178: 731
36. Gronenborn, A. M., Clore, G. M., Hobbs, L., Jeffery, J. 1984. *FEBS Lett.* 172: 219
37. Gronenborn, A. M., Clore, G. M., Jeffery, J. 1984. *J. Mol. Biol.* 172: 559
38. Hammond, S. J., Birdsall, B., Feeney, J., Searle, M. S., Roberts, G. C. K., Cheung, H. T. A. 1987. *Biochemistry* 26: 8585
39. Ito, W., Nishimura, M. N., Sakato, N., Fujio, H., Arata, Y. 1987. *J. Biochem.* 102: 643
40. Keepers, J. W., James, T. L. 1984. *J. Magn. Reson.* 57: 404
41. Kohda, D., Kawai, G., Yokoyama, S., Kawakami, M., Mizushima, S., Miyazawa, T., 1987. *Biochemistry* 26: 6531
42. Kuroda, Y., Kitamura, K. 1984. *J. Am. Chem. Soc.* 106: 1
43. Landy, S. B., Rao, B. D. N. 1989. *J. Magn. Reson.* 81: 371
44. Lefevre, J. F., Lane, A. N., Jardetsky, O. 1987. *Biochemistry* 26: 5076
45. Levy, H. R., Assulin, O., Scherf, T., Levitt, M., Anglister, J. 1989. *Biochemistry* 28: 71684
46. Levy, H. R., Ejchart, A., Levy, G. C. 1983. *Biochemistry* 22: 2792
47. Lipari, G., Szabo, A. 1982. *J. Am. Chem. Soc.* 104: 4546
48. Live, D. H., Cowburn, S., Breslow, E. 1987. *Biochemistry* 26: 6415

49. London, R. E., Perlman, M. E., Davis, D. G. 1992. *J. Magn. Reson.* 97: 79
50. Macura, S., Ernst, R. R. 1980. *Mol. Phys.* 41: 95
51. Marshall, A. G., Schmidt, P. G., Sykes, B. D. 1972. *Biochemistry* 11: 3875.
52. McConnell, H. M. 1958. *J. Chem. Phys.* 28: 430
53. Meyer, E. F., Clore, G. M., Gronenborn, A. M., Hansen, H. A. S. 1988. *Biochemistry* 27: 725
54. Milon, A., Miyazawa, T., Higashijima, T. 1990. *Biochemistry* 29: 65
55. Ni, F. 1992. *J. Magn. Reson.* 96: 651
56. Ni, F., Konoshi, Y., Bullock, C. D., Rivetna, M. N., Scheraga, H. A. 1989. *Biochemistry* 28: 3106
57. Ni, F., Konoshi, Y., Frazier, R. B., Scheraga, H. A., Lord, S. T. 1989. *Biochemistry* 28: 3082
58. Ni, F., Konoshi, Y., Scheraga, H. A. 1990. *Biochemistry* 29: 4479
59. Ni, F., Meinwald, Y. C., Vasquez, M., Scheraga, H. A. 1989. *Biochemistry* 28: 3094
60. Olejniczak, E. T. 1989. *J. Magn. Reson.* 81: 392
61. Rosevear, P. R., Bramson, H. N., O'Brian, C., Kaiser, E. T., Mildvan, A. S. 1983. *Biochemistry* 22: 3439
62. Rosevear, P. R., Fox, T. L., Mildvan, A. S. 1987. *Biochemistry* 26: 3487
63. Rosevear, P. R., Powers, V. M., Dowhan, D., Mildvan, A. S., Kenyon, G. L. 1987. *Biochemistry* 26: 5338
64. Solomon, I. 1955. *Phys. Rev.* 99: 559
65. Wakamatsu, K., Okada, A., Higashijima, T., Miyazawa, T. 1986. *Biopolymers* 25: S193
66. Wakamatsu, K., Okada, A., Miyazawa, T., Masui, Y., Sakakibara, S., Higashijima, T. 1987. *Eur. J. Biochem.* 163: 331
67. Wüthrich, K. 1986. *NMR of Proteins and Nucleic Acids.* New York: Wiley and Sons
68. Yip, P. F. 1990. *J. Magn. Reson.* 90: 382
69. Zilber, B., Scherf, T., Levitt, M., Anglister, J. 1990. *Biochemistry* 29: 10032

# PROLYL ISOMERASE: Enzymatic Catalysis of Slow Protein-Folding Reactions

*Franz X. Schmid*

Laboratorium für Biochemie, Universität Bayreuth, D-W–8580 Bayreuth, Germany

KEY WORDS: kinetics of protein folding, peptide bond isomerization, ribonuclease T1

## CONTENTS

| | |
|---|---|
| PERSPECTIVES AND OVERVIEW | 124 |
| PROLYL ISOMERIZATION | 125 |
|    *Isomerization of Peptide Bonds* | *125* |
|    *Prolyl Isomerizations in Protein Folding* | *126* |
| PROLYL ISOMERASES | 128 |
|    *Substrate Specificity* | *128* |
|    *Three-Dimensional Structure* | *129* |
| CATALYSIS OF SLOW STEPS IN FOLDING BY PROLYL ISOMERASES | 129 |
|    *Folding Mechanism of Ribonuclease T1* | *131* |
|    *Catalysis of RNase T1 Folding by Prolyl Isomerases* | *133* |
|    *Synergism of Prolyl Isomerase and Protein Disulfide Isomerase as Catalysts of Folding* | *135* |
|    *Efficiency of Catalysis* | *136* |
| ROLE OF PROLYL ISOMERASE IN CELLULAR FOLDING | 137 |
| PROLYL ISOMERASES AS TOOLS IN PROTEIN FOLDING | 138 |
| MECHANISTIC ASPECTS OF PROLYL ISOMERASE ACTION | 138 |
| CONCLUDING REMARKS | 139 |

## PERSPECTIVES AND OVERVIEW

*Cis/trans* isomerizations of prolyl peptide bonds constitute slow steps in the folding of many proteins. These isomerizations and the formation of ordered structure are closely interrelated processes because polypeptide chains with incorrect prolyl bonds can undergo partial folding only. The acquisition of the native conformation requires the correct prolyl isomers, and therefore, prolyl peptide-bond isomerization often limits the rate of the final steps of folding.

Prolyl isomerizations can be catalyzed by peptidyl prolyl *cis/trans* isomerases. Two unrelated classes of these isomerases are presently known, the cyclophilins and the FK 506 binding proteins, which bind with very high affinity to the immunosuppressants cyclosporin A and FK 506, respectively. The functional correlation between the tight binding of immunosuppressants and the catalysis of prolyl isomerization is still unclear. Cyclophilins are ubiquitous enzymes that occur in all species and in all cellular compartments. This review focuses on these prolyl isomerases and their role in in vitro protein folding. Apparently, the accessibility of the prolyl bonds in the refolding polylpeptide chains is a major factor that determines the efficiency of catalysis. In favorable cases, such as in ribonuclease T1, the time range of folding can be accelerated from hours to seconds.

Unlike molecular chaperones, prolyl isomerases act as classical enzymes in protein folding. They accelerate prolyl isomerization in refolding and in unfolding, and they do not affect the equilibrium between the native and the unfolded state. *Cis/trans* isomerizations during oxidative folding are also catalyzed, and the formation of the correct disufide bonds is facilitated. Prolyl isomerases may influence the folding pathway when the efficiency of catalysis varies for the different prolyl bonds in the folding protein. In such a case, the rank order of isomerization can be changed, thus altering the folding mechanism.

The cellular functions of the prolyl isomerases are still largely unclear. There is preliminary evidence that cyclosporin A, a potent inhibitor of prolyl isomerases, can interfere with protein maturation in the endoplasmic reticulum. The catalysis of prolyl *cis/trans* isomerizations may indeed be a major physiological function of the prolyl isomerases. Their natural substrates are, however, not known. In addition to a role in cellular protein folding, prolyl isomerases could conceivably also bind to prolyl bonds in folded proteins and catalyze transitions between conformations with alternative prolyl isomers and differing biological activity. Experimental evidence is rapidly accumulating that prolyl isomerases and their complexes with immunosuppressants play a role in signal-transduction pathways. The high diversity and ubiquitous occurrence of prolyl isomerases suggest that they have multiple cellular functions.

## PROLYL ISOMERIZATION

### Isomerization of Peptide Bonds

The peptide bond has appreciable double bond character, and consequently, the isomerization between the *cis* and *trans* forms is a slow reaction. The *trans* conformation is strongly favored for peptide bonds that do not involve prolyl residues. Their *cis/trans* ratio is very difficult to determine experimentally, and the *cis* contents of short peptides were estimated using energy calculations to be as low as 1.0% or 0.1% (43, 93). *Cis* peptide bonds can be formed, however, in folded proteins, provided that sufficient folding energy can be utilized to stabilize the *cis* isomer. About 30 kJ/mol would be necessary to shift the equilibrium distribution of *cis/trans* from 1/1000, as present in the unfolded chain, to e.g. 100/1 in a folded protein. This value is similar to the total free energies of stabilization of small proteins. Accordingly, X-ray crystallographers have found very few nonproline *cis* peptide bonds (about 0.05% of all examined peptide bonds) in native proteins (37, 77, 112).

For peptide bonds that precede proline (Xaa-Pro bonds, Figure 1) the difference in Gibbs free energy between the *cis* and the *trans* conformation is much smaller, and both forms are significantly populated at equilibrium, unless structural constraints such as those in folded proteins stabilize one of the two isomers. In short oligopeptides, *cis* contents of 10–30% are frequently found (10, 28–30). *Cis/trans* isomerizations about Xaa-Pro peptide bonds are slow because of the high activation energy ($E_A \approx 85$ kJ/mol),

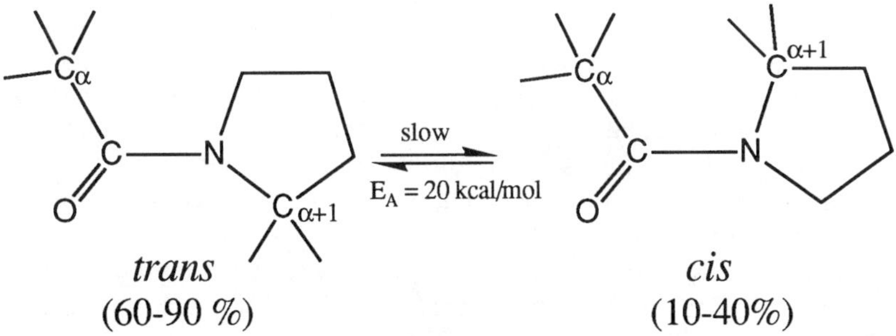

trans (60-90 %) ⇌ [slow, $E_A = 20$ kcal/mol] cis (10-40%)

Figure 1  Isomerization between the *cis* and the *trans* forms of an Xaa-Pro peptide bond.

and time constants of 10–100 s are observed for isomerization in short peptides near 25°C (10, 28–30).

## Prolyl Isomerizations in Protein Folding

FAST AND SLOW PROTEIN-FOLDING REACTIONS   In 1973, Garel & Baldwin (26) discovered that unfolded ribonuclease (RNase) A consists of a mixture of molecules that differ vastly in their rates of refolding. The respective fast-folding ($U_F$) and slow-folding ($U_S$) species coexist in a slow equilibrium and give rise to parallel fast (in the time range of milliseconds) and slow (in the time range of minutes) refolding reactions. $U_F$ and $U_S$ species have since been detected for many other proteins (see 59, 60, 95 for reviews). The proline hypothesis of Brandts et al (5) provided a plausible molecular explanation for this phenomenon. They suggested that the fast- and slow-folding molecules differ in the *cis/trans* isomeric state of one or more Xaa-Pro peptide bonds (cf Figure 1).

THE PROLINE MODEL   Normally, in the native protein, N, each prolyl peptide bond has a unique conformation, either *cis* or *trans*. However, during unfolding, (N → $U_F$; Equation 1),

$$N \underset{\phantom{folding transition}}{\overset{\text{folding transition}}{\rightleftharpoons}} U_F \underset{\phantom{prolyl isomerization}}{\overset{\text{prolyl isomerization}}{\rightleftharpoons}} U_S^i, \qquad 1.$$

the conformational restraints of the native state vanish and these bonds become free to isomerize slowly in the $U_F \rightleftharpoons U_S^i$ reactions as in short oligopeptides. These isomerizations lead ultimately to an equilibrium mixture of a single unfolded species with correct prolyl isomers, $U_F$, and one or more unfolded species with incorrect prolyl isomers, $U_S^i$. Refolding of the $U_F$ molecules with correct prolyl isomers is fast. Refolding of the $U_S^i$ molecules is usually not inhibited by nonnative isomers, but it is decelerated because it is coupled with the reisomerizations of the incorrect prolyl bonds. Under native solvent conditions, refolding does not occur by a reversal of the unfolding mechanism (i.e. reisomerization, followed by folding; cf Equation 1). The mechanism of Equation 1 is valid only under unfolding conditions and within the transition region, where partially folded species are usually not formed. Theoretical models for the kinetic analysis of folding under such conditions and the interrelationship with one or two prolyl isomerization reactions have been worked out and tested for several model proteins (52, 55).

Under solvent conditions that strongly favor folded structure (strongly native conditions), chains with incorrect isomers, $U_S^i$, can form partially folded intermediates, $I_S^i$ (Equation 2) prior to the reisomerization of the prolyl peptide bonds

$$U_S^i \xrightarrow{\text{fast folding}} I_S^i \xrightarrow[\text{isomerization}]{\text{slow proline}} N \qquad 2.$$

(13, 27, 50, 54, 83, 97). The rates of the rapid $U_S^i \to I_S^i$ reactions and the structures of the intermediates $I_S^i$ depend on the location of the nonnative prolyl isomers in the protein chain and on the solvent conditions selected for folding. Generally, incorrect prolyl isomers at the surface of the folding protein or in flexible chain regions will allow extensive structure formation prior to reisomerization. Correspondingly, solvent conditions that strongly stabilize folded proteins will also allow the formation of partially folded intermediates even with incorrect isomers.

Proteins for which prolyl isomerizations have been detected and characterized in unfolding and in refolding include pancreatic RNases (13, 69, 96, 98), the $C_L$ fragment of the Ig light chain (27), thioredoxin (48, 49), yeast iso–1 and iso–2 cytochromes c (83, 94, 119, 120), RNase T1 (51, 54), barnase (79), staphylococcal nuclease (66), carbonic anhydrase (22, 104), the chymotrypsin inhibitor CI2 (41, 42), and pancreatic trypsin inhibitor (39, 44). In almost all cases, prolyl reisomerizations are preceded by rapid structure formation. Probably not all proline residues are equally important, and nonessential prolines may exist that do not affect protein folding (2, 40, 62–65, 71, 83).

Isomerizations of peptide bonds that do not contain proline were discussed as potential sources for $U_S$ species as well (5). Even though the correct *trans* state is strongly favored for nonprolyl peptide bonds, their large number could nevertheless lead to a significant fraction of protein molecules with wrong (*cis*) peptide bonds. Experimental evidence for a role in protein folding of *cis* peptide bonds not involving proline is still missing. Possibly *cis* → *trans* isomerizations at normal peptide bonds are sufficiently rapid to not be rate-limiting events in folding.

PROLYL ISOMERIZATIONS IN FOLDED PROTEINS   About 7% of all prolyl peptide bonds are *cis* in native proteins of known three-dimensional structure (77, 112). The *cis* or *trans* state is usually clearly defined in the native protein. In a few cases, simultaneous occurrence of the *cis* and *trans* form could be detected by using $^1$H-NMR (9, 17, 38, 113). The respective prolyl bonds are all exposed to solvent in the folded proteins. In staphylococcal nuclease, the equilibrium at Pro117 can be shifted towards the *cis* isomer by adding $Ca^{2+}$ ions and an inhibitor that binds at the active site. Prolyl *cis/trans* isomerizations in folded proteins could possibly be a widespread phenomenon and of importance for the biological function of a protein. The detection of *cis/trans* equilibria with NMR spectroscopy is, however, difficult. Prolyl isomerizations have been proposed to occur in native concanavalin A (6) and prothrombin (78). Distinctive experimental support for these suggestions is still lacking (cf 72).

## PROLYL ISOMERASES

Prolyl isomerases catalyze prolyl peptide bond isomerizations in oligopeptides with high efficiency. They also accelerate slow, proline-limited steps in the folding of numerous proteins. They were discovered 1984 by Fischer and his coworkers (19) who used an assay that exploits the conformational specificity of chymotrypsin. This protease cleaves the assay peptide succinyl-Ala-Ala-Pro-Phe-4-nitroanilide only when the Ala-Pro bond is *trans*. Ninety percent of the peptide molecules contain such a *trans* bond and are hydrolyzed rapidly in the presence of chymotrypsin at high concentration. The remaining 10% are cleaved slowly, limited in rate by the *cis* → *trans* isomerization of the Ala-Pro bond. This slow hydrolysis reaction is strongly accelerated in the presence of prolyl isomerase activity. Kofron et al (61) improved this assay and developed a direct fluorometric assay that functions in the absence of chymotrypsin (25). They determined the $K_M$ value of human 18-kDa prolyl isomerase for the *cis* form of the assay peptide as 1 mM and $k_{cat}$ as 13,200 s$^{-1}$ (at 5°C).

The 18-kDa prolyl isomerase from porcine kidney is identical to cyclophilin (20, 114). Cyclophilin was identified in 1984 by virtue of its very high affinity for the immunosuppressant cyclosporin A (CsA) (31). Prolyl isomerases of the cyclophilin type are ubiquitous proteins. They are found in all organisms and subcellular compartments. CsA, a cyclic undecapeptide, binds tightly to most cyclophilins and inhibits their prolyl isomerase activity in a competitive manner. The inhibition constants are in the nanomolar range (20, 61).

Interestingly, the binding protein for another immunosuppressant, FK 506, also has prolyl isomerase activity (32, 105) and catalyzes proline-limited steps in protein folding (117). FK 506 is a macrolid and is not structurally related to CsA. Similarly, the amino acid sequences of the 12-kDa FK 506 binding protein, known as FKBP12, and cyclophilin show no sequence similarities. Nevertheless the FKBPs share the prolyl isomerase activity and the ubiquitous distribution with the cyclophilins. Additional members of the cyclophilin and FKBP families were recently discovered (12, 24, 58, 86, 100, 107, 115, 116, 121).

### Substrate Specificity

The specificities of prolyl isomerases of the cyclophilin type were examined, with regard to the proline ring itself and to the amino acid Xaa that precedes proline, by employing tetrapeptides of the general sequence succinyl-Ala-Xaa-Yaa-Phe–4-nitroanilide. The specificity for proline at position Yaa is high. Replacement with four- or six-membered homologues of proline or with sarcosine (N-methyl-glycine) resulted in an approximately $10^4$-fold decrease in activity (G. Fischer, personal communication). In contrast, the enzyme tolerates many amino acids at position Xaa. The $k_{cat}/K_M$ values varied by less than one order of magnitude when nine different amino acids were inserted at position Xaa of the peptide (33, 34). When a similar set of peptides was used to assay the specificity of a mammalian FKBP12, a more than $10^3$-fold variation in $k_{cat}/K_M$ was observed. The $k_{cat}/K_M$ values increase with increasing hydrophobicity of the residue that precedes proline (34, 92). Apparently, FKBP12 and cyclophilin differ substantially in sequence specificity with regard to residue Xaa. Data for other positions are not yet available. Clearly, it is too early to speculate how this difference is related to the biological functions of these two classes of prolyl isomerases.

At present, there is growing evidence that the immunosuppressive effects of CsA and FK 506 are mediated by interactions of the respective prolyl isomerase/inhibitor complexes with calcineurin, a $Ca^{2+}$-dependent protein phosphatase (11, 21, 23, 73, 74, 85, 103). It will be interesting to learn whether additional proteins with high affinity for immunosuppressants can be found and whether they also display enzymatic activities as prolyl isomerases.

## Three-Dimensional Structure

The three-dimensional structures of both 18-kDa cyclophilin and FKBP12 were solved in 1991. They are shown in Figures 2 and 3. There are no readily apparent similarities in the backbone structure of the two proteins. FKBP12 is composed of a five-stranded antiparallel $\beta$ sheet, a short $\alpha$ helix, and a high amount of aperiodic structure (80, 81, 118). The protein chain of cyclophilin is arranged in an eight-stranded antiparallel $\beta$-barrel structure and in two short $\alpha$ helices (45–47, 106). Cyclophilin was cocrystallized with a proline-containing *cis* peptide, which is bound in a hydrophobic groove on the surface of the $\beta$ barrel (45, 46). The structure of FKBP12 was solved with and without the inhibitor FK 506. The pipecolic amide moiety of FK 506, which is thought to mimic the proline residue of the substrate, is also bound in a hydrophobic pocket.

# CATALYSIS OF SLOW STEPS IN FOLDING BY PROLYL ISOMERASES

A catalysis of slow protein-folding reactions by a prolyl isomerase of the cyclophilin family was initially observed for the immunoglobulin light

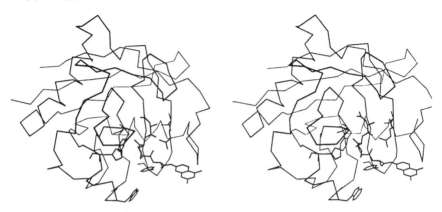

*Figure 2* Stereo representation of the three-dimensional structure of human cyclophilin A complexed with the tetrapeptide N-acetyl-Ala-Ala-Pro-Ala-methylcoumarin. Note that in the refined structure the Ala-Pro bond of the peptide is in the *cis* conformation (taken from 118a, with permission).

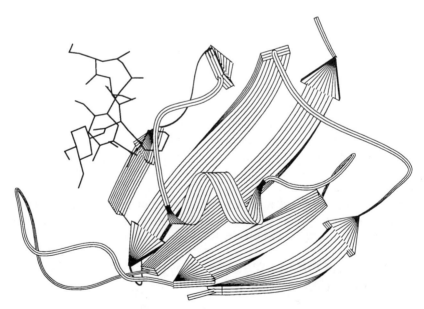

*Figure 3* Three-dimensional structure of the complex between human FKBP and FK 506 (118) (courtesy of J. Clardy).

chain, porcine RNase, and the S-protein fragment of bovine RNase A (70). Intriguingly, folding of intact RNase A and of thioredoxin could not be catalyzed (67), although the folding reactions of both proteins involve proline-limited steps (48, 98). Possibly, the rapid formation of ordered structure early in folding renders the respective prolyl peptide bonds inaccessible to prolyl isomerase. Prolyl and hydroxyprolyl isomerizations are rate-limiting steps in the maturation of the collagen triple helix and are also accelerated by prolyl isomerase (3, 14). The folding of collagen in vitro and in vivo is discussed later.

Prolyl isomerases of the cyclophilin type as well as of the FKBP type can catalyze slow steps in protein folding (70, 117). Very little is known, however, about the function of FKBP as a catalyst of in vitro folding. Therefore, this review discusses only the role of the cyclophilins for folding, and the term prolyl isomerase is used when the function of this protein family in folding reactions is described.

Not all slow steps in protein folding involve prolyl isomerization. The very slow refolding of large proteins is often limited in rate by other events, such as slow conformational rearrangements, domain pairing, or subunit associations. Whether prolyl isomerizations occur at earlier stages in the folding of such proteins remains to be demonstrated.

## Folding Mechanism of Ribonuclease T1

RNase T1 has been a major model for investigating the role of prolyl isomerases for protein folding. The folding kinetics of this protein are largely determined by the slow trans → cis isomerizations of two prolyl bonds that are cis in the native protein, but predominantly trans in the unfolded chains. The following section briefly describes the properties and the folding mechanism of this protein, to give an understanding of the observed effects of prolyl isomerase in this system.

RNase T1 is a small single-domain protein of 104 amino acids (for a review see 91) with two disulfide bonds (Cys2-Cys10 and Cys6-Cys103). It contains four prolyl peptide bonds; two are trans (Trp59-Pro60 and Ser72-Pro73) and the other two are cis (Tyr38-Pro39 and Ser54-Pro55) in the native protein. The thermodynamic stability of RNase T1 shows a maximum near pH 5 (89) and is strongly enhanced by the addition of NaCl (87, 88, 90). In the absence of the disulfide bonds, RNase T1 can still fold reversibly to a native-like, catalytically active form in the presence of 2 M NaCl (56, 82, 87, 88, 90).

After unfolding of RNase T1, the two cis prolyl peptide bonds at Pro39 and Pro55 isomerize largely to the incorrect trans state (53, 54). Results obtained for the wild-type protein and a variant with substitutions at positions 54 and 55 (51) suggest that in denatured RNase T1 80–90%

of both Pro39 and Pro55 are in the incorrect *trans* conformation. The isomerizations at these two prolines lead to four different unfolded species. Three to four percent of all molecules contain the correct isomers ($U_{55c}^{39c}$) and fold rapidly. In addition, three slow-folding species are present (cf Figure 4): two with one incorrect proline isomer each ($U_{55t}^{39c}$ and $U_{55c}^{39t}$) and another, dominant species with two incorrect prolines ($U_{55t}^{39t}$).

Figure 4 shows the kinetic mechanism for the slow refolding of RNase T1 (53, 54). Early in the process of refolding, all slow-folding molecules regain most of their secondary structure and presumably part of their tertiary structure in the time range of milliseconds (the $U_i \rightarrow I_i$ steps, Figure 4) (57). The subsequent slow folding steps involve the reisomerizations of the incorrect prolyl isomers. These reactions are coupled with further

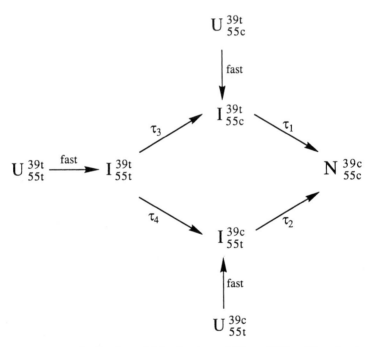

*Figure 4* Scheme of a kinetic model for the slow refolding of RNase T1 under strongly native conditions. U stands for unfolded species, I for intermediates of refolding, and N for the native protein. The superscript and the subscript indicate the isomeric states of prolines 39 and 55, respectively, in the correct *cis* (c) and the incorrect *trans* (t) isomeric states. For example, $U_{55c}^{39t}$ is an unfolded species with Pro55 in the correct *cis* and Pro39 in the incorrect *trans* state.

folding and can be monitored by observing changes in absorbance and fluorescence. The major unfolded species has two incorrect isomers ($U_{55t}^{39t}$) and it can enter two alternative folding pathways (the upper or the lower pathway in Figure 4), depending upon which reisomerization occurs first. The partitioning between these two pathways is determined by the relative rates of reisomerization at Pro39 and at Pro55 in the partially folded intermediate $I_{55t}^{39t}$. Though complex, the scheme in Figure 4 nevertheless is still a simplification. It is valid only for folding under strongly native conditions, where intermediates are populated and reverse reactions are not significant. Also, the (probably minor) contributions from the isomerizations of the *trans* prolines 60 and 73 are not accounted for.

According to the scheme in Figure 4, protein folding and reisomerization of prolyl peptide bonds are interrelated processes. Intermediates with native-like secondary structure can form very rapidly in the presence of nonnative proline isomers (54, 57). Correct isomers are required for the final events of folding, and hence these slow steps of folding are limited in rate by reisomerization. The rapid formation of ordered structure can affect the isomerization kinetics. Unlike the situation in RNase A, where prolyl isomerization in a folding intermediate was accelerated (13, 97), one isomerization in RNase T1 (at Pro39 in the $I_{55c}^{39t} \rightarrow N$ step) is strongly decelerated in refolding (54).

## *Catalysis of RNase T1 Folding by Prolyl Isomerases*

The slow refolding of RNase T1, as measured by the changes in absorbance or fluorescence, can roughly be approximated as a sum of two phases, an intermediate phase with a time constant, $\tau \approx 400$ s, and a very slow phase with $\tau \approx 3000$ s (at pH 8, 10°C) (53). The intermediate phase is heterogeneous, with contributions from the reisomerizations of Pro55 (in the $I_{55t}^{39t} \rightarrow I_{55t}^{39t}$ and the $I_{55t}^{39c} \rightarrow N$ steps; Figure 4) and of Pro39 in molecules that also contain an incorrect Pro55 (the $I_{55t}^{39t} \rightarrow I_{55t}^{39c}$ step; Figure 4). This assignment was supported by results on the Ser54Gly/Pro55Asn variant in which the intermediate phase was missing (51). The very slow phase originates from the $I_{55c}^{39t} \rightarrow N$ step (Figure 4) and reflects the *trans* $\rightarrow$ *cis* reisomerization at Pro39 in molecules that had already undergone *trans* $\rightarrow$ *cis* isomerization at Pro55. All slow phases of RNase T1 folding are catalyzed by prolyl isomerases, albeit with a strongly different efficiency (20, 54, 101). Catalysis at Pro55 is generally efficient. Good catalysis of Pro39 isomerization is only observed as long as a second incorrect isomer (at Pro55) is present. $I_{55c}^{39t}$ molecules with a single incorrect *trans* isomer (at Pro39) can apparently form extensive structure. This rapid folding has two effects: it strongly decelerates *trans* $\rightarrow$ *cis* isomerization at Pro39 ($I_{55c}^{39t} \rightarrow N$; Figure 4), and it renders catalysis by prolyl isomerase very poor.

The heterogeneous intermediate phase of folding is dominated by Pro55 isomerization and is well catalyzed by prolyl isomerase. All isomerizations and their catalysis occur after the formation of partially folded intermediates (Figure 4). Pro39, but not Pro55, appears to be partly buried at this stage of folding already and is not readily accessible to prolyl isomerase. As a matter of fact, in native RNase T1, Pro55 is located at the surface; however, Pro39 is in the interior of the protein.

The slow refolding reactions of RNase T1 are catalyzed by prolyl isomerases of the cyclophilin family from a variety of different organisms (101). In general, their activities are similar. The best catalysis of Pro39 isomerization in the very slow reaction was observed with the cytoplasmic prolyl isomerase from *Escherichia coli* (36). In the presence of a physiological concentration of this enzyme, the folding rate is increased 300-fold (101).

CATALYSIS OF FOLDING IN THE ABSENCE OF DISULFIDE BONDS   The function of prolyl isomerases in folding was initially investigated by using proteins with intact disulfide bonds because unfolding of these proteins is reversible under a wide variety of conditions and because there was good evidence that prolyl isomerizations were involved as rate-limiting steps in their refolding. Proteins without disulfides should, however, be better models for nascent proteins. The two disulfide bonds of RNase T1 can be cleaved by reduction and subsequent carboxymethylation of the free thiols. The corresponding variant, RCM-RNase T1, is unfolded at pH 8.0 and 15°C. In the presence of 2 M NaCl, however, folding to a native-like and enzymatically active form can be induced (87, 88, 90). The midpoint of the reversible NaCl-dependent folding transition is near 1 M NaCl (82). Refolding induced by transfer of the protein from 0 M to 2 M NaCl is a very slow reaction ($\tau = 1820$ s) that is accelerated in the presence of prolyl isomerase. This observation indicates that the slow refolding of RNase T1 without crosslinks is also limited in rate by prolyl isomerization. The catalysis by prolyl isomerase is more efficient in the absence of the disulfide bonds than in their presence when folding is compared under identical conditions (82). When the disulfide bonds have not yet formed, refolding proteins cannot easily acquire ordered structure and are thus better substrates for prolyl isomerases. The intriguing consequence of this is that rapid structure formation may not always be advantageous in protein folding.

CATALYSIS OF PROLYL ISOMERIZATION IN UNFOLDING PROTEINS   Prolyl isomerases function in protein folding as enzymes. They catalyze the reaction in either direction and they do not determine the isomeric states of the prolyl peptide bonds in the protein substrates. The distribution of *cis* and/or *trans* isomers depends only on the stability of the folding protein under the particular solvent conditions. Under refolding conditions, the native isomers are stabilized by further rapid folding; under unfolding conditions, prolyl isomerase catalyzes the *cis/trans* equilibration reaction.

Catalysis of prolyl isomerization in an unfolded protein cannot be measured easily, since prolyl isomerase is rapidly inactivated by denaturants and at elevated temperatures. These problems could be overcome by using RCM-RNase T1 as a substrate. As outlined above, this protein is only folded in the presence of $\geqslant 2$ M NaCl, and its unfolding can be induced by a dilution to 0.4 M NaCl at pH 8 and 15°C. Prolyl isomerases are stable under these conditions and retain their full enzymatic activity. Unfolding, or to be more precise, the equilibration of $U_F$ and $U_S$ species of RCM-RNase T1 in the unfolding chains, is indeed catalyzed, and the efficiency of prolyl isomerase is the same in unfolding and refolding experiments under identical conditions near the midpoint of the unfolding transition. The final values reached in these kinetic experiments with and without prolyl isomerase are identical, indicating that this enzyme does not affect the equilibrium between native and unfolded protein (82). Clearly, prolyl isomerases have no directional information, and their enzymatic function in protein folding is comparable to the action of protein disulfide isomerase, which can catalyze the reduction, the oxidation, or the scrambling of disulfide bonds, depending on the redox conditions and on the stability of the disulfide bonds in the substrate protein.

## *Synergism of Prolyl Isomerase and Protein Disulfide Isomerase as Catalysts of Folding*

Both the *cis/trans* isomerizations of Xaa-Pro peptide bonds and the formation of the correct disulfide bonds are slow processes in the folding of disulfide-containing proteins. Disulfide-bond formation is catalyzed by protein disulfide isomerase, an enzyme required for the de novo folding of nascent proteins in the endoplasmic reticulum (ER) (7, 18). Prolyl isomerases of the cyclophilin type are also present in the ER (35) and in other compartments of the secretory pathway (1, 8). In vitro, prolyl isomerase accelerates the oxidative folding of reduced RNase T1, i.e. folding coupled with formation of the disulfide bonds (102). This is probably an indirect effect. Partially native-like structure and thus the correct disulfides are presumably formed most rapidly in protein molecules that contain the native-like isomers of the prolyl peptide bonds. These molecules would be analogous to the fast-folding $U_F$ species of unfolded RNase T1 with intact disulfides (53, 54). Most unfolded species have incorrect prolyl isomers, however, and cannot form native-like structure easily. Prolyl isomerase catalyzes their isomerizations and thereby indirectly facilitates the formation of the correct disulfide bonds.

An analogous effect of prolyl isomerase occurs when disulfide isomerase catalyzes disulfide-bond formation. In the absence of prolyl isomerase, addition of 1.6 $\mu$M disulfide isomerase increases the rate of oxidative folding of RNase T1 by a factor of 2.9. However, when 6.9 $\mu$M prolyl isomerase is present in addition to the disulfide isomerase, this same concentration of disulfide isomerase leads to a 12-fold acceleration of reoxidation (102). This synergistic effect of the two enzymes is most pronounced under conditions where prolyl isomerization and disulfide-bond formation are similar in rate. No acceleration of disulfide-bond formation by prolyl isomerase was found during the folding of pancreatic RNase A under similar conditions (68, 72). The simplest explanation is that, in the reoxidation of pancreatic RNase A, incorrect disulfides form rapidly during folding unless the concentration of glutathione is very small. The rate-limiting slow reshuffling of the incorrect disulfide bonds is not affected by prolyl isomerase.

Clearly, prolyl isomerase can also accelerate the oxidative folding of reduced protein chains, which resemble the nascent polypeptide chains in the ER. The simultaneous presence of prolyl isomerase markedly enhances the efficiency of disulfide isomerase as a catalyst of disulfide-bond formation. It remains to be demonstrated whether a similar synergistic action of the two enzymes occurs during the de novo synthesis and folding of proteins in the cell.

## Efficiency of Catalysis

The efficiency of prolyl isomerases as catalysts of slow folding varies strongly. It depends on the kind of prolyl isomerase, on the substrate protein, and on the folding conditions. A recent review (99) summarizes the relative efficiencies of various prolyl isomerases in the folding of different proteins. It is difficult to assign numerical values to the efficiency of prolyl isomerases in folding reactions. In the previous review (99), we used the ratio of the activities towards the assay peptide and towards refolding proteins as a measure for the efficiency of catalysis. In both cases, the activities were expressed as acceleration factors of the uncatalyzed reactions. The highest activity was observed when cytoplasmic prolyl isomerase from *E. coli* was used to catalyze the slow refolding of RNase T1. In this case, one of the slow reactions was catalyzed with a 40% efficiency relative to the enzyme assay employing the oligopeptide succinyl-Ala-Ala-Pro-Phe–4-nitroanilide. Catalytic efficiencies of prolyl isomerases in protein folding, expressed as $k_{cat}/k_{uncat}$ under saturating-substrate conditions cannot be given, because the $K_M$ values for protein substrates cannot be measured. In the concentration range accessible in protein-folding experiments ($\leqslant 100$ $\mu$M substrate protein) there is no evidence for saturation behavior. If we assume the same $K_M$ value of 1 mM for protein substrates as was found for the assay peptide (61), then ratios for $k_{cat}/k_{uncat}$ as high as $10^5$ would result for the most active prolyl isomerases in RNase T1 refolding. Clearly, this number should be considered with care, since the kinetic mechanisms underlying catalysis are not well known.

## ROLE OF PROLYL ISOMERASE IN CELLULAR FOLDING

The cellular functions of prolyl isomerases are largely unclear. Immunosuppressants bind at the active site with very high affinity and inhibit the prolyl isomerase activity competitively. Indirect evidence for a possible function of prolyl isomerases in cellular protein folding appears in recent studies reporting that the in vivo maturation of two proteins, collagen and transferrin, is slightly retarded in the presence of CsA.

The folding of procollagen, the precursor of collagen, starts at the carboxyterminal end of the molecules and proceeds unidirectionally in a zipper-like manner to the N terminus (16). This zero-order helix propagation reaction is limited in rate by the successive *cis* → *trans* isomerization of Xaa-proline and/or Xaa-hydroxyproline bonds. Bächinger (3) showed that this in vitro annealing of the collagen triple helix is indeed accelerated by a factor of about three in the presence of 0.03 mg/ml prolyl isomerase. Steinmann et al (111) followed the folding of procollagen I in fibroblasts and found that 8.5 min are required for half-completion of the triple helix. When 5 $\mu$M CsA was diffused into the cells, this value increased to 13.5 min. This result is most easily explained by assuming that the rate-limiting prolyl and hydroxyprolyl isomerizations during the in vivo folding of collagen are accelerated by prolyl isomerase and that this catalysis can be inhibited by CsA.

Transferrin is a large secretory protein with 19 disulfide bonds. Secretion is slow, with a half time of about three hours, and is sensitive to CsA. The level of transferrin secretion drops about threefold in the presence of $10^{-5}$ M CsA. Other secretory proteins, such as serum albumin or $\alpha 1$ antitrypsin, are transported more rapidly, and their secretion is not affected by CsA (75). The overall time course of transferrin maturation is very slow both in the absence and in the presence of CsA. In the presence of CsA, however, intracellular folding shows an initial lag that is absent in the experiment without CsA. The authors conclude that this constitutes evidence for the involvement of a prolyl isomerase in an initial step in transferrin folding.

In both the maturation of collagen and the folding of transferrin, the inhibitory effects of CsA on cellular folding were found to be fairly small. However, even a small retardation of intracellular folding could be deleterious, because proteins in a partially folded conformation are much more susceptible to nonproductive side reactions, such as aggregation, unwanted covalent modification, or proteolytic digestion.

## PROLYL ISOMERASES AS TOOLS IN PROTEIN FOLDING

Previously, proline-limited steps in protein folding were identified by comparing their kinetic properties with those of prolyl isomerizations in short peptides (5, 84, 96). Unfortunately, these properties are often changed significantly when they are coupled with structure formation, making a molecular assignment very difficult. Prolyl isomerases simplify the identification of folding reactions that involve prolyl isomerization as a rate-limiting step. The results are only significant, however, when catalysis by prolyl isomerase is actually found. A lack of catalysis could originate either from a rate-limiting event other than prolyl isomerization, or from prolyl peptide bonds that are not accessible for the enzyme because ordered structure forms rapidly in folding. In the latter case, prolyl isomerases with lower steric requirements, such as the cytosolic enzyme from *E. coli*, could be used. An increase in the residual concentration of denaturant in the refolding solution to destabilize intermediates is not advisable, because prolyl isomerases are sensitive to low concentrations of denaturant.

## MECHANISTIC ASPECTS OF PROLYL ISOMERASE ACTION

The reaction catalyzed by prolyl isomerases differs from most other enzymatic reactions. It is not correlated with the formation or breakage of covalent bonds; intermediates do not accumulate; and no cofactors are known to participate in catalysis. Therefore, as Stein points out, no simple experimental "handles" are available to help elucidate the mechanism of catalysis (108, 109). On the other hand, prolyl isomerization provides us with a rare opportunity: both the uncatalyzed and the catalyzed reactions occur at measurable rates and can be studied. A direct comparison of the properties of these reactions should give valuable information on the mechanisms of the reactions themselves and on the catalytic strategy used by the enzyme (34).

In oligopeptides, the free energy of activation, $\Delta G^*$, of uncatalyzed prolyl isomerization is largely dominated by the enthalpic contribution,

$\Delta H^*$. The entropy of activation, $\Delta S^*$, is almost zero for several different Xaa-Pro bonds (34), indicating that the transition from the ground state to the activated state is not correlated with a significant rearrangement of solvent structure.

Prolyl isomerizations are accelerated in the presence of strong acids (4, 110) or nonpolar solvents (15, 34). In acid catalysis, the imide nitrogen of proline is protonated and the partial double-bond character of the peptide bond is abolished. For some peptides, acid catalysis is thought to bring about a $10^6$-fold acceleration of *cis/trans* isomerization (34). Protonation of the imide nitrogen is difficult to achieve, however, because its pKa is presumably below $-2$. The decrease in activation energy for isomerization in nonpolar solvents suggests that the activated state is less polar than the ground state. In attaining such a transition state, however, the ordering of the solvent water molecules near the prolyl bond should change. This is difficult to reconcile with an activation entropy that is close to zero.

An excellent discussion of the mechanistic aspects of enzyme-catalyzed prolyl isomerization is found in the recent work of Stein and colleagues (34, 108, 109). These authors suggest that the solvent-mediated and the enzymic acceleration of prolyl isomerization are brought about by a similar stabilization of a nonpolar transition state, in which the single-bond character of the peptide bond is increased. The substrate binding sites of both prolyl isomerases, cyclophilin and FKBP12, could provide a nonpolar environment for the Xaa-Pro bond, and the enzyme could induce strain to distort the prolyl peptide bond in the substrate. Strain is not necessarily purely geometric but can be mediated by desolvation and/or electrostatic effects that reduce the resonance of the amide bond. Electrostatic effects could include the stabilization of a $sp^3$ configuration at the imide nitrogen of proline by a hydrogen-bond donor or a positively charged group of the enzyme. This would be analogous to acid catalysis.

# CONCLUDING REMARKS

The formation of the correct disulfide bonds and the isomerization of prolyl peptide bonds are slow processes that determine the overall rate of the folding of many small proteins. They are catalyzed by the enzymes protein disulfide isomerase and prolyl isomerase, respectively. Disulfide isomerase is required for the de novo folding of nascent-protein chains in the ER, but the role of prolyl isomerases for cellular folding is still unclear.

The question of whether catalysts for prolyl isomerization are needed to ensure the productive folding of nascent proteins is tied to the question of whether prolyl isomerizations are in fact slow steps in cellular folding as well. The conformation of proteins during or immediately after biosynthesis is not known. Completely *trans* polypeptide chains are probably formed. They rapidly adopt a condensed conformation, because the native cellular environment is a poor solvent for extended unfolded proteins. In such newly synthesized molecules, all prolyl bonds that are *cis* in the folded protein would be in the wrong conformational state. For collagen, there is enough time for *cis/trans* equilibrations to occur prior to triple helix formation in vivo. This may represent a rather unique situation since the annealing of collagen is preceded by the correct folding, association, and disulfide-bonding of the registration peptides. For other proteins, it is not known whether sufficient time is available in vivo for *cis/trans* equilibrations to occur before folding gets started. In any case, prolyl isomerizations must be involved in cellular folding.

If we accept that prolyl isomerizations do occur during cellular folding, then the next question is whether enzymatic catalysis would be advantageous. In vitro folding experiments are usually performed at low temperature where prolyl isomerizations take place in the range of minutes. At 37°C, however, these isomerizations occur in the time range of 5–10 s. Also, for most large proteins, other folding steps are slower and determine the overall rate of folding. Protein folding both in the test tube and in the cell is plagued by unspecific aggregation, a major competing reaction. Such aggregations could be suppressed by two strategies: (*a*) the transient binding to molecular chaperones, such as GroE, and/or (*b*) the acceleration of folding steps that lead to an aggregation-resistant form. Examples for the first mechanism are abundant (76), but the second mechanism is still largely speculative. In the in vitro folding of small single-domain proteins, intermediates with incorrect prolyl isomers are only marginally stable, and prolyl isomerases accelerate their conversion to the stable folded state. In cellular folding, prolyl isomerases could act in a similar way and ensure rapid folding of individual domains of large proteins to a conformation that is resistant to aggregation and can undergo the final steps of folding.

The multitude of prolyl isomerases suggests that their functions are clearly not restricted to protein folding. Growing evidence indicates that the complexes of prolyl isomerases with immunosuppressants block signal-transduction pathways (11, 23, 73, 74, 85). This finding leads one to ask whether prolyl isomerizations and their catalysis in peptides or folded proteins could be involved in cellular regulation. Proline residues occur preferentially in exposed turn regions. This good accessibility together with the intrinsically slow rate of *cis/trans* isomerizations would render prolyl bonds attractive candidates for the construction of molecular switches to transiently activate or inactivate proteins. Such switching mechanisms could be coupled with phosporylation or dephoshorylation reactions. Unfortunately, evidence for such processes and for an involvement of prolyl isomerases is still lacking. The respective experiments will be very difficult because good handles for the identification of prolyl isomerizations in cellular processes are not available.

## ACKNOWLEDGMENTS

I thank Gunter Fischer, Christian Frech, Thomas Kiefhaber, Lorenz Mayr, Matthias Mücke, E. Ralf Schönbrunner, and Stefan Walter for many helpful discussions and comments. Our own work described in this review was supported by the Deutsche Forschungsgemeinschaft, the Fonds der Chemischen Industrie, and the Volkswagenstiftung.

*Literature Cited*

1. Arber, S., Krause, K.-H., Caroni, P. 1992. *J. Cell. Biol.* 116: 113–25
2. Babul, J., Nakagawa, A., Stellwagen, E. 1978. *J. Mol. Biol.* 126: 117–21
3. Bächinger, H.-P. 1987. *J. Biol. Chem.* 262: 17144–48
4. Berger, A., Loewenstein, A., Meiboom, S. J. 1959. *J. Am. Chem. Soc.* 81: 62–67
5. Brandts, J. F., Halvorson, H. R., Brennan, M. 1975. *Biochemistry* 14: 4953–63
6. Brown, R. D. III, Brewer, C. F., Koenig, S. H. 1977. *Biochemistry* 16: 3883–96
7. Bulleid, N. J., Freedman, R. B. 1988. *Nature* 335: 649–51
8. Caroni, P., Rothenfluh, A., McGlynn, E., Schneider, C. 1991. *J. Biol. Chem.* 266: 10739–42
9. Chazin, W. J., Kördel, J., Drakenberg, T., Thulin, E., Brodin, P., et al. 1989. *Proc. Natl. Acad. Sci. USA* 86: 2195–98
10. Cheng, H. N., Bovey, F. A. 1977. *Biopolymers* 16: 1465–72
11. Clipstone, N. A., Crabtree, G. R. 1992. *Nature* 357: 695–97
12. Colley, N. J., Baker, E. K., Stamnes, M. A., Zuker, C. S. 1991. *Cell* 67: 255–63
13. Cook, K. H., Schmid, F. X., Baldwin, R. L. 1979. *Proc. Natl. Acad. Sci. USA* 76: 6157–61
14. Davis, J. M., Boswell, B. A., Bächinger, H.-P. 1989. *J. Biol. Chem.* 264: 8956–62
15. Drakenberg, T., Dahlquist, H.-I., Forsén, S. 1972. *J. Phys. Chem.* 76: 2178–83
16. Engel, J., Prockop, D. J. 1991. *Annu. Rev. Biophys. Biophys. Chem.* 20: 137–52
17. Evans, P. A., Dobson, C. M., Kautz, R. A., Hatfull, G., Fox, R. O. 1987. *Nature* 329: 266–68
18. Farquhar, R., Honey, N., Murant, S. J., Bossier, P., Schultz, L., et al. 1991. *Gene* 108: 81–89
19. Fischer, G., Bang, H., Mech, C. 1984. *Biomed. Biochim. Acta* 10: 1101–11
20. Fischer, G., Wittmann-Liebold, B., Lang, K., Kiefhaber, T. Schmid, F. X. 1989. *Nature* 337: 476–78
21. Flanagan, W. M., Corthésy, B., Bram, R. J., Crabtree, G. R. 1991. *Nature* 352: 803–7
22. Fransson, C., Freskgård, P.-O., Herbertsson, H., Johansson, Å., Jonasson, P., et al. 1992. *FEBS Lett.* 296: 90–94
23. Friedman, J., Weissman, I. 1991. *Cell* 66: 799–806
24. Galat, A., Lane, W. S., Standaert, R. F., Schreiber, S. L. 1992. *Biochemistry* 31: 2427–34
25. Garcia-Echeverria, C., Kofron, J. L., Kuzmic, P., Kishore, V., Rich, D. H. 1992. *J. Am. Chem. Soc.* 114: 2758–59
26. Garel, J. R., Baldwin, R. L. 1973. *Proc. Natl. Acad. Sci. USA* 70: 3347–51
27. Goto, Y., Hamaguchi, K. 1982. *J. Mol. Biol.* 156: 891–910
28. Grathwohl, C., Wüthrich, K. 1976. *Biopolymers* 15: 2025–41
29. Grathwohl, C., Wüthrich, K. 1976. *Biopolymers* 15: 2043–57
30. Grathwohl, C., Wüthrich, K. 1981. *Biopolymers* 20: 2623–33
31. Handschumacher, R. E., Harding, M. W., Rice, J., Drugge, R. J., Speicher, D. W. 1984. *Science* 226: 544–47
32. Harding, M. W., Galat, A., Ueling, D. E., Schreiber, S. L. 1989. *Nature* 341: 758–60

33. Harrison, R. K., Stein, R. L. 1990. *Biochemistry* 29: 3813–16
34. Harrison, R. K., Stein, R. L. 1992. *J. Am Chem. Soc.* 114: 3464–71
35. Hasel, K. W., Glass, J. R., Godbout, M., Sutcliffe, J. G. 1991. *Mol. Cell. Biol.* 11: 3484–91
36. Hayano, T., Takahashi, N., Kato, N., Maki, N., Suzuki, M. 1991. *Biochemistry* 30: 3041–48
37. Herzberg, O., Moult, J. 1991. *Proteins: Struct. Funct. Genet.* 11: 223–29
38. Higgins, K. A., Craik, D. J., Hall, J. G., Andrews, P. R. 1988. *Drug Des. Deliv.* 3: 159–70
39. Hurle, M. R., Anderson, S., Kuntz, I. D. 1991. *Protein Eng.* 4: 451–55
40. Ihara, S., Ooi, T. 1985. *Biochim. Biophys. Acta* 830: 109–12
41. Jackson, S., Fersht, A. R. 1991. *Biochemistry* 30: 10428–35
42. Jackson, S., Fersht, A. R. 1991. *Biochemistry* 30: 10436–43
43. Jorgensen, W. J., Gao, J. 1988. *J. Am. Chem. Soc.* 110: 4212–16
44. Jullien, M., Baldwin, R. L. 1979. *J. Mol. Biol.* 145: 265–80
45. Kallen, J., Spitzfaden, C., Widmer, H., Wüthrich, K., Walkinshaw, M. D. 1991. *Nature* 353: 276–79
46. Kallen, J., Walkinshaw, M. D. 1992. *FEBS Lett.* 300: 286–90
47. Ke, H., Zydowsky, L. D., Liu, J., Walsh, C. T. 1991. *Proc. Natl. Acad. Sci. USA* 88: 9483–87
48. Kelley, R. F., Richards, F. M. 1987. *Biochemistry* 26: 6765–74
49. Kelley, R. F., Stellwagen, E. 1984. *Biochemistry* 23: 5095–5102
50. Kelley, R. F., Wilson, J., Bryant, C., Stellwagen, E. 1986. *Biochemistry* 25: 728–32
51. Kiefhaber, T., Grunert, H.-P., Hahn U., Schmid, F. X. 1990. *Biochemistry* 29: 6475–79
52. Kiefhaber, T., Kohler, H.-H., Schmid, F. X. 1992. *J. Mol. Biol.* 224: 217–29
53. Kiefhaber, T., Quaas, R., Hahn, U., Schmid, F. X. 1990. *Biochemistry* 29: 3053–60
54. Kiefhaber, T., Quaas, R., Hahn, U., Schmid, F. X. 1990. *Biochemistry* 29: 3061–70
55. Kiefhaber, T., Schmid, F. X. 1992. *J. Mol. Biol.* 224: 231–40
56. Kiefhaber, T., Schmid, F. X., Renner, M., Hinz, H.-J., Hahn, U., Quaas, R. 1990. *Biochemistry* 29: 8250–57
57. Kiefhaber, T., Schmid, F. X., Willaert, K., Engelborghs, Y., Chaffotte, A. 1992. *Protein Sci.* 1: In press
58. Kieffer, L. J., Thalhammer, T., Handschumacher, R. E. 1992. *J. Biol. Chem.* 267: 5503–7
59. Kim, P. S., Baldwin, R. L. 1982. *Annu. Rev. Biochem.* 51: 459–89
60. Kim, P. S., Baldwin, R. L. 1990. *Annu. Rev. Biochem.* 59: 631–60
61. Kofron, J. L., Kuzmic, P., Kishore, V., Colón-Bonilla, E., Rich, D. 1991. *Biochemistry* 30: 6127–34
62. Kördel, J., Drakenberg, T., Forsén, S., Thulin, E. 1990. *FEBS Letters* 263: 27–30
63. Kördel, J., Forsén, S., Drakenberg, T., Chazin, W. J. 1990. *Biochemistry* 29: 4400–9
64. Krebs, H., Schmid, F. X., Jaenicke, R. 1983. *J. Mol. Biol.* 169: 619–33
65. Krebs, H., Schmid, F. X., Jaenicke, R. 1985. *Biochemistry* 24: 3846–52
66. Kuwajima, K., Okayama, N., Yamamoto, K., Ishihara, T., Sugai, S. 1991. *FEBS Lett.* 290: 135–38
67. Lang, K. 1988. *Molekularer Mechanismus und enzymatische Katalyse langsamer Proteinfaltungsreaktionen.* PhD thesis. Univ. Regensburg, Germany. 158 pp.
68. Lang, K., Schmid, F. X. 1988. *Nature* 331: 453–55
69. Lang, K., Schmid, F. X. 1990. *J. Mol. Biol.* 212: 185–96
70. Lang, K., Schmid, F. X., Fischer, G. 1987. *Nature* 329: 268–70
71. Levitt, M. 1981. *J. Mol. Biol.* 145: 251–63
72. Lin, L.-N., Hasumi, H., Brandts, J. F. 1988. *Biochim. Biophys. Acta* 956: 256–66
73. Liu, J., Albers, M. W., Wandless, T. J., Luan, S., Alberg, D. G., et al. 1992. *Biochemistry* 31: 3896–3901
74. Liu, J., Farmer, J. D. Jr., Lane, W. S., Friedman, J., Weissman, I., Schreiber, S. L. 1991. *Cell* 66: 807–15
75. Lodish, H. F., Kong, N. 1991. *J. Biol. Chem.* 266: 14835–38
76. Lorimer, G. 1992. *Curr. Opin. Struct. Biol.* 2: 26–34
77. MacArthur, M. W., Thornton, J. M. 1991. *J. Mol. Biol.* 218: 397–412
78. Marsh, H. C., Hiskey, R. G., Kochler, K. A. 1979. *Biochem. J.* 183: 513–17
79. Matouschek, A., Kellis, J. T., Serrano, L., Bycroft, M., Fersht, A. R. 1990. *Nature* 346: 440–45
80. Michnick, S. W., Rosen, M. K., Wandless, T. J., Karplus, M., Schreiber, S. L. 1991. *Science* 252: 836–39
81. Moore, J. M., Peattie, D. A., Fitzgibbon, M. J., Thomson, J. A. 1991. *Nature* 351: 248–50
82. Mücke, M., Schmid, F. X. 1992. *Biochemistry* 31: 7848–54
83. Nall, B. T. 1990. In *Protein Folding*, ed.

L. M. Gierasch, J. King, pp. 198–207. Washington: AAAS Press.
84. Nall, B. T., Garel., J.-R., Baldwin, R. L. 1978. *J. Mol. Biol.* 118: 317–30
85. O'Keefe, S. J., Tamura, J., Kincaid, R. L., Tocci, M. J., O'Neill, E. A. 1992. *Nature* 357: 692–94
86. Ondek, B., Hardy, R. W., Baker, E. K., Stamnes, M. A., Shieh, B.-H., Zuker, C. A. 1992. *Nature*. In press
87. Oobatake, M., Takahashi, S., Ooi, T. 1979. *J. Biochem.* 86: 55–63
88. Oobatake, M., Takahashi, S., Ooi, T. 1979. *J. Biochem.* 86: 65–70
89. Pace, C. N. 1990. *Trends Biochem. Sci.* 15: 14–17
90. Pace, C. N., Grimsley, G. R., Thomson, J. A., Barnett, B. J. 1988. *J. Biol. Chem.* 263: 11820–25
91. Pace, C. N., Heinemann, U., Hahn, U., Saenger, W. 1991. *Angew. Chem. Int. Ed. Eng.* 30: 343–60
92. Park, S. T., Aldape, R. A., Futer, O., DeCenzo, M. T., Livingston, D. J. 1992. *J. Biol. Chem.* 267: 3316–24
93. Ramachandran, G. N., Mitra, A. K. 1976. *J. Mol. Biol.* 107: 85–92
94. Ramdas, L., Sherman, F., Nall, B. T. 1986. *Biochemistry* 25: 6952–58
95. Schmid, F. X. 1992. In *Protein Folding*, ed. T. E. Creighton, pp. 197–241. San Francisco: Freeman
96. Schmid, F. X., Baldwin, R. L. 1978. *Proc. Natl. Acad. Sci. USA* 75: 4764–68
97. Schmid, F. X., Blaschek, H. 1981. *Eur. J. Biochem.* 114: 111–17
98. Schmid, F. X., Grafl, R., Wrba, A., Beintema, J. J. 1986. *Proc. Natl. Acad. Sci. USA* 83: 872–76
99. Schmid, F. X., Mayr, L. M., Mücke, M., Schönbrunner, E. R. 1992. *Adv. Protein Chem.* In press
100. Schneuwly, S., Shortridge, R. D., Laarrivee, D. C., Ono, T., Ozaki, M., Pak, W. L. 1989. *Proc. Natl. Acad. Sci. USA* 86: 5390–94
101. Schönbrunner, E. R., Mayer, S., Tropschug, M., Fischer, G., Takahashi, R., Schmid, F. X. 1991. *J. Biol. Chem* 266: 3630–35
102. Schönbrunner, R., Schmid, F. X. 1992. *Proc. Natl. Acad. Sci. USA* 89: 4510–13
103. Schreiber, S. L. 1991. *Science* 251: 283–87
104. Semisotnov, G. V., Uversky, V. N., Sokolovsky, I. V., Gutin, A. M., Razgulyaev, O. I., Rodionova, N. A. 1990. *J. Mol. Biol.* 213: 561–68
105. Siekierka, J. J., Hung, S. H. Y., Poe, M., Lin, C. S., Sigal, N. H. 1989. *Nature* 341: 755–57
106. Spitzfaden, C., Weber, H.-P., Braun, W., Kallen, J., Wider, G., et al. 1992. *FEBS Lett.* 300: 291–300
107. Stamnes, M. A., Shieh, B.-H., Chuman, L., Zuker, C. S. 1991. *Cell* 65: 219–27
108. Stein, R. L. 1991. *Curr. Biol.* 1: 234–36
109. Stein, R. L. 1992. *Adv. Protein Chem.* In press
110. Steinberg, I. Z., Harrington, W. F., Berger, A., Sela, M., Katchalski, E. 1960. *J. Am. Chem. Soc.* 82: 5263–79
111. Steinmann, B., Bruckner, P., Supertifurga, A. 1991. *J. Biol. Chem.* 266: 1299–1303
112. Stewart, D. E., Sarkar, A., Wampler, J. E. 1990. *J. Mol. Biol.* 214: 253–60
113. Svensson, L. A., Thulin, E., Forsén, S. 1992. *J. Mol. Biol.* 223: 601–6
114. Takahashi, N., Hayano, T., Suzuki, M. 1989. *Nature* 337: 473–75
115. Tai, P.-K., Albers, M. W., Chang, H., Faber, L. E., Schreiber, S. L. 1992. *Science* 256: 1315–18
116. Thalhammer, T., Kieffer, L. J., Jiang, T., Handschumacher, R. E. 1992. *Eur. J. Biochem.* 206: 31–37
117. Tropschug, M., Wachter, E., Mayer, S., Schönbrunner, E. R., Schmid, F. X. 1990. *Nature* 346: 674–77
118. Van Duyne, G. D., Standaert, R. F., Karplus, P. A., Schreiber, S. L., Clardy, J. 1991. *Science* 252: 839–42
118a. Walkinshaw, M., Kallen, J., Weber, H.-J., Widmer, A., Widmer, H., Zurini, M. 1992. *Transplant. Proc.* 24(Suppl. 2): 8–13
119. White, T. B., Berget, P. B., Nall, B. T. 1987. *Biochemistry* 26: 4358–66
120. Wood, L. C., Muthukrishnan, K., White, T. B., Ramdas, L., Nall, B. T. 1988. *Biochemistry* 27: 8554–61
121. Yem, A. W., Tomasselli, A. G., Heinrikson, R. L., Zurcher-Neely, H., Ruff, V. A., et al. 1992. *J. Biol. Chem.* 267: 2868–71

# MODELS OF LIPID-PROTEIN INTERACTIONS IN MEMBRANES

## Ole G. Mouritsen

Canadian Institute of Advanced Research, and Department of Physical Chemistry, The Technical University of Denmark, Building 206, DK–2800 Lyngby, Denmark

## Myer Bloom

Canadian Institute of Advanced Research, and Department of Physics, University of British Columbia, Vancouver, British Columbia, V6T 1W5 Canada

KEY WORDS:  lipid bilayers, phase transition, hydrophobic matching, integral membrane protein, amphiphilic polypeptides, cholesterol

CONTENTS

| | |
|---|---|
| PERSPECTIVES AND OVERVIEW | 146 |
|     Integral Membrane Proteins and the Concept of Hydrophobic Matching | 146 |
|     The Membrane as a Many-Particle System | 147 |
| MODELS OF LIPID-PROTEIN INTERACTIONS AND THE HYDROPHOBIC MATCHING CONDITION | 150 |
|     Phenomenological Thermodynamic Models | 150 |
|     Microscopic Interaction Models | 155 |
| PROTEIN-INDUCED LATERAL MEMBRANE ORGANIZATION | 157 |
|     Lipid Transition Temperature, Phase Diagrams, and Phase Separation | 157 |
|     Lateral Distribution of Proteins | 159 |
| LIPID STRUCTURE AND COMPOSITION NEAR INTEGRAL MEMBRANE PROTEINS | 160 |
|     Lipid Order Parameter Profiles | 160 |
|     Lipid Selectivity | 162 |
| A MODEL SYSTEM FOR LIPID-PROTEIN INTERACTIONS: SYNTHETIC AMPHIPHILIC POLYPEPTIDES | 164 |
|     Short and Long Polypeptides in Thin and Thick Lipid Bilayers | 164 |
|     Controlling Membrane Thickness by Cholesterol and Polypeptides | 166 |
| MEMBRANE FUNCTION AND LIPID-PROTEIN HYDROPHOBIC MATCHING | 167 |
| FUTURE PERSPECTIVES | 168 |

## PERSPECTIVES AND OVERVIEW

### Integral Membrane Proteins and the Concept of Hydrophobic Matching

The plasma membranes of eukaryotic cells are composite, stratified structures (5, 22) consisting of a lipid-bilayer matrix sandwiched between a cytosol polymeric filament (the cytoskeleton) and an extramembraneous glycocalyx. Figure 1 shows a schematic illustration of this complex macromolecular assembly. The lipid bilayer of the membrane provides the cell with a permeability barrier and it mediates many cellular functions that involve the proteins associated with the membrane that act as structural elements, enzymes, and/or receptors. Numerous membrane-bound proteins are integral membrane proteins, which implies that they have at least one membrane-spanning domain (17). The membrane-spanning domain may consist of a single α-helical polypeptide or several structured strands of amino acids. In order for the membrane-spanning domain to be compatible with the amphiphilic nature of the lipid-bilayer matrix, this domain itself must be amphiphilic with a high degree of hydrophobic matching between its hydrophobic part and the hydrophobic core of the lipid bilayer. This observation has led to the concept of hydrophobic matching (19, 48, 50), which is known to play a major role not only in determining the

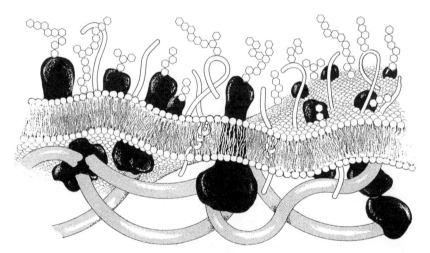

*Figure 1* Schematic illustration of a eukaryotic cell plasma membrane, which highlights the membrane as a stratified composite of a fluid lipid bilayer sandwiched between the carbohydrate glycocalyx on the outside and the cytoskeleton on the inside. Integral membrane proteins and polypeptides are indicated as large molecular objects penetrating the lipid bilayer matrix (illustration by Ove Broo Sorensen).

thermodynamics of lipid-protein interactions but also in controlling the activity of integral membrane-bound enzymes and receptors (50).

In this review, we critically evaluate the concept of hydrophobic matching in membranes and its use in modeling lipid-protein interactions. We describe the current status of statistical thermodynamic modeling of lipid-protein interactions in membranes built on the hydrophobic-matching concept and show how the concept has been used to interpret and guide experimental studies of model membrane systems. Moreover, we provide some perspectives in the use of hydrophobic matching to study the effect on membrane function of various molecular agents, such as cholesterol and drugs, that interact with membranes.

## The Membrane as a Many-Particle System

A conventional approach in molecular membranology is to focus attention on membrane phenomena of singular importance, e.g. active transport and drug-receptor interactions, by mainly considering the active molecule or its active site and how these entities interact with a few other molecules. The many-particle character of the entire membrane assembly is often neglected. In particular, the lipid bilayer is often considered to be quiescent and to play only a passive role in providing anchoring sites for the membrane proteins. The conventional view of the biological membrane is embodied in the fluid-mosaic model of Singer & Nicolson (64). This model supports and encourages a particular physical intuition about lateral membrane structure and organization in terms of an agglomerate of lipids and proteins in a fluid-like random structure. An often overlooked fact that should be added to this picture is that the lipid membrane is a highly dynamic system that, despite its fluid character, has a high degree of both global and local coherent structure in its organization: the membrane is heterogeneous statically (16, 22) as well as dynamically (4, 49). In the case of eukaryotic cell plasma membranes, the heterogeneity of the membrane assembly is controlled by the couplings between the lipid membrane and the cytoskeleton, as well as by the many-particle character of the lipid membrane itself. The mutual lipid-lipid interactions and the lipid-protein interactions make the membrane a correlated and cooperative system associated with properties that cannot be fully understood in terms of only a few molecules.

Among the more striking cooperative phenomena in the lipid membrane are lipid phase transitions (33), protein-induced phase separation (65), protein aggregation and crystallization (35), and lipid selectivity of proteins (69). Many of these phenomena need not be induced by specific biochemical reactions but can be controlled and triggered by purely physical forces (61). The cooperative phenomena underlie not only membrane organization but also the various other physical properties of lipid membranes (5), such as passive permeability, mechanical cohesion and strength, elastic flexibility, and fluidity.

Several structural transitions occur in lipid membranes, of which the main gel-to-fluid lipid-bilayer phase transition (33) is believed to be of particular interest for understanding the organization and functioning of biological membranes (32). The fluid lipid-bilayer phase is relevant at physiological conditions in biological membranes. The very different manifestations of lipid-protein interactions in the different lipid phases are probably most clearly seen in NMR studies of reconstituted membranes (7). As illustrated in Figure 2, the fluid phase exhibits almost no effect, whereas the NMR orientational order parameters are strongly modified in the gel phase because of the presence of integral proteins (such as rhodopsin) (3). This apparent lack of influence on the lipid structure in the fluid phase may indicate the fulfillment of a hydrophobic-matching condition (48). In this light, the main lipid transition can be seen as a unique phenomenon by which one can probe the physical effects of the interaction between lipids and integral membrane proteins because the

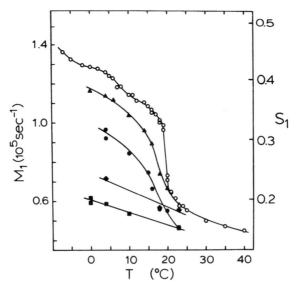

*Figure 2*  Temperature dependence of the first moment, $M_1$, and the average acyl chain orientational order parameter, $S$, of the deuterium NMR spectra for DMPC-$d_{54}$ lipid bilayers incorporated with different amounts of rhodopsin. From top to bottom the five sets of data correspond to lipid/protein ratios of $\infty$ (pure DMPC), 150, 50, 30, and 12. Reproduced from Ref. 3 with permission.

transition is associated with a dramatic change in lipid-bilayer hydrophobic thickness. The hydrophobic matching condition is therefore significantly altered in the transition region, as illustrated in Figures 3a and b. As this transition is extremely sensitive to the molecular interactions of the composite membrane system, its appearance should provide insight into the fundamental principles underlying lipid-protein interactions in model membranes and ultimately in biological membranes. Interpretation of physical measurements of membrane properties in terms of molecular properties and, conversely, derivation of macroscopic membrane properties from theoretical molecular-interaction models relies on a route from microscopics to macroscopics. The present review is about this route.

In recent years, investigators have used mainly model membrane systems, such as simple lipid bilayers and reconstituted lipid-protein systems, to obtain solid information on the physical properties of membranes with physical experimental measurements as well as theoretical modeling and simulation. Important experimental techniques include calorimetry, magnetic resonance spectroscopy, neutron and X-ray

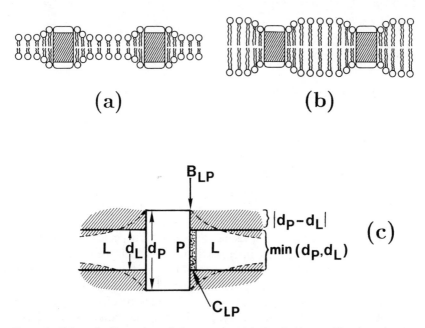

*Figure 3* Schematic illustration of the concept of hydrophobic matching between the hydrophobic length of the transmembrane domain of a protein (P) and the hydrophobic thickness of a lipid bilayer (L). (*a*) A thin membrane; (*b*) a thick membrane; (*c*) hydrophobic-thickness parameters, $d_L$ and $d_P$, and interaction parameters, $B_{LP}$ and $C_{LP}$, of the mattress model of Equation 8.

diffraction, and micromechanics. The physical properties of membranes in relation to different time and length scales obtained using these techniques are discussed in a recent review by Bloom et al (5). Other reviews discuss the various theoretical approaches to lipid-membrane physical properties and the physical effects of lipid-protein interactions using statistical mechanical molecular-interaction models (1, 47, 51) and computer-simulation techniques (46, 50).

## MODELS OF LIPID-PROTEIN INTERACTIONS AND THE HYDROPHOBIC MATCHING CONDITION

### Phenomenological Thermodynamic Models

The Landau theory provides a standard theoretical prescription for a phenomenological model of the thermodynamic free energy of a system with a phase transition perturbed by another component (27). The Landau free energy is an expansion in terms of an appropriate order parameter that for the lipid membrane usually is taken to be the lipid-bilayer hydrophobic thickness, $d_L$, the bilayer area, $a_L$, or the average acyl-chain orientational order parameter, $S$,

$$S = \left\langle \frac{1}{2(N-1)} \sum_{i=2}^{N} (3\cos^2\theta_i - 1) \right\rangle, \qquad 1.$$

where the summation extends over all $N$ segments of the acyl chain and $\theta_i$ is the angle between the bilayer normal and the normal to the plane spanned by the $i$th $CH_2$ group. Provided that the orientational order arises from conformational and rotational motions of axial symmetry about the bilayer normal, the bilayer thickness and the average orientational order parameter are linearly related (25):

$$d_L = d_L^{\max}(\beta_1 S + \beta_2), \qquad 2.$$

where $\beta_1$ and $\beta_2$ are geometrical parameters. Equation 2 should be regarded as a semiempirical relation between $d_L$ and $S$ that is expected to be approximately correct when no large intercalation effects occur between the lipids of the two leaflets of the bilayer. Up to now, Equation 2 has been found to be in agreement with available measurements of $d_L$ and $S$.

Several different strategies have been adopted to incorporate the effects of proteins into the Landau theory. In all cases, the protein is introduced as a perturbing impurity in the pure lipid-bilayer matrix. In the approaches by Owicki et al (53, 54) and Jähnig (27–29), the proteins are assumed to be fixed in the membrane and only acting to provide a boundary condition for the lipid order parameter. Minimization of the free energy then leads to an exponentially varying lipid-order parameter profile near the protein that can be written:

$$d_L(r) = d_L^0 + (d_P - d_L^0)\exp(-r/\xi_P),\qquad 3.$$

where $\xi_P$ is the coherence length of the protein perturbation, $d_P$ is the hydrophobic length of the protein, $d_L^0$ is the unperturbed lipid-bilayer thickness, and $r$ is a distance parameter. The term $\xi_P$ is a function of temperature and composition of the membrane. Since the proteins in this approach are assumed to be fixed in the membrane, their presence does not induce thermodynamic phase separation of the lipid gel and fluid phases. The free energy for the lipid-protein system can then be written (27):

$$G = G^0 + \frac{2\pi b}{(d_L^{0,f} - d_L^{0,g})^2}\left(\frac{\rho_P}{\pi\xi_L} + 1\right)(d_P - d_L)^2,\qquad 4.$$

where $G^0$ is the free energy of the unperturbed membrane and $b$ is a phenomenological constant related to an elastic energy of spatial density fluctuations in the membrane. The term $\xi_L$ is the coherence length of the lipid-bilayer fluctuations, which is assumed to be proportional to the decay length $\xi_P$, an assumption that only holds approximately. The term $\rho_P$ is the circumference of the protein, which is assumed to have a cylindrical shape. The superscripts on the pure lipid bilayer thicknesses refer to the gel (g) and the fluid (f) phases. The dependence on protein concentration, $x_P$, is now implicitly given through the equilibrium bilayer thickness $d_L(x_P)$, which is determined by minimization of the total free energy. Within this theory, the full excess free energy is attributed to an elastic distortion, and this excess free energy is, according to Equation 4, proportional to the square of the hydrophobic mismatch,

$$\text{mismatch} = |d_P - d_L|,\qquad 5.$$

between the lipid-bilayer and the protein hydrophobic thicknesses. Equation 4 shows that the shift in phase-transition temperature, relative to the pure lipid-bilayer transition temperature $T_m$, resulting from the presence of the proteins, is (55):

$$T - T_m = \Delta T(x_P) = 16\xi_P^2\left(\frac{\rho_P}{\pi\xi_L} + 1\right)\left(\frac{\overline{d_L} - d_P}{d_L^{0,f} - d_L^{0,g}}\right)x_P,\qquad 6.$$

where the mean lipid-bilayer hydrophobic thickness is defined as

$$\overline{d_L} = \frac{1}{2}(d_L^g + d_L^f).\qquad 7.$$

Hence, within this type of Landau theory, $\Delta T$ equals zero for $d_P = \overline{d_L}$, which is a consequence of the neglect of the direct lipid-protein interactions.

Although we have presented the Landau theory here using the bilayer hydrophobic thickness as an order parameter, this theory has a wider sphere of applicability and can take into account other types of protein-induced perturbation on the lipids, e.g. in terms of constraints on lipid-chain tilting or as a specific boundary condition on the lipid-protein interface caused by the roughness of this interface (29). The advantage of this type of Landau approach is that it is simple and transparent. Among its drawbacks are that it does not allow for phase separation and that the phenomenological parameters of the model are in principle unknown.

The mattress model of lipid-protein interactions in membranes (48, 65–68) is also a phenomenological theory of the Landau type. However, the mattress model differs from the conventional Landau-theory approach in that it is a two-component real-solution theory that allows for phase separation. Moreover, the mattress model assumes an expansion of the free energy that requires an experimental input for the thermodynamic properties of the pure lipid bilayer and then formulates the phenomenological expansion parameters describing the effects of the proteins in terms of physically identifiable properties and interaction parameters. The mattress model uses the hydrophobic thickness as an order parameter and in its simplest version, which applies at low protein concentrations, the free energy is written

$$G = x_L \mu_L^0 + x_P \mu_P^0 + RT(x_L \ln x_L + x_P \ln x_P)$$
$$+ x_L x_P [B_{LP} |d_L - d_P| + C_{LP} \min(d_L, d_P)], \qquad 8.$$

where $x_L$ and $x_P$ are the composition variables, $x_L + x_P = 1$, $\mu_L^0$ and $\mu_P^0$ are the standard chemical potentials of the lipids and proteins, respectively, and $B_{LP}$ and $C_{LP}$ are interaction parameters. $B_{LP}$ is related to the hydrophobic effect of the mismatch, and $C_{LP}$ is the direct lipid-protein van der Waals–like interaction associated with the interfacial contact of the two species. The geometry and the interactions of the mattress model are illustrated in Figure 3c.

The standard chemical potential of the lipids refers to the pure lipid system, and its difference in the two lipid phases can therefore be related to the transition enthalpy, $\Delta H_L$. By choosing the standard state of the proteins as the infinite-dilution limit, it is possible to express the standard chemical potential of the proteins in terms of the geometrical and interaction parameters of the model. If it is assumed that the hydrophobic part of the transmembrane proteins has the form of a cylinder with a smooth surface, of which a cross-section is uniform along the long axis of the protein, it follows that $B_{LP}$ and $C_{LP}$, and thus $\Delta\mu_P^0$, are proportional to the perimeter, $\rho_P$, of the protein cross-section:

$$\Delta\mu_P^0 = -\rho_P\Gamma, \quad B_{LP} = \rho_P\gamma, \quad C_{LP} = \rho_P\nu. \qquad 9.$$

The terms $\gamma$ and $\nu$ are reduced energy parameters of the model. The parameter $\nu$, being related to the direct interaction between the hydrophobic parts of the lipids and the proteins, depends explicitly on the lipid phase, i.e. $\nu = \nu^\alpha$. The parameter $\gamma$ measures the interaction between the hydrophobic part of the longer species and the hydrophilic material to which this species is exposed owing to the mismatch. $\Gamma$ is a function of $\gamma$, $\nu^\alpha$, and the hydrophobic thickness variables. In the dilute regime, the phase boundaries, $T_m^g$ and $T_m^f$, denoting the solidus and liquidus lines in the phase diagram $(T, x_P)$, are linear in $x_P$, and the two scenarios shown in Figures 4a and b can arise. With the definition of the midpoint transition temperature

$$\overline{T}(x_P) = \frac{1}{2}[T^g(x_P) + T^f(x_P)], \qquad 10.$$

the shift in midpoint transition temperature relative to the pure melting transition temperature as obtained from the model in Equation 8 can be expressed in the form (66):

$$\Delta T(x_P) = \overline{T}(x_P) - T_M \cong \frac{RT_m^2}{\Delta H_L} x_P \sinh\left(\frac{\rho_P\Gamma}{RT_m}\right). \qquad 11.$$

$\Delta T(x_P)$ is a convenient quantity to compare with experimental measurements. The $\Delta T$ of Equations 6 and 11 are in principle different quantities because a sharp phase transition remains at $x_P \neq 0$ in the conventional Landau theory, whereas the mattress-model solution theory allows for a two-phase coexistence region.

In its full version (48), the mattress model represents the bilayer as a planar elastic sheet by including a term in the free energy expression of the form

$$G_{el} = A_L(d_L - d_L^0)^2, \qquad 12.$$

where $A_L$ is the elastic constant that can be related to the elastic area compressibility modulus at constant temperature. In equilibrium, Equation 12 added to Equation 8 leads to

$$d_L = d_L^0 + \left(\frac{\rho_P\gamma}{2A_L}\right)x_P, \qquad 13.$$

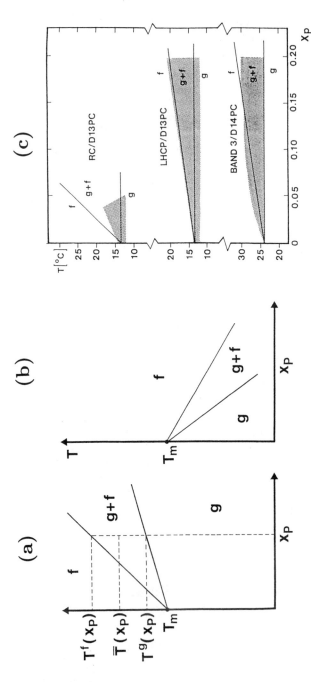

*Figure 4* Phase diagrams in temperature, $T$, versus protein concentration, $x_P$, of lipid-protein bilayers in the dilute regime, $x_P \ll 1$. The designations g and f denote gel and fluid lipid phases, and f+g is the two-phase region. Panels *a* and *b* are generic phase diagrams derived from the mattress model in Equation 8. Panel *a* corresponds to $\Delta\mu_P^0 < 0$ and Panel *b* to $\Delta\mu_P^0 > 0$. $\overline{T}(x_P)$ is the midpoint phase transition temperature, Equation 10, and $T^f$ and $T^g$ are the liquidus and solidus lines, respectively. (*c*) Specific phase diagrams of lipid membranes reconstituted with different proteins. The solid lines are the theoretical prediction of the phase boundaries of the phenomenological mattress model in the dilute regime. The experimental coexistence regions are shown in grey. RC/D13PC: photosynthetic reaction center protein in D13PC lipid bilayers (experimental data from 55). LHCP/D13PC: photosynthetic antenna protein in D13PC lipid bilayers (experimental data from 55). Band 3/D14PC: human erythrocyte band 3 protein in D14PC (DMPC) lipid bilayers (experimental data from 43). Reproduced from Ref. 65 with permission.

which implies that the elastic energy contribution enters to second order in $x_P$. Hence, the term $G_{el}$ becomes important at higher protein concentrations. For $x_P \lesssim 10^{-3}$, with realistic values of the model parameters $\gamma$ and $v^\alpha$ (66), the elastic interactions can be neglected.

The determination of the parameters of the mattress model has been discussed at length (48, 65–68). Here it suffices to mention that the input regarding the pure lipid-bilayer properties is obtained from thermodynamic and thermomechanic measurements. The protein parameters are in most cases obtained from estimates based upon knowledge of the geometry of the three-dimensional protein structure to the extent this is known. The interaction parameters $\gamma$ and $v^\alpha$ are assumed to be universal (66) and independent of the particular protein in question.

## Microscopic Interaction Models

Formulation of a microscopic interaction model for lipid-protein interactions and their effects on membrane thermodynamics and phase equilibria requires the availability of a microscopic model that describes the pure lipid bilayer. One of the first models proposed to describe the lipid-bilayer main transition was that of Marčelja (39), who later extended the model to describe lipid-protein interactions, treating the protein as a cylindrical boundary condition on the lipid-orientational order parameter (40). The lipid-protein interactions are nonspecific within this approach, which is very similar to that of the conventional Landau theory. The Marčelja model does not account for phase separation, and it focuses on calculating the lipid-order parameter profile, $S(r)$, around the protein. One of the most important results of the Marčelja model is its prediction of effective lipid-mediated attractive protein-protein interactions when the order-parameter profiles of neighboring proteins overlap. The implications of these results for protein aggregation are somewhat unclear, because the assumption of immobile proteins in the Marčelja model results in an attractive force between the proteins of purely enthalpic origin and does not include the counteracting forces from the entropy of mixing.

Another useful approach is based on the Pink model (57) for the lipid-bilayer main transition. The Pink model is a statistical mechanical lattice model that takes detailed account of the lipid-acyl chain interactions. Several extensions of this model have been put forward to account for the dependence of the lipid-protein interactions on the conformational states of the lipid chains in a specific manner (37, 38, 56, 58, 72). The advantage of these approaches is that they allow for phase separation. One of their drawbacks, however, is that the values of the lipid-protein interaction parameters are not simply related to molecular properties. The interaction constants remain unknown model parameters.

The Pink model allows for a series of conformational states of the lipid-acyl chains. In one of its more elaborate versions, the Pink model incorporates 10 states. This 10-state Pink model has been extended to include lipid-protein interactions (67, 68) in the spirit of the mattress model (48). In this microscopic mattress-model approach, the Pink model provides the pure lipid-bilayer phase transition and its properties. In the phenomenological mattress-model approach, the phase transition and its properties were fed in from experimental data as described in the previous section. The lipid-protein interactions are included in the microscopic model by assuming that the hydrophobic membrane-spanning part of the protein molecule is a stiff, rod-like, and hydrophobically smooth object with no appreciable internal flexibility. In this way, the protein is characterized only by a cross-sectional area, $a_P$ (or circumference $\rho_P$). The model is formulated in terms of a Hamiltonian, $\mathcal{H} = \mathcal{H}_{L-L} + \mathcal{H}_{L-P}$, which expresses the internal energy associated with the lipid-lipid and the lipid-protein interactions in terms of the microscopic variables of the system. Direct protein-protein interactions are neglected in this approach but could in principle also be taken into account. Statistical mechanics then relates the microscopic Hamiltonian function to the macroscopic thermodynamic free energy as a functional:

$$G = -k_B T \ln \int_\Omega \exp[-\mathcal{H}(\Omega)/k_B T] \, d\Omega. \qquad 14.$$

The integral in Equation 14 is over all possible microstates, $\Omega$, of the system.

In contrast to the phenomenological mattress model, whose solution in terms of thermodynamic properties can be derived by straightforward minimization of the free energy, the microscopic mattress model poses a computational problem because it is formulated in terms of microscopic variables rather than macroscopic averages. It is possible to solve this computational problem, and hence provide a route from microscopics to macroscopics, by using modern computer-simulation, e.g. Monte Carlo, techniques (46). A statistical equilibrium ensemble of microstates of the model can be generated using these techniques. From this ensemble of states one can calculate virtually any macroscopic thermodynamic quantity that is a derivative of the free energy, as well as any microscopic correlation function. The simulation provides the typical equilibrium configurations of the model from which the lateral structure and organization of the membrane can be determined. Specifically, the membrane state in terms of lipid phase, protein-induced phase separation, and state of aggregation of the proteins can be assessed. Moreover, the structure of the lipid membrane at the protein-lipid interface can be studied.

## PROTEIN-INDUCED LATERAL MEMBRANE ORGANIZATION

*Lipid Transition Temperature, Phase Diagrams, and Phase Separation*

In the limit of low protein concentrations, where the properties of protein-dense phases can be neglected, two of the possible generic types of phase diagrams are shown schematically in Figures 4a and b. These two cases represent two situations in which the mismatch interaction is dominant. The transition region is then suppressed or enhanced depending approximately on whether or not the value of $d_P$ is smaller or larger than the mean lipid-bilayer thickness, $d_L$ in Equation 7. In other words, the solubility of the protein in the two lipid phases in these cases is larger in the phase to which the protein is better matched. If we now also take into account the elastic distortion of the lipid bilayer caused by incorporation of the protein (cf Equation 12), a more complicated phase diagram can result with an upper azeotropic point (48). The elastic effect tends to suppress the phase boundaries because $A_L^g > A_L^f$ for lipid bilayers.

Using freeze-fracture and light-scattering experiments, Möhwald and collaborators (55, 60) have carried out a systematic experimental study of the effect of hydrophobic mismatch on lipid-protein interactions. These authors studied the phase behavior of photosynthetic reaction center (RC) proteins and antenna proteins (light-harvesting chlorophyll protein; LHCP) reconstituted into lipid bilayer vesicles of saturated 1,2-diacyl-*sn*-glycero-3-PCs (diacyl-glycero-phosphatidylcholine with $n_C$ acyl-chain carbon atoms; $Dn_CPC$) with different acyl-chain lengths, $n_C = 12, 13, 14, 15, 16$. These proteins are parts of the photosynthetic apparatus of *Rhodopseudomonas sphaeroides*; they are of very different size; and they both have hydrophobic transmembrane segments. The results of the experiments can be rationalized within the framework of the phenomenological mattress model. The comparison between the experimental and the theoretical phase diagrams is given in Figure 4c. In the light of the uncertainties associated with deriving the phase diagrams from the experimental data (65), the accordance between theory and experiment is satisfactory and provides some support to the hydrophobic-matching hypothesis.

Also shown in Figure 4c is a comparison between experimental data and theoretical predictions for band 3 protein. The family of human erythrocyte band 3 proteins are large transmembrane glycoproteins, some of which act as receptors for concanavalin A. The receptor portion of band 3 is well characterized on the molecular level in terms of receptor function and active ion transport. A careful experimental study of the phase behavior of phospholipid bilayers incorporated with the concanavalin A receptor of band 3 protein has been performed (43). This study combines information obtained from calorimetry with that from deuterium NMR difference spectroscopy. The experimental phase diagram obtained is probably one of the most reliable phase diagrams for lipid-protein systems presently available.

By detecting transmission changes due to light scattering, Möhwald and collaborators (55, 60) monitored the shift in phase transition temperature, $\Delta T(x_P)$ in Equation 10, as a function of $\overline{d_L}$ and $x_P$ for photosynthetic reaction center proteins and antenna proteins. Figures 5a and b reproduce the experimental data. Obviously, hydrophobic mismatch strongly influences the phase behavior. The theoretical predictions obtained from the phenomenological mattress model are also shown in Figure 5. It is of interest for this system to compare the theoretical prediction of $\Delta T(\overline{d_L})$ from the mattress model, Equation 11, with that of the conventional

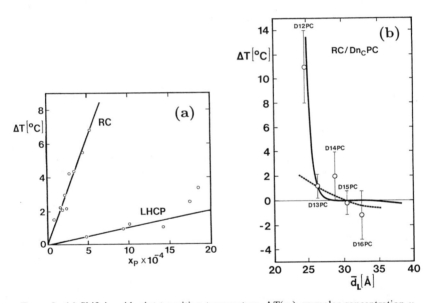

*Figure 5* (a) Shift in midpoint transition temperature, $\Delta T(x_P)$, vs molar concentration $x_P$, of RC protein and LHCP protein respectively reconstituted into D13PC lipid bilayers. Open circles indicate the experimental data from Ref. 55, and solid curves show the linear theoretical predictions of Equation 11 based on the phenomenological mattress model in the dilute regime. (b) Shift in midpoint transition temperature, $\Delta T(\overline{d_L})$, vs mean lipid hydrophobic thickness $\overline{d_L}$ for a series of D$n_C$PC lipid bilayers incorporated with RC protein in a molar concentration of $x_P = 10^{-4}$. Open circles indicate experimental data from Refs. 55, 60. The solid line is the theoretical prediction from the phenomenological mattress model of Equation 11. The dashed line is the prediction from the conventional Landau theory including elastic distortions, Equation 6. Reproduced from Ref. 66 with permission.

Landau theory in Equation 6 based on elastic interactions. As explained earlier, the $\Delta T$s of Equations 6 and 11 are in principle different quantities since a sharp phase transition remains at $x_P \neq 0$ in the Landau theory (with immobile impurities), whereas a two-phase coexistence region is a natural consequence of the mattress-model solution theory. The comparison made in Figure 5b shows that the two theories account equally well for the experimental data involving relatively small changes in $\Delta T$ but produce significantly different results at more extreme changes. In particular, the elastic Landau theory fails to reproduce the dramatic increase in $\Delta T$ for D12PC.

The hyperbolic curve, $\Delta T(d_L)$ in Equation 11, passes through $\Delta T = 0$ for a value of $d_L$ that in general is different from $d_P$. Only in the case $v^g = v^f$ or $v^\alpha \ll \gamma$ does one obtain $d_L = d_P$. This implies that it is not strictly correct, in dealing with a protein having unknown hydrophobic thickness, to assume that the value of $d_P$ is the mean hydrophobic thickness of that lipid bilayer that does not exhibit a shift in transition-temperature in the presence of the protein.

## Lateral Distribution of Proteins

In addition to protein concentration and temperature, the lateral distribution of proteins in the lipid membrane plane is also controlled by (a) the lipid-protein interactions and lipid-mediated protein-protein interactions and (b) the direct protein-protein interactions that may be of long range because of extramembrane moieties. The microscopic version of the mattress model (68) has been used to systematically study the lateral protein distribution in model membranes as controlled by temperature and a. The results suggest that the formation of protein aggregates in the membrane plane is predominantly controlled by the strength of the direct van der Waals–like lipid-protein interaction. Whereas the hydrophobic mismatch is of prime importance for determining the phase equilibria, a mismatch may not be the sole reason for protein aggregation within each of the individual phases: depending on the strength of the van der Waals–like interaction associated with the direct lipid and protein hydrophobic contact, the proteins may remain dispersed in the fluid phospholipid bilayer, even if the mismatch between the protein and the bilayer thickness is as high as 12 Å. This observation is consistent with the findings of Lewis & Engelman (36) on the state of aggregation of bacteriorhodopsin in fluid lipid bilayers of varying thickness.

The Monte Carlo simulation results (68) indicate that, when the direct protein-lipid interaction parameter is sufficiently small, protein aggregates form in the fluid region of the phase diagram just above the phase boundary because of dynamic aggregation induced by the lipid-density fluctuations.

This effect is therefore a consequence of the way the protein couples to the dynamic membrane heterogeneity. By this mechanism, lipid fluctuations can induce dynamic protein aggregation, which should be most pronounced close to the phase boundaries. As the strength of the direct protein-lipid interaction is increased, the tendency for formation of protein aggregates via this mechanism is diminished. Larger proteins would, however, induce stronger lipid-mediated attractive protein-protein interactions, which in turn would enhance the tendency for aggregation. This effect is likely to be of importance close to the phase boundaries where the coherence length of the lipid-mediated force is maximal. For larger proteins, a further complication is that the aggregates would be complexes of proteins with lipids trapped in the interstitial regions between several proteins.

## LIPID STRUCTURE AND COMPOSITION NEAR INTEGRAL MEMBRANE PROTEINS

### Lipid Order Parameter Profiles

A microscopic version of the mattress model has been studied using computer simulation (67) to determine the coherence length, $\xi_P(T, x_P)$, for the spatial fluctuations of the lipid order parameter profiles around integral membrane proteins for a fixed distribution of proteins in a large lipid bilayer array in the transition region. The model is studied at low protein concentrations where the overlap between the lipid profiles from neighboring proteins is negligible. The protein is characterized by a hydrophobic length, $d_P$, and a cross-sectional area. Schematically, the lipid order parameter profile may look as shown in Figure 6a, in the case of a protein whose hydrophobic length is larger than the hydrophobic thickness of the unperturbed lipid bilayer.

The coherence length, which can be determined from the profile with an exponential shape (cf Equation 3), has a strong temperature dependence with a sharp peak at the transition, as illustrated in Figure 6b for a protein with a hydrophobic length close to the hydrophobic thickness of the fluid lipid bilayer. The protein-induced disturbance of the lipid bilayer is seen to extend beyond the first few molecular layers over a wide range of temperatures. The coherence length becomes very large close to the transition. The effect of the protein on the local structure of the lipid bilayer depends in a detailed manner on the temperature, on the size of the protein, and on the protein hydrophobic length relative to the hydrophobic thickness of the lipid-bilayer phases (67). The long coherence length found in these calculations provides a mechanism for indirect lipid-mediated

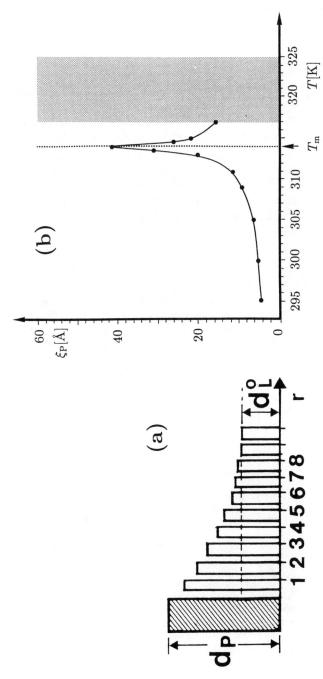

*Figure 6.* (a) Schematic drawing of the lipid acyl-chain length profile near an integral membrane protein. The term $r$ is the distance parameter in units of lattice spacings. (b) Temperature dependence of decay length, $\xi_P$, for a protein with a hydrophobic length $d_P = 24$ Å and a cross-sectional area corresponding to seven sites on the triangular lattice of the Pink model. The shaded area indicates a temperature region where $d_L^0 \approx d_P$ very close to the protein, and the decay cannot be resolved within the numerical accuracy in order to yield a reliable value of $\xi_P$. Reproduced from Ref. 67 with permission.

protein-protein long-range attraction and may hence play an important role in regulating protein segregation.

## Lipid Selectivity

The microscopic mattress model for lipid-protein interactions in a lipid membrane incorporated with a very dilute static dispersion of proteins has been extended to membranes with two different lipid species characterized by different acyl-chain lengths (69). This extended model was considered with a view to determining to what extent bare physical effects may be responsible for lipid selectivity and lipid specificity of membrane proteins (14). The basic idea behind the model study is that lipid chains of varying length are perturbed by the protein surface to different extents via the hydrophobic matching condition. The lipid species best adapted to the matching condition will be selected, on a statistical basis, and thereby have an increased probability of being close to the protein-lipid interface. This is an example of interface enrichment.

The fact that such a selectivity can be a consequence of the hydrophobic matching condition is demonstrated by the data in Figure 7, which is derived from computer simulations on the microscopic mattress model, now appropriately extended to account for two different lipid species, dimyristoyl phophatidylcholine (DMPC) and distearoyl phosphatidylcholine (DSPC). The data in Figure 7 refer to a protein with a large cross-

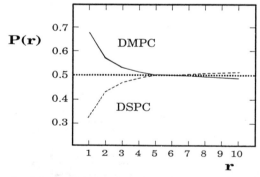

*Figure 7* Example of protein-lipid interface enrichment and physical lipid specificity in a binary lipid mixture as described by the microscopic mattress model. Lipid concentration profiles, $P(r)$, for the two lipid species are shown as a function of distance, $r$, from a very large integral membrane protein. The data refer to computer simulation on an equimolar binary mixture of DMPC and DSPC at 340 K, which is well above the coexistence region. The protein hydrophobic thickness, $d_P = 20$ Å, is close to twice the acyl-chain length in a fluid DMPC lipid bilayer. Reproduced from Ref. 50 with permission.

section and a hydrophobic length $d_P = 20$ Å embedded in an equimolar lipid mixture at a temperature well above the coexistence region of the mixture, i.e. in the fluid lipid phase. The value of $d_P$ is chosen to be close to the hydrophobic thickness of fluid DMPC bilayers. Lipid-concentration profiles of DMPC and DSPC are shown in Figure 7 as a function of distance, $r$, from the protein. The protein selects the lipid species (in this case DMPC) that most easily wets the hydrophobic surface of the protein. Conversely, the protein lipid interface is depleted in the other species. Hence, the lipid-protein interactions lead to local compositional demixing of the two lipid species. As a result of the hydrophobic matching condition, the lipid-protein interactions couple to the compositional fluctuations of the binary lipid mixture in the fluid phase.

A most striking observation made from the model simulations (69) of protein-induced compositional heterogeneity is related to a nonequilibrium transient effect found in the compositional different concentration profiles of the two lipid species as these profiles establish themselves in the course of time. This effect, which may have some important consequences for steady-state membrane organization, refers to a situation in which a thermally equilibrated binary lipid mixture is prepared in the fluid phase and then suddenly is made subject to the boundary condition imposed by the presence of the proteins. In response to the presence of the proteins, the mixture has to reorganize itself laterally and decompose locally as illustrated by the equilibrium concentration profiles in Figure 7. This reorganization proceeds via long-range diffusional processes. The interdiffusion of the species is, however, limited by the conservation law imposed by the global composition of the mixture. As a consequence, on its way towards equilibrium in the presence of proteins, the mixture displays a pronounced oscillatory behavior in the concentration profiles.

The results presented above refer to immobile model proteins, such as proteins bound to specific positions of the membrane, for example via the cytoskeleton, or to proteins that diffuse very slowly relative to the lipids. However, in the case of mobile proteins, the general nature of the results for static proteins indicate that structured concentration profiles of the type shown in Figure 7 will facilitate a medium-range lipid-mediated indirect protein-protein attraction that will influence the state of protein aggregation. This observation may have biological relevance for those proteins whose biological activity depends on their aggregational state. For a nonequilibrium system, say a protein-lipid membrane driven by external sources of energy that couple to protein conformational changes, the oscillatory profile may be dynamically maintained. The mobile proteins may in the driven system organize themselves laterally to fit into the part of the profile that is enriched in the lipid species with the higher affinity for the protein. This picture can be generalized to systems of different proteins with different lipid selectivity.

## A MODEL SYSTEM FOR LIPID-PROTEIN INTERACTIONS: SYNTHETIC AMPHIPHILIC POLYPEPTIDES

*Short and Long Polypeptides in Thin and Thick Lipid Bilayers*

The possibility of synthesizing specific amphiphilic membrane-spanning polypeptides has added an exciting dimension to quantitative model studies of lipid-protein interactions in membranes. Davis et al (12) were the first to propose the use of synthetic amphiphilic polypeptides specifically designed to span a lipid bilayer by having hydrophilic amino-acid residues at the ends and a central core of hydrophobic amino-acid residues organized into an α-helical arrangement. The use of polypeptides with different lengths of the hydrophobic core and the incorporation of these into phospholipid bilayers of different hydrophobic thickness led to the proposal that such a family of model membranes would constitute a unique model system for investigating consequences of the hydrophobic matching condition. Since the original proposal, several studies (24, 44, 45, 52) have emerged for this model system that use a variety of techniques, including calorimetry and X-ray scattering, as well as NMR and infrared spectroscopy. The polypeptides used in these studies were of the type $Lys_2$-Gly-$(Leu)_n$-$Lys_2$-Ala-amide, with $n = 16$, 20, or 24.

Figure 8a shows selected deuterium NMR spectra for the peptide with $n = 24$ incorporated into dipalmitoyl phosphatidylcholine (DPPC)-$d_{62}$ bilayers. Figure 8b shows the first moment, $M_1$, of such spectra as a function of temperature for different polypeptide concentrations and for two peptide lengths, $n = 16$ and $n = 24$. Because $M_1$ is a measure of the degree of acyl-chain order and is related to the bilayer hydrophobic thickness, Figure 8b demonstrates that both peptides tend to disorder and thin the gel bilayer phase and to order and thicken the fluid bilayer phase. However, the ordering effect in the fluid phase is stronger for the long peptide, and the disordering effect in the gel phase is more pronounced for the short peptide. These observations can be rationalized within the mattress model: since the hydrophobic lengths of the two peptides are $d_P \simeq 24$ and 36 Å for $n = 16$ and 24, respectively, the short peptide is rather closely matched to the hydrophobic thickness of the fluid DPPC phase, whereas the long peptide is close to the mean bilayer thickness, $\overline{d_L}$, and hence has a distorting effect on both phases. The difference between the experimental phase diagrams, as determined by NMR difference spectroscopy and calorimetry (24, 44), for these two peptides in DPPC is not

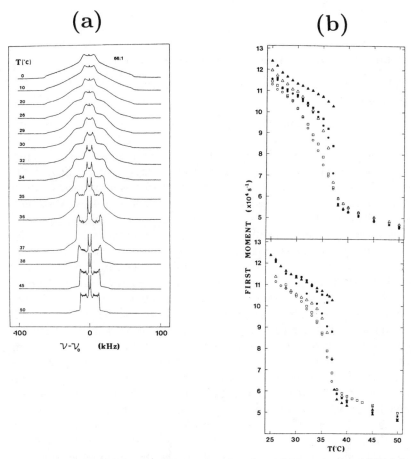

*Figure 8* (*a*) Temperature dependence of the deuterium NMR spectrum of DPPC-$d_{62}$ bilayers incorporated with the amphiphilic membrane-spanning polypeptide Lys$_2$-Gly-(Leu)$_n$-Lys$_2$-Ala-amide, with $n = 24$, in a molar concentration of $x_P = 0.0149$ (24). (*b*) Temperature dependence of the first moment, $M_1$, of the deuterium NMR spectrum of DPPC-$d_{62}$ bilayers incorporated with the amphiphilic membrane-spanning polypeptides Lys$_2$-Gly-(Leu)$_n$-Lys$_2$-Ala-amide, with $n = 16$ (*upper panel*) and $n = 24$ (*lower panel*) in molar concentration ranging from $x_P = 0$ to 0.0323 (*upper panel*) and to 0.0243 (*lower panel*). Reproduced from Ref. 24 with permission.

very large, but the systematics are consistent with expectations from the mattress-model considerations. The experiments have hence provided strong qualitative support to the idea that the degree of hydrophobic matching is an important feature of lipid-peptide interactions in membranes.

It would be interesting to investigate the physical properties of lipid-bilayer membranes incorporated with polypeptides that more closely resemble those occurring in real integral proteins, e.g. α-helical polypeptides with an anisotropic hydrophobic surface (13).

## Controlling Membrane Thickness by Cholesterol and Polypeptides

Instead of merely manipulating the bilayer hydrophobic thickness by varying the temperature as in previous studies, Nezil & Bloom (52) used the well-known membrane-thickening effect of cholesterol (25, 73) to change the hydrophobic thickness of 1-palmitoyl–2-oleoyl phosphatidylcholine (POPC)-$d_{62}$ bilayers. The change in bilayer thickness upon incorporation of amphiphilic polypeptides of the type $Lys_2$-Gly-$(Leu)_n$-$Lys_2$-Ala-amide, with either $n = 16$ or $n = 24$, was then determined using the measured NMR order-parameter profiles. Two types of membranes were considered: a thin membrane consisting of pure POPC-$d_{31}$ and a thick membrane consisting of POPC-$d_{31}$ mixed with 30 mol% cholesterol. This amount of cholesterol makes the hydrophobic membrane thickness increase from 27 to 32 Å. These membranes were then incorporated with one of the two polypeptides that are characterized by hydrophobic lengths of $d_P = 24$ Å and $d_P = 36$ Å, respectively. The resulting membrane thicknesses as determined experimentally are shown in Figure 9 as a function of temperature. As shown, both polypeptides strongly perturb these bilayers. The perturbations are fully consistent with the hydrophobic matching condition of the mattress model: addition of the short peptide to the thin membrane or the long peptide to the thick membrane results in little change in the bilayer thickness. In both cases the peptide hydrophobic length is rather closely matched to the bilayer hydrophobic thickness. In contrast, addition of the long peptide to the thin membrane leads to a substantial increase in membrane thickness, and similarly, addition of the short peptide to the thick membrane induces a significant reduction in membrane thickness. Nezil & Bloom made the interesting observation from their study (cf Figure 9) that the perturbation in membrane thickness resulting from mismatch is not symmetric with respect to the sign of the mismatch. The perturbation is relatively larger when the short peptide is incorporated into the thick bilayer than when the long peptide is incorporated into the thin bilayer. This asymmetry is in fact predicted by the phenomenological mattress model, Equation 8, because the value of the mismatch parameter $\gamma$ depends on which is longer: the lipid bilayer or the protein/polypeptide. When the peptide is the shorter species, the effective value of $\gamma$ is about twice that corresponding to the opposite situation, simply because the hydrophobic effect is stronger when the lipid-acyl

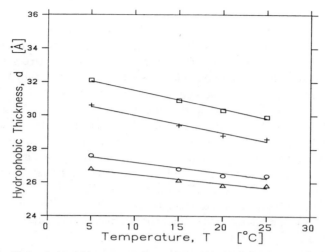

*Figure 9* Hydrophobic thickness vs temperature for different POPC-$d_{31}$ bilayers as measured by deuterium NMR spectroscopy. Symbols represent: △, pure POPC bilayer; ○, POPC bilayer incorporated with 4.8 mol% long polypeptide (Lys$_2$-Gly-(Leu)$_{24}$-Lys$_2$-Ala-amide); □ POPC bilayer with 30 mol% cholesterol; +, POPC bilayer with 30 mol% cholesterol incorporated with 4.8 mol% short peptide [Lys$_2$-Gly-(Leu)$_{16}$-Lys$_2$-Ala-amide]. Reproduced from Ref. 52 with permission.

chains are exposed to water than when the side groups of the polypeptide are exposed (71). A similar asymmetry has been observed experimentally for bacteriorhodopsin in fluid lipid bilayers (36).

The experimental study of Nezil & Bloom (52) isolates the influence of hydrophobic matching on membrane thickness in a particularly direct manner and provides strong support for the general importance of hydrophobic matching in membrane-protein (polypeptide) systems.

## MEMBRANE FUNCTION AND LIPID-PROTEIN HYDROPHOBIC MATCHING

Several authors have suggested (32, 61, 63) that the physiological functioning of membrane-bound proteins, receptors, and enzymes is controlled to a large extent by the structure and dynamics of the lipid-bilayer matrix with which they are associated. In order to put the hydrophobic-matching hypothesis in perspective, we refer to a few illustrative cases that strongly indicate that physiological membrane functions are controlled by hydrophobic-membrane thickness.

Enzymes like cytochrome $c$ oxidase (41), ($Na^+$-$K^+$)-ATPase (30), and $Ca^{2+}$-ATPase (8, 26) all exhibit an optimal enzymatic activity when incorporated into lipid bilayers of a given thickness. This observation strongly indicates that the hydrophobic matching criterion is operative since the optimum occurs at hydrophobic thicknesses that are in the neighborhood of the hydrophobic length that can be estimated for these proteins. It has been shown quite clearly that in the case of $Ca^{2+}$-ATPase (8, 15, 26) the dominant factor is the hydrophobic lipid-bilayer thickness, rather than the degree of acyl-chain saturation or concepts such as fluidity. In one of these experiments (26), the bilayer hydrophobic thickness was manipulated by incorporation of $n$-alkanes into the bilayer. These alkanes are known to increase the hydrophobic thickness and they are often said to fluidize the bilayer. In thin bilayers, where the $Ca^{2+}$-ATPase enzyme activity is low, high activity could be regained by incorporation of $n$-alkanes. In contrast, introduction of $n$-alkanes into a thick bilayer, where the initial activity is low, leads to a further decrease in the activity (26).

In the case of the acetylcholine receptor, which is an integral membrane protein involved in neural activity, there are also some experimental indications that membrane thickness and receptor function are correlated (11, 21). The action of certain drugs, such as local and general anesthetics (2, 20, 23), on the acetylcholine receptor may be mediated by a structural change in the lipid-protein interface. Because many drugs are known to change lipid-bilayer thickness, it is tempting to speculate that the drug action may be lipid-mediated via changes in the hydrophobic-matching condition at the lipid-protein interface (49).

## FUTURE PERSPECTIVES

The nature of the physical forces underlying lipid-protein interactions in membranes is still a subject of considerable controversy. In this review, we have put forward the hypothesis that the degree of hydrophobic matching between lipid-bilayer and integral-membrane protein/polypeptide hydrophobic thicknesses constitutes an important contribution to lipid-protein interactions in membranes. We have explored the consequences of this hypothesis for membranes containing integral proteins and polypeptides, using simple thermodynamic and statistical theories and models including the so-called mattress model. The theoretical predictions are in qualitative and sometimes quantitative accordance with experiments. The specific systems described in this review included phospholipid bilayers of varying thickness incorporated with transmembrane molecules, such as rhodopsin, photosynthetic-reaction-center and antenna proteins, synthetic amphiphilic polypeptides, and band 3 protein. Numerous other systems not discussed here have also been investigated in the light of the hydrophobic matching hypothesis (50), including bacteriorhodopsin (9, 36, 59), the transferrin receptor (34, 70), the insulin receptor (62, 70), and gramicidin A (10, 18, 31, 42, 74). Many aspects of the physical effects of lipid-protein interactions in these systems indicate that the degree of hydrophobic matching is an important element of the physical interactions. More reliable experimental data are, however, required to perform a stringent and critical test of the mattress model and the hydrophobic-matching hypothesis.

The apparent, overall success of the hydrophobic-matching hypothesis in accounting for existing experimental data on lipid-protein and lipid-polypeptide systems suggests that the concept should be useful for guiding and suggesting further experiments geared towards a fuller understanding of the physical forces involved in lipid-protein interactions, and how these interactions ultimately control biological function. Integral proteins ought to be approximately matched to the fluid-bilayer thickness of their natural membranes in order to function optimally. Local or global changes in the conditions for matching, e.g. as induced by changes in lateral composition, by drugs, by external chemical gradients, or by electric fields, may dramatically affect protein activity and could indeed form the basis for trigger processes (61). The concept of hydrophobic matching may have played an important role in the evolution of biological membranes (5, 6). Hence, approaches that use a physical concept like hydrophobic matching, together with renewed insight into the material-design principles Nature has exploited during the evolution of biological membranes, should provide a fruitful and constructive framework for further experimental and theoretical progress in the area of lipid-protein interactions in membranes.

ACKNOWLEDGMENTS

The research work by O. G. M. is supported by the Danish Natural Science Research Council under grant No. 11–7785 and by the Danish Technical Research Council under grant No. 16–4890–1. That of M. B. is supported by the National Sciences and Engineering Research Council of Canada. The review was prepared under a program on the Science of Soft Surfaces and Interfaces under the aegis of the Canadian Institute of Advanced Research. We are indebted to the stimulating interaction with several colleagues, including Martin J. Zuckermann, John Hjort Ipsen, Maria M. Sperotto, Kent Jørgensen, Rodney L. Biltonen, Clare Morrison, Frank Nezil, and Jennifer Thewalt.

## Literature Cited

1. Abney, J. R, Owicki, J. C. 1985. See Ref. 75, p. 1
2. Arias, H. R., Sankaram, M. B., Marsh, D., Barrantes, F. J. 1990. *Biochim. Biophys. Acta* 1027: 287
3. Bienvenue, A., Bloom, M., Davis, J. H., Devaux, P. F. 1982. *J. Biol. Chem.* 257: 3032
4. Mouritsen, O. G., Biltonen, R. L. 1993. In *New Comprehensive Biochemistry. Protein-Lipid Interactions*, ed. A. Watts. Amsterdam: Elsevier. In press
5. Bloom, M., Evans, E., Mouritsen, O. G. 1991. *Q. Rev. Biophys.* 24: 293
6. Bloom, M., Mouritsen, O. G. 1988. *Can. J. Chem.* 66: 706
7. Bloom, M., Smith, I. C. P. 1985. See Ref. 75, p. 61
8. Caffrey, M., Feigenson, G. W. 1981. *Biochemistry* 20: 1949
9. Cherry, R. J., Müller, U., Holenstein, C., Heyn, M. P. 1980. *Biochim. Biophys. Acta* 596: 145
10. Cornell, B. A., Separovic, F., Thomas, D. E., Atkins, A. R., Smith, R. 1989. *Biochim. Biophys. Acta* 985: 229
11. Criado, M., Eibl, H., Barrantes, F. J. 1984. *J. Biol. Chem.* 259: 9188
12. Davis, J. H., Clare, D. M., Hodges, R. S., Bloom, M. 1983. *Biochemistry* 22: 5298
13. DeGrado, W. F., Wasserman, Z. R., Lear, J. D. 1989. *Nature* 243: 622
14. Devaux, P. F., Seigneuret, M. 1985. *Biochim. Biophys. Acta* 822: 63
15. East, J. M., Jones, O. T., Simmonds, A. C., Lee, A. G. 1984. *J. Biol. Chem.* 259: 8070
16. Edidin, M. 1990. *Curr. Topics Membr. Transp.* 36: 81
17. Eisenberg, D. 1984. *Annu. Rev. Biochem.* 53: 595
18. Elliott, J. R., Needham, D., Dilger, J. P., Haydon, D. A. 1983. *Biochim. Biophys. Acta* 735: 95
19. Engelman, D. M., Zaccai, G. 1980. *Proc. Natl. Acad. Sci. USA* 77: 5894
20. Fraser, D. M., Louro, S. R. W., Horváth, L. I., Miller, K. W., Watts, A. 1990. *Biochemistry* 29: 2664
21. Fong, T. M., McNamee, M. G. 1986. *Biochemistry* 25: 830
22. Gennis, R. B. 1989. *Biomembranes. Molecular Structure and Function.* Heidelberg: Springer-Verlag. 533 pp.
23. Horváth, L. I., Arias, H. R., Hankovszky, H. O., Hideg, K., Barrantes, F. J., Marsh, D. 1990. *Biochemistry* 29: 8707
24. Huschilt, J. C., Hodges, R. S., Davis, J. H. 1985. *Biochemistry* 24: 1377
25. Ipsen, J. H., Mouritsen, O. G., Bloom, M. 1990. *Biophys. J.* 57: 405
26. Johannsson, A., Keithley, C. A., Smith, G. A., Richards, C. D., Hesketh, T. R., Metcalfe, J. C. 1981. *J. Biol. Chem.* 256: 1643
27. Jähnig, F. 1981. *Biophys. J.* 36: 329
28. Jähnig, F. 1981. *Biophys. J.* 36: 347
29. Jähnig, F., Vogel, H., Best, L. 1982. *Biochemistry* 21: 6790
30. Johannsson, A., Smith, G. A., Metcalfe, J. C. 1981. *Biochim. Biophys. Acta* 641: 416
31. Killian, J. A., Prasad, K. U., Urry, D. W., de Kruijff, B. 1989. *Biochim. Biophys. Acta* 978: 341
32. Kinnunen, P. K. J. 1991. *Chem. Phys. Lipids* 57: 375
33. Kinnunen, P., Laggner, P., eds. 1991. *Phospholipid Phase Transitions. Chem. Phys. Lipids* 57: 109–408
34. Kurrle, A., Rieber, P., Sackmann, E. 1990. *Biochemistry* 29: 8274
35. Kühlbrandt, W. 1992. *Q. Rev. Biophys.* 25: 1
36. Lewis, B. A., Engelman, D. M. 1983. *J. Mol. Biol.* 166: 203
37. Lookman, T., Pink, D. A., Grundke, E. W., Zuckermann, M. J., de Verteuil, F. 1982. *Biochemistry* 21: 5593
38. MacDonald, A. L., Pink, D. A. 1987. *Biochemistry* 26: 1909
39. Marčelja, S. 1974. *Biophys. Biochim. Acta* 367: 165
40. Marčelja, S. 1976. *Biophys. Biophys. Acta* 455: 1
41. Montecucco, C., Smith, G. A., Dabbeni-Sala, F., Johannsson, A., Galante, Y. M., Bisson, R. 1982. *FEBS Lett.* 144: 145
42. Morrow, M. R., Davis, J. H. 1988. *Biochemistry* 27: 2024
43. Morrow, M. R., Davis, J. H., Sharom, F. J., Lamb, M. P. 1986. *Biochim. Biophys. Acta* 858: 13
44. Morrow, M. R., Huschilt, J. C., Davis, J. H. 1985. *Biochemistry* 24: 5396
45. Morrow, M. R., Whitehead, J. P. 1988. *Biochim. Biophys. Acta* 941: 271
46. Mouritsen, O. G. 1990. In *Molecular Description of Biological Membrane Components by Computer Aided Conformational Analysis*, ed. R. Brasseur, 1: 3. Boca Raton, FL: CRC Press. 336 pp.
47. Mouritsen, O. G. 1991. *Chem. Phys. Lipids* 57: 179
48. Mouritsen, O. G., Bloom, M. 1984. *Biophys. J.* 46: 141
49. Mouritsen, O. G., Jørgensen, K. 1992. *BioEssays* 14: 129
50. Mouritsen, O. G., Sperotto, M. M. 1992.

In *Thermodynamics of Cell Surface Receptors*, ed. M. Jackson, pp. 127–81. Boca Raton, FL: CRC Press
51. Nagle, J. F. 1980. *Annu. Rev. Phys. Chem.* 31: 157
52. Nezil, F. A., Bloom, M. 1992. *Biophys. J.* 61: 1176
53. Owicki, J. C., McConnell, R. M. 1979. *Proc. Natl. Acad. Sci. USA* 76: 5750
54. Owicki, J. C., Springgate, M. W., McConnell, H. M. 1978. *Proc. Natl. Acad. Sci. USA* 75: 1616
55. Peschke, J., Riegler, J., Möhwald, H. 1987. *Eur. Biophys. J.* 14: 385
56. Pink, D. A., Chapman, D. 1979. *Proc. Natl. Acad. Sci. USA* 76: 1542
57. Pink, D. A., Green, T. J., Chapman, D. 1980. *Biochemistry* 19: 349
58. Pink, D. A., Hamboyan, H. 1990. *Eur. Biophys. J.* 18: 245
59. Rehorek, M., Dencher, N. A., Heyn, M. P. 1985. *Biochemistry* 24: 5980
60. Riegler, J., Möhwald, H. 1986. *Biophys. J.* 49: 1111
61. Sackmann, E. 1984. In *Biological Membranes*, ed. D. Chapman, 5: 105. New York: Academic
62. Sackmann, E., Sui, S.-f., Wirthensohn, K., Maksymiw, R., Uromow, T. 1987. In *Biomembranes and Receptor Mechanisms*, ed. E. Bertoli, D. Chapman, 7: 97. Padova: Fidia Res. Ser. Liviana Press
63. Sanderman, H. 1978. *Biochim. Biophys. Acta* 515: 209
64. Singer, S. J., Nicolson, G. L. 1972. *Science* 175: 720
65. Sperotto, M. M., Ipsen, J. H., Mouritsen, O. G. 1989. *Cell Biophys.* 14: 79
66. Sperotto, M. M., Mouritsen, O. G. 1988. *Eur. Biophys. J.* 16: 1
67. Sperotto, M. M., Mouritsen, O. G. 1991. *Biophys. J.* 59: 262
68. Sperotto, M. M., Mouritsen, O. G. 1991. *Eur. Biophys. J.* 19: 157
69. Sperotto, M. M., Mouritsen, O. G. 1992. *Biochim. Biophys. Acta*. Submitted
70. Sui, S.-f., Urumow, T., Sackmann, E. 1988. *Biochemistry* 27: 7463
71. Tanford, C. 1973. *The Hydrophobic Effect. Formation of Micelles and Biological Membranes*. New York: Wiley and Sons. 233 pp.
72. Tessier-Lavigne, M., Boothroyd, A., Zuckermann, M. J., Pink, D. A. 1982. *J. Chem. Phys.* 76: 4587
73. Vist, M. R., Davis, J. H. 1990. *Biochemistry* 29: 451
74. Watnick, P. I., Chan, S. I., Dea, P. 1990. *Biochemistry* 29: 6215
75. Watts, A., de Pont, J. J. H. H. M., eds. 1985. *Progress in Protein-Lipid Interactions*, Vol. 1. Amsterdam: Elsevier. 291 pp.

# FUNCTIONAL BASES FOR INTERPRETING AMINO ACID SEQUENCES OF VOLTAGE-DEPENDENT K$^+$ CHANNELS

*Arthur M. Brown*

Department of Molecular Physiology and Biophysics, Baylor College of Medicine, One Baylor Plaza, Houston, Texas 77030

KEY WORDS: voltage sensor, K$^+$ pore, gate, K$^+$:Rb$^+$ selectivity filter

CONTENTS

| | |
|---|---|
| PERSPECTIVES AND OVERVIEW | 174 |
| CLASSIFICATION OF VOLTAGE-DEPENDENT K$^+$ CHANNELS | 175 |
| FUNCTIONAL CLASSIFICATION OF VOLTAGE-DEPENDENT K$^+$ CURRENTS | 176 |
| STRUCTURAL CLASSIFICATION OF K$^+$ CHANNELS | 179 |
|     *Subfamilies Identified by Similarities of Core Sequences* | 179 |
|     *Hetero- and Homomultimeric Assemblies* | 179 |
|     *Slowpoke and Other Ligand-Gated K$^+$ Channels* | 181 |
| CHANNEL TOPOGRAPHY | 183 |
| INTERPRETING THE FUNCTION OF AMINO ACID SEQUENCES IN K$^+$ CHANNELS | 185 |
| AMINO ACID SEQUENCES INVOLVED IN INACTIVATION | 186 |
|     *N-Terminus Inactivation: the Ball-and-Chain Model* | 186 |
|     *C-Terminus Inactivation* | 187 |
|     *Pore or P-Type Inactivation* | 187 |
| AMINO ACIDS FORMING THE P REGION OF K$^+$ CHANNELS | 189 |
|     *Localization of the Pore* | 189 |
| THE VOLTAGE SENSOR | 192 |
| CONCLUDING REMARKS | 194 |

I have yet to see any problem, however complicated, which when you looked at it in the right way did not become still more complicated.

P. Anderson, 1978 (7)

## PERSPECTIVES AND OVERVIEW

Voltage-dependent $K^+$ channels are the most numerous and diverse members of a larger family of voltage-dependent ion channels that includes $Na^+$ and $Ca^{2+}$ channels. Voltage-dependent ion channels are quaternary proteins with a major polypeptide of 1500–2000 amino acids. This peptide senses changes in membrane potential, usually depolarizations, and responds by changing the flux of ions from 0 to about $10^7 \, s^{-1}$ (1, 50). The $K^+$-channel gene encodes a subunit of about 400–700 amino acids, and four of these subunits are thought to assemble as homo- or heterotetramers. For $Na^+$ and $Ca^{2+}$ channels, a single gene encodes four similar repeats, each of which corresponds to a single $K^+$-channel subunit. One or two additional peptides may be associated with $Na^+$ channels (21), as many as four may be associated with $Ca^{2+}$ channels (21), and an additional peptide may be associated with $K^+$ channels (96). Although the quaternary structure is clearly important to function, this review considers only the main peptide of $K^+$ channels.

Historically, functional studies in the virtual absence of structural data have provided a treasure of information on voltage-dependent activation and inactivation, gating currents, ionic permeation, and transition rates between closed, open, and inactivated states. The key elements responsible for these functions have been given names like voltage sensor, gate, pore, and selectivity filter and have been arranged in electromechanical models (50) of channel proteins (Figure 1). As we shall see, these metaphors have strongly influenced the interpretation of structurally based experiments.

In recent years, the emphasis has shifted from strictly functional experiments because many voltage-dependent ion channels have been cloned, their nucleotide sequences determined, and from them, the primary amino acid sequences deduced. The problem now is to find the location and composition of the key elements. This task is complicated by the fact that while the linear amino acid sequences are known, the macromolecular structures are not. At present, the preferred way of examining the problem is to use the combined approach of mutational analysis (59) and electrophysiological measurements (44) of the resultant currents expressed in heterologous systems. The results have certainly been complicated—hence the premise quoted above (7). Thus we can settle into the comfortable recognition that mutational analysis and electrophysiological measurement must be the correct way of examining the problem.

This review evaluates results obtained using the combined approach up to July, 1992. The main conclusion is that specific stretches of the linear amino acid sequence of $K^+$ channels determine key functional components such as the inactivation gate, voltage sensor, and pore.

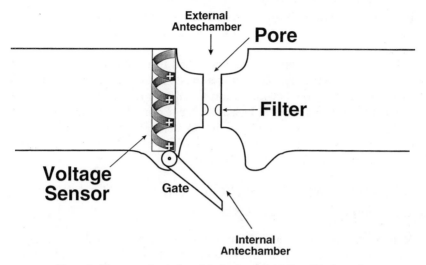

*Figure 1* Electromechanical model of voltage-dependent K⁺ channels.

## CLASSIFICATION OF VOLTAGE-DEPENDENT $K^+$ CHANNELS

The successes of cloning have not only produced a change in the experimental paradigm but have introduced into classification of voltage-dependent ion channels the entirely new consideration of structure. Because ion channels are ubiquitous and of central importance to cellular function, a wide variety of channel genes have evolved. Among $Na^+$ and $Ca^{2+}$ channels, the differences have been reflected mainly in tissue specificity; $Na^+$- and $Ca^{2+}$-channel genes are classified by the tissues, such as neurons and cardiac and skeletal muscle, in which they are primarily expressed.

Confusion has arisen in the classification of $K^+$ channels, however. The original Shaker gene from *Drosophila melanogaster* (14, 63, 93) gave rise to five alternatively spliced variants (103), four of which produced identifiable phenotypes (118) when expressed heterologously in oocytes of *Xenopus laevis*. Subsequently, homology screening was used to isolate clones from a variety of tissues including brain, muscle, and lymphocytes, and the different tissues often expressed the same clones. Names were attached that reflected the tissue source, and as a result, identical cDNAs often have different names. Complicating the situation more, significant differences in tissue distribution may occur between closely related genes (56). In addition, various species have been used, and the minor differences reflecting species rather than functional differences have received undue emphasis in the nomenclature. A structural classification based on sequence homology and chronology of isolation has been proposed (22) and will in general be used here.

Up to 1984, $K^+$ channels were classified from functional studies that used mainly electrophysiological methods. As the molecular structure is unknown, a functional classification remains practical because the functional features and models based on them are more certain than any present interpretation based on amino acid sequences. Furthermore, considerations of functional results have greatly influenced structurally based $K^+$-channel models (16, 32).

## FUNCTIONAL CLASSIFICATION OF VOLTAGE-DEPENDENT $K^+$ CURRENTS

Voltage-dependent $K^+$ currents may be grouped into three classes: delayed rectifiers (DR K) (52), transient currents (T K) (25, 43), and inward rectifiers (IR K) (33, 105). Numerous DR and T Ks have been cloned, but currently no IR Ks have been cloned. A fourth class requiring ligands as well as changes in membrane potential for activation is referred to as ligand-gated $K^+$ channels (L K). Channels in this class have the properties of DR, T, or IR Ks, but membrane potential alone is not sufficient for activation; the appropriate ligand, which may be an ion such as $Ca^{2+}$ (86) or $Na^+$ (64), a nucleotide such as ATP (90), or a G protein (75, 126), must also be present. In the L K class, only the $K^+_{Ca}$ channel has been cloned (3, 10) and expressed (3).

Delayed rectifiers were reported first in squid giant axon (52). The adjective *delayed* was used to describe the slow activation process relative to that of $Na^+$ currents. The characteristic outward rectification is caused by: (*a*) voltage- and time-dependent gating, (*b*) transmembrane ion gradients, and (*c*) asymmetrical ion transport through the open pore. The latter two properties give rise to time-independent rectification as measured in single-channel current-voltage (i-v) relationships. In whole-cell I-V curves, rectification results mainly from the voltage dependence of activation. The single-channel i-v curves under physiological conditions also rectify outwardly because of the outwardly directed concentration gradient for $K^+$. However, even in symmetrical isotonic $K^+$ solutions, the single channel i-v values may show inward or outward rectification, suggesting an underlying asymmetry in the structure of the open pore (124).

Unlike $Na^+$ currents, which are transient because of rapid inactivation, DR K currents remain steady during depolarization (52). Subsequently, a class of $K^+$ currents with the transient properties of $Na^+$ currents was identified (25, 43). These T K currents are sometimes referred to as A currents (25). Both T Ks and DR Ks are activated from closed states at the resting potential to an open state at depolarized potentials, but T Ks move quickly from the open state to a closed, inactivated state during maintained depolarizations. For T Ks, like DR Ks, activation is time and depolarization dependent, with the latter giving an outward rectification to peak current-voltage (I-V) curves recorded from whole cells. This rectification is exaggerated by the time-independent net flux of $K^+$ ions from their much higher internal concentration through the open pores.

The $K^+$ channels responsible for the Shaker phenotype in *D. melanogaster* are prototypical T $K^+$ channels. Like $Na^+$ currents (4), inactivation was interpreted as being coupled to activation (129). The inactivated state is absorbing, and for macroscopic currents, the apparent voltage dependence of inactivation is thought to originate from the voltage dependence of activation. At the single-channel level, mean open time does not reflect the voltage dependence of macroscopic inactivation and in fact may actually increase at depolarized potentials where macroscopic $K^+$ currents are inactivating most quickly.

IR Ks pass current more efficiently in the inward rather than the outward direction; hence, their i-v curves show inward rectification. Following the convention of high occupancy of closed states at the resting potential and high occupancy of open and inactivated states at depolarized potentials, IR Ks may be thought of as inactivated at resting potentials, the inactivation being removed by hyperpolarization. In response to hyperpolarizing test pulses, the removal of inactivation causes a rapid, time-dependent increase in current, and upon return to the holding potential, deactivation causes a time-dependent decrease in the tail currents. The current flowing through the open pore has, under physiological conditions, the opposite rectification from that expected from the asymmetry of $K^+$ concentrations. Inward rectification is attributed to block by intracellular $Mg^{2+}$ (82, 120) of a pore containing multiple ion-binding sites (51). Inward rectification occurs in whole-cell currents even in the absence of $Mg^{2+}$ as a result of depolarization-induced inactivation (62, 82). Another interesting property of IR Ks is that they are regulated by external $K^+$ ion concentration; the higher the external $K^+$, the more depolarized the potential at which removal of inactivation occurs (24). This behavior may have relevance for recent observations on the effects of external $K^+$ on expressed DR K currents (28, 95). The single-channel conductance is determined by external $K^+$ such that conductance increases with a square-root dependence on the concentration of external $K^+$.

The best-studied example of an L K is the $Ca^{2+}$-activated ($K_{Ca}$) channel, in particular the large conductance or maxi-$K_{Ca}$ channel (72, 92). For this channel, the time-independent rectification is similar to that of DR Ks. Activation is time dependent and is determined by the levels of both $Ca^{2+}$ and membrane potential. At physiological potentials, activation is absent at $Ca_i$s of $10^{-8}$ M or less (13), and at physiological $Ca_i$s, membrane potential alone is not sufficient to activate the channels. For threshold levels of activation, both $Ca^{2+}$ and depolarization are necessary, but either may be sufficient above threshold. Maxi-$K_{Ca}$ channels are usually blocked by the scorpion toxin Charybdotoxin (CTX) (80), while another toxin from bee venom, Apamin, blocks a different, smaller conductance $K_{Ca}$ channel (101).

The requirement of both ligand and membrane potential for threshold activation appears to be similar in other types of L Ks. ATP-inhibited $K^+$ ($K_{ATP}$) channels (90) are regulated by levels of ATP and membrane potential. However, the I-V relationships resemble those of IR Ks rather than DR Ks. The time-independent inward rectification results from $Mg^{2+}$ blockade, but the blockade occurs at much higher concentrations of $Mg^{2+}$ because of a faster dissociation rate. $Na^+$ and $Ca^{2+}$ ions also may block this channel. Depolarization-induced inactivation is minor when present.

G protein–activated $K^+$ channels such as the muscarinic cholinergic $K^+$ channel in atrium (18) and brain (122) are members of the same class of IR Ks. For the heart channel, not only is the time-independent rectification similar to that of IR Ks but the time-dependent rectification also shows depolarization-induced inactivation (53). The G protein–gated atrial $K^+$ channel may be spontaneously active in the absence of agonist; nevertheless basal activity is determined by G-protein activation of the channel (91).

Another muscarinically activated $K^+$ current is the M current in neurons. The M current is activated at resting potentials, has the properties of a DR K, and is inactivated after acetylcholine occupies its associated muscarinic receptor (2). The coupling between receptor and M channel is complex, involving cytoplasmic mediators and possibly phosphorylation. The S channel in *Aplysia* neurons (106) has analogous properties. A serotonin receptor appears to be coupled via A kinase to a $K^+$ channel that is inhibited by phosphorylation.

Abundant evidence indicates that phosphorylation modulates the activity of DR Ks, possibly from the negative charge that is added to the channel protein (11). The DR K in heart is regulated both by phosphorylation and G proteins (37), but this channel is thought to be the slow $K^+$, or $I_{SK}$, channel originally cloned from kidney (114) and subsequently isolated from heart (34). The primary sequence of $I_{SK}$ is completely different from the class of voltage-dependent $K^+$ channels under consideration in this chapter.

# STRUCTURAL CLASSIFICATION OF K$^+$ CHANNELS

## Subfamilies Identified by Similarities of Core Sequences

The standard for classification is the one used for *Drosophila* K$^+$ channels. The K$^+$-channel gene at the Shaker locus is Kv1.0. Subsequently, three other closely related genes have been identified in *Drosophila* using homology screening with *Shak*-derived oligonucleotide probes. These were named *Shab*, *Shaw*, and *Shal* (20, 125) and in the present classification are *Kv2*, *Kv3*, and *Kv4*, respectively. Mammalian homologs of *Shak*, *Shab*, *Shaw*, and *Shal* have been isolated and expressed (see below), and the four classes may be considered as subfamilies of the larger family of voltage-dependent K$^+$-channel genes.

The proteins encoded by these genes have three domains, N and C termini that show great variety and are mainly hydrophilic, and a conserved hydrophobic core (Figure 2). The core is about 70% similar for members of each subfamily and about 40% similar among subfamilies. Shal channels, like Shak channels, express currents that inactivate with a relatively rapid rate (125); Shab channels express inactivating currents that, by comparison with Shak and Shal, are much slower. Shaw channels (85) in comparison to Shab channels express currents with even slower inactivation with one exception (97). Shaw channels are activated at much more depolarized potentials than any of its relatives (20, 125).

Homology screening with *Shak*-based probes has revealed homologs of each of these four genes in rat (59, 60), mouse (116), and human (115). An exception to isolation by homology screening was the isolation from rat brain of a Shab-like channel Kv2.1 by expression cloning in *Xenopus* oocytes (36). Unlike *Drosophila Shak*, some mammalian genes are intronless (116). As in *Drosophila*, alternative splicing, exemplified by the rat *Shaw* gene in mammals, has added to the variety of expressed K$^+$ channels (78). A puzzle that has arisen is that only one of the mammalian homologs of *Shak*, *Kv1.4*, had the inactivation properties expected of Shaker-like T K$^+$ currents (110). Moreover, the inactivation rate was too slow for that shown by most native T K channels. A solution may be provided by considering another possible mechanism by which diversity in K$^+$ channels may arise.

## Hetero- and Homomultimeric Assemblies

Diversity may be introduced not only at the DNA and RNA levels, but also at the subunit-assembly level. Co-injection of two cRNAs of the *Kv1* subfamily, each of which expressed very different gating and pharmacological (98) properties, led to the expression of single channels with mixed properties that is expected from expression of both cRNAs (98). The

## Hydropathy Profiles and Topography of Voltage-Dependent K⁺ Channels

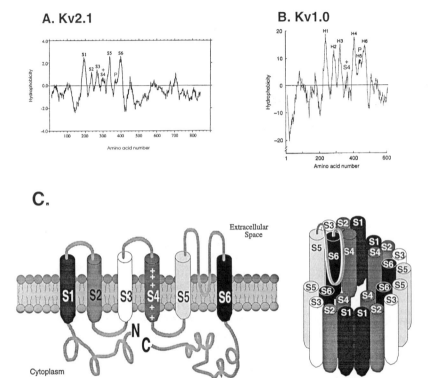

*Figure 2* Hydropathy profiles and topography of voltage-dependent $K^+$ channels. (*A*, *B*) Kyte-Doolittle plots. The transmembrane segments are numbered. P identifies the pore region. (*C*) The two-dimensional topology for a homotetrameric $K^+$ channel subunit (*left*) and subunit assembly into a tetramer (*right*). In the assembled channel, an outer ring consists of S1–S3 and S5, a middle ring consists of S4 and S6, and an inner ring consists of the pore shown as a loop between S5 and S6 in one subunit. The positively charged S4 is thought to be the voltage sensor, and most of the pore is formed by the loop between S5 and S6. *A* is reprinted with permission from Ref. 117 (Copyright 1987 by the AAAS). *B* is reprinted with permission from *Nature* (36; Copyright 1989 Macmillan Magazines Ltd.).

interpretation was that heteromultimeric channels had formed. Whole-cell currents expressed by co-injection of cRNAs with different sensitivities toward blockade by external tetraethylammonium (TEA) were interpreted similarly (66). Even more convincing were experiments in which dimer tandems of two different cDNAs, again from the *Kv1* subfamily, expressed single-channel currents with properties expected from heteromultimeric channels (57). Heteromultimer formation is not without constraints; heteromultimers did not form following co-injection of cRNAs from different $K^+$-channel subfamilies (26).

These results have two important implications: the structure of $K^+$ channels expressed in native tissues for the moment remains unknown, and subunits may contribute in a predictable way to the function of the assembled heteromultimer. Inferences about native $K^+$ channels are limited by the possibility that they may be heteromultimers. This limitation does not apply to native $Na^+$ (89) and $Ca^{2+}$ channels (87) in which a single type of cDNA has been shown to encode the functional protein (87, 89). The composition of native $K^+$ channels awaits definitive characterization, which will require a combination of biochemical and structural measurements.

The possibility that different subunits may contribute in a predictable manner to the heteromultimeric phenotype is consistent with the interpretation that the $K_d$ for external blockade by the symmetrical organic cation blocker TEA was the geometric mean of the $K_d$s of each subunit in a homotetrameric channel. A point mutation in which Y at position 449 in Shak was substituted for T (Figure 3) greatly enhanced blockade and abolished its weak voltage dependence, leading to the suggestion that TEA was coordinated by the $\pi$ electrons of a ring of aromatic Ys (47, 65). On the other hand, for the asymmetrical channel blocker CTX, the $K_d$ was the arithmetic mean of the $K_d$s of each subunit of the homotetrameric channel (79).

Substitution of N for D at position 431 in Shak B channels (Figure 3) greatly reduced CTX blockade for homomultimeric channels but had much less effect on heteromultimeric channels, even when the concentration of mutated cRNA relative to wild-type cRNA was low. This observation was exploited to determine the number of subunits in a $K^+$ channel (79). Along with the assumptions that wild-type and mutant channels expressed equally and assembled randomly, binomial statistics were used to argue for a tetrameric $K^+$ channel. This interpretation is supported by the expectation that voltage-dependent $K^+$ channels, just like their relations, voltage-dependent $Na^+$ and $Ca^{2+}$ channels, should have a tetrameric structure.

## *Slowpoke and Other Ligand-Gated $K^+$ Channels*

Recently, the $K_{Ca}$ gene responsible for the Slowpoke phenotype in *Drosophila* (3, 10) was cloned. Alternative splicing of this gene could produce numerous phenotypes (3). The conserved core, especially the pore, or P region (107), is homologous with other $K^+$ channels except that part of

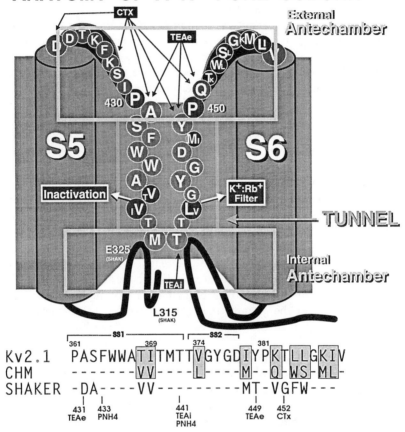

*Figure 3* Anatomy of a $K^+$ pore subunit. Antechambers and tunnel or pore formed by the S5–S6 loop are shown. The numbering scheme at the bottom is for Shak, but some of the numbering for Kv2.1 and the Kv2.1–Kv3.1 chimeric channel, CHM, is shown above the Kv2.1 alignment. The prolines are structure breakers and delimit the $\beta$-hairpin lining the pore. Residues are indicated that are involved in external CTX and TEA blockade, internal TEA blockade, P-type inactivation, a $K^+$:$Rb^+$ selectivity filter, and changes in $NH_4^+$ permeability. The inner part of the pore is formed in part by the S4–S5 loop and the C-terminus end of S6, both of which are shown as black loops. Amino acid residues are labeled using the single letter code: S, serine; T, threonine; P, proline; A, alanine; G, glycine; M, methionine; I, isoleucine; L, leucine; V, valine; D, aspartate; E, glutamate; N, asparagine; Q, glutamine; K, lysine; R, arginine; H, histidine; F, phenylalanine; Y, tyrosine; W, tryptophan; C, cysteine.

S6 is absent. The phenotype that was expressed in *Xenopus* oocytes lacked the $Ca^{2+}$ sensitivity of vertebrate $K_{Ca}$ channels; possibly one of the spliced variants may have this property or it may appear in a different heterologous-expression system. However, a completely different gene may express the vertebrate phenotype (123).

With the cloning of Slowpoke, the first of the L Ks was identified. A possible lead on L Ks with IR K properties, as well as the voltage-gated IR K, has come from experiments using complementation of a yeast null mutant for $K^+$ flux with *Arabidopsis*-derived cDNA libraries. Two cDNAs have been isolated. These had stretches resembling the P and S4 regions of $K^+$ channels (6, 104). In yeast, the resting potential is about $-140$ mV so that the gene product may be active at this potential. This would be consistent with the behavior of a voltage-dependent IR K. In addition, the complementation was abolished by TEA and barium ion, both of which block IR Ks.

An unrelated gene, $I_{SK}$, apparently produces a phenotype similar to DR Ks (114). This gene encoding 129 amino acids has been isolated from human kidney, heart, and myometrium (34, 114), and heterologous expression produced a slowly activating current remarkably similar to the DR K of cardiac ventricular myocytes (100). Hydropathy plots yield a single transmembrane-spanning segment that when mutagenized changes unitary conductance (39, 46). This argues against the possibility initially raised that the gene product was simply a modulator of DR Ks. Furthermore, not only the kinetics and voltage dependence of activation but the pharmacology of the phenotype expressed in oocytes is that of the cardiac DR K.

## CHANNEL TOPOGRAPHY

The deduced amino acid sequences have been interpreted using hydropathy plots and secondary-structure predictions. The hydropathy plots are quite characteristic (Figure 2) and show six transmembrane segments S1–S6, which from considerations of secondary structure, are thought to be α-helical. A seventh peak arising from the segment between S5 and S6 is usually present, and different considerations (see below) suggest that this may be a β-hairpin (45, 127, 128). From the two approaches, a common topographical model has been derived for the superfamily of voltage-dependent ion channels (Figure 2). A difference between $K^+$ channels on the one hand and $Na^+$ and $Ca^{2+}$ channels on the other is that homotetrameric $K^+$ channels have rotational symmetry around a central pore whereas $Na^+$ and $Ca^{2+}$ channels are rotationally pseudosymmetrical.

The hydrophobic central core region is the most conserved structure among K$^+$ channels. Experiments showed that large deletions could be made simultaneously in the N and C termini without altering gating or conduction (30, 121), although deletions at either terminus alone altered gating (121). The results (121) indicate that the N and C termini are on the same side of the membrane and may interact with each other and the central core. Extending the deletions into the transmembrane segments produced nonfunctional channels (121).

Within the core, the most strikingly conserved sequences are S4 and the SS1–SS2 segments (Figure 3) in the loop connecting S5 and S6. S4 has an Arg or Lys–X-X–Arg or Lys repeat where -X-X- is usually a hydrophobic pair, and the repeat may occur from four to eight times per helix. Based on its charge and its high degree of conservation, S4 is thought to be the voltage sensor. Even more conserved in K$^+$ channels are the SS1–SS2 segments. Point mutations in SS1–SS2 (81, 127, 128) and mutagenesis of a stretch including SS1–SS2 (45) have shown that 20 amino acids in this region span the membrane twice to form roughly the outer three-fourths of the pore. A structure satisfying this requirement would be an antiparallel pair of $\beta$-strands ($\beta$-hairpin) with 10 residues per strand (16, 45, 127, 128).

A detailed structural model of K$^+$ channels has been formulated using the topographical structures shown in Figure 2 and the functional results obtained from mutagenesis experiments (32). The model has three concentric rings. An outer ring consists of 16 $\alpha$-helices formed by S1–S3 and S5. A middle ring consists of eight $\alpha$-helices formed by S4 and S6. An inner ring forms the pore and consists in its outer part of a $\beta$-hairpin derived from the SS1–SS2 stretch (Figure 3) and in its inner part of an $\alpha$-helix derived from the S4–S5 loop. Part of the S4–S5 loop has a Leu heptad (83), which may interact with conserved residues in S5.

The outer ring was established by placing nonconserved hydrophobic residues in S1–S3 and S5 in contact with the membrane lipids. Salt bridges between ion pairs determined the deactivated position of the S4 voltage sensor, and an important disulfide bridge was postulated between S6 and S2. A propagating helical-screw model (41) produced an outward movement of S4 upon depolarization. To account for the estimated two electronic charges per subunit moving through the membrane field (129), the movement must have been greater than 10 Å (32). New ion pairs were formed, and the S4–S5 $\alpha$-helix moved outward, revealing an E to enhance outward K$^+$ flow. The narrowest part of the pore was formed by 449Y in Shak (Figure 3), and it was suggested that K$^+$ ions may be coordinated by $\pi$ electrons from the ring of aromatic residues at this position.

A more restricted model involving only the P region has been proposed (16, 17). This has a long $\beta$-barrel composed of SS1 and SS2. The preferred orientation is right-handed with a counter-clockwise rotation. The short $\beta$-barrel was selected (32) because it allowed the P residues of Slowpoke to be aligned in a manner that was conserved among $K^+$ channels. The long $\beta$-barrel included F433, which seemed to regulate $NH_4^+$ and $Rb^+$ permeability (128), a pore property. Both models clearly have heuristic value and deserve to be tested.

## INTERPRETING THE FUNCTION OF AMINO ACID SEQUENCES IN $K^+$ CHANNELS

The strategy of these experiments is simple; the interpretation is not. Guesses based upon the models that have been discussed are made as to which sequences are relevant to a particular function; mutations are introduced into the relevant cDNA; and the transcribed cRNAs are expressed in heterologous cells. The principle methods are site-directed mutagenesis, in which one or more point mutations are introduced into the channel protein, or large-scale mutagenesis, in which peptide fragments are deleted, added, or exchanged. The resultant currents are measured from either the whole cell, macropatches containing many channels, or single channels. Transient current expression in *Xenopus* oocytes following injection of cRNA is by far the most widely used method. Stable (27, 71) or transient (23) transfection of cell lines is becoming more popular, and results to date suggest that the essential biophysical characteristics of $K^+$ channels are maintained despite marked differences among the various heterologous cells that have been used (27).

For large proteins such as $K^+$ channels, this approach might at first glance be dismissed because the macromolecular structure is unknown. Any results are certain to be hoisted on the petard of remote effects. However, a modular design for these proteins appears to save the day. As we shall see, one can delete stretches of amino acids that greatly influence inactivation (54, 130) or exchange stretches of amino acids that completely alter the pore phenotype (45) with little effect on the voltage sensor (112). On the other hand, point mutations can be made (28, 55) that produce circumscribed effects limited to changes in inactivation, TEA blockade of the pore (81, 127), and ionic permeation (128). This is not to say that remote effects do not occur nor that functions are precisely circumscribed by specific sequences, but a correlation between different functional requirements and different linear sequences of $K^+$ channels clearly exists.

## AMINO ACID SEQUENCES INVOLVED IN INACTIVATION

Three regions of the amino acid sequence of $K^+$ channels are important for different types of inactivation. These are the N terminus, the sixth transmembrane segment S6, and the P region.

### N-Terminus Inactivation: the Ball-and-Chain Model

A comparison of the inactivation properties expressed by alternatively spliced Shaker cRNAs led to the conclusion that differences in the N or C termini were responsible for differences in rates of inactivation (118). Kinetic studies showed that the mechanism of Shaker inactivation (129) resembled that of $Na^+$-channel inactivation (4, 67) in being intrinsically voltage independent (4, 67). Extending the comparison, trypsin, which removed inactivation of $Na^+$ channels (9), similarly affected Shak B channels (54). Attention was then focused on the N terminus because of its many potential trypsin cleavage sites.

An N-terminus stretch of 19 residues was shown to be necessary for fast inactivation, and a more distal stretch of residues modified the rate at which inactivation occurred (54). In the context of the ball-and-chain metaphor used for inactivation of $Na^+$ channels (9), a similar metaphor was adopted for $K^+$ channels (54, 130).

Subsequently, a ball peptide was synthesized and reproduced fast inactivation in Shak channels from which the N-terminus ball region had been previously deleted. Measurements of on and off rates indicate that positive charge steers the ball to its receptor, whereas hydrophobic residues determine its release (88). When the ball contains cysteines as it does in some fast-inactivating mammalian homologs of Shak, oxidation of the cysteines may abolish fast inactivation (99). The ball peptide produced inactivation in related rat Shak channels that otherwise showed little or no inactivation. The peptide also produced inactivation in the slowly inactivating Shab channel Kv2.1 (58) and $K_{Ca}$ channels of *Necturus* smooth muscle (31), rat brain or skeletal muscle (35), and porcine coronary artery (119). The results can be interpreted in two ways: either the channels have receptors for the inactivation ball or the peptide has nonspecific, probably hydrophobic, effects.

These elegant experiments, along with experiments on an exchange of the P region (45), demonstrate that consecutive stretches of amino acids in an ion channel may form complete functional domains. Unfortunately, the structure of the synthetic ball peptide could not be determined with NMR spectroscopy (73), making literal testing of the ball metaphor impossible.

That the ball plugs the inner mouth of the pore seems likely because increases in *trans* $K^+$ concentration increased the rate of recovery from inactivation in a voltage-dependent manner (29). The results suggested interaction at the inner mouth of the pore between the ball peptide and $K^+$ ions flowing inward under the influence of the membrane field. The ball seemed to plug the pore in the open state because recovery from inactivation following removal of the ball correlated strongly with reopenings. The ball peptide also appeared to interact with portions of the S4–S5 loop (58). In one model (32), the ball interacts with the inner antechamber formed by the S4–S5 α-helix and the C terminus of S6.

## C-Terminus Inactivation

Inactivation was different in alternatively spliced Shaker channels that shared common N termini but differed in their C termini (118). Shak A and B channels with the same ball-removing N-terminal deletions differed markedly in their inactivation rates (55). These differences could be attributed to an Ala-Val difference at position 463, which in the topographical map for Shak is at the N-terminus end of S6. The V at 463 in Shak A allows rapid inactivation to persist even after deletion of the N-terminus ball.

TEA blockade distinguished C and N types of inactivation (23). Internal but not external TEA blocked N-type inactivation, an effect that was explained as a simple competition between TEA and the inactivating ball for a shared internal binding site. By contrast, external but not internal TEA blocked C-type inactivation. Competition for a common binding site was again suggested, but in this case a foot-in-the-door model was used as an explanation. In this model, the C-type inactivation gate cannot close in the presence of external TEA, and the TEA receptor is barely located in the pore because external TEA produces blockade with little or no voltage dependence.

## Pore or P-Type Inactivation

The A463V difference in S6 responsible for C-type inactivation in Shak was thought to be near the external mouth of the pore (55). In lymphocytes, an H at a position corresponding to 449 in Shak (Figure 3) forms part of the external TEA receptor, and when protonated produced large changes in inactivation (19). Thus, residues near the pore appeared to contribute to inactivation of the nonball type. A study involving point reversions of a chimeric $K^+$ channel showed that a Val → Ile reversion at a position in the tunnel or deep part of the pore (Figure 3) produced a novel phenotype with the unique property of rapid inactivation (68). Unitary conductance was unchanged, however. Subsequently, a Val → Ser substitution at the same position produced more complete inactivation and a marked reduction in conductance, confirming that this position was in the pore (28).

P-type inactivation differed from C-type inactivation because external TEA rather than slowing inactivation (23) increased channel availability (28). Only at higher concentrations did external TEA actually produce blockade (28). It was unlikely that the pore mutations introduced a receptor for N-terminus inactivation, because internal TEA scaled the currents downward but did not slow inactivation (28). The effects of external TEA were interpreted by a model in which a transition between closed and inactivated states was impeded, although blockade of open channels still occurred (28).

Despite their differences, both P and C types of inactivation involve residues in or near the pores, and therefore, they may share a similar mechanism. They should be distinguished from the N terminus ball (54, 130) and from deletions of the C terminus that in Kv2.1 increased the rate of inactivation (121). Thus, the distinction appears to lie between inactivation mechanisms involving the nonconserved, hydrophobic core; a more suitable nomenclature might be core and noncore inactivation.

The enhanced currents produced by external TEA may be compared with the observation that extracellular $K^+$ increased availability of Kv1 channels (95). The $K_o$-induced increase in availability was blocked by a Thr → Lys substitution that also abolished block by external TEA (95). However, the changes in sensitivity to external TEA and $K_o$ were not accompanied by changes in the rate of inactivation, single-channel conductance, mean open time, or gating-charge displacement. Apparently, $K_o$ modulated the coupling between gating charge and channel opening.

On the other hand, in inactivation produced by substitutions of pore residues, external $K^+$ slowed inactivation and increased recovery from inactivation. These results indicate a highly complex situation in which TEA and $K^+$ receptors near the external mouth of the pore couple to an inactivation process within the pore. TEA apparently produces blockade at an external site (70, 113) and has an enhancing effect on $K^+$ currents (28) at another site, possibly the same site as the one responsible for the enhancing effect of $K_o$ (95). In addition, there may be a site at which increased $K_o$ slows inactivation (28).

# AMINO ACIDS FORMING THE P REGION OF $K^+$ CHANNELS

Several permeant and impermeant ions and a number of organic and peptide toxins have been used in electrophysiological studies to probe $K^+$ pores. The results have been systematized using an electromechanical model in which the pore has inner and outer antechambers connected by a narrower tunnel (Figure 1). The inner antechamber is covered by an activation gate that is coupled to the voltage sensor.

The antechambers are sites at which blockade may be produced by impermeant ions and toxins too large to enter the tunnel. Within the tunnel, blockade may be voltage dependent, and one may determine the electrical distance of the blocking site in the membrane field from the voltage-dependence. Blockade may be sufficiently slow for the blocking and unblocking rates to be determined. This type of blockade often interferes with normal channel gating. Faster blockade may produce noisy openings, or if blockade is sufficiently fast relative to the bandwidth of the recording system, may produce a reduction in amplitude.

The tunnel or deep pore allows ions to move in a single file. For $K_{Ca}$ channels, two to four ions may occupy the tunnel at one time (72). In addition to the sites they occupy, the ions have to pass through selectivity filters in the deep pore. A major goal is to identify the structures that give rise to the wells and barriers that $K^+$ ions encounter as they traverse the deep pore.

## *Localization of the Pore*

The first evidence that the S5–S6 loop was near the pore came from a study on $Na^+$ channels in which the point mutation E374Q in repeat I reduced the blockade produced by tetrodotoxin (TTX) (referenced in 48). The TTX receptor is thought to be located near the external mouth of the pore. Consistent with a reduction in charge at the site, a small reduction in inward $Na^+$ conductance was observed. In early topographical models, the S5–S6 loop of $Na^+$ channels was divided into two shorter α-helical segments SS1 (40) and SS2 (42), which were placed in the membrane. The S5–S6 loops are not similar between $Na^+$ and $K^+$ channels, but for $K^+$ channels, the hydropathy plots placed the highly conserved S5–S6 loop in the membrane, giving the initial topographical model seven identifiable transmembrane segments (117).

More pertinent observations relating the S5–S6 loop of $K^+$ channels to the P region were soon made. The scorpion peptide toxin CTX acted by blocking the external mouth of Shak channels, and cytoplasmic $K^+$ ions relieved the blockade in a voltage-dependent manner (80). Point mutations in the S5–S6 loop that neutralized negative charge reduced blockade by CTX. Another blocker of Shak channels at an external site was TEA, and point mutations in the S5–S6 loop that neutralized negative charge reduced TEA affinity (81). In fact, two neutralizing mutations separated by 18 residues reduced blockade by both TEA and CTX.

The extent to which the S5–S6 loop defined the pore was more firmly established in three papers that appeared simultaneously. In one (127), internal TEA blockade of Shak was reduced 10-fold by a T → S substitution at a position midway between two of the more widely separated residues that affected external TEA blockade. The internal site was about 20% in from the cytoplasmic surface in terms of electrical distance so that 80% of the membrane field fell across a stretch of nine amino acids (Figure 3).

Two other point mutations in the S5–S6 loop of Shak affected pore properties. A T → S mutation at the same site that reduced internal blockade by TEA (127) and a nearby F → S mutation (Figure 3) increased $Rb^+$ and $NH_4$ permeability and conductance (128) while leaving $K^+$ permeability and conductance unchanged.

Another study did not use point mutations but used large-scale mutagenesis instead to show that a stretch of 21 amino acids defined $K^+$ conductance and TEA blockade of the pore (45). Two related $K^+$ channels, Kv2.1 and Kv3.1, that have different pore phenotypes were identified. Kv3.1 had a $K^+$ conductance that was three times greater than Kv2.1 and was blocked by external TEA, whereas Kv2.1 was blocked by internal TEA. Silent, unique restriction sites were engineered into Kv2.1, and 21 of 837 amino acids were exchanged with a corresponding stretch that had been synthesized using Kv3.1 as a template in a polymerase chain reaction. The chimeric channel CHM thus formed had the pore phenotype of the donor Kv3.1 channel, establishing that this stretch of amino acids embodied a significant extent of the P region. When Kv2.1, Kv3.1, and Shak are aligned (Figure 3), the stretch including the sites for TEA blockade and $NH_4^+$ permeability in Shak overlaps the stretch including both TEA sites and the region responsible for $K^+$ conductance in Kv2.1 and Kv3.1. For Kv3.1, the external TEA site is located at the outer end of the pore (45, 70, 113), and for Kv2.1, the internal site is located about 75% of the way through the pore (45, 70, 113). Therefore, the chimeric exchange defined approximately the outer three-fourths of the pore. The inner one-fourth may come from the S4–S5 loop and the C-terminus end of S6 (32).

The simplest interpretation of these three sets of results was for the K pore to span three-fourths of the membrane twice within about 20 residues as a $\beta$-hairpin (45, 127, 128). The pore would then be a $\beta$-barrel formed by four rotationally symmetrical $\beta$-hairpins. The model is far from definitive, and as noted two structural models of the pore have been proposed that differ in the lengths of their $\beta$-barrels (16, 32).

The swap of pore phenotypes in the chimeric channel was accompanied by only small changes in activation, suggesting a modular arrangement for the pore and voltage sensor. Such an arrangement was already proposed in the electromechanical models of ion pores (Figure 1). Considering the success of the ball-and-chain metaphor for N-terminus inactivation, a sense of deja vu surrounds the predictive powers of the electromechanical models. For the chimeric exchange, extensive differences in pore structure and function might be associated with few differences in structure or function of the voltage sensor. This was tested by comparing gating currents between Kv2.1 and CHM L374V, which differed from Kv2.1 at eight positions in the pore but was identical at S4, the presumed voltage sensor. Despite large differences in $K^+$ conductance and TEA blockade, the gating currents did not differ (112).

Kirsch et al (69) further studied the $K^+$ pore with mutational analysis, taking advantage of the phenotypic and genotypic differences between CHM and Kv2.1. Point reversions in CHM outside the deep pore, and downstream from P381 (Figure 3), either, in the case of the conservative reversions, resulted in no effect, or, in the case of a change in side-chain charge, produced effects consistent with the change in charge and a change in the local concentration of $K^+$. Thus, Q382K reduced inward and outward current; the reduction in inward current was greater, resulting in a decrease in outward rectification of the single-channel current. Similar though less marked results were obtained for M387K.

Within the deep pore or tunnel were four differences between CHM and Kv2.1, three of which were conservative (Figure 3). Surprisingly, conservative reversions at 369 and 374 introduced novel phenotypes. In both cases, open times were shortened almost 10-fold, but for V369I, the shortening resulted from stabilization of an inactivated state that was apparent in the whole-cell currents producing the P-type inactivation discussed earlier (28). For L374V, the reduction in open time was caused by rapid departure from the open state to a proximal closed state, and the currents remained steady rather than inactivating during standard test pulses. Channel availability was unaltered by increased $K_o$, and single-channel $K^+$ conductance was greatly reduced. Combined reversions at 369 and 374 restored most of the host-pore phenotype. None of the other five possible double reversions had this effect, indicating that positions 369 and 374 interacted with each other. The likelihood of interaction among separate subunits was confirmed when it was established that co-injection of V369I and L374V cRNAs produced channels that had the properties of the double reversion (G. E. Kirsch, J. A. Drewe, M. DeBiasi, H. A.

Hartmann & A. M. Brown, submitted). This result has the important implication that different subunits may not simply contribute their recognizable properties to heterotetrameric channel assemblies but that preexisting differences in pore properties may be cancelled by suitable interactions between different subunits.

The substitution L374V in CHM and Kv2.1 switched $K^+:Rb^+$ selectivity (69). To analyze this selectivity filter, extensive substitutions were introduced at this position. The results showed that hydrophobic residues favored $Rb^+$ whereas polar residues favored $K^+$ (111). Internal TEA blockade was enhanced by hydrophobic substitutions (111) consistent with a hydrophobic component in the interaction between TEA and the pore (8, 38). Interactions between preferred ions and TEA blockade, as well as anomalous effects of L at this position, led to an interpretation in which both the side chains at 374 and the peptide backbone were involved in selectivity and internal TEA blockade.

More support for localizing the pore to the S5–S6 loop has come from recent experiments on voltage-dependent $Na^+$ channels (48). Alignment of this region for the four repeats of $Na^+$ and $Ca^{2+}$ channels identified a $K \rightarrow E$ difference in repeat III and an $A \rightarrow E$ difference in repeat IV. When these substitutions were introduced into $Na^+$ channels, the currents expressed in *Xenopus* oocytes became nonselective among cations in the absence of $Ca^{2+}$, were blocked by micromolar concentrations of $Ca^{2+}$, and at higher concentrations conducted $Ca^{2+}$ ions. A $Na^+$-selective pore had been converted to a pore that was similar to a $Ca^{2+}$-selective pore. However, neither the gating nor the pharmacology reflected a $Ca^{2+}$-channel phenotype. Once again, localized changes in the S5–S6 loop could change pore function without affecting other functions of the protein.

## THE VOLTAGE SENSOR

The positive charge and conservation of S4 have led to the idea that S4 is the voltage sensor. However, cGMP- and cAMP-gated ion channels that are not voltage sensitive have regions homologous to S4. If S4 is the voltage sensor, then neutralization or reversal of positive charge should change voltage sensitivity; in general this is the case. The first tests of this idea were made on $Na^+$ channels. Changes in charge of S4 in repeat I produced changes in the slope of the opening probability-membrane potential, or Po-V relationship. An inverse relationship was observed between the effective valence, which determined the slope, and the charge removed (109). Substitutions at different Rs or Ks also produced shifts in the Po-V curves that were not easily reconciled with the changes in voltage sensitivity. The S4 in repeat I was not dominant, since similar results were reported for substitutions of the S4 in repeat II. Therefore, changes in charge of a single S4 also affected the work required to open the channel by the voltage sensors of all four subunits. This result implies cooperativity among subunits. It also shows that the shifts in gating plus the differences in charge among the S4s of $Na^+$ channels greatly complicate quantitative interpretations of the effects of changes in charge. The homomultimeric nature of $K^+$ channels made them more attractive for studies on gating, but as we shall see, the complications of shifts in gating again clouded interpretation.

For $K^+$ channels, the lower limit for the gating charge estimated from the Po-V curves may be considerably less than the charge that is actually measured (102). This raises two possibilities: gating-current measurements are contaminated by charge movement unrelated to gating, and gating currents include movements of the voltage sensor that are not directly translated into channel opening. The burden of correlating the changes in charge of the sensor with functional effects falls upon the Po-V curve. But since Po-Vs arise from multiple kinetic states, the Po-V curve must be interpreted by multiple Boltzmann relationships. At very hyperpolarizing potentials, the complex Boltzmann equations become greatly simplified and the limiting slope becomes a measure of an equivalent gating charge that is a lower limit of the actual charges that may move (5). However, such measurements are difficult to make because they require a range of Po values between 0.001 and 0.100.

In work on Shak, little correlation of effective valence with changes in charge in S4 was observed (94). However, an inactivating channel was used, and this limited measurements of effective valence. When inactivation was removed, the correlation was better (74, 76). Nevertheless, the same problems present for $Na^+$ channels persisted; namely, the voltage shifts are difficult to explain and the correlation between amount of charge changed and the change in voltage sensitivity is not straightforward when shifts occur.

An interesting observation was that a Leu heptad (83) began towards the C-terminus end of S4 and extended to the S4–S5 loop. Two groups showed that substitutions of these Leu proteins produced large shifts in Po-V, although the effective valence of the Po-V curves was not affected (77, 84). A leucine zipper was proposed, although the other half of the zipper was not identified. A similar mutation in $Na^+$ channels had produced a marked shift in Po-V, also without any change in voltage sensitivity (12).

Recently, direct measurements on gating currents in $K^+$ channels have led to conflicting interpretations on the correlation between charge immobilization and inactivation. Evidence for charge immobilization was reported for both noninactivating and inactivating Shak channels in one laboratory (108), whereas another laboratory found charge immobilization only in the inactivating channels (15). Furthermore, in the noninactivating Shak channels, application of internal TEA induced charge immobilization. The differences can be reconciled partly by the effects of TEA (15) and partly by differences in phenotypes because the apparent charge immobilization in noninactivating channels may have resulted from a slow deactivation process in those channels.

## CONCLUDING REMARKS

Despite the successes of the combined approaches of mutational engineering and electrophysiological measurements, we are still ignorant of the three-dimensional structure of voltage-dependent $K^+$ channels. No ready solution to this problem is apparent, which is particularly ironic considering that few proteins have been probed functionally in as many revealing ways as voltage-dependent $K^+$ channels. At present, the structure of only one membrane protein, the photosynthetic reaction center, has been solved at atomic resolution—the resolution required to understand ionic permeation. Significant progress has been made with the channel-forming proteins of bacterial walls, porins, whose structures have been resolved to better than 3.0 Å (61). However, porins are weakly ion selective, have unusual voltage dependence, and may offer few insights into voltage-dependent ion channels of excitable membranes. The structure of bacteriorhodopsin is known at 3 Å (49) in the membrane plane but the structure normal to the membrane is known only at 10 Å. This protein has the structural motif of G-protein receptors and is unlikely to shed much light on membrane proteins through which ions permeate. The data base is therefore too small for computer modeling to provide much confidence regarding theoretical predictions based upon derived amino acid sequences.

For voltage-dependent $K^+$ channels, the rate at which structures may be solved is also limited by the inability to accumulate sufficient functional protein for X-ray crystallography. This in turn is limited by the relatively few membranes in cells into which channel proteins can be inserted, even if the protein is overproduced by genetic engineering.

Nevertheless, great strides have been made in analyzing the functional significance of various regions of the amino acid sequence that defines voltage-dependent $K^+$ channels. The P region has been delimited to a considerable degree. An N-terminus inactivating ball-and-chain region has been established. Residues in the P region, in nearby S6, and in the S4–S5 loop are involved in forms of core inactivation other than that due to the noncore inactivation. S4 appears to be part of the voltage sensor. Partial identification of the external receptor for CTX and the external and internal receptors for TEA has been achieved. Salvation has come in the form of a modular design for $K^+$-channel components in which large parts of each module are happily ensconced in stretches of the linear sequence. This model will not be the whole story, however, because folding ensures that separate regions of the protein will come together to completely specify any particular function. For this next stage, many more experiments using point mutations or large-scale mutagenesis combined with electrophysiological measurements will be required.

ACKNOWLEDGMENTS

I thank Diana Kunze, Tony Lacerda, Maurizio Taglialatela, and Glenn Kirsch for their comments; Judy Breedlove and Marianne Anderson for their secretarial assistance; Brooke Summers for drawing the figures; and Mariella De Biasi, John Drewe, Hali Hartmann, Glenn Kirsch, and Maurizio Taglialatela for our collaboration. This work was supported in part by National Institutes of Health Grants NS23877, HL37044, and HL36930 to A. M. Brown.

Literature Cited

1. Adams, D. J., Nonner, W. 1990. In *Potassium Channels Structure, Classification, Function and Therapeutic Potential*, ed. N. S. Cook, pp. 40–69. Chichester, UK: Ellis Horwood
2. Adams, P. R., Brown, D. A., Constanti, A. 1982. *J. Physiol.* 330: 537–72
3. Adelman, J. P., Shen, K.-Z., Kavanaugh, M. P., Warren, R. A., Wu, Y.-N., et al. 1992. *Neuron* 9: 209–16
4. Aldrich, R. W., Corey, D. P., Stevens, C. F. 1983. *Nature* 306: 436–40
5. Almers, W. 1978. *Rev. Physiol. Biochem. Pharmacol.* 82: 97–190
6. Anderson, J. A., Huprikar, S. S., Kochian, L. V., Lucas, W. J., Gaber, R. F. 1992. *Proc. Natl. Acad. Sci. USA* 89: 3736–40
7. Anderson, P. 1978. *The Washitonian* 14: 152
8. Armstrong, C. M. 1971. *J. Gen. Physiol.* 58: 413–37
9. Armstrong, C. M., Bezanilla, F., Rojas, E. 1973. *J. Gen. Physiol.* 62: 375–91
10. Atkinson, N. S., Robertson, G. A., Ganetzky, B. 1991. *Science* 253: 551–55
11. Augustine, C. K., Bezanilla, F. 1990. *J. Gen. Physiol.* 95: 245–71
12. Auld, V. J., Goldin, A. L., Krafte, D. S., Marshall, J., Dunn, J. M., et al. 1988. *Neuron* 1: 449–61
13. Barrett, J. N., Magleby, K. L., Pallotta, B. S. 1982. *J. Physiol.* 331: 211–30
14. Baumann, A., Krah-Jentgens, I., Mueller, R., Mueller-Holtkamp, F., Seidel, R., et al. 1987. *EMBO J.* 6: 3419–29
15. Bezanilla, F., Perozo, E., Papazian, D. M., Stefani, E. 1991. *Science* 254: 679–83
16. Bogusz, S., Boxer, A., Busath, D. 1992. *Protein Eng.* 5: 285–93
17. Bogusz, S., Busath, D. 1992. *Biophys. J.* 62: 19–21
18. Brown, A. M., Birnbaumer, L. 1990. *Annu. Rev. Physiol.* 52: 197–213
19. Busch, A. E., Hurst, R. S., North, R. A., Adelman, J. P., Kavanaugh, M. P. 1991. *Biochem. Biophys. Res. Commun.* 179: 1384–90
20. Butler, A., Wei, A., Baker, K., Salkoff, L. 1989. *Science* 243: 843–947
21. Catterall, W. A. 1988. *Science* 242: 50–61
22. Chandy, K. G. 1991. *Nature* 352: 26
23. Choi, K. L., Aldrich, R. W., Yellen, G. 1991. *Proc. Natl. Acad. Sci. USA* 88: 5092–95
24. Ciani, S., Krasne, S., Miyazaki, S.,

Hagiwara, S. 1978. *J. Membr. Biol.* 14: 103–34
25. Connor, J. A., Stevens, C. F. 1971. *J. Physiol.* 213: 21–30
26. Covarrubias, M., Wei, A., Salkoff, L. 1991. *Neuron* 7: 763–73
27. Critz, S., Wible, B. A., Lopez, H. S., Brown, A. M. 1993. *J. Neurochem.* 60: 1175–78
28. De Biasi, M., Hartmann, H. A., Drewe, J. A., Taglialatela, M., Brown, A. M., et al. 1992. *Pfluegers Arch.* In press
29. Demo, S. D., Yellen, G. 1991. *Neuron* 7: 743–53
30. Drewe, J. A., Verma, S., Frech, G., Joho, R. H. 1992. *J. Neurosci.* 12: 538–48
31. Dubinsky, W. P., Mayorga-Wark, O., Schultz, S. G. 1992. *Proc. Natl. Acad. Sci. USA* 89: 1770–74
32. Durell, S. R., Guy, H. R. 1992. *Biophys. J.* 62: 243–52
33. Fatt, P., Katz, B. 1953. *J. Physiol. (London)* 120: 374–89
34. Folander, K., Smith, J. S., Antanavage, J., Bennett, C., Stein, R. B., et al. 1990. *Proc. Natl. Acad. Sci. USA* 87: 2975–79
35. Foster, C. D., Chung, S., Zagotta, W. N., Aldrich, R. W., Levitan, I. B. 1992. *Neuron* 9: 229–36
36. Frech, G. C., VanDongen, A. M. J., Schuster, G., Brown, A. M., Joho, R. H. 1989. *Nature* 340: 642–45
37. Freeman, L. C., Kwok, W.-M., Kass, R. S. 1992. *Am. J. Physiol.* 31: H1298–H1302
38. French, R. J., Shoukimas, J. J. 1981. *Biophys. J.* 34: 271–91
39. Goldstein, S. A. N., Miller, C. 1991. *Neuron* 7: 403–8
40. Greenblatt, R. E., Blatt, Y., Montal, M. 1985. *FEBS Lett.* 193: 125–34
41. Guy, H. R., Conti, F. 1990. *TINS* 13: 201–6
42. Guy, H. R., Seetharamulu, P. 1986. *Proc. Natl. Acad. Sci. USA* 83: 508–12
43. Hagiwara, S., Kusano, K., Sato, R. 1961. *J. Physiol. (London)* 155: 470–89
44. Hamill, O. P., Marty, A., Neher, E., Sakmann, B., Sigworth, F. J. 1981. *Pfluegers Arch.* 391: 85–100
45. Hartmann, H. A., Kirsch, G. E., Drewe, J. A., Taglialatela, M., Joho, R. H., et al. 1991. *Science* 251: 942–44
46. Hausdorff, S. F., Goldstein, S. A. N., Rushin, E. E., Miller, C. 1991. *Biochemistry* 30: 3341–46
47. Heginbotham, L., Mackinnon, R. 1992. *Neuron* 8: 483–91
48. Heinemann, S. H., Terlau, H., Stühmer, W. 1992. *Nature* 356: 441–43
49. Henderson, R., Baldwin, J. M., Ceska, T. A., Zemlin, F., Beckmann, E., et al. 1990. *J. Mol. Biol.* 213: 899–929
50. Hille, B. 1992. In *Ionic Channels of Excitable Membranes*, pp. 83–114. Sunderland: Sinauer
51. Hille, B., Schwarz, W. 1978. *J. Gen. Physiol.* 72: 409–42
52. Hodgkin, A. L., Huxley, A. F. 1952. *J. Physiol.* 117: 500–44
53. Horie, M., Irisawa, H. 1987. *Am. J. Physiol.* 22: H210–14
54. Hoshi, T., Zagotta, W. N., Aldrich, R. W. 1990. *Science* 250: 533–38
55. Hoshi, T., Zagotta, W. N., Aldrich, R. W. 1991. *Neuron* 7: 547–56
56. Hwang, P. M., Glatt, C. E., Bredt, D. S., Yellen, G., Snyder, S. H. 1992. *Neuron* 8: 473–81
57. Isacoff, E. Y., Jan, Y. N., Jan, L. Y. 1990. *Nature* 345: 530–34
58. Isacoff, E. Y., Jan, Y. N., Jan, L. Y. 1991. *Nature* 353: 86–90
59. Jan, L. Y., Jan, Y. N. 1989. *Cell* 56: 13–25
60. Jan, L. Y., Jan, Y. N. 1992. *Annu. Rev. Physiol.* 54: 537–55
61. Jap, B. K., Walian, R. J., Gehring, K. 1991. *Nature* 350: 167–70
62. Josephson, I. R., Brown, A. M. 1986. *J. Membr. Biol.* 94: 19–35
63. Kamb, A., Iverson, L. E., Tanoye, M. A. 1987. *Cell* 50: 405–13
64. Kameyama, M., Kakei, M., Sato, R., Shikasaki, T., Matsuda, H., et al. 1984. *Nature* 309: 354–56
65. Kavanaugh, M. P., Hurst, R. S., Yakel, J., Varnum, M. D., Adelman, J. P., et al. 1992. *Neuron* 8: 493–97
66. Kavanaugh, M. P., Varnum, M. D., Osborne, P. B., MacDonald, J. C., Busch, A. E., et al. 1991. *J. Biol. Chem.* 266: 7583–87
67. Kirsch, G. E., Brown, A. M. 1989. *J. Gen. Physiol.* 93: 85–99
68. Kirsch, G. E., Drewe, J. A., Hartmann, H. A., Taglialatela, M., De Biasi, M., et al. 1992. *Neuron* 8: 499–505
69. Kirsch, G. E., Drewe, J. A., Taglialatela, M., Joho, R. H., De Biasi, M., et al. 1992. *Biophys. J.* 62: 136–44
70. Kirsch, G. E., Taglialatela, M., Brown, A. M. 1991. *Am. J. Physiol.* 30: 583–C90
71. Koren, G., Liman, E. R., Logothetis, D. E., Nadal-Ginard, B., Hess, P. 1990. *Neuron* 4: 39–51
72. Latorre, R., Miller, C. 1983. *J. Membr. Biol.* 71: 11–30
73. Lee, C. W. B., Aldrich, R. W., Gierasch, L. M. 1992. *Biophys. J.* 61: A379 (Abstr.)
74. Liman, E. R., Hess, P., Weaver, F., Koren, G. 1991. *Nature* 353: 752–56

75. Logothetis, D. E., Kurachi, Y., Galper, J., Neer, E. J., Clapham, D. E. 1987. *Nature* 325: 321–26
76. Logothetis, D. E., Movahedi, S., Satler, C., Lindpaintner, K., Nadal-Ginard, B. 1992. *Neuron* 8: 531–40
77. Lopez, G. A., Jan, Y. N., Jan, L. Y. 1991. *Neuron* 7: 327–36
78. Luneau, C. J., Williams, J. B., Marshall, J., Levitan, E. S., Oliva, C., et al. 1991. *Proc. Natl. Acad. Sci. USA* 88: 3932–36
79. MacKinnon, R. 1991. *Nature* 350: 232–35
80. MacKinnon, R., Miller, C. 1989. *Science* 245: 1382–85
81. MacKinnon, R., Yellen, G. 1990. *Science* 250: 276–79
82. Matsuda, H. 1988. *J. Physiol.* 397: 237–58
83. McCormack, K., Campanelli, J. T., Ramaswami, M., Mathew, M. K., Tanouye, M. A. 1989. *Nature* 340: 103
84. McCormack, K., Tanouye, M. A., Iverson, L. E., Lin, J.-W., Ramaswami, M., et al. 1991. *Proc. Natl. Acad. Sci. USA* 88: 2931–35
85. McCormack, T., Vega-Saenz de Miera, E. C., Rudy, B. 1990. *Proc. Natl. Acad. Sci. USA* 87: 5227–31
86. Meech, R. W., Standen, N. B. 1975. *J. Physiol.* 249: 211–39
87. Mikami, A., Imoto, K., Tanabe, T., Niidome, T., Mori, Y., et al. 1989. *Nature* 340: 230–33
88. Murrell-Lagnado, R. D., Aldrich, R. W. 1992. *Biophys. J.* 61: A290 (Abstr.)
89. Noda, M., Ikeda, T., Kayano, T., Sujuki, H., Takeshima, H., et al. 1986. *Nature* 320: 188–92
90. Noma, A. 1983. *Nature* 305: 147–48
91. Okabe, K., Yatani, A., Brown, A. M. 1991. *J. Gen. Physiol.* 97: 1279–93.
92. Pallotta, B. S., Magleby, K. L., Barrett, J. N. 1981. *Nature* 293: 471–74
93. Papazian, D. M., Schwarz, T. L., Tempel, B. L., Jan, Y. N., Jan, L. Y. 1987. *Science* 237: 749–53
94. Papazian, D. M., Timpe, L. C., Jan, Y. N., Jan, L. Y. 1991. *Nature* 349: 305–10
95. Pardo, L. A., Heinemann, S. H., Terlau, H., Ludewig, U., Lorra, C., et al. 1992. *Proc. Natl. Acad. Sci. USA* 89: 2466–70
96. Rehm, H., Tempel, B. L. 1991. *FASEB J.* 5: 164–70
97. Rettig, J., Wunder, F., Stocker, M., Lichtinghagen, R., Mastiaux, F., et al. 1992. *EMBO J.* 11: 2473–86
98. Ruppersberg, J. P., Schröter, K. H., Sakmann, B., Stocker, M., Sewing, B., et al. 1990. *Nature* 345: 535–37
99. Ruppersberg, J. P., Stocker, M., Pongs, O., Heinemann, S. H., Frank, R., et al. 1991. *Nature* 352: 711–14
100. Sanguinetti, M. C., Jurkiewicz, N. K. 1990. *J. Gen. Physiol.* 96: 195–215
101. Schmid-Antomarchi, H., Hugues, M., Lazdunski, M. 1986. *J. Biol. Chem.* 261: 8633–37
102. Schoppa, N. E., McCormack, K., Tanouye, M. A., Sigworth, F. J. 1992. *Science* 255: 1712–15
103. Schwarz, T. L., Tempel, B. L., Papazian, D. M., Jan, Y. N., Jan, L. Y. 1988. *Nature* 331: 137–42
104. Sentenac, H., Bonneaud, N., Minet, M., Lacroute, F., Salmon, J.-M., et al. 1992. *Science* 256: 663–65
105. Shimoni, Y., Clark, R. B., Giles, W. R. 1992. *J. Physiol.* 448: 709–27
106. Siegelbaum, S. A., Camardo, J. S., Kandel, E. R. 1982. *Nature* 299: 413–17
107. Stevens, C. F. 1991. *Nature* 349: 657–58
108. Stühmer, W., Conti, F., Stocker, M., Pongs, O., Heinemann, S. H. 1991. *Pfluegers Arch.* 418: 423–29
109. Stühmer, W., Conti, F., Suzuki, H., Wang, X., Noda, M., et al. 1989. *Nature* 339: 597–603
110. Stühmer, W., Ruppersberg, J. P., Schroter, K., Sakmann, B., Stocker, M., et al. 1989. *EMBO J.* 8: 3235–44
111. Taglialatela, M., Drewe, J., Kirsch, G. E., De Biasi, M., Hartmann, H., et al. 1992. *Pfluegers Arch.* In press
112. Taglialatela, M., Kirsch, G. E., VanDongen, A. M. J., Drewe, J. A., Hartmann, H. A., et al. 1992. *Biophys. J.* 62: 34–36
113. Taglialatela, M., VanDongen, A. M. J., Drewe, J. A., Joho, R. H., Brown, A. M., et al. 1991. *Mol. Pharmacol.* 40: 299–307
114. Takumi, T., Ohkubo, H., Nakanishi, S. 1988. *Science* 242: 1042–45
115. Tamkum, M. M., Knoth, K. M., Walbridge, J. A., Kroener, H., Roden, D. M., et al. 1991. *FASEB J.* 5: 331–37
116. Tempel, B. L., Jan, Y. N., Jan, L. Y. 1988. *Nature* 332: 837–39
117. Tempel, B. L., Papajian, D. M., Schwartz, T. L., Jan, Y. N., Jan, L. Y. 1987. *Science* 237: 770–75
118. Timpe, L. C., Jan, Y. N., Jan, L. Y. 1988. *Neuron* 1: 659–67
119. Toro, L., Stefani, E., Latorre, R. 1992. *Neuron* 9: 237–45
120. Vandenberg, C. A. 1987. *Proc. Natl. Acad. Sci. USA* 84: 2560–64
121. VanDongen, A. M. J., Frech, G. C., Drewe, J. A., Joho, R. H., Brown, A. M. 1990. *Neuron* 5: 433–43

122. Van Dongen, T., Codina, J., Olate, J., Mattera, R., Joho, R., et al. 1988. *Science* 242: 1433–37
123. Vazquez, J., Feigenbaum, P., King, V. F., Kaczorowski, G. J., Garcia, M. L. 1990. *J. Biol. Chem.* 265: 15564–71
124. Wagoner, P. K., Oxford, G. S. 1987. *J. Gen. Physiol.* 90: 261–90
125. Wei, A., Covarrubias, M., Butler, A., Baker, K., Pak, M., et al. 1990. *Science* 248: 599–603
126. Yatani, A., Codina, J., Brown, A. M., Birnbaumer, L. 1987. *Science* 235: 207–11
127. Yellen, G., Jurman, M., Abramson, T., Mackinnon, R. 1991. *Science* 251: 939–42
128. Yool, A. J., Schwarz, T. L. 1991. *Nature* 349: 700–4
129. Zagotta, W. N., Aldrich, R. W. 1990. *J. Gen. Physiol.* 95: 29–60
130. Zagotta, W. N., Hoshi, T., Aldrich, R. W. 1990. *Science* 250: 568–71

# THE EFFECTS OF PHOSPHORYLATION ON THE STRUCTURE AND FUNCTION OF PROTEINS

## L. N. Johnson

Laboratory of Molecular Biophysics, Rex Richards Building, University of Oxford, Oxford, OX1 3QU, United Kingdom

## D. Barford

W. M. Keck Structural Biology Laboratory, Cold Spring Harbor Laboratory, PO Box 100, Cold Spring Harbor, New York 11724

KEY WORDS: protein phosphorylation, allosteric mechanism, phosphate recognition sites, glycogen phosphorylase, isocitrate dehydrogenase, cyclic AMP–dependent protein kinase

CONTENTS

| | |
|---|---|
| PERSPECTIVES AND OVERVIEW | 200 |
| PHOSPHATE-RECOGNITION SITES | 201 |
| GLYCOGEN PHOSPHORYLASE | 202 |
|     Structural Changes upon Phosphorylation | 204 |
|     Signal Transduction from the Phosphorylation to the Allosteric Sites and to the Catalytic Site | 207 |
| E. COLI ISOCITRATE DEHYDROGENASE | 209 |
| CYCLIC AMP–DEPENDENT PROTEIN KINASE | 213 |
|     Autophosphorylation Site | 216 |
|     Relevance for Other Kinases | 217 |
| PHOSPHOENOLPYRUVATE SUGAR PHOSPHOTRANSFERASE SYSTEM | 218 |
|     The Structure of HPr | 219 |
|     Glucose Permease IIA Domains | 222 |
| BACTERIAL CHEMOTAXIS AND CheY | 223 |
| SILENT PHOSPHORYLATION | 226 |
| DISCUSSION | 227 |

## PERSPECTIVES AND OVERVIEW

Regulation of protein function through selective phosphorylation of certain amino acids was first observed in mammalian cells and tissue extracts but is now known to occur in both eukaryotic and prokaryotic cells. Protein phosphorylation is the most ubiquitous intracellular control mechanism. It regulates certain metabolic pathways, gene transcription and translation, membrane transport, hormonal response, muscle contraction, light harvesting and photosynthesis, cell division and cell growth, viral oncogene response, and learning and memory. In eukaryotic cells, reversible phosphorylation catalyzed by protein kinases and phosphatases occurs predominantly on serine, threonine, and tyrosine residues (54), although recently a histone H4 protein histidine kinase was reported (37). Signal transduction in prokarotes is also mediated by phosphorylation on serine, threonine, aspartate, and histidine residues (107), but no instances of tyrosine phosphorylation have yet been observed (16). In this review, we ask the question: what are the structural consequences of the addition of a phosphate group to a protein and how may this alter the protein's structure and function?

Protein phosphorylation can result in either activation or inhibition of enzyme activity. It can promote conformational changes that may be local to, or more remote from, the site of phosphorylation, and these changes can also alter the surface properties of the protein that affect self-association or recognition by other proteins. To date, two systems are understood in depth from protein crystallographic studies: glycogen phosphorylase and isocitrate dehydrogenase. In phosphorylase, phosphorylation on a single serine near the amino terminus of the polypeptide chain is associated with significant local conformational changes that are accompanied by long range tertiary and quaternary conformational shifts that lead to allosteric activation of the enzyme. In isocitrate dehydrogenase, however, the enzyme is inhibited by serine phosphorylation in an electrostatic steric blocking mechanism so that the phosphate group prevents the binding of the anionic substrate without recourse to conformational change.

The structural details of several other systems are available. Although a comparison of the phosphorylated and dephosphorylated forms has not yet been possible, the likely structural consequences of phosphorylation can be deduced from other evidence. In the cyclic AMP (cAMP)–dependent protein kinase, phosphorylation of threonine 197 appears to play a conformational role in forming a contact across two domains and promoting the correct conformation for recognition of substrate and the regulatory subunit. The bacterial phosphoenolpyruvate sugar phosphotransferase proteins, HPr and glucose permease (which have histidine phosphorylation at the phosphotransfer sites and a regulatory serine phosphorylation for HPr), are an example in which phosphorylation may change surface properties that influence recognition by other proteins. The bacterial protein of chemotaxis, CheY, is phosphorylated on an aspartate in an acidic region of the protein, and this could result in conformational changes from electrostatic repulsion. Finally, in four structural examples, phosphorylation apparently does nothing; these are the so-called silent phosphorylation sites (14a). These phosphorylation sites are located on the surface and make no defined contacts. They may be involved with some as yet unidentified interactions with other proteins.

## PHOSPHATE-RECOGNITION SITES

Before the structural work on proteins controlled by phosphorylation, considerable data were available on phosphate-binding sites from numerous proteins. A summary of the major contacts established by X-ray crystallography for some 30 phosphate sites in 18 different proteins was published in 1984 (43). To date, the Protein Data Bank contains coordinates for over 50 different proteins that exhibit phosphate recognition, and this number will increase substantially from the addition of the proteins whose coordinate entries are in preparation to the data base.

When the phosphate-recognition site is involved solely in binding, the interactions displayed are usually numerous and well defined. These characteristics should dominate the phosphate regulatory sites where conformational changes are involved. Most often tight binding is achieved by hydrogen bonds and ionic interactions to arginine side chains, as for example in the recent structure of a phosphotyrosine peptide bound to the SH2 domain of v-src (115a). The guanidinium group is well suited for such interactions by virtue of its planar geometry and ability to form multiple hydrogen bonds to the phosphate oxygens. As indicated by the high $pK_a$ ( > 12) of the guanidinium group in the free amino acid, the group almost invariably carries a positive charge unless there is a very unusual environment. Lysine side chains may also be involved in phosphate binding, but not all phosphate recognition sites contain positively charged groups. In several enzymes, the phosphate site is located near the N-terminal end of an $\alpha$-helix, and the phosphate oxygens are directed towards the main-chain NH groups. The rationale was provided by Hol et al (36), who noted that the alignment of peptide dipoles parallel to the helix axis gave rise to a macrodipole with approximately one half unit of positive charge located at the amino terminal and one half unit of negative charge located at the carboxy terminal ends of the helix. Phosphate groups may also hydrogen bond to side chains of serine, threonine, asparagine, and glutamine residues as well as main-chain NH groups of peptides not involved in helices.

Phosphate-binding sites in protein crystal structures have often been recognized by sulfate-binding sites because of the frequent use of ammonium sulfate in crystallization of proteins. Recent work with triose-phosphate isomerase has shown that sulfate- and phosphate-binding sites may not be identical (114). A nice distinction between interactions for sulfate and phosphate has been made by Quiocho and colleagues from the comparison of the dianions bound to the sulfate- and the phosphate-binding proteins (61, 80). Both dianion-recognition sites exhibit multiple hydrogen-bond interactions with donor groups, but the phosphate site also makes contact with an aspartate. A role for the aspartate has been proposed in the discrimination against sulfate or other tetrahedral fully ionized divalent oxyanions that cannot contribute as donors to hydrogen bonding.

The phosphate catalytic sites are generally less well defined, and the local regions of the protein may require conformational changes to stabilize the transition state on the catalytic pathway. Such phosphate catalytic recognition sites are represented by the $\gamma$ phosphate sites of ATP in the kinases, or the substrate phosphate–attacking site in glycogen phosphorylase, or the site for phosphate ester cleavage in ribonuclease.

## GLYCOGEN PHOSPHORYLASE

Glycogen phosphorylase catalyzes the degradation of glycogen to glucose–1-P. In muscle, glucose–1-P is utilized through glycolysis to provide ATP to meet the energy requirements of the cell and in liver is metabolized to glucose and exported to provide fuel for other tissues. Glycogen phosphorylase was the first enzyme recognized as being controlled by reversible phosphorylation. In 1955, Fischer & Krebs observed phosphorylase kinase activity in crude skeletal muscle extracts after they had been passed through unwashed filter paper as a consequence of extraction of calcium from the paper (25, 55). The site of phosphorylation was established in 1959 (24), and the authors speculated correctly on the likely significance of "the accumulation of positive charges on this exposed segment of the phosphorylase molecule." Later studies showed that phosphorylase kinase is also controlled by phosphorylation (for review, see 77); this was the second enzyme recognized to be controlled by phosphorylation. These reactions of phosphorylase kinase and phosphorylase are part of the integral cascade system whereby in response to nervous (e.g. calcium) or hormonal (e.g. adrenalin or insulin) signals glycogen metabolism may be controlled through effective activation or inhibition of the enzymes involved in glycogen degradation and glycogen synthesis (e.g. reviewed in 15). Activation of the nonphosphorylated phosphorylase b (GPb) is achieved by addition of a phosphate group from $Mg^{2+}$ ATP to a single serine residue, Ser14, catalyzed by phosphorylase kinase to give the activated phosphorylated form phosphorylase a (GPa). The reverse process of inactivation through dephosphorylation was first demonstrated by Sutherland & Wosilait in 1955 (110, 121) and is achieved through the action of protein phosphatase 1 (28).

In the liver, phosphorylase activation is achieved solely through phosphorylation, but in muscle, phosphorylase is also controlled by intracellular signals that act through noncovalent allosteric mechanisms. The nonphosphorylated muscle phosphorylase can be activated by AMP and to a lesser extent by IMP and is inhibited by ATP and glucose-6-P. Both phosphorylated and nonphosphorylated forms of both muscle and liver phosphorylase are inhibited by glucose and purine analogs such as caffeine and activated by glycogen. Several major reviews have discussed glycogen phosphorylase (e.g. 29, 46, 64, 68, 72), and several short reviews focus on allostery and control (3, 10, 44, 45).

To a first approximation following the model of Monod et al (67), the activation mechanisms can be understood in terms of interconversions between alternative structural states; the inactive/low affinity T state and the active/high affinity R state, where these interconversions are modulated by allosteric interactions and reversible phosphorylation. X-ray diffraction studies on rabbit muscle glycogen phosphorylase have resulted in structural representations of each of the four forms of the enzyme. Crystals of the low affinity, low activity T-state GPb were obtained by crystallization of GPb in the presence of IMP, a weak activator, under low salt conditions (10 mM magnesium acetate) (1). Similar conditions were used to produce crystals of T-state GPa, except that IMP was replaced by glucose. Glucose is an inhibitor competitive with substrate that promotes the T-state conformation (101). Growth of crystals of the R state was more problematic, possibly because of the difficulty in obtaining a homogeneous population of R-state molecules (the allosteric constant $L_o$ is 3000). R-state crystals were achieved through crystallization of both GPb and GPa in the presence of 1 M ammonium sulfate (4, 6). The T-state crystals are well ordered and the structures have been refined at 1.9-Å (GPb) and 2.1-Å (GPa) resolution with inclusion of water molecules. The R-state crystals diffract less well and the structures have been refined to 2.8-Å resolution without inclusion of water molecules.

The large protein (842 amino acids) folds into a relatively compact structure that has the same topology in both T and R states and a and b forms. The protein is an $\alpha/\beta$ protein that can be divided into two domains, residues 10–484 and 485–842. The first nine residues have not been located in the electron-density maps (except in T-state GPa in which residues 5–9 have been located). The catalytic site is at the center of the molecule where the two domains come together and is buried approximately 15 Å from the surface of the molecule and includes the essential cofactor pyridoxal phosphate (Figure 1). The N-terminal domain contains the allosteric effector sites for Ser14-P, AMP, ATP, glucose–6-P, and glycogen. The inhibitor site for purines is located at the entrance to the catalytic-site cleft where the two domains come together. The physiologically active form of the enzyme is a dimer. In the T-state structures, the two subunits of the dimer are related by a crystallographic two-fold axis of symmetry. In all R-state structures, the dimers aggregate to form tetramers, a phenomenon recognized from solution studies to be associated with activation (5).

## Structural Changes upon Phosphorylation

In GPb, the N terminal residues 10 to 18 are disordered. They have been located in the electron density map but exhibit the highest temperature factors of the whole structure. They adopt an irregular extended conformation and make intrasubunit contacts with three helical regions. The Ser14 side chain is turned in towards Glu501, and Arg16 hydrogen bonds to Glu105, Gln96, and Val483 (main-chain O). The sequence around the Ser14 (Arg-Lys-Gln-Ile-Ser*-Val-Arg) contains a number of positively charged residues, and these are located close to a cluster of acidic residues on the protein surface. Upon phosphorylation, the conformation of the N-terminal residues changes dramatically (4, 6, 100). Evidently, the acidic environment encountered by the N-terminal residues in GPb is an inhospitable environment for the phosphate group. The first half-turn of the $\alpha 1$ helix is unwound at residue 23, and residues 10–22 are rotated up through approximately 120° relative to their position in GPb (Figure 1). There is a turn of $3_{10}$-helix between residues 10 and 13 and between residues 15 and 18 so that, overall, residues 10–18 have the appearance of a distorted helix, although the conformation of residues 13 and 14 is $\beta$. Ser14 shifts more than 36 Å between GPb and GPa. The Ser14-P makes hydrogen bonds and ionic contacts to Arg69 and Arg43' from the other subunit and to main-chain N of Val15 and Arg16 (Figure 2). The two arginines undergo substantial movement upon phosphorylation. (Residues of the other subunit are denoted by primes.)

The localization of the Ser14-P at this site is accompanied by interdigitation of the N-terminal residues to sites on the protein surface. Arg10 occupies a position previously occupied by Arg43' in T-state GPb and hydrogen bonds with main-chain carbonyls at the end of a helix of the other subunit. The side chain of Ile13 docks into a nonpolar pocket on

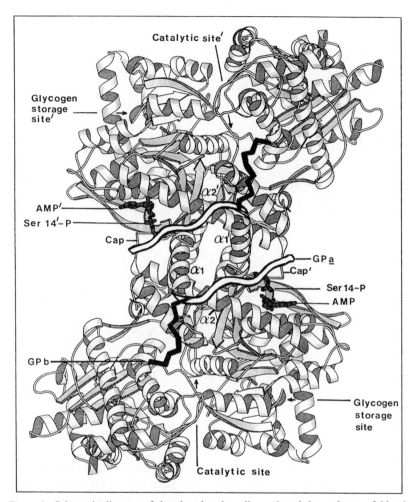

*Figure 1* Schematic diagram of the phosphorylase dimer viewed down the two-fold axis with the phosphorylation (Ser14-P) and allosteric (AMP) sites towards the viewer. Access to the catalytic site is from the far side of the molecule. The diagram illustrates the major change in the N-terminal residues upon phosphorylation. Residues 10–23 of GPb, which make intrasubunit contacts, are shown as solid black lines. Residues 10–23 of GPa, which make intersubunit contacts, are shown as white lines with thick borders. In the fold of the chain from residues 24–80, the α1-cap-α2 region is indicated. For further details, see text. Diagram produced with MOLSCRIPT (53) by M. S. P. Sansom.

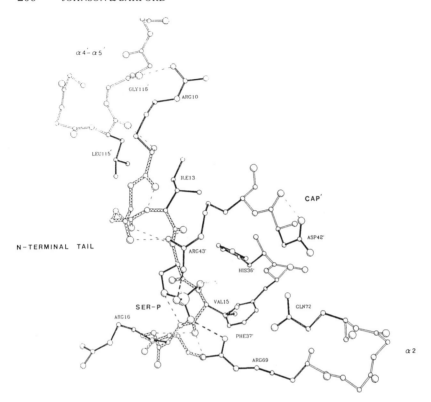

*Figure 2* Details of the interaction of Ser14-P in R-state GPa. The main-chain residues for residues 10–16 are dotted. The phosphate interacts with Arg43' and 69 and with main-chain N of residues 15 and 16. For further details, see text (from 6).

the surface formed mostly by nonpolar side chains of the other subunit. Likewise, the side chain of Val15 packs into a hydrophobic pocket formed mostly from residues Ile68, Arg69, and His36'. Arg16 is orientated towards Gln12 but is external and makes no strong interactions. On binding to the subunit surface, the phosphorylated N terminus displaces five C-terminal residues 838–842, which are disordered in GPa, and Arg16 fills the space previously occupied by these residues. The displacement of the C terminus disrupts an ion pair between Asp838 and His36', and the movement of the histidine leads to the recognition site for Val15. Thus, phosphorylation is accompanied by a concerted disorder-to-order transition of the N-terminal residues and an order-to-disorder transition of the C-terminal residues. These interactions, a combination of electrostatic interactions made by the

Ser-P and polar and nonpolar interactions of the surrounding residues, are crucial to the allosteric response.

The major incentive for the dramatic change in conformation of the N terminal residues on phosphorylation appears to be electrostatic. Sulfate ions at high concentrations can activate GPb, and this was utilized to produce crystals of R-state GPb (4). Ammonium sulfate produces properties of GPa in GPb such as enhanced affinity for AMP, a reversal of glucose–6-P inhibition and enzyme tetramerization (22, 58, 99). The crystal structure of R-state GPb in the presence of ammonium sulfate showed that sulfate ions occupied the Ser-P site of GPa, and the N-terminal residues were ordered also as in GPa. Sulfate ions also bound at the catalytic site and the AMP phosphate recognition site.

Phosphorylase b' (GPb') is a proteolytic product obtained by limited proteolysis with subtilisin and lacks the first 16 residues. The modified enzyme can be activated by AMP, showing that the N-terminal residues are not essential for AMP activation, but of course GPb' cannot be activated by phosphorylation [see Graves & Wang (29) for review]. Recent studies (59) have shown that GPb' exhibits no activation by sulfate in the absence of AMP, and the AMP activation is inhibited by sulfate. These results indicate that activation by the dianion can only be achieved with the involvement of the N-terminal residues and the conformational changes that these residues promote. Bacterial expression systems for mammalian phosphorylases have been developed, and these have allowed site-directed mutagenesis experiments to probe the roles of individual residues (11, 14). Interestingly, the mutation Ser14 to Asp does not result in activation (M. F. Browner & R. J. Fletterick, personal communication), which could indicate that a monoanionic group is insufficient to promote the movement of both arginines, although structural data are required to test this hypothesis.

## Signal Transduction from the Phosphorylation to the Allosteric Sites and to the Catalytic Site

Phosphorylation of GP is accompanied by significant changes in tertiary and quaternary conformations that connect the phosphorylation site with the other allosteric effector sites and with the catalytic site. One subunit interface that is composed mainly of the interactions between the $\alpha 1$-cap-$\alpha 2$ helix-loop-helix region (residues 20–80) and the cap' (residues 36–45) region of the other subunit is tightened by the movements that accompany phosphorylation (Figure 1). As the two arginines move to contact the Ser-P, the cap' region is brought closer to the $\alpha 2$ helix. These shifts create a high affinity AMP site that is located some 12 Å from the Ser14-P site on the inner side of the $\alpha 2$ helix (6, 100).

The structure of an R-state GPb in the presence of AMP and the absence of activating sulfate has been solved (102). Despite the different crystallization conditions, the quaternary and tertiary structures are similar to the previously determined R-state structures, and the AMP binds in similar conformation and makes similar contacts to the structures observed in the other activated AMP complexes. Interestingly, the N-terminal residues are reported to be partially ordered in a conformation similar to that observed in GPa, but the crucial arginine side chains Arg43' and Arg69 that have no compensating phosphate are not well ordered. The concerted movements around the cap'/α2 interface show how phosphorylation leads to tighter AMP binding, although it is not clear that AMP binding should necessarily lead to order of the N-terminal residues.

The central core of the phosphorylase molecule, comprising about 60% of the structure, does not change conformation between T-state GPb and R-state GPa (5, 6). (The overall root mean square shift in Cα coordinates from T-state GPb to R-state GPa is 3.6 Å, or 1.3 Å if residues 10–21 are excluded and 66% of the Cα atoms superimpose to within 0.5 Å.) The signals from the phosphorylation and AMP sites are transmitted via the quaternary structural changes and extensive changes in tertiary structure at the subunit interface (4, 6). On the T-to-R transition, the quaternary structural change is represented by a rotation of one subunit of 10° about an axis normal to the two-fold axis of symmetry of the dimer that intercepts this axis near the cap'/α2 interface. As a result, the cap'/α2 interface is tightened and the other major interface of the tower helices is altered. The tower helix, residues 262–278, makes a major excursion from the subunit and packs against the tower helix of the symmetry-related subunit. In the T state, the two tower helices are antiparallel with an angle of tilt about $-20°$. In the R state, the tower helices change their angle of tilt to about $-70°$ and pull apart by two turns of helix. In the T state, Tyr262' from the top of the tower' helix packs against Ile165. Ile165 in turn packs with Pro281 and Asn133. In the R state, Tyr262' shifts so that this cluster is disrupted. Ile165, which is at the end of the parallel strand (residues 162–164) of β-sheet with strand residues 276–279, shifts. These shifts appear to form the link between changes at the interface and changes at the catalytic site (for illustrations of these shifts, see 4, 6)

In the T state, access from the bulk solvent to the catalytic site is blocked mostly by the loop of chain residues 282–286 (the 280s loop). The side chain of Asp283 is directed towards the catalytic site and is hydrogen bonded via two water molecules to the 5' phosphate of the pyridoxal phosphate. The guanidinium group of a nearby arginine, Arg569, is almost completely buried, and its charged group is compensated by hydrogen bonds to the carbonyl oxygens of Pro281, Asn133, and Lys608. The T-to-

R transition results in (a) a replacement of the acidic residue (Asp283) by a basic residue (Arg569) to create the substrate recognition site, (b) displacement of the loop that blocks the access to the catalytic site, and (c) movements of subdomains that line either side of the access channel. The shifts in Arg569 as it breaks its T-state contacts with Ile165, Pro281, and Asn133 are correlated with the conformational changes in these residues so that events at the catalytic site are linked to events at the tower/tower interface and, through the quaternary structural changes, to the phosphorylation and AMP recognition sites. In the T-state GPa structure, these subunit-interface interactions are uncoupled. Glucose bound at the catalytic site favors the T state through its interactions with Asp283 and Asn284, which tie the 280s loop in the T-state conformation. However at the Ser-P site, the tertiary structure and the subunit interactions are almost identical to those of R-state GPa. The catalytic mechanism is discussed elsewhere (44, 46, 74).

Finally, the allosteric transition promoted by phosphorylation in phosphorylase is accompanied by association of a pair of functional dimers to form a tetramer. As a result of tertiary and quaternary structural changes, a new protein/protein interface is created that involves part of the glycogen storage subdomain, a C-terminal subdomain, and the tower helices (5). These surface subdomains shift as rigid bodies, and the cores of the two major domains remain unchanged. The structural properties of the protein/protein tetramer interface are atypical of protein oligomerization sites and are more similar to those properties observed for protease/inhibitor or antibody/antigen interactions. In muscle, most phosphorylase is associated with glycogen particles, and glycogen promotes dissociation of the less active tetramers to active dimers (29) so that the dimer/tetramer association for phosphorylase is probably not physiologically relevant. However, the structural results provide a model that explains how protein phosphorylation can be accompanied by changes in surface properties of a protein that lead to changes in protein/protein association.

## *E. COLI* ISOCITRATE DEHYDROGENASE

Control of the enzymic activity of *Escherichia coli* isocitrate dehydrogenase (IDH) illustrates a control mechanism different from that occurring in glycogen phosphorylase. Biochemical, genetic, and structural studies have demonstrated that inhibition of the catalytic activity by phosphorylation is a result of direct inhibition of substrate binding by a phosphorylated catalytic-site serine residue. Allosteric conformational changes are not involved [reviewed by Stroud (108)].

The enzyme catalyzes a step in the citric acid cycle, namely the oxidative decarboxylation of isocitrate-producing α-ketoglutarate, NADPH and $CO_2$:

$NADP^+ + isocitrate \rightarrow NADPH + \alpha\text{-ketoglutarate} + CO_2$.

The reaction lies on a branch point in carbohydrate metabolism regulating flux between the citric acid cycle and the glyoxylate by-pass pathway. Inhibition of IDH by phosphorylation allows flux through the glyoxylate pathway, enabling cells to grow on an acetate carbon source (57). Phosphorylation of the enzyme causes a complete inactivation of its activity (112), and the site of phosphorylation was established as Ser113 (7). A bifunctional enzyme, IDH kinase/phosphatase, catalyzes both the phosphorylation and dephosphorylation of IDH (57, 69). An understanding of the nature of the inhibitory effect caused by protein phosphorylation was obtained by studies of mutants of IDH containing amino-acid substitutions at position 113. Substitution of Ser113 with either aspartate or glutamate resulted in a reduction in activity by $10^6$-fold (18), whereas substitution by threonine, tyrosine, cysteine, or alanine only slightly reduced enzymic activity. Also shown was that phosphorylation of Ser113 or substitution with aspartate or glutamate reduces activity by blocking binding of isocitrate to the enzyme, while having no effect on the binding affinity of $NADP^+$ (18).

The structures of phosphorylated and unphosphorylated *E. coli* isocitrate dehydrogenase have been determined to a resolution of 2.5 Å and refined to R factors of 0.197 and 0.169, respectively (41, 42). The structure is composed of a homodimer, each subunit comprising 416 amino acid residues. Consistent with the absence of sequence similarity to other dehydrogenases (112), no structural homology to the Rossmann nucleotide fold exists. Serine 113 is situated on helix D at the edge of the catalytic-site pocket located at the interface of the major domains of IDH and at the dimer interface. The Ser-P makes no direct contacts to the enzyme (Figure 3*a*). There are hydrogen bonds through waters to Thr105 (OG2 and N) and to Leu103 (O) and a long (3.42 Å) contact to Asn232' of the other subunit. Several positively charged groups reside in the vicinity of the catalytic site that include Arg119, Arg153, and Lys230', but these side chains are over 6 Å from the phosphate oxygens and play other roles in the substrate-recognition site. The serine is located at the start of an α-helix, and the phosphate oxygens are directed to the main chain N of residues 115 and 116 so that there is a contribution from the helix dipole.

A comparison of the structures showed that both forms of the enzyme share identical global conformations with small conformational changes localized to the phosphoserine site (41, 42). The conformations of the Ser113 for aspartate and glutamate mutant structures also resemble the unphosphorylated isocitrate dehydrogenase structure, suggesting that allosteric conformational changes are not responsible for enzymic regulation (40). Support for the notion that phosphorylation prevents isocitrate binding to the enzyme because of a direct interaction between the Ser-P residue and the substrate-binding site was obtained following structural studies on the binary complex of unphosphorylated IDH and isocitrate (40). The structure of the complex determined to a resolution of 2.5 Å indicated that the α-carboxylate group of isocitrate bound to the catalytic site forms a hydrogen bond to the hydroxy side chain of Ser113. Binding of $Mg^{2+}$ isocitrate to the catalytic site induces side-chain rotation of Arg119 and Arg129 in order to form favorable interactions with the substrate. Figure 3b shows a representation of $Mg^{2+}$ isocitrate bound to the catalytic site. The structure of the binary complex of IDH and $NADP^+$ shows that $NADP^+$ binds to the large domain of the enzyme close to the C2 hydrogen of isocitrate so that $NADP^+$ is dehydrogenated and remote from the site of phosphorylation (39).

Electrostatic calculations based on the model-built structure of isocitrate bound to phosphorylated IDH and aspartate and glutamate mutants indicated that phosphorylation of Ser113 or replacement by aspartate or glutamate prevents isocitrate binding by a combination of a loss of a hydrogen bond between the substrate and enzyme, and repulsive electrostatic and steric effects. These predictions of the mechanism of inactivation of IDH by phosphorylation based on structural and kinetic data were elegantly confirmed by a kinetic analysis of the rate of wild-type and mutant enzyme activity against a substrate analogue 2R malate, which is isologous to isocitrate but lacks a γ-carboxylate group (17). 2R Malate acts as a substrate for IDH at a substantially reduced rate; $V_{max}$ is reduced $10^4$-fold, and $K_m$ is increased $10^3$-fold. Phosphorylation of isocitrate dehydrogenase causes a 70-fold reduction in $V_{max}$ and a doubling of $K_m$ towards 2R malate compared with a complete inactivation of the enzyme towards isocitrate. Substitution of serine for aspartate or glutamate, which reduces $V_{max}/K_m$ for isocitrate by $10^5$, reduces $V_{max}/K_m$ only two- to three-fold for 2R malate. The activity of mutant forms of IDH towards isocitrate allowed a quantitative assessment of the contribution of loss of a hydrogen bond and electrostatic and steric repulsive effects to be determined. Loss of a hydrogen bond in the Ser113-Ala mutant results in a reduction of $V_{max}/K_m$ by 17-fold. Steric effects were assessed by determining the activities of serine for tyrosine and lysine mutants, producing a reduction in $V_{max}/K_m$ by $8.5 \times 10^3$ and $3.5 \times 10^3$, respectively. Aspartate and glutamate mutants, mimicking the electrostatic repulsive effects, reduce $V_{max}/K_m$ by $7.5 \times 10^5$ and $2.0 \times 10^5$, respectively. Phosphoserine, which is a bulky residue, will produce steric effects equivalent to those of tyrosine in addition to electro-

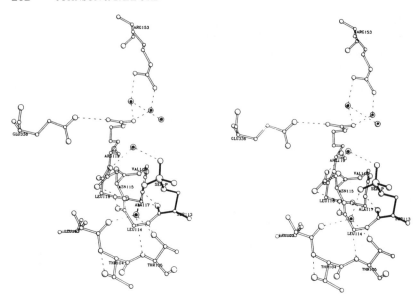

static repulsive effects leading to a net reduction in activity of $10^9$-fold. This observation easily accounts for the complete loss of activity of the phosphorylated enzyme towards isocitrate.

## CYCLIC AMP–DEPENDENT PROTEIN KINASE

The discovery in 1968 (116) of the cAMP-dependent protein kinase (PKA) provided the final piece for solving the puzzle of the hormone-stimulated cascade that uses cAMP as the second messenger. PKA was the second protein kinase to be purified, and its broader specificity compared with phosphorylase kinase led to an explosion in the number of proteins recognized to be controlled by phosphorylation. The protein kinase family now contains over 100 mammalian protein kinases, and if cloned kinases from other species are included, then the total number is over 200 (38). Although these enzymes differ in size, substrate specificity, activation and inhibition mechanisms, and subunit composition, they all show sufficient similarity in their catalytic cores to allow sequence alignment and identification of key conserved groups (31, 32). PKA has become the prototype protein kinase. The crystallographic structure solution in the summer of 1991 (51, 52) has led to insights into the protein-fold, recognition, and catalytic properties of PKA and also has provided a framework for understanding many other serine, threonine, and tyrosine kinases. The *Saccharomyces cerevisae* cAMP-dependent protein kinase has also been crystallized (56), and the structure was determined recently (J. W. Pflugrath & J. Kuret, personal communication).

The mechanism of activation of PKA is somewhat unusual in that it involves dissociation of the regulatory subunit. In the absence of cAMP, the protein exists as an $R_2C_2$ tetramer where R and C are the regulatory and catalytic subunits, respectively. Activation is achieved through binding of cAMP to the R subunits, which promotes dissociation into an $R_2$ dimer and two free active C subunits. The general consensus sequence recognized by PKA is Arg-Arg-X-Ser[Thr]-Y, where X and is any small residue and Y is a large hydrophobic residue. Chemical analysis of the C subunit

---

*Figure 3* The contacts at the catalytic site of isocitratedehydrogenase. (*a*) A stereo diagram of the Ser113-P contacts. The view is down the αD helix and shows the contacts through waters to Thr103, Leu105, and Asn115. Asn232′ from the other subunit is close but is not shown. Drawing produced from coordinate set 4ICD available from the Protein Data Bank from the structure of Hurley et al (41). (*b*) $Mg^{2+}$ isocitrate bound in the active site of IDH. Hydrogens are shown for illustrative purposes where hydrogen bonds between the enzyme and substrate are thought to occur. Ser113 at the bottom of the figure forms a hydrogen bond with the γ carboxylate of the substrate (from 40).

revealed the presence of an acid–stable protein–bound phosphate that could be removed with *E. coli* alkaline phosphatase. The phosphate could be replaced in an autocatalytic process (13), and sequence analysis showed that the sites of phosphorylation occurred at Thr197 and Ser338 (96). The properties of the dephosphorylated enzyme were not reported. The Thr197 autophosphorylation site occurs 10 residues before the sequence APE, a sequence that is highly conserved in all protein kinases and may be a motif associated with autophosphorylation (75).

The crystals of PKA were obtained as a binary complex of the kinase with a tight binding oligopeptide inhibitor ($K_i$ = 2.3 nM), and formation of this complex was a key factor in obtaining X ray–quality crystals. The inhibitor was derived from the amino-terminal region (residues 5–24) of a naturally occurring thermostable protein kinase inhibitor in which the target serine was replaced by an alanine. The crystal structure of the mouse recombinant PKA catalytic subunit crystallized from polyethylene glycol in the presence of the 20–amino acid inhibitor has been solved to 2.7-Å resolution (51, 52). The chain has been traced for residues 15–350 and the 20 residues of the inhibitor.

The molecule is bilobal in shape and is comprised of a small N-terminal domain (residues 40–125) and a larger C-terminal domain (Figure 4). The cleft between the domains is filled by a portion of the inhibitor. The small domain is mostly antiparallel $\beta$-sheet and contains the residues important in binding ATP. These include the Gly-X-Gly-X-X-Gly motif (residues 50–55), which in PKA occurs at a sharp turn between two antiparallel strands. The larger domain is mostly $\alpha$-helical and contains residues important for peptide recognition and catalysis. The N-terminal residues 15–31 form a helix that contacts the large domain, and the C-terminal residues 281–350 extend over a large region from the large domain to the small domain. The highly conserved loop, residues 165–171, has been identified as the catalytic loop. It contains the invariant residues Asp166 and Asn171, which interact with each other, and the highly conserved residues Arg165 and Leu167. Knowledge of the binding mode of the inhibitor peptide indicates that Asp166 is the most likely candidate for the catalytic base. Another conserved aspartate, Asp184, participates in chelation of $Mg^{2+}$.

The inhibitor peptide-recognition site extends over a large area and includes widely separated regions of the sequence located on the enzyme surface. The amino terminal portion of the inhibitor is in an $\alpha$-helical conformation and its interactions are dominated by hydrophobic interactions by a phenylalanine from the inhibitor peptide and electrostatic interactions with an arginine from the inhibitor peptide. In the vicinity of the catalytic site, the inhibitor peptide is in an extended conformation. Residues Arg-Arg-Asn-Ala\*-Ile-His of the inhibitor comprise the con-

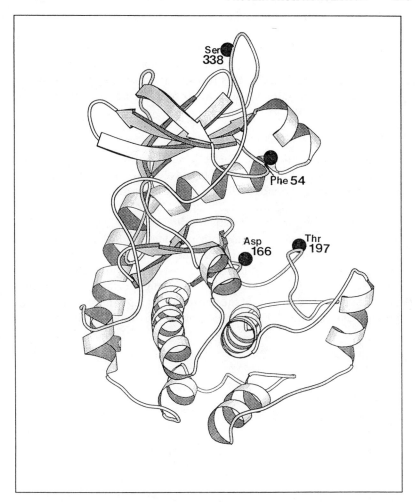

*Figure 4* A shematic diagram of the fold of cAMP-dependent protein kinase (PKA) showing the positions of the phosphorylation sites (Thr197 and Ser338), the catalytic site (Asp166), and the ATP phosphate anchor loop marked by Phe54. Diagram produced with coordinates from the Protein Data Bank from the structure of Knighton et al (51, 52) with the program MOLSCRIPT (53) by M. S. P. Sansom.

sensus peptide-recognition site; these residues are denoted $P-3$, $P-2$, $P-1$, P, $P+1$, $P+2$, where P denotes the phosphorylation-site or pseudo-phosphorylation-site residue (here occupied by Ala). The interactions in the immediate vicinity of the catalytic site involve electrostatic interactions with the two arginine residues and hydrophobic interactions for the nonpolar Ile at the P+1 site. The P+1 site lies at the edge of the cleft and is important for proper orientation of the site of phosphorylation. The isoleucine side chain fits into a hydrophobic pocket containing residues Leu198, Pro202, and Leu205. It is this site and probably the orientation of the catalytic aspartate that appear to be most affected by the phosphorylation of Thr197.

## Autophosphorylation Site

The autophosphorylation site Thr197 occurs on an external loop. The phosphate group is partially accessible to solvent on one side and on the other makes hydrogen bonds and ionic contacts to residues His87, Arg165, Lys189, and Thr195 and main chain N of Thr197 (Figure 5). Arg165 is immediately adjacent to the catalytic residue Asp166. The Thr-phosphate draws together side chains from both the small and large domains and appears to contribute to localization of this region. The interactions of the phosphate with Thr195 and the main-chain N of 197 may also locate the loop residues 192–217 that include Leu 198, an important determinant of the P+1 site. The precise role of the phosphate group cannot be established in absence of structural data on the conformation of the dephosphorylated

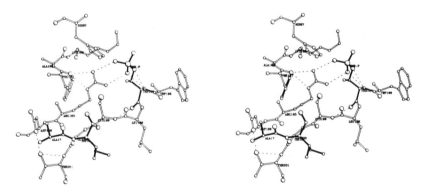

*Figure 5* A stereo diagram of the contacts at the autophosphorylation Thr197 site in PKA. Thr197 is shown in heavy lines. The phosphate group makes direct contacts to Arg165 and Thr197 N and to Thr195 OG2 (not shown). The contact to His87 is 3.4 Å and is just outside the cutoff for hydrogen bonds (3.3 Å). Lys189 NZ is 4 Å away. Part of the 20–amino acid inhibitor is also shown in thick lines (Ala17 and Ile18) and the packing of the inhibitor Ile18 side chain with Leu198 enzyme side chain and the disposition of the catalytic Asp166 to the active-site residue Ala17 are also shown. Figure drawn from coordinates from the Protein Data Bank from the work of Knighton et al (51, 52).

form of the enzyme, but from the understanding of the substrate recognition and the disposition of catalytic residues obtained from the PKA structure, the evidence suggests that the role of the phosphate at Thr197 is primarily conformational. Experimental evidence indicates the phosphate may be important for recognition and not for catalysis. In yeast PKA, mutation of the corresponding residue (Thr241) to alanine resulted in an enzyme that showed a threefold decrease in affinity for substrate and negligible change in activity (59a). The Thr241 → Ala mutant displayed a dramatic decrease in affinity for the regulatory subunit. Substitution of residue 241 with Glu, Asp, or Ser-phosphate partially restored affinity for the regulatory subunit (59b).

The second autophosphorylation site at Ser338 has also been located. This site is on the surface at a site removed from the catalytic site, and the phosphate group makes only one rather long ionic contact with Arg336 (Figure 4). Phosphorylation at this site does not appear to influence the protein.

*Relevance for Other Kinases*

Several other kinases are known to have phosphorylation sites that from sequence alignments correspond to the region of Thr197 in PKA. The sequences in this region are often not conserved, so some variation in optimal alignments may occur (31; G. J. Barton & D. J. Owen, personal communication). Two of these of considerable biological importance are the cell division–control kinase and the mitogen-activated protein kinase.

The genetic characterization of cell division–control mutants in yeast demonstrated the importance of a single protein kinase in cell-division control [reviewed by Nurse (71)]. The *cdc2* gene encodes a 34-kDa phosphoprotein with serine/threonine protein kinase activity that is controlled by association with other proteins, notably the cyclins, and by phosphorylation. Phosphorylation on Thr14 or Tyr15 leads to inhibition, and these residues occur in the sequence GEGT*Y*G (where * indicates phosphorylation site), which maps to the GTGSFG region involved in ATP recognition in PKA. The expression of a mutated *cdc2* gene with the Tyr15-Phe mutation resulted in cells entering mitosis prematurely (27). One postulate is that phosphorylation of this tyrosine may impede ATP binding and lead to loss of activity, whereas the mutant *cdc2* kinase would exhibit no such control. In PKA, the residue equivalent to Tyr15 is a Phe. Examination of the PKA structure in this region shows that the sequence is indeed involved in ATP binding but that phosphorylation may also act by promoting conformational changes that destroy the ATP recognition site.

A recent series of papers showed that Thr167 in *Schizosaccharomyces pombe* or Thr161 in human p34$^{cdc2}$ is phosphorylated and that phosphorylation is essential for kinase activity and for tight binding of cyclin (20, 26). Sequence alignments show that this threonine can align with Thr197 of PKA. It is therefore tempting to conclude that phosphorylation of this residue plays a similar conformational role in the *cdc2* kinases. Replacement of Thr161 in human *cdc2* kinase (or Thr167 in *S. pombe*) with a glutamate causes uncoordination of mitosis and multiple cytokinesis, results that suggest the action of the phosphate group can be partially mimicked by the negatively charged glutamate leading to uncontrolled activation (20, 26).

In the phosphorylase kinase catalytic subunit, the site equivalent to Thr197 in PKA is occupied by a glutamate. Although both the $\alpha$ and $\beta$ subunits of phosphorylase kinase are regulated by phosphorylation, the $\gamma$ catalytic subunit shows no regulation by phosphate (77). Hence, one could speculate that the replacement of the threonine by a glutamate allows the phosphorylase kinase catalytic-subunit active site to be correctly formed in the absence of posttranslational modification, although this has yet to be confirmed with structural studies.

Mitogen-activated protein kinases (MAP kinases) are activated by dual tyrosine and threonine phosphorylation in response to various stimuli, including insulin and phorbol esters (90). In quiescent cells capable of division, MAP kinase activation corresponds with reentry into the cell cycle. The complex roles and history of these enzymes have been reviewed (109). The threonine and tyrosine residues that are phosphorylated occur at positions 183 and 185 in the sequence FLT*EY*VATRW, and both phosphorylations are required for full enzymic protein kinase activity. Sequence alignment (31) shows that the similar protein kinase from yeast KSS1 and the MAP kinase sequences align with these phosphorylation sites close to the corresponding position of Thr197 in PKA. Hence again, the addition of phosphate to the tyrosine and the threonine may act in a role similar to the autophosphorylation of Thr197 in PKA. MAP kinase, however, contains the extraordinary cluster of negatively charged groups (five units of charge in three residues) resulting from phosphorylation that may contribute a more dramatic effect.

## PHOSPHOENOLPYRUVATE SUGAR PHOSPHOTRANSFERASE SYSTEM

The phosphoenolpyruvate (PEP) sugar phosphotransferase system (PTS) mediates exogenous sugar accumulation in bacterial cells and regulates certain non-PTS sugar permeases, transcription from some operons, and chemotaxis towards PTS sugars (reviewed in 66, 78, 84, 89, 91, 92). A phosphoryl group derived from PEP is transferred to the transported sugar via four phosphoprotein intermediates where three of the phosphorylation sites are histidine residues. The system consists of two energy-coupling components (enzyme I and the histidine-containing phosphocarrier protein, HPr) and of a sugar-specific permease enzyme II. PTS permeases consist of either a single three-domain polypeptide of approximately 68 kDa (with domains termed IIA, IIB, and IIC) or two or three distinct polypeptide chains (92). A total of five phosphoryl group transfers occur along the pathway:

$$PEP \rightarrow \text{Enzyme I-(His-N}\varepsilon) \sim P \rightarrow \text{HPr-(His-N}\delta) \sim P \rightarrow \text{Enzyme}$$

$$\text{IIA-(His-N}\varepsilon) \sim P \rightarrow \text{Enzyme IIB} \sim P \rightarrow \text{sugar} \sim P(\text{in}).$$

Domain IIC forms the sugar-specific binding site and transmembrane channel (92). Phosphorylation of IIB leads to an opening of the channel, allowing sugar translocation across the membrane. The activity of the PTS of gram-positive bacteria is regulated by phosphorylation on Ser46 of HPr (81, 83). Phosphorylation inhibits the interaction between HPr and enzyme I 100-fold, thus preventing phosphoryl-transfer (86). Replacement of Ser46 by an aspartate residue mimics the effects of Ser46 phosphorylation (85).

## The Structure of HPr

HPr of *E. coli* has been crystallized in the presence of 68% lithium sulfate at pH 3.8 and the structure determined with isomorphous replacement at a resolution of 2.8 Å and refined to an R factor of 0.23 (21). HPr of *Bacillus subtilis* was crystallized from 63% ammonium sulfate and 0.5% polyethylene glycol 1000 at pH 6.0 (49). The structure was determined independently and refined to an R factor of 0.15 at 2.0-Å resolution (33). Secondary-structure assignments and a proposal for the topology have been made using two-dimensional $^1$H NMR spectroscopy for HPr of *E. coli* (50), *B. subtilis* (119), and *Staphylococcus aureus* (48). Wittekind et al (118) have analyzed differences in chemical-shift resonances of $^1$H between wild-type *B. subtilis* HPr and HPr(Ser-P) and a Ser-to-Asp mutant at position 46. Three-dimensional NMR studies on *E. coli* HPr have also been reported (30, 113).

The structures of *B. subtilis* HPr determined using X-ray crystallography and of *E. coli*, *B. subtilis*, and *S. aureus* HPr determined using NMR are similar, exhibiting an open faced antiparallel β-sheet flanked on one face by three α-helices (Figure 6). The sequence of *E. coli* HPr shares 34 and 32% identity with the sequences of *B. subtilis* and *S. aureus*, respectively, and therefore similarities in tertiary structure are to be expected. The crystal structure of *E. coli* HPr, while exhibiting the same secondary structure, shows a markedly different tertiary structure because the pairs of β strands AD and BC do not form a β-sheet. Epitope mapping results on *E. coli* HPr are consistent only with the structure determined using NMR spectroscopy, which correctly predicts a contiguous epitope, and are inconsistent with the crystal structure (95). The more open conformation of the crystal structure of *E. coli* HPr could represent a partially unfolded conformation caused by the relatively harsh conditions necessary for crystallization (68% lithium sulfate at pH 3.9). In contrast, the NMR structures and the crystal structure of *B. subtilis* HPr were determined at pH 6, close to the physiological pH.

The protein fold consists of a single domain with the catalytic site His15 located at the surface of the molecule at the N terminus of the first α-helix, and the histidine is accessible to solvent (33). A hydrogen bond is formed between Ser12 Oγ and Nε of His15; this may be important in orientating the histidine side chain. Phosphoryl transfer occurs through Nδ of His15 (49). In the crystal structure of *B. subtilis* HPr, a sulfate ion of crystallization is located 3.9 Å from the Nδ atom of His15, and a salt bridge

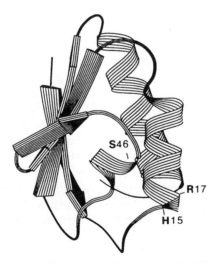

*Figure 6* A ribbon diagram showing the fold of *B. subtilis* HPr. The positions of the phosphorylated residues His15 and Ser46 together with Arg17 are shown (from 33).

is formed between the sulfate ion and the side chain of an invariant arginine at position 17. Sulfate also forms crystal contacts with two neighboring protein molecules. These interactions consist of contacts to a positively charged N terminus, a main-chain nitrogen from Ile47 and O$\gamma$ of Ser66 of one molecule, and a main-chain nitrogen from Met48 and the O$\gamma$ of Ser46 of a second.

The sulfate-binding site in HPr may represent a site similar to that occupied by the phosphate of a phosphohistidine residue (33). Modeling of a phosphohistidine residue with the phosphorous atom in the plane of the imidazole ring of His15 suggests that only small adjustments in the side chains of His15 and Arg17 are needed to accomodate the phosphate group. The phosphoryl group hydrogen bonds to the main chain nitrogens of Ala16 and Arg17 at the N terminus of the first $\alpha$-helix and a salt bridge to the side chain of Arg17.

Ser46, the regulatory site, is located at the N terminus of the second helix at the surface of *B. subtilis* HPr. Modeling shows that the $\chi_1$ angle should be rotated by 60° when Ser46 is phosphorylated to avoid close contact between the phosphoryl group and the helix (33). A surface hydrophobic region formed by Ile47, Met48, and Met51 and flanked by His15 and Ser46 is suggested to form the recognition site for enzyme I and the permease (33, 34). The proximity of His15 and Ser46 suggests a mechanism for the reduction in affinity between HPr and enzyme I caused by phosphorylation of Ser46. Introduction of an electrostatic charge onto the side chain of Ser46 would alter the steric and electrostatic properties of the protein surface close to His 15, and this may be sufficient to impede the interaction between HPr and enzyme I through formation of unfavorable interactions. Thus, phosphorylation on Ser46 could regulate protein interactions by directly altering the structure of the region of the molecule responsible for these interactions.

Perturbations in the chemical-shift positions for NH-C$\alpha$H resonances are observed in the two-dimensional $^1$H NMR spectrum of *B. subtilis* HPr resulting from phosphorylation of Ser46 (118). Similar perturbations are noted for the mutant HPr where Ser46 is replaced by an aspartate. Backbone resonances of residues in the primary sequence near the phosphorylation site exhibit chemical-shift differences in the modified forms. Residues close to His15, including the first two turns of helix 1, also exhibit changes in chemical shift, a result consistent with the proximity of Ser46 to the catalytic site (33). Chemical-shift perturbations are not straightforward to interpret and cannot be directly understood in terms of the nature and magnitude of structural changes. However, the localization of the chemical-shift perturbations to the catalytic and phosphorylation sites suggests that structural changes are restricted to these regions and that global conformational changes do not occur.

## Glucose Permease IIA Domains

*B. subtilis* glucose permease consists of a single polypeptide chain with the domain order IICBA. The IIA$^{glc}$ domain consisting of 162 amino acid residues has been overexpressed (111) and crystallized in the presence of 57–62% ammonium sulfate and 0.5% polyethylene glycol 1000 at pH 8.0 (49). The crystal structure was determined using isomorphous replacement and has been refined to an R factor of 0.2 at 2.2-Å resolution (60). The protein shares 41% sequence identity to *E. coli* IIA$^{glc}$ (also termed Enzyme III$^{glc}$), which exists as a separate polypeptide chain of 168 amino acid residues. Crystals of *E. coli* IIA$^{glc,fast}$ [a form of IIA$^{glc}$ with an N-terminal hexapeptide proteolytically removed (65)], were grown in the presence of 20% polyethylene glycol 8000 at pH 6.6. The crystal structure was determined by isomorphous and molecular replacement methods using the IIA$^{glc}$ domain of *B. subtilis* as a search object (120). The structure has been refined to an R factor of 0.166 for data between 6.0- and 2.1-Å resolution. NMR studies have also been performed on *E. coli* (76) and *B. subtilis* IIA$^{glc}$ (23). The crystallization of IIA$^{lac}$ of *S. aureus* has been reported (12).

The protein fold comprises an antiparallel $\beta$-barrel incorporating Greek Key and jelly-roll topological motifs (60). Approximate two-fold symmetry has been noted in the protein topology (120). The histidine residue that is phosphorylated by HPr (His83 of IIA of *B. subtilis*, His90 of *E. coli*; *E. coli* residue numbers are seven higher than the *B. subtilis* equivalents) is located in a shallow depression at the C terminus of the seventh strand of the barrel and adjacent to a second conserved histidine at position 68 (the equivalent is His75 in *E. coli*), forming a histidine pair. The orientation of the histidine side chains are determined by hydrogen bonds between N$\delta$ of His83 to a main-chain carbonyl of Gly85 and between N$\delta$ of His68 to O$\gamma$ of Thr66. The phosphorylated atom, N$\varepsilon$ (19), is exposed to solvent and is close (3.2 Å) to N$\varepsilon$ of His68. In the *B. subtilis* structure, a water molecule bridges both histidines through hydrogen bonds to the side-chain N$\varepsilon$ atoms. The active site is surrounded by a ring of hydrophobic residues in *B. subtilis* including Phe34, Phe64, Val33, Met38, and Met39, with a similar constellation of side chains in *E. coli* IIA$^{glc}$. Two evolutionarily conserved aspartate residues (Asp31 and 87 in *B. subtilis*) are located some 7 Å from His83 and are orientated towards each other across a narrow groove. Their distance from His83 suggests that they may play roles in forming the protein-recognition interface with HPr and enzyme IIB rather than participating directly in catalysis. Replacement of one of these conserved aspartates to alanine in the homologous $\beta$-glucoside permease of E. coli causes a decrease in the catalytic rate (94). A model proposed for IIA(His-P) of B. subtilis, where the phosphoryl atom is built coplanar with the histidine ring, suggests that an oxygen on the phosphate would hydrogen bond to Nε of His68 and to a main-chain nitrogen of Val89 with no conformational changes. A protonated His68 may function to stabilize the phosphorylated histidine 83. Amide nitrogens of 94 and 95 project towards the pocket, forming an oxyanion hole, reminiscent of the oxyanion hole occurring in serine proteases. Replacement of His75 in E. coli IIA$^{glc}$ with glutamine or substitution of His68 in B. subtilis with alanine results in proteins that are competent to accept phosphate from HPr but cannot transfer a phosphoryl group to IIB$^{glc}$ (79, 82). This implicates the second histidine in catalysis, and Presper et al (79) interpreted the results to suggest that a phosphoryl group is transferred from the catalytic histidine to the adjacent histidine prior to transfer to IIB$^{glc}$.

From knowledge of the three-dimensional structures of HPr and the glucose permease IIA domain of B. subtilis, O. Herzberg (submitted) has proposed a model for the transient complex formed during the phosphoryl-transfer reaction. Docking is achieved with minimal protein conformational changes, and the interface is formed by two complementary surfaces on both protein molecules, comprising the surface hydrophobic regions surrounding the catalytic sites. An electrostatic interaction is proposed between Arg17 of HPr and Asp31 and 87 of IIA$^{glc}$. The model proposes that phosphoryl transfer is catalyzed by the switching of a salt bridge formed between Arg17 and the phosphohistidine residue of HPr to a new position with Asp31 and 87 of IIA$^{glc}$.

## BACTERIAL CHEMOTAXIS AND CheY

Bacterial cells respond to changes in their environment by regulating motility and gene expression. These processes are controlled by a family of proteins including a membrane receptor that detects environmental changes, an autophosphorylating protein histidine kinase triggered by the receptor, a response regulator phosphorylated by the histidine kinase, and a target for regulator action. The chemotactic response to attractants and repellents has been characterized at a molecular and structural level in E. coli and Salmonella typhimurium. The proteins regulating this process are related to those regulating other processes such as osmoregulation, sporulation, virulence, and response to nitrogen, phosphorous, and oxygen depletion (105, 107). Chemotactic behavior is achieved by regulating the frequency of bacterial tumbling caused by the reversal of the rotation of the flagellar motor (8, 107). Transduction of a signal from a membrane receptor (e.g. the aspartate receptor encoded by the Tar gene) to the flagellar motor is mediated by a set of four proteins in a process involving protein phosphorylation and phosphoryl group transfer. CheA (a protein histidine kinase) catalyzes phosphoryl-group transfer from a histidine residue to CheY (a response-regulator protein) (70). The phosphorylated residue is Asp57 (93). Phospho-CheY interacts with the flagellar switch proteins promoting clockwise flagellar rotation and bacterial tumbling.

Insight into the molecular mechanism for the basis of the control of bacterial tumbling has come from X-ray crystallographic studies of CheY from *S. typhimurium* (105–107) and *E. coli* (115), together with site-directed mutagenesis experiments. CheY is a protein of 128 residues. The sequence of *E. coli* and *S. typhimurium* differ at only four positions. CheY from both species crystallizes isomorphously in the presence of 2 M ammonium sulfate; however, at different pH values (*E. coli* at pH 8.0 and *S. typhimurium* at 4.6). The structure of *S. typhimurium* has been reported at 2.7 Å and refined to an R factor of 0.29, while the structure of *E. coli* CheY has been determined at 1.7-Å resolution, aided by knowledge of the *S. typhimurium* structure, and subsequently refined to an R factor of 0.15. The protein is a monomer folding into a single domain composed of a doubly wound five-stranded parallel β-sheet surrounded by five α-helices, a topology similar to p21 *ras* (Figure 7). An acidic active site pocket is

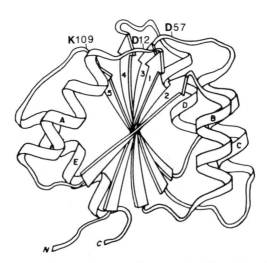

*Figure 7* Ribbon trace of the *S. typhimurium* CheY protein. The positions of Asp12, Asp57, and Lys109 are shown (from 107).

formed by the side chain of the invariant Asp57 residue at the C terminus of $\beta 3$ and the highly conserved residues Asp12 and 13 at the C terminus of $\beta 1$. A divalent metal ion–binding site is formed by the carboxylate side chains of these residues (63). Electron density corresponding to a sulfate anion is located at the catalytic site of *E. coli* CheY. The sulfate ion forms salt-bridges to N$\delta$ of the Asn59 side-chain and to the $\varepsilon$-amino nitrogen of Lys109. The Lys109 side chain also forms a salt bridge to the carboxyl oxygen of Asp57 and via a H$_2$O bridge to the carboxyl group of Asp12. Although the density corresponding to Lys109 in *S. typhimurium* CheY crystals is poorly ordered, it is suggested to form a salt bridge to the carboxyl side chain of Asp57 (62).

Phosphorylation of CheY produces structural changes, which result in 100-fold stimulation of the interaction of this protein with the flagellar motor (2). The instability of phosphoaspartate prevents structural studies on the phospho-CheY [the half-life of phospho-CheY is two minutes (35)], and therefore, the structure of the phosphorylated form of CheY cannot be directly determined. The conformation of side chains of Asp12, 13, and 57 and Lys109 and the proposed Mg$^{2+}$ ion site forms a network of salt bridges and hydrogen bonds in the active site of CheY. The residues and solvent molecules are tightly packed, and all possible interactions involving Asp57 and Lys109 are satisfied. This would suggest that phosphorylation of Asp57 would result in a concerted conformational change of these side chains. The side chain of Lys109 (in the *E. coli* structure) is well placed to form a salt bridge to the phosphoaspartate group, and the position of the sulfate ion and the Lys109 side chain in the crystal structure of *E. coli* CheY may represent a conformation close to the activated one. The occurrence of weak electron density corresponding to Lys109 in the crystals of *S. typhimurium* CheY may suggest that in the inactivated state of CheY, Lys109 is poorly ordered but becomes ordered on phosphorylation of Asp57 or binding of a sulfate ion close to the Asp57 side chain.

Features of the activated conformation of CheY resulting from phosphorylation of Asp57 have been proposed on the basis of site-directed substitution of Asp12 and 13 to Asn and Lys109 to Arg (9, 62). The role of Lys109 in regulating tumbling was suggested from two findings. First, although it did not affect the phosphorylation of CheY, replacing Lys109 with Arg abolished the ability of phospho-CheY to reverse flagellar motor rotation. Moreover, dephosphorylation, catalyzed by CheZ, is inhibited, suggesting that the conformation of mutant phospho-CheY, inactive in regulating tumbling, cannot be recognized by CheZ. A second finding implicating Lys109 is that substitution of Asp13 for Lys, while abolishing Asp57 phosphorylation, mimics the wild-type phosphoprotein in stimulating tumbling behavior. This result suggests that the Lys13 side chain competes with Lys109 for a salt bridge to Asp57, causing a conformational change in the side chain of Lys109 similar to that caused by phosphorylation of Asp57 in the wild-type protein. The result also demonstrates that phosphorylation of Asp57 causes conformational changes remote from the site of phosphorylation because the activated conformation of CheY, induced by Asp57 phosphorylation, may be produced by modifications remote from Asp57. Substitution of Asp13 or 57 for Asn abolishes regulation of tumbling because phosphorylation of Asp57 is prevented.

The function of CheY has been likened to the *ras* family of GTP binding proteins in eukaryotic cells (105, 107). These proteins show similarities in the process of transducing signals from receptors to effectors, and their tertiary structures share a common topology (73). The molecular mechanism of switching is similar in that conversion to the inactive state requires hydrolysis of a phosphoanhydride bond, a reaction stimulated by activating proteins CheZ and GAP, respectively. Such similarities suggest a common ancestor for CheY and *ras*-mediated signal transduction processes in bacteria and eukaryotes, respectively.

## SILENT PHOSPHORYLATION

Pepsinogen and pepsin are phosphorylated at Ser68. Enzymatic removal of phosphate has no effect on activity of either pepsinogen or pepsin, and the function of the phosphate is unknown. The structure of porcine pepsin at 1.8-Å resolution (97, 98, 117) shows that Ser68-P is located within a $\beta$ bulge in a strand that forms the bend leading into an extended $\beta$ hairpin loop. The residue is located on the surface and is exposed to solvent.

Ovalbumin is a noninhibitory member of the serine protease inhibitor (serpin) superfamily. The protein is present as 60–65% of the protein in egg white but its physiological function is unknown. Three main forms have been distinguished by their electrophoretic behavior with, respectively, zero, one, or two phosphate groups. The phosphorylation positions have been established as Ser87c and Ser350. In the crystal structure (103, 104), the electron density for both phosphate sites is weak. Both sites are external and associated with helices. The position of Ser350 is at the start of an external, rather mobile, helix. In the serpins, this loop is the reaction center and the bond between residues 358 and 359 is cleaved by proteases, which leads to a dramatic change in structure. The role of the phosphate in ovalbumin is not known, but interestingly one phosphorylation site is close to the analogous reaction center of the serpins.

## DISCUSSION

The structural results discussed above on proteins controlled by phosphorylation have given rise to at least five possible mechanisms. (*a*) In glycogen phosphorylase, the mechanism of activation by phosphorylation is allosteric; local conformational changes are associated with long-range effects transmitted by subunit-subunit interactions so that phosphorylation at Ser14 leads to activation at the catalytic site over 35 Å away. (*b*) In isocitrate dehydrogenase, the mechanism is based on an electrostatic blocking process in which phosphorylation of Ser113 prevents binding of the anionic substrate without promoting any conformational change. (*c*) In PKA and CheY, the effects of phosphorylation appear to be mostly conformational: in PKA the assumption is that phosphorylation of Thr197 localizes a loop that forms part of the substrate recognition site and regulatory-subunit recognition site and maintains the catalytic-site residues in their correct conformation; in CheY, the assumption is that phosphorylation of an aspartic acid residue Asp57 causes conformational changes in nearby residues (e.g. Lys109). (*d*) In HPr, phosphorylation of Ser46 is proposed to inhibit protein/protein interactions through changes in the surface properties and conformation so that the protein is not recognized by the enzyme that transfers phosphate to the catalytic His15. (*e*) Finally, phosphorylation may not do anything to the protein. In this category of silent phosphorylation sites we include pepsin, ovalbumin, and probably the Ser338 site in PKA.

The electrostatic properties of the covalently linked (assumed) dianionic phosphate group are important in mechanisms *a–d*. In phosphorylase, the electrostatic interactions are important but not sufficient. Phosphorylase requires the N-terminal residues for transmission of the allosteric response, and a single, negative-charge Ser14-to-Asp mutant is insufficient to trigger response. In isocitrate dehydrogenase, addition of a single negative-charge (Ser-to-Asp) mutant is almost as effective (rate reduction $10^5$) as phosphorylation (rate reduction $10^6$) in inactivation of the enzyme, but calculations show that steric effects are also important. Concomitant disorder/order transitions in response to charge localization are also a feature of some of these proteins. In phosphorylase, the N-terminal 20 residues become localized on phosphorylation. In CheY the structural transitions observed for Lys109 have been identified as likely to be important in transmission of conformational changes. And in HPr the movements of Arg17 upon binding phosphate and the subsequent movements and interactions proposed when HPr is recognized by glucose permease IIA also follow order/disorder transitions.

Table 1 summarizes the topology and contacts at the phosphorylation

**Table 1** Phosphorylation sites in proteins determined from X-ray evidence

| Enzyme | Site of phosphorylation | Topology | Polar contacts to phosphate <3.4 Å | Comment |
|---|---|---|---|---|
| GPa | Ser14 | Start of a 1 turn $3_{10}$ helix, residues 15–17 | O1 ... Val15 N<br>O2 ... Arg43′ NH1<br>O3 ... Arg69 NH1, NH2<br>O3 ... Arg16 N | Conformation of unphosphorylated protein different in this region |
| IDH | Ser113 | Start of α-helix D, residues 113–132 | O2 ... Wat ... Thr105 OG1, N<br>O3 ... Wat ... Asn115 ND2<br>O1 ... Asn232′ ND2 | No change on phosphorylation |
| PKA | (i) Thr197 | Surface loop; β conformation but no defined secondary structure | O2 ... Arg165 NH1<br>O2 ... Thr197 N<br>O2 ... Thr195 OG1<br>O3 ... His87 ND1<br>O3 ... Arg165 NH2 | |
| PKA | (ii) Ser338 | Surface loop | No contacts | |
| HPr | (i) His15 | N-cap position; start of α-helix residues 16–26 | O1 ... Arg17 NH1, NH2<br>O2 ... Arg17 N<br>O2 ... Ala16 N | Catalytic site<br>Phosphate modeled from sulphate |
| HPr | (ii) Ser46 | N-cap position; start of α-helix residues 47–51 | ND[a] | Inhibitory site |
| Glucose permease IIA | His83 | C-terminal end of β-strand | O1 ... His68 NE2<br>O2 ... His68 NE2<br>O2 ... Val89 N | Phosphotransfer site<br>Phosphate modeled |
| CheY | Asp57 | C terminal end of β-strand | Lys109 | From sulphate site |
| Pepsin | Ser68 | External loop β-bulge | No contacts | Weak electron density |
| Ovalbumin | (i) Ser87c | In 1 turn helix C2, residues 87b–87e | No contacts | Weak electron density |
| | (ii) Ser350 | Start of helix R, residues 350–359 | No contacts | Weak electron density |

[a] ND: not determined.

sites. At present, there is no easy relationship between the topology of the phosphorylation site and the resulting mechanism. All the sites are on the surface and at least partially accessible, as one would expect if the proteins are to be recognized by kinases and phosphatases. Representation of phosphorylation sites is more frequent at the start of an α-helix (in phosphorylase, isocitrate dehydrogenase, both sites in HPr, and both sites in ovalbumin). Certainly such a position is favorable for a phosphate group because it allows the phosphate oxygens to interact with the partial positive charge at the N terminus of the helix and to hydrogen bond to the mainchain NH groups. However, serine tends to occupy the N-cap positions at the start of α-helices (87; see also 1) and more examples of other phosphoproteins are needed to see whether phosphorylation at these sites is more general. We note that the conformation adopted by the inhibitor of PKA at the catalytic site is a $\beta$ structure. If it is assumed that the inhibitor mimics the substrate, then $\beta$ conformations around the phosphorylation site must be adopted by many proteins, at least in the recognition process. In two proteins, CheY and glucose permease IIA, the phosphate site is at the C-terminal end of a $\beta$ strand.

Examination of the contacts made at the phosphorylation sites (Table 1) shows a dramatic difference between GPa and PKA, in one group in which the role of phosphorylation is to promote conformational changes, and between IDH in another group in which the role is to block. In the conformational group (GPa and PKA), the contacts are strong, involving arginine residues and other side chains. In the blocking group, the contacts are weak and are made only to polar groups through water molecules.

The past few years have seen remarkable progress in understanding some of the basic responses of enzymes to phosphorylation. Likely, other variations on the mechanisms described here will account for the marvellous variety of proteins controlled by phosphorylation. In the structural examples described above, we have definite evidence for regulation by a single phosphate site. An important feature of control by phosphorylation is multiple phosphorylation in which several sites may be phosphorylated or dephosphorylated in response to different signals by different enzymes (88). To date, we have no evidence for the structural basis of these mechanisms.

ACKNOWLEDGMENTS

We are grateful to Dr. M. S. P. Sansom and Mrs. A. Roper for help with the diagrams and to all who sent us material in advance of publication.

## Literature Cited

1. Acharya, K. R., Stuart, D. I., Varvill, K. M., Johnson, L. N. 1991. *Glycogen Phosphorylase b: Description of the Protein Structure.* Singapore/London: World Scientific. 123 pp.
2. Barak, R., Eisenbach, M. 1992. *Biochemistry* 31: 1821–26
3. Barford, D. 1991. *Biochim. Biophys. Acta* 113: 55–62
4. Barford, D., Johnson, L. N. 1989. *Nature* 340: 609–16
5. Barford, D., Johnson, L. N. 1992. *Protein Sci.* 1: 472–93
6. Barford, D., Hu, S.-H., Johnson, L. N. 1991. *J. Mol. Biol.* 218: 233–60
7. Borthwick, A. C., Holms, W. H., Nimmo, H. G. 1984. *FEBS Lett.* 174: 2956–60
8. Bourret, R., Boikovich, K. A., Simon, M. I. 1991. *Annu. Rev. Biochem.* 60: 401–41
9. Bourret, R. B., Hess, J. F., Simon, M. I. 1990. *Proc. Natl. Acad. Sci. USA* 87: 41–45
10. Browner, M. F., Fletterick, R. J. 1992. *TIBS* 17: 66–71
11. Browner, M. F., Rasor, P., Tugendreich, S., Fletterick, R. J. 1991. *Protein Eng.* 4: 351–57
12. Celikel, R., Dai, X., Stewart, G. C., Sutrina, S. L., Saier, M. H. Jr., et al. 1991. *J. Mol. Biol.* 222: 857–59
13. Chiu, Y. S., Tao, M. 1978. *J. Biol. Chem.* 253: 7145–48
14. Coats, W. S., Browner, M. F., Fletterick, R. J., Newgard, C. B. 1991. *J. Biol. Chem.* 266: 16113–19
14a. Cohen, P. 1982. *Nature* 296: 613–20
15. Cohen, P. 1991. *Biochim. Biophys. Acta* 1094: 292–99
16. Cozzone, A. J. 1988. *Annu. Rev. Microbiol.* 42: 97–125
17. Dean, A. M., Koshland, D. E. Jr. 1990. *Science* 249: 1044–46
18. Dean, A. M., Lee, M. H. I., Koshland, D. E. Jr. 1989. *J. Biol. Chem.* 264: 20482–86
19. Dorschug, M., Frank, R., Kalbitzer, H. R., Hengstenberg, W., Deutscher, J. 1984. *Eur. J. Biochem* 144: 113–19
20. Ducommun, B., Brambilla, P., Felix, M.-A., Franza, B. R., Kransenti, E., Draetta, G. 1991. *EMBO J.* 10: 3311–19
21. El-Kabbani, O. A. L., Waygood, E. B., Delbaere, L. T. J. 1987. *J. Biol. Chem.* 27: 12926–29
22. Engers, H. D., Madsen, N. B. 1968. *Biochem. Biophys. Res. Commun.* 33: 49–54
23. Fairbrother, N., Cavangh, J., Dyson, H. J., Palmer, A. G., Sutrina, L .S., et al. 1991. *Biochemistry* 30: 6896–6907
24. Fischer, E. H., Graves, D. J., Crittenden, E. R. S., Krebs, E. G. 1959. *J. Biol. Chem.* 234: 1698–1704
25. Fischer, E. H., Krebs, E. G. 1955. *J. Biol. Chem.* 216: 121–32
26. Gould, K., Moreno, S., Owen, D. J., Sazer, S., Nurse, P. 1991. *EMBO J.* 10: 3297–3309
27. Gould, K., Nurse, P. 1989. *Nature* 342: 39–45
28. Gratecos, D., Detwiler, T. C., Hurd, S., Fischer, E. H. 1977. *Biochemistry* 16: 4812–17
29. Graves, D. J., Wang, J. H. 1972. *Enzymes* 7: 435–82
30. Hammen, P. K., Waygood, E. B., Klevit, R. E. 1991. *Biochemistry* 30: 11842–50
31. Hanks, S. K., Quinn, A. M. 1991. *Methods Enzymol.* 200: 38–62
32. Hanks, S. K., Quinn, A. M., Hunter, T. 1988. *Science* 241: 42–52
33. Herzberg, O., Reddy, P., Sutrina, S., Saier, M. H. Jr., Reizer, J., Kapadia, G. 1992. *Proc. Natl. Acad. Sci. USA* 89: 2499–2503
34. Herzberg, O. 1992. *J. Biol. Chem.* 267: 24819–23
35. Hess, J. F., Oosawa, K., Kaplan, N., Simon, M. I. 1988. *Cell* 53: 79–88
36. Hol, W. G. J., van Duijnen, P. T., Berendsen, H. J. C. 1978. *Nature* 273: 443–46
37. Huang, J. M., Wei, Y. F., Kim, Y. H., Osterberg, L., Matthews, H. R. 1991. *J. Biol. Chem.* 266: 9023–31
38. Hunter, T. 1991. *Methods Enzymol.* 200: 3–37
39. Hurley, J. H., Dean, A. M., Koshland, D. E. Jr., Stroud, R. M. 1991. *Biochemistry* 30: 8671–78
40. Hurley, J. H., Dean, A. M., Sohl, J. L., Koshland, D. E. Jr., Stroud, R. M. 1990. *Science* 249: 1012–16
41. Hurley, J. H., Dean, A. M., Thorsness, P. E., Koshland, D. E. Jr., Stroud, R. M. 1990. *J. Biol. Chem.* 265: 3599–3602
42. Hurley, J. H., Thorsness, P. E., Ramalingham, V., Helmers, N. H., Koshland, D. E. Jr., Stroud, R. M. 1989. *Proc. Natl. Acad. Sci. USA* 86: 8635–39
43. Johnson, L. N. 1984. In *Inclusion Compounds*, ed. J. L. Atwood, J. E. D. Davies, D. MacNicol, pp. 509–69. London: Academic
44. Johnson, L. N. 1992. *FASEB J.* 6: 2274–82
45. Johnson, L. N., Barford, D. 1990. *J. Biol. Chem.* 265: 2409–12

46. Johnson, L. N., Hajdu, J., Acharya, K. R., Stuart, D. I., McLauchlin, P. J., et al. 1989. In *Allosteric Enzymes*, ed. G. Herve, pp. 81–127. Boca Raton, FL: CRC Press
47. Kalbitzer, H. R., Hengstenberg, W., Rosch, P., Muss, P., Bernmann, P., et al. 1982. *Biochemistry* 21: 2879–85
48. Kalbitzer, H. R., Neidig, K. I., Hengstenberg, W. 1991. *Biochemistry* 30: 11186–92
49. Kapadia, G., Chen, C., Reddy, P., Siaer, M. H. Jr, Reizer, J., Herzberg, O. 1991. *J. Mol. Biol.* 221: 1079–80
50. Klevit, R. E., Waygood, E. B. 1986. *Biochemistry* 25: 7774–81
51. Knighton, D. R., Zheng, J., Ten Eyck, L. F., Ashford, V. A., Xuong, N.-H., et al. 1991. *Science* 253: 407–13
52. Knighton, D. R., Zheng, J., Ten Eyck, L. F., Xuong, N.-H., Taylor, S. S., Sowadski, J. M. 1991. *Science* 253: 414–20
53. Kraulis, P. J. 1991. *J. Appl. Crystalogr.* 24: 946–50
54. Krebs, E. G. 1986. *Enzymes* 17: 3–20
55. Krebs, E. G., Fischer, E. H. 1956. *Biochim. Biophys. Acta* 20: 150–57
56. Kuret, J., Pflugrath, J. W. 1991. *Biochemistry* 30: 10595–10600
57. LaPorte, D. C., Koshland, D. E. Jr. 1982. *Nature* 300: 458–60
58. Leonidas, D. D., Oikonomakos, N. G., Papageorgiou, A. C., Xenakis, A., Cazianis, C. T., Bem, F. 1990. *FEBS Lett.* 261: 23–27
59. Leonidas, D. D., Oikonomakos, N. G., Papageorgiou, A. C. 1991. *Biochim. Biophys. Acta* 1076: 305–7
59a. Levin, L. R., Kuret, J., Johnson, K. E., Powers, S., Cameron, S., et al. 1988. *Science* 240: 68–70
59b. Levin, L. R., Zoller, M. J. 1990. *Mol. Cell. Biol.* 10: 1066–75
60. Liao, D.-I., Kapadia, G., Reddy, P., Saier, M. H. Jr, Reizer, J., Herzberg, O. 1991. *Biochemistry* 30: 9583–94
61. Luecke, H., Quicho, F. A. 1990. *Nature* 347: 402–6
62. Lukat, G. S., Lee, B. H., Mottonen, J. M., Stock, A. M., Stock, J. B. 1991. *J. Biol. Chem.* 266: 8348–54
63. Lukat, G. S., Stock, A. M., Stock, J. B. 1990. *Biochemistry* 29: 5436–44
64. Madsen, N. B. 1986. *Enzymes* 17: 366–94
65. Meadow, N. D., Coyle, P., Komoryia, A., Anfinsen, C. B., Roseman, S. 1986. *J. Biol. Chem.* 261: 13504–9
66. Meadow, N. D., Fox, D. K., Roseman, S. 1990. *Annu. Rev. Biochem.* 59: 497–542
67. Monod, J., Wyman, J., Changeux, J.-P. 1965. *J. Mol. Biol.* 12: 88–118
68. Newgard, C. B., Hwang, P. K., Fletterick, R. J. 1989. *Crit. Rev. Biochem. Mol. Biol.* 24: 66–99
69. Nimmo, G. A., Borthwick, A. C., Holms, W. H., Nimmo, H. G. 1984. *Eur. J. Biochem.* 141: 401–8
70. Ninfa, E. G., Stock, A., Mowbray, S., Stock, J. 1991. *J. Biol. Chem.* 266: 9674–9770
71. Nurse, P. 1990. *Nature* 344: 503–8
72. Oikonomakos, N. G., Acharya, K., Johnson, L. N. 1991. In *Post-Translational Modification of Proteins*, ed. J. Crabbe, J. Harding, pp. 81–151. Boca Raton, FL: CRC Press
73. Pai, E. F., Kabsch, W., Krengel, U., Holmes, K. C., John, J., Wittinghofer, A. 1989. *Nature* 341: 209–14
74. Palm, D., Klein, H. W., Schinzel, R., Buehner, M., Helmreich, E. J. M. 1990. *Biochemistry* 29: 1099–1107
75. Payne, D. M., Rossomando, A. J., Martino, P., Erickson, A. K., Her, J.-H., et al. 1991. *EMBO J.* 10: 885–92
76. Pelton, J. G., Torchia, D. A., Meadow, N. D., Wong, C.-Y., Roseman, S. 1991. *Proc. Natl. Acad. Sci. USA* 88: 3479–83
77. Pickett-Gies, C. A., Walsh, D. 1986. *Enzymes* 17: 395–459
78. Postma, P. W., Lengeler, J. W. 1985. *Microbiol. Rev.* 49: 232–69
79. Presper, K. A., Wong, C.-Y., Liu, L., Meadow, N. D., Reseman, S. 1989. *Proc. Natl. Acad. Sci. USA* 86: 4052–55
80. Quiocho, F. A., Sack, J. S., Vyas, N. K. 1987. *Nature* 329: 561–64
81. Reizer, J., Novotny, M. J., Hengstenberg, W., Saier, M. H. 1984. *J. Bacteriol.* 160: 333–40
82. Reizer, J., Pao, G. M., Saier, M. H. Jr. 1992. *J. Mol. Evol.* In press
83. Reizer, J., Peterkosky, A. 1987. In *Sugar Transport and Metabolism in Gram-Positive Bacteria*, ed. J. Reizer, A. Peterkosky, pp. 333–64. Chichester, UK: Ellis Horwood
84. Reizer, J., Saier, M. H. Jr., Deutscher, J., Grenier, F., Thompson, J., Hengstenberg, W. 1988. *CRC Crit. Rev. Microbiol.* 15: 297–338
85. Reizer, J., Sutrina, S. L., Saier, M. H., Stewart, G. C., Peterkofsky, A., Reddy, P. 1989. *EMBO J.* 8: 2111–20
86. Reizer, J., Sutrina, S. L., Wu, L.-F., Deutscher, J., Reddy, P., Saier, M. H. Jr. 1992. *J. Biol. Chem.* 267: 9158–69
87. Richardson, J. S., Richardson, D. C. 1988. *Science* 240: 1648–52
88. Roach, P. J. 1990. *FASEB J.* 4: 2961–68

89. Roseman, S., Meadow, N. D. 1990. *J. Biol. Chem.* 265: 2993–96
90. Rossomando, A., Wu, J., Weber, M. J., Sturgill, T. W. 1992. *Proc. Natl. Acad. Sci. USA* 89: 5221–25
91. Saier, M. H. Jr. 1985. *Mechanism, Regulation of Carbohydrate Transport in Bacteria.* New York: Academic
92. Saier, M. H. Jr., Reizer, J. 1990. *Res. Microbiol.* 141: 1033–38
93. Sanders. D. A., Gilleco-Castro, B. L., Stock, A. M., Burlingame, A. L., Koshland, D. E. Jr. 1989. *J. Biol. Chem.* 264: 21770–78
94. Schnetz, K., Sutrina, S. L., Saier, M. H. Jr., Rak, B. 1990. *J. Biol. Chem.* 265: 13464–71
95. Sharma, S., Georges, F., Delbaere, L. T. J., Lee, J. S., Klevit, R. E., Waygood, E. B. 1991. *Proc. Natl. Acad. Sci. USA* 88: 4877–81
96. Shoji, S., Titani, K., Demaille, J. G., Fischer, E. H. 1979. *J. Biol. Chem.* 254: 6211
97. Sielecki, A. R., Fedorov, T., Boodhoo, A., Andreeva, N. S., James, M. N. G. 1990. *J. Mol. Biol.* 214: 143–70
98. Sielecki, A. R., Fujinagu, M., Read, R. J., James, M. N. G. 1991. *J. Mol. Biol.* 219: 671–92
99. Sotiroudis, J. G., Oikonomakos, N. G., Evangelopoulos, A. E. 1978. *Biochem. Biophys. Res. Commun.* 90: 234–39
100. Sprang, S. R., Acharya, K. R., Goldsmith, E. J., Stuart, D. I., Varvill, K. M., et al. 1988. *Nature* 336: 215–21
101. Sprang, S. R., Fletterick, R. J. 1979. *J. Mol. Biol.* 131: 523–51
102. Sprang, S. R., Withers, S. G., Goldsmith, E. J., Fletterick, R. J. Madsen, N. B. 1991. *Science* 254: 1367–71
103. Stein, P. E., Leslie, A. G. W., Finch, J. T., Turnell, W. G., McLaughlin, P. M., Carrel, R. W. 1990. *Nature* 347: 99–102
104. Stein, P. E., Leslie, A. G. W., Finch, J. T., Carrell, R. W. 1991. *J. Mol. Biol.* 221: 941–59
105. Stock, J. B., Lukat, G. S., Stock, A. M. 1991. *Annu. Rev. Biophys. Chem.* 20: 109–36
106. Stock, A. M., Mottonen, J. M., Stock, J. B., Schutt, C. E. 1989. *Nature* 337: 745–49
107. Stock, J. B., Stock, A. M., Mottonen, J. M. 1990. *Nature* 344: 395–400
108. Stroud, R. M. 1991. *Curr. Opin. Struct. Biol.* 1: 826–35
109. Sturgill, T. W., Wu, J. 1991. *Biochem. Biophys. Acta* 1092: 350–57
110. Sutherland, E. W., Wosilait, W. D. 1955. *Nature* 175: 169–71
111. Sutrina, S. L., Reddy, P., Saier, M. H. Jr., Reizer, J. 1990. *J. Biol. Chem.* 265: 18581–89
112. Thorsness, P. E., Koshland, D. E. Jr. 1987. *J. Biol. Chem.* 262: 10422–25
113. van Nuland, N. A., van Dijk, A. A., Dijkstra, K., van Hoessel, F. H., Scheek, R. M., Robilland, G. T. 1992. *Eur. J. Biochem.* 203: 483–91
114. Verlinde, C. L. M. J., Noble, M. E. M., Kalk, K. H., Groendijk. H., Wierenga, R. K., Hol, W. G. J. 1991. *Eur. J. Biochem.* 198: 53–57
115. Volz, K., Matsumura, P. 1991. *J. Biol. Chem.* 266: 15511–19
115a. Waksman, G., Kominos, D., Robertson, S. C., Pant, N., Baltimore, D., et al. 1992. *Nature* 358: 646–53
116. Walsh, D. A., Perkins, J. P., Krebs, E. G. 1968. *J. Biol. Chem.* 243: 3763–74
117. Williams, S. P., Bridger, W. A., James, M. N. G. 1986. *Biochemistry* 25: 6655–59
118. Wittekind, M., Reizer, J., Deutscher, J. Saier, M. H., Klevit, R. E. 1989. *Biochemistry* 28: 9908–12
119. Wittekind, M., Reizer, J., Klevit, R. E. 1990. *Biochemistry* 29: 7191–7200
120. Worthylake, D., Meadow, N. D., Roseman, S., Liao, D.-I., Herzberg, O., Remington, S. J . 1991. *Proc. Natl. Acad. Sci. USA* 88: 10382–86
121. Wosilait, W. D., Sutherland, E. W. 1956. *J. Biol. Chem.* 218: 469–81

# WHAT DOES ELECTRON CRYOMICROSCOPY PROVIDE THAT X-RAY CRYSTALLOGRAPHY AND NMR SPECTROSCOPY CANNOT?

## W. Chiu

Verna and Marrs McLean Department of Biochemistry and The W. M. Keck Center for Computational Biology, Baylor College of Medicine, One Baylor Plaza, Houston, Texas 77030

KEY WORDS: image reconstruction, three-dimensional macromolecule structure

CONTENTS

PERSPECTIVES AND OVERVIEW ............................................................. 233
BIOPHYSICS OF NMR SPECTROSCOPY, X-RAY CRYSTALLOGRAPHY, AND ELECTRON
    MICROSCOPY ....................................................................... 234
    How to Do Three-Dimensional Reconstruction .................................... 236
    How to Prepare Cryospecimens and Obtain Electron Images ........................ 237
    Structural Resolution Attainable with Electron Cryomicroscopy and
        Three-Dimensional Reconstruction ............................................ 238
    Biological Specimens Most Amenable to Study with Electron Cryomicroscopy and
        Three-Dimensional Reconstruction ............................................ 239
FUTURE PROSPECTS ........................................................................ 251

## PERSPECTIVES AND OVERVIEW

The three-dimensional structure of a biomolecule provides a visual framework for understanding its mechanism of action. X-ray crystallography is an established technique that can provide the atomic positions of biomolecules that can be crystallized (42). Nuclear magnetic resonance (NMR) spectroscopy is an emerging technique that can determine in solution atomic structures of biomolecules with moderately small molecular weights (29, 118, 125). Both of these techniques are now regarded as essential structural tools in biochemistry and molecular biology.

DeRosier & Klug (35) were the first to determine a three-dimensional structure from electron images. In their work with the T4 bacteriophage tail, they laid the conceptual and computational foundation of image reconstruction in macromolecular electron microscopy. Later, Henderson & Unwin (59) reached a major breakthrough in protein electron crystallography; they determined the three-dimensional structure of a periodic array of bacteriorhodopsin in *Halobacterium halobium*.

During the past decade, numerous advances have been made in electron cryomicroscopy. These advances include improved methods for making two-dimensional periodic arrays of membrane and soluble proteins (72, 75), better techniques for preserving specimens in the microscopic vacuum (38, 39), improved cryospecimen holders (37), the advent of computer-guided imaging (17, 19, 36), and the application of correlation analysis for noisy and imperfect images (45, 56, 101). These advances have made electron imaging a useful tool for discerning the structures of two-dimensional crystals and specimens with helical or icosahedral symmetries, thus spanning a broad range of resolution from 3.5 Å to 35 Å. Significant developments in computational procedures that employ multivariate statistical analysis have also created new opportunities for determining the structures of single molecules that have no internal symmetry (48, 116).

Here, after briefly reviewing the principles of electron cryomicroscopy and image reconstruction and their current technology, I give several examples to support the rationale for using these techniques instead of X-ray crystallography and NMR spectroscopy to derive structural information about biomolecules.

## BIOPHYSICS OF NMR SPECTROSCOPY, X-RAY CRYSTALLOGRAPHY, AND ELECTRON MICROSCOPY

The experimental measurements derived from X-ray crystallography, electron microscopy, and NMR spectroscopy differ because of the physical interactions between atoms and the types of electromagnetic radiation. NMR spectroscopy usually measures spectra resulting from the resonance signals of protons in a molecule, whereas electron microscopy and X-ray crystallography measure the scattering from entire atoms and valence electrons, respectively. To determine protein structure with NMR spectroscopy, resonances must be assigned to specific amino acids in a known sequence of the protein, and the crosspeaks in the two-dimensional spectra are interpreted in terms of the geometrical constraints of the atomic groups (126). This method does not generate a unique structure but a family of structures consistent with the constraints of NMR spectroscopy. Comparisons of structures derived from NMR spectroscopy and X-ray crystallography of the same proteins have affirmed the consistency of both techniques while pointing out their respective advantages and pitfalls (118). A unique feature of NMR spectroscopy is its ability to probe the dynamics of molecular conformations not readily detectable using crystallographic methods. Because the interpretation of complex spectra is difficult, NMR spectroscopy can be used to study the structure of relatively small molecules (up to 150 amino acid residues in a single protein). However, the prospect of applying this technique to structural studies of larger molecules is quite optimistic (29).

In X-ray crystallography, a diffraction pattern measures the amplitudes of the scattering but not of the phases. The electron density of an object can be represented by a sum of sinusoidal waves, each of which is characterized by its frequency, amplitude, and phase. Whereas the phases contribute to the positional accuracy in the electron density of the molecule, the amplitudes affect the relative weights of the electron density. X-ray crystallography obtains its amplitudes from diffraction intensities. The major task in solving a molecular structure derived from X-ray crystallography is to determine the phases in various ways, including multiple isomorphous replacement, molecular replacement, noncrystallographic symmetry averaging, anomalous scattering, and multiple-wavelength anomalous dispersion (12). The accuracy of the X-ray structure is measured in terms of the figure of merit and the R factor comparing the measured and calculated diffraction intensities. The number of new structures derived from X-ray crystallography has increased during the past few years because of advances in data collection with area detectors and synchrotron light sources and the availability of faster and more affordable computers and graphics workstations (15, 60). The macromolecules now studied by X-ray crystallography range from small peptides to large assemblies, including membranes and viruses.

Because of their charged properties, high-energy electrons can be focused by electromagnetic lenses to form a diffraction pattern or a high-resolution image of a specimen. The electron energy primarily used in electron microscopy is 100 keV, although new high-resolution instruments can operate with electron energies up to 400 keV. Electron diffraction patterns and images of a thin biological object can be interpreted with the weak phase approximation that relates the object's density function, diffraction pattern, and images by Fourier transformation (114). An electron-diffraction pattern of a thin protein crystal is similar to an X-ray diffraction pattern that loses the phases. A unique advantage of electron microscopy over X-ray diffraction is its ability to record images from which amplitude and phase information essential for three-dimensional reconstruction of an object can be retrieved by computing the Fourier transform of an image (50).

As with any imaging apparatus, an electron image is imperfect because of a combination of instrumental factors as well as statistical noise. This image imperfection implies an inaccuracy in the computed amplitudes and phases derived from the raw image data. For a crystalline specimen, the amplitudes for reconstruction can be calculated more accurately and more directly from electron diffraction intensities than from images (9, 114, 120). Currently, the lack of an electron diffraction pattern in specimens other than the crystalline type makes the amplitude determination less reliable. However, various computer restoration schemes have been attempted to correct the amplitudes derived directly from images (25, 68, 78, 106, 110). Phase information can be readily corrected if the defocus and beam-tilt parameters of the image are known (56, 114). The accuracy of phase determination is evident in the phase agreement among the symmetry-related reflections in the computed structural factors of crystalline specimens (54, 114). Several computational procedures have been developed to correct image distortion or to select images that are highly similar before averaging many of them to enhance the signal-to-noise ratio and to restore a high-resolution image that resembles the original object (47, 55, 101).

An electron micrograph provides a two-dimensional projection of a three-dimensional object along the pathway of an electron beam. However, the data derived from electron microscopy are not free from distortions. In order to retrieve an object's three-dimensional structure, multiple projections of an object must be recorded. Computer reconstruction, therefore, is an integral and essential step for combining different projections and retrieving the three-dimensional mass density of an object (32, 44, 71).

## How to Do Three-Dimensional Reconstruction

Image reconstruction is a two-step process—collecting three-dimensional data and merging them computationally. The ways of collecting the data differ depending on the type of specimen. It is useful to classify the various specimen types according to symmetry (i.e. crystalline, helical, and icosahedral) and asymmetry. Different reconstruction algorithms have been written to take advantage of the symmetric or asymmetric properties of these different objects (see 5, 44, 89, 108). Basically, two computational approaches have been used successfully to perform three-dimensional reconstructions: Fourier and weighted back-projection methods. The Fourier method has mainly been used to reconstruct images of symmetric objects (32), whereas the weighted back-projection method has been used to reconstruct images of asymmetric objects (99).

For any of these reconstruction algorithms to be used successfully, the investigator must sample sufficient points of different views of an object and must know the angular orientation of each view. The required amount of three-dimensional data and the tolerance of errors in angular parameters depend on the resolution of the reconstruction ($d$) and the size of the objects ($T$) (31, 97, 99, 105). Based on a geometrical argument, the required number of evenly sampled views is about $\pi T/d$. For example, for an asymmetric object that is 250 Å in diameter, 150 evenly sampled views are needed to reconstruct its three-dimensional structure to a resolution of 5 Å. Of course, if the object has definite symmetry, the number of independent views can be reduced. Because the data are generally quite noisy and the data sampling is not likely to be evenly distributed, more data are usually recorded for the final reconstruction. For an object and the expected resolution of its reconstruction to be the same size, the tolerable error in the tilt angle determination at 60° must be less than 0.05° (97).

Different angular projection views of an object can be obtained in several ways. For a crystalline specimen, different tilts of the specimen can be collected with a continuous-tilt microscope specimen holder or with specimen holders that have different fixed tilt angles. This approach results in a cone of missing data because the tilt angles accessible in the microscope are limited to $\pm 60°$; however, these missing data may not affect the final reconstruction (51, 55). In a helical specimen, the molecule presents different views in a single projection, and hence in theory, only one image is needed for most reconstructions at low resolution (108). Because of their symmetry, icosahedral particles present up to 60 different orientations of the identical unit for every projection (31). Because the particles are usually oriented randomly on the grid with respect to the electron beam, very few images are needed to provide sufficient data points for reconstruction. Asymmetric particles may or may not have a preferred orientation along the electron microscopic grid. For the conical reconstruction algorithm, a minimum of two images has been used for particles with preferred orientations but random azimuthal distributions on the grid (44, 99).

## How to Prepare Cryospecimens and Obtain Electron Images

During the past decade, methods for preserving specimens in a microscopic vacuum have significantly improved. The most effective procedure is to embed the specimen in a thin layer of vitreous ice or glucose (114; see 26, 38). In either procedure, the specimen is kept at a temperature below $-150°C$ for observation in order to reduce damage (65). In the method involving vitreous ice, the low temperature also keeps the ice from subliming in the vacuum. These preparative procedures can maintain the specimen's native structure in a state similar to that in solution without the use of chemical fixatives or negative stains in the microscope vacuum. Evidence for the effectiveness of these preparative procedures is the structural resolution of a crystalline specimen revealed in electron diffraction patterns or computed Fourier transforms of electron images.

Biological specimens are more difficult to study with electron microscopy than other objects. The interaction of the high-energy electrons can either destroy the specimen with a chemical reaction (i.e. the transfer of energy from the irradiating electrons to the specimen) (129) or make the specimen move (57). Fortunately, the damage caused by the electron beam can be effectively reduced by using a low dose, low temperature procedure (65, 114). Furthermore, a computer-controlled spot scan of the electron beam also reduces the beam-induced movement of specimens (19, 36). A successful application of these imaging procedures has facilitated the routine yield of high-quality, high-resolution images (16, 17, 55).

## Structural Resolution Attainable with Electron Cryomicroscopy and Three-Dimensional Reconstruction

The wavelength of 100 keV of electron energy is 0.034 Å, and hence, the resolution of images is not diffraction limited. Image quality can be affected by chromatic aberration of the microscope lenses, the monochromaticity of electron energy, and the partial spatial coherence of the electron-illumination source (107). Under the most optimal conditions for electron optical imaging, projection images of protein crystals have been resolved beyond 3.5 Å (10, 17, 67). This structural resolution is confirmed by crystallographic criteria, i.e. the signal-to-noise ratio of the amplitudes and the phase coherence of the Fourier coefficients around the expected reciprocal lattice positions calculated from images.

The highest resolution achieved to date is that of the three-dimensional structure of bacteriorhodopsin in which the polypeptide backbone can be traced. In this study, the resolution was 3.5 Å in plane and about 7 Å normal to the plane of the membrane (55). The poor resolution in the $z$ direction may be caused by the disorder of the crystal in this direction, by the difficulty of preparing a flat specimen, or both. However, progress has been made in overcoming the flatness problem in different specimens (21, 52). Studies of several crystalline specimens are now underway, and diffraction and image data of tilted crystals are being collected mainly with a 400-kV microscope at a resolution around 3 Å (27, 119).

Many biological specimens do not appear in crystalline arrays in vivo. For a helical specimen, the highest resolution attained has been about 9 Å, the resolution at which $\alpha$-helices of coat-protein subunits can be resolved as exemplified by electron microscopic studies of the tobacco mosaic virus (TMV) (68). An electron diffraction pattern of a nearly parallel aggregate of TMV has been recorded to a resolution of 3 Å (66), so imaging this specimen to similar resolution should be possible. More recent images of TMV have been recorded with a 400-kV cryomicroscope and spot scanning to a resolution better than 9 Å (K. H. Downing, personal communication; J. Jakana, unpublished results). Reconstruction of TMV's three-dimensional structure at even higher resolutions would require the following: (a) computational straightening of the helical array in both the horizontal and vertical directions, (b) separation of the overlapping Bessel functions in the high-resolution layer lines in the Fourier transform of the image, and (c) correction of the phases with an accurate determination of the contrast transfer function of the micrographs caused by defocus, spherical abberation, and astigmatism.

Other biological specimens with icosahedral symmetry or asymmetry have been studied at resolutions between 40 and 25 Å (8, 13, 46, 49, 98). These structures were studied at these low resolutions because the images used in their reconstruction were taken at a relatively high defocus in order to optimize the contrast in the images of the unstained and hydrated specimens. Highly defocused images simplify defining the positions and orientations of the particles in the initial stages of image reconstruction. However, the trade-off is the loss of resolution due to the contrast-damping effect of the spatial coherence of highly defocused images. This loss of resolution can be overcome if an electron microscope with a more coherent electron source is used (130, 131). Of course, improved procedures for computerized reconstruction of these types of objects are also required.

## Biological Specimens Most Amenable to Study with Electron Cryomicroscopy and Three-Dimensional Reconstruction

In practice, any highly purified biological macromolecule or macromolecular assembly in a size range from $10^3$ to $10^6$ kDa can be studied with electron cryomicroscopy and reconstructed in three dimensions. As mentioned above, the level of structural resolution currently attainable depends to some extent on the symmetry of an object. For X-ray crystallography, the macromolecule must be crystallized, and the minimum size of the crystal must be 0.2–0.5 mm (84). For NMR spectroscopic studies, the molecule must be small, that is, from 15 to 25 kDa (125). The following section presents several biological specimens that are classified by their types of symmetry. Using both biophysical and biological rationales, I explain why electron cryomicroscopy and three-dimensional reconstruction are the best techniques for studying these specimens.

CRYSTALS   Some thin protein crystals naturally occur in vivo or can be induced to form in vitro. The best example of a naturally occurring two-dimensional crystal is bacteriorhodopsin, which is 45 Å thick and 1–2 $\mu$m on edge. Bacteriorhodopsin is a membrane protein that has seven $\alpha$-helices spanning its lipid bilayer. It is the first protein for which a three-dimensional atomic model was reconstructed from electron microscopic data (59). This work yielded the first high-resolution structural motif of a membrane protein, namely, transmembrane helices. This motif was subsequently found in the X-ray crystallographic structure of a membrane-protein complex in a photosynthetic reaction center (34). Based on this motif and protein-sequence information, transmembrane helical models of other membrane proteins have been constructed (22, 43, 82, 83). Using several cumulative technical improvements in data collection and processing, Henderson et al (55) traced the polypeptide of bacteriorhodopsin and derived some side-chain conformations (Figure 1) (55). Serious attempts at making three-dimensional crystals of bacteriorhodopsin have not been successful (58, 86). A three-dimensional crystal relies significantly on the molecular contacts of the protein surface. Attaining the molecular interactions necessary for a stable crystal formation may be difficult for a membrane protein that does not have much of its polypeptide domain exposed outside the membrane lipid (85).

Another in vivo crystal that has been studied successfully with electron cryomicroscopy and three-dimensional reconstruction is the acrosomal bundle of *Limulus* sperm. This macromolecular assembly contains several tens of filaments of two proteins: actin and scruin in a 1:1 molar ratio in a bundle 1000 Å in diameter and several tens of microns in length. Although the bundle is twisted, small segments of the bundle can be analyzed each as a single crystal (104). This assembly is being used as a model system to probe for the detailed interaction between actin and actin-binding proteins. A 400-kV electron image of an ice-embedded acrosomal process shows remarkably fine details, and its computed diffraction pattern suggests structural data extending to a resolution of at least 7 Å (103). Since these crystals are not two-dimensional, the strategy for collecting and processing their structural data will be different from that used for bacteriorhodopsin. Because the crystals are too small and too disordered (i.e. twisted) for X-ray crystallography and much too large for NMR spectroscopy, electron cryomicroscopy is the best tool for analyzing their structure.

Attempts have been made to prepare crystals of membrane proteins of 1 to 2 $\mu$m$^2$ and only a single layer ($<100$ Å) that are suitable for electron crystallographic analysis. The most successful preparations, which use appropriate choices of lipid and detergent mixtures, have yielded high-

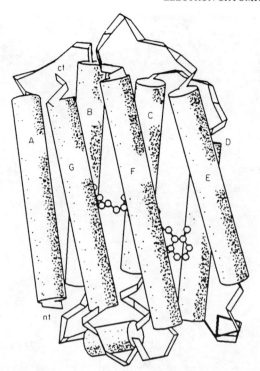

*Figure 1* A polypeptide chain trace of bacteriorhodopsin with seven α-helices (A through G) represented by the cylinders and the retinal position, which occupies a binding pocket formed by amino acid residues from all seven helices (55) (courtesy of Dr. K. Downing; reproduced with permission from Academic Press).

resolution (3 Å) crystals (63, 75). Two examples are *Escherichia coli* porin (PhoE) and plant light-harvesting complex. In the case of porin, β-strands at an angle tilted about 35° from the axis normal to the membrane are lined up around the β-sheet wall of each monomer as illustrated in Figure 2 (40, 64). More recent structural refinements of porins have yielded a density map to which a few large porin side chains have been assigned (40). During the course of this work, the X-ray analysis of other porins was solved to atomic resolution (30, 123). The structural features seen by electron crystallographic analysis were consistent with those seen with X-ray crystallography. Continual refinement of the electron crystallographic data will be useful for confirming the potential power of the technique. Structural studies of light-harvesting protein complex has led to a three-

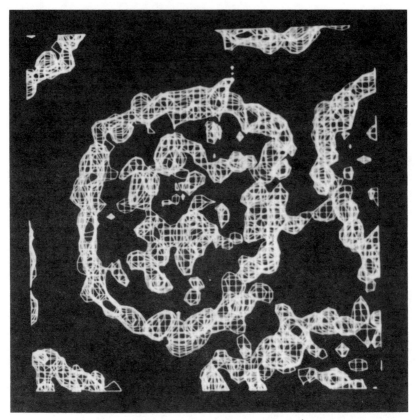

*Figure 2* A monomeric unit of *E. coli* PhoE porin shown in a 10-Å thick central section of the three-dimensional density map at 3.4-Å resolution. This map was generated after phase refinement and extension using molecular averaging about the noncrystallographic threefold solvent flattening, and inclusion of partial model information during phase refinement (40). The $\beta$-barrel lines up at the perimeter of the monomer (courtesy of Drs. Thomas Earnest & Bing Jap).

dimensional density map that has been interpreted, not only in terms of the protein's three transmembrane $\alpha$-helices, but also with respect to the disposition of several chlorophyll pigment molecules as shown in Figure 3 (76). A recent map of this structure at a resolution of 4 Å has been interpreted in terms of identification of some side chains (119). Though the structures of these two membrane systems are not fully determined, continuing efforts to get better tilted data should eventually yield the entire three-dimensional structure of both systems so their polypeptide backbones can be traced.

A recent advance in the preparation of true two-dimensional crystals of

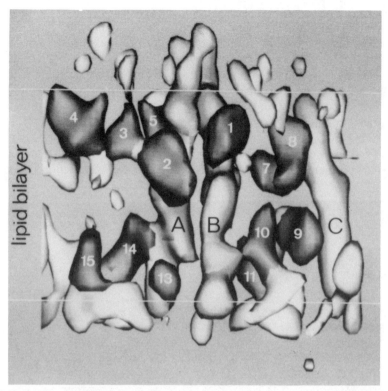

*Figure 3* The side view of plant light-harvesting complex seen from a central position in the lipid bilayer outside the trimer. This reconstruction was done from the two-dimensional crystal of this membrane complex at 6 Å in which 3 membrane-spanning α-helices (labeled A-C) and the 15 chlorophyll molecules were interpreted (76, 119) (courtesy of Drs. D. N. Wang & W. Kühlbrandt).

soluble proteins is the use of a phospholipid monolayer formed at the water-air interface (for review, see 72). Proteins at optimal concentration, pH, and ionic strength in the aqueous phase can be induced to crystallize via their interactions either with a ligand molecule chemically engineered to the phospholipid head group or with the charged group of the phospholipid. The most encouraging result was obtained with the streptavidin molecule that was crystallized on a biotinylated phospholipid film. This two-dimensional crystal diffracts beyond a 3-Å resolution (28, 33). This technique appears to be quite promising for a variety of proteins, particularly those that have an affinity for membrane surfaces.

Other soluble proteins or nucleic acids have been crystallized with a thickness of several hundreds of Ångstroms and over 1 μm on edge. These diffract in an electron cryomicroscope at resolutions better than 3 Å. One example is an electron diffraction pattern of the glucose-embedded crotoxin complex crystal shown in Figure 4 (27). Examples of other protein crystals for which high-resolution diffraction data have been reported are: T4 DNA-helix destabilizing protein (54), α-helical coiled-coil protein from the ootheca of the praying mantis (20), influenza neuraminidase-antibody fragment complex (20), and tropomyosin from pig cardiac tissue (7). The reasons why only thin, rather than thick, crystals form are not known. Under the same or different crystallization conditions, some of these proteins can be made into large crystals suitable for X-ray diffraction analysis

*Figure 4* Four hundred–kilovolt electron diffraction pattern of glucose-embedded crotoxin complex crystal tilted at 50° and recorded in a 1024 × 1024 Gatan slow-scan charge coupled device camera attached to a JEOL4000 electron cryomicroscope. The crystal was kept at −160°C during data collection and received a dose of 0.01 electrons Å$^{-2}$. The two-dimensional crystal unit cell is **a** = **b** = 38.8 Å. This pattern has been background subtracted and has reflections corresponding to a 3-Å resolution (courtesy of Dr. J. Brink).

(1, 124); however, none has yet been solved by X-ray analysis because of various technical problems. It is possible that many proteins might be induced to form thick as well as thin crystals. However, studying thin crystals is difficult because of the variable thicknesses either within or among single crystals. Therefore, an optimal experimental and computational procedure is needed to collect and merge the tilted data from crystals with the same thickness under the constraints of radiation damage. Several approaches have been attempted to solve this relatively difficult problem (18, 27, 91), but when the detailed procedure is worked out, electron crystallography could be an alternate tool for high-resolution structure determination for a broad spectrum of soluble proteins.

Questions have often been raised about the short-term and long-term need of electron crystallography in structural biology because of the overwhelming successes with X-ray crystallography. Given its current technology, electron crystallography cannot compete with X-ray crystallography in terms of resolution and speed of structural solution. As seen from structural analyses of porin, X-ray crystallography can solve a structure much more readily than electron crystallography whenever a suitable crystal can be prepared and isomorphous heavy atoms can be obtained. On the other hand, three-dimensional, rather than two-dimensional, crystals of some proteins are harder to make, as illustrated by the above examples. The phase problems cannot always be solved quickly in X-ray crystallography. In some instances, studying a thin crystal is preferable because the molecule may be in a more native state as in the living system. No theoretical limitations to future technical improvements in electron crystallography prohibit it from becoming as automated and routine as X-ray crystallography. In view of the continuing demand that macromolecular structures be solved more rapidly and more accurately, more tools should be available to biologists so they can try different techniques and use whichever works best at deriving a particular macromolecular structure.

Another potential use of electron crystallography is to provide a low resolution (i.e. 5–7 Å) three-dimensional model of a protein from which the X-ray crystallographer can derive an initial model and improve it by using phase extension to achieve high resolution. In this way, complex structures could be solved more quickly. An obvious example of the possible advantages of combining these technologies is the ribosome. The X-ray crystal structure of the ribosome diffracts up to resolutions of 3.5 Å (127). Because this particle has a large mass and internal asymmetry, phasing its X-ray intensities is a difficult task (122). A model built using electron microscopy therefore would be helpful in phasing the X-ray diffraction intensities. Hence, it is realistic to expect that electron cryomicroscopy and three-dimensional reconstruction will play a significant role in solving this complex structure at high resolution.

HELICAL ARRAYS  Macromolecular assemblies occur as helical arrays in the living cell. Examples include helical viruses, such as TMV (68, 115); filamentous viruses (93); the T4 bacteriophage tail (4); viral nucleocapsid (24, 41); cytoskeleton assemblies, such as actin filaments complexed with actin-binding proteins (87, 88); microtubule assemblies (81); and bacterial flagella (90, 112). In addition, some macromolecular assemblies can also form helical arrays in certain chemical environments. Examples are the RecA-protein complex (23, 128), acetylcholine receptor (110, 111), and botulism toxin (102).

X-ray solution scattering is often applied to studies of filamentous structures such as microtubules (6, 11). However, X-ray fiber diffraction is a better technique for specifically studying the high-resolution structures of helical filaments, including viruses and flagella. For example, the atomic structure of TMV has been determined with X-ray fiber diffraction (92). Nevertheless, this approach can be problematic because it requires well-oriented and well-ordered fiber and the phasing of the fiber diffraction intensities (80). Alternatively, electron cryomicroscopy and three-dimensional reconstruction can be used to image individual helical filaments and to reconstruct their three-dimensional structure as shown in Figure 5. This figure shows how four core $\alpha$-helices of the TMV coat protein can be visualized from a 9-Å density map (68). This structural analysis was undertaken to illustrate the current technical feasibility of obtaining the secondary structure of a helical assembly. A recent study of the flagellar filament from *Salmonella typhimurium* has also suggested the feasibility of visualizing high-resolution features of the flagellar protein (D. Morgan & D. DeRosier, personal communication).

Another important aspect of electron cryomicroscopic studies of helical structures is the need for information from the naturally occurring helical array so that the biological function of the system can be understood. A case in point is the structure of the actin filament. The atomic structure of g-actin has been solved with X-ray crystallography (69). However, the f-actin filament and its complex, which has various actin-binding proteins that are the functional entities in the cytoplasm, are the structures of ultimate interest because of their biological activities (70). These filamentous structures are hard to study because they are flexible and have relatively disordered structure. Nevertheless, the structures derived from electron cryomicroscopy and three-dimensional reconstruction can be taken as useful constraints for model building based on the structure of the g-actin crystal (62).

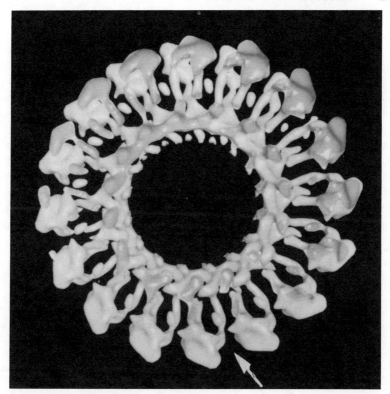

*Figure 5* Fifteen-Ångstrom-thick cross-sectional slice from a three-dimensional density map of tobacco mosaic virus reconstructed from electron images of ice-embedded virus particles at 9-Å resolution (68). In this display, the two pairs of core α-helices of the coat protein are most easily visible in the bottom coat protein subunit (*arrow*). The helices themselves appear as overlapping tubes directed radially from the outer region of the coat protein (courtesy of Dr. M. F. Schmid). The map is similar to that obtained from X-ray fiber diffraction (92).

ICOSAHEDRAL PARTICLES  Many viral particles have icosahedral symmetry. These particles are large, ranging from 150 to 2500 Å in diameter. Solution NMR cannot be used to study their structures. With X-ray diffraction, the largest viral particle solved is SV40, which is about 500 Å in diameter (79). In theory, there is no reason why X-ray crystallography cannot be used to study larger viral particles, but in practice, obtaining high-quality crystals of large viruses has been difficult. Furthermore, the typically large unit-cell size in the crystal, which is often 1000 Å, presents difficulties with the currently available technology for collection and processing of data.

During the past few years, electron cryomicroscopy and three-dimensional reconstruction have been applied to very productive studies of the molecular structures of a broad spectrum of viruses from animals, insects, plants, and bacteria. These activities are flourishing partly because of improved procedures for preparing frozen, hydrated viral particles and the general applicability of the three-dimensional reconstruction algorithm first introduced by Crowther (31). Because of the technical constraints in studying large viral particles, as mentioned above, these successful structural studies were carried out only at low resolution. However, the biological information derived from them has helped molecular virologists understand the structural basis of biological processes during the replication cycle of viruses, including receptor binding (94), antibody neutralization (94, 96, 121), virion maturation and assembly (98), and viral genome packaging (14). For example, the first structural study of the virus-Fab complex (Figure 6) revealed the binding sites and two different conformations of the Fabs of a neutralizing antibody that are bound to each

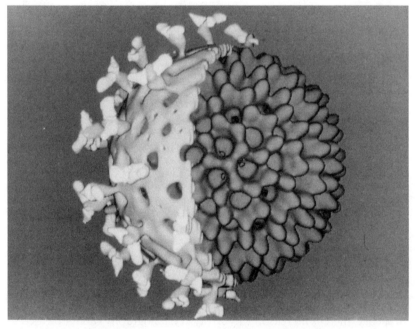

*Figure 6* Three-dimensional surface density map of rotavirus-Fab complex determined to 40-Å resolution (96). The diameter of this viral assembly is over 1000 Å. The map shows the binding sites of the neutralizing monoclonal Fab against the vp4, which are the spike proteins. The two Fab densities per spike suggest that each spike is dimeric, a supposition that was recently confirmed by independent biochemical measurements. The reconstruction also reveals the positions of the 132 channels per virion and the interaction between two shells of proteins as revealed in this cut-away view (courtesy of Dr. B. V. V. Prasad).

spike of the rotavirus (96). The density map suggested that the spike is a dimer, and this interpretation was subsequently confirmed by biochemical studies (F. Ramig, unpublished data). Another striking example of imaging virus is the direct visualization of the dsDNA in a bacteriophage (Figure 7), the interpretation of which supports the liquid crystalline model of this viral DNA (13, 77).

Another very fruitful study that used X-ray crystallography and electron cryomicroscopy in combination is exemplified by an investigation of the separate binding of the cellular receptor and of the neutralization antibodies to the human rhinovirus (Figure 8). The structures of these complexes together with the X-ray crystallographic structure of the rhinovirus (100) provides a foundation for model building in order to determine the interaction between these molecules (94). Another sensible approach for using these techniques in combination is to use X-ray structure analysis of a single viral protein and the image reconstruction of a large virus such as adenovirus to elucidate the architecture of a complex viral assembly much greater than 1000 Å in diameter (109).

ASYMMETRIC PARTICLES  A very intriguing aspect of electron microscopy is its use in visualizing single asymmetric macromolecules. Although X-ray scattering can be used to estimate the size and shape of single macromolecules, there is a practical limit to the structural detail that it can

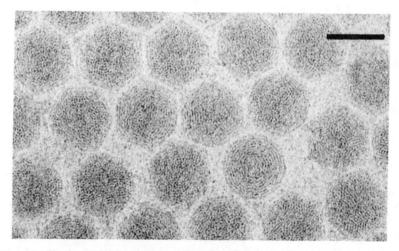

*Figure 7*  One hundred–kilovolt electron image of ice-embedded bacteriophage T7 virions in which different motifs of packaged dsDNA due to different virus orientations are visible. This genome configuration supported the model of the viral DNA as a liquid crystal (13) (courtesy of Drs. F. Booy, A. Steven & P. Serwer). Bar = 500 Å.

250  CHIU

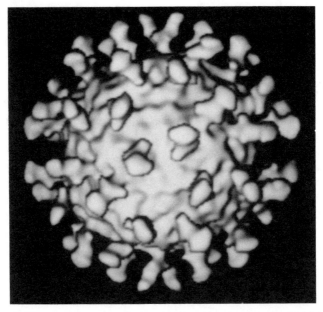

*Figure 8* Three-dimensional surface display of human rhinovirus complexed with 16/D1D2, a cellular receptor embedded in vitreous ice. The receptor is 75 Å long and is bilobed as expected for the two-domain ICAM-1 fragment. The proposed binding site for the receptor is at the center of the canyon seen in the X-ray structure of the rhinovirus (94) (courtesy of Drs. T. S. Baker & N. H. Olson).

determine. For a small molecule, NMR solution spectroscopy is obviously the structural technique of choice. However, the development of correlation and multivariate statistical analysis by van Heel & Frank (116) has revolutionized electron cryomicroscopic study of asymmetric macromolecules. This computational tool allows the electron microscopist to sort out electron images of the same view from thousands of images and combine them to enhance the statistical definition of features not readily visible in the original micrographs. So far, the tool has been applied to electron microscopic studies of various macromolecular assemblies embedded either in negative stain or in vitreous ice. Examples include ribosomal particles of different sizes (46), a nuclear pore (2, 3, 61), the calcium-release channel (117), and a protein-nucleic acid complex (53). This technique recently was extended to analyses of low contrast images of ice-embedded particles (95). Some of these biological macromolecules contain several different polypeptides, and the low resolution three-dimensional structures have revealed several interesting features crucial to our understanding of the complex structure. For example, the ribosome structure shown in Figure 9 led to a hypothesis about the binding sites of tRNA and mRNA in the 70S particle (95). Using a monoclonal antibody that binds to specific molecular fragments may help to delineate the spatial organization of the subunits in these assemblies. Furthermore, cryomicroscopy can also be used to visualize the three-dimensional structures of macromolecular assemblies in different functional states that may not be easily obtained by X-ray diffraction or other biophysical techniques (26, 113).

## FUTURE PROSPECTS

The importance of structural results is that they provide or stimulate the design of subsequent experiments that lead to a better understanding of the biological system under investigation. This contribution can occur not only at high resolution but also at low resolution, depending on the nature of the biological questions that need to be answered. During the past decade, electron cryomicroscopy and three-dimensional reconstruction have provided important structural information to many investigators, particularly those studying biological problems in membrane biology, molecular virology, cytoskeletal protein assembly, and protein-nucleic acid complexes.

For many biological problems, amino acid side chains must be visualized to answer questions regarding their function in terms of their chemical

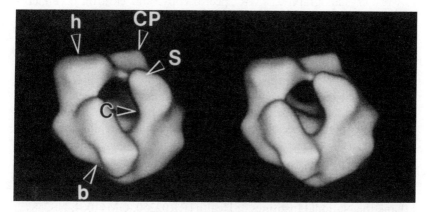

*Figure 9* Stereo-pair image of ice-embedded 70S ribosome particle reconstructed from correlation and multivariate statistical analysis. In this particular view, the 30S subunit is on the left described as head (h) and main body (b), and the 50S is on the right described as central protuberance (cp) and stalk base (s) (95) (courtesy of Dr. J. Frank).

behavior. The ultimate goal of electron crystallography is to determine the atomic structures of all protein specimens that can form two-dimensional crystals. In principle, there is no fundamental reason why this imaging technique cannot be extended to the atomic resolution of specimens with helical or icosahedral symmetry or of asymmetric particles. This goal can be reached with further development in the electron optics of the microscope, refinement of imaging conditions and data-collection procedures, and application of more advanced computational analysis. The combined use of a field emission gun and an intermediate high-voltage electron microscope could enhance the image contrast by improving spatial and temporal coherence (130, 131). An elimination of the inelastically scattered electrons also can improve the image contrast (106). Future generations of instruments with all of these features can yield images of crystalline and noncrystalline specimens from which high-resolution structural factors could more readily be extracted for three-dimensional reconstruction. An on-line analysis with data captured from a charged couple device camera and a high-speed computer would enhance the efficiency of data collection and make the complex data-collection scheme more feasible (18, 27, 73, 74). In addition, computer software for restoration and reconstruction from low contrast and high-resolution images will be needed to extract the structural information after proper corrections. Essentially, this technique has the potential to mature further and reach its theoretical limit at atomic resolution and, therefore, can become an attractive tool of choice for solving three-dimensional structures of symmetric and asymmetric biological specimens.

X-ray crystallography, NMR spectroscopy, and electron cryomicroscopy differ significantly in terms of the specimen required and the method needed to analyze the data derived from them. Clearly, electron cryomicroscopy and three-dimensional reconstruction allow us to study the three-dimensional structures of many specimens in states that X-ray crystallography or NMR spectroscopy cannot. And in some cases, electron imaging is the vital link between cell biology and X-ray crystallography. This link will facilitate better understanding of the crystal structure in its biological context, especially when macromolecular assemblies and their interrelationships are involved.

ACKNOWLEDGMENTS

I thank the National Institutes of Health (RR02250, GM41064, NS25877) and the W. M. Keck Foundation for their support of this research; Drs. T. S. Baker, F. Booy, J. Brink, K. Downing, T. Earnest, J. Frank, B. Jap, W. Kühlbrandt, N. H. Olson, B. V. V. Prasad, M. F. Schmid, P. Serwer,

A. C. Steven, and D. N. Wang for providing the figures; Drs. T. S. Baker, S. Hamilton, G. N. Phillips, Jr., B. V. V. Prasad, and M. F. Schmid for helpful comments regarding the manuscript; Ms. P. P. Powell for excellent editorial assistance; and Mr. W. Scoular for organizing the literature citations.

*Literature Cited*

1. Achari, A., Radvanyi, F. R., Scott, D., Cassian, B., Sigler, P. B. 1985. *J. Biol. Chem.* 260: 9385
2. Akey, C. W. 1989. *J. Cell Biol.* 109: 955
3. Akey, C. W., Goldfarb, D. S. 1989. *J. Cell Biol.* 109: 971
4. Amos, L., Klug, A. 1975. *J. Mol. Biol.* 99: 51
5. Amos, L. A., Henderson, R., Unwin, P. N. T. 1982. *Prog. Biophys. Mol. Biol.* 39: 183
6. Andreu, J. M., Borda, J., Diaz, J. F., deAncos, J. G., Gil, R., et al. 1992. *J. Mol. Biol.* 226: 169
7. Avila-Sakar, A. J., Schmid, M. F., Lee, L. S., Whitby, F., Phillips, G. N. Jr., Chiu, W. 1993. *J. Struct. Biol.* In press
8. Baker, T. S., Newcomb, W. W., Olson, N. H., Cowsert, L. M., Olson, C., Brown, J. C. 1991. *Biophys. J.* 60: 1445
9. Baldwin, J., Henderson, R. 1984. *Ultramicroscopy* 14: 319
10. Baldwin, J. M., Henderson, R., Beckman, E., Zemlin, F. 1988. *J. Mol. Biol.* 202: 585
11. Beese, L., Stubbs, G., Thomas, J., Cohen, C. 1987. *J. Mol. Biol.* 196: 575
12. Blundell, T. L. Johnson, L. N. 1976. *Protein Crystallography*. London: Academic
13. Booy, F., Trus, B. L., Newcomb, W. W., Brown, J. C., Serwer, P., Steven, A. C. 1992. In *Proc. Annu. Meet. Electron Microsc. Soc. Am.*, 50th, 1: 452. Boston: San Franscico Press
14. Booy, F. P., Newcomb, W. W., Trus, B. L., Brown, J. C., Baker, T. S., Steven, A. C. 1991. *Cell* 64: 1007
15. Branden, C., Tooze, J. 1991. *Introduction to Protein Structure*. New York: Garland
16. Brink, J., Chiu, W. 1991. *J. Microsc.* 161: 279
17. Brink, J., Chiu, W., Dougherty, M. 1992. *Ultramicroscopy.* 46: 229
18. Brink, J., Chiu, W. 1992. In *Proc. Eur. Congr. Electron Microsc.*, *10th*, 3: 15. Granada, Spain: Secretariado de Publicaciones de la Universidad de Granada
19. Bullough, P. A., Tulloch, P. A. 1990. *J. Mol. Biol.* 215: 161
20. Bullough, P. A., Tulloch, P. A. 1991. *Ultramicroscopy* 35: 131
21. Butt, H. J., Wang, D. N., Hansma, P. K., Kühlbrandt, W. 1991. *Ultramicroscopy* 36: 307
22. Chabre, M. 1985. *Annu. Rev. Biophys. Biophys. Chem.* 14: 331
23. Chang, C. F., Rankert, D. A., Jeng, T. W., Morgan, D. G., Schmid, M. F., Chiu, W. 1988. *J. Ultrastruct. Mol. Struct. Res.* 100: 166
24. Charest, P. M., Jakana, J., Schmid, M. F., Vidal, S., Kolakofsky, D., Chiu, W. 1992. In *Proc. Annu. Meet. Electron Microsc. Soc. Am.*, 50th, 1: 870. Boston: San Francisco Press
25. Cheng, R. H. 1992. In *Proc. Annu. Meet. Electron Microsc. Soc. Am.*, 50th, 1: 996. Boston: San Francisco Press
26. Chiu, W. 1986. *Annu. Rev. Biophys. Biophys. Chem.* 15: 237
27. Chiu, W., Brink, J., Soejima, T., Schmid, M. F. 1992. In *Proc. Annu. Meet. Electron Microsc. Soc. Am.*, 50th, 2: 1054. Boston: San Francisco Press
28. Chiu, W., Guan, T. L., Soejima, T., Jakana, J., Schmid, M. F. 1992. In *Asia-Pacific Electron Microsc. Conf.*, 5th, p. 134. Beijing: World Scientific
29. Clore, G. M., Gronenborn, A. M. 1991. *Annu. Rev. Biophys. Biophys. Chem.* 20: 29
30. Cowan, S. W., Schirmer, T., Rummel, G., Steiert, M., Ghosh, R., et al. 1992. *Nature* 358: 727
31. Crowther, R. A. 1971. *Philos. Trans. R. Soc. London Ser. B* 261: 221
32. Crowther, R. A., DeRosier, D. J., Klug, A. 1970. *Proc. R. Soc. London* 317: 319
33. Darst, S. A., Ahlers, M., Meller, P. H., Kubalek, E. W., Blankenburg, R., et al. 1991. *Biophys. J.* 59: 387
34. Deisenhofer, J., Epp, O., Miki, K., Huber, R., Michel, H. 1985. *Nature* 318: 618
35. DeRosier, D. J., Klug, A. 1968. *Nature* 217: 130
36. Downing, K. H. 1991. *Science* 251: 53

37. Downing, K. H., Chiu, W. 1990. *J. Electron Microsc. Rev.* 3: 213
38. Dubochet, J., Adrian, M., Chang, J. J., Homo, J. C., Lepault, J., et al. 1988. *Q. Rev. Biophys.* 21: 129
39. Dubochet, J., McDowall, A. W. 1981. *J. Microsc.* 124: RP3
40. Earnest, T. N., Walian, P. J., Gehring, K., Jap, B. K. 1992. *Trans. Am. Crystallogr. Assoc.* In press
41. Egelman, E. H., Wu, S. S., Amrein, M., Portner, A., Murti, G. 1989. *J. Virol.* 63: 2233
42. Eisenberg, D., Hill, P. C. 1989. *TIBS* 14: 260
43. Engelman, D. M., Steitz, T. A., Goldman, A. 1986. *Annu. Rev. Biophys. Biophys. Chem.* 15: 321
44. Frank, J. 1990. *Q. Rev. Biophys.* 23: 281
45. Frank, J., Chiu, W., Degn, L. 1988. *Ultramicroscopy* 26: 345
46. Frank, J., Penczek, P., Grassucci, R., Srivastava, S. 1991. *J. Cell Biol.* 115: 597
47. Frank, J., van Heel, M. 1982. *J. Mol. Biol.* 161: 134
48. Frank, J., Verschoor, A., Boublik, M., van Heel, M. 1982. *J. Mol. Biol.* 161: 107
49. Fuller, S. D. 1987. *Cell* 48: 923
50. Glaeser, R. M. 1982. In *Methods of Experimental Physics*, pp. 391. New York: Academic
51. Glaeser, R. M., Tong, L., Kim, S. H. 1989. *Ultramicroscopy* 27: 307
52. Glaeser, R. M., Zilker, A., Radermacher, M., Gaub, H. E., Hartmann, T., Baumeister, W. 1991. *J. Microsc.* 161: 21
53. Gogol, E. P., Seifried, S. E., von Hippel, P. H. 1991. *J. Mol. Biol.* 221: 1127
54. Grant, R. A., Schmid, M. F., Chiu, W. 1991. *J. Mol. Biol.* 217: 551
54a. Hawkes, P. W., Valdre, U., eds. 1990. *Biophysical Electron Microscopy*. London: Academic
55. Henderson, R., Baldwin, J. M., Ceska, T. A., Zemlin, F., Beckmann, E., Downing, K. H. 1990. *J. Mol. Biol.* 213: 899
56. Henderson, R., Baldwin, J. M., Downing, K. H., Lepault, J., Zemlin, F. 1986. *Ultramicroscopy* 19: 147
57. Henderson, R., Glaeser, R. M. 1985. *Ultramicroscopy* 16: 139
58. Henderson, R., Shotton, D. 1980. *J. Mol. Biol.* 139: 99
59. Henderson, R., Unwin, P. N. T. 1975. *Nature* 257: 28
60. Hendrickson, W. A., Wüthrich, K. 1992. *Macromolecular Structures*. London: Current Biology
61. Hinshaw, J. E., Carragher, B. O., Milligan, R. A. 1992. *Cell* 69: 1133
62. Holmes, K. C., Tirion, M., Popp, D., Lorenz, M., Kabsch, W., Milligan, R. A. 1992. In *Mechanism of Myofilament Sliding in Muscle Contraction*, ed. H. Sugi, G. H. Pollack. New York: Plenum. In press
63. Jap, B. K. 1988. *J. Mol. Biol.* 199: 229
64. Jap, B. K., Walian, P. J., Gehring, K. 1991. *Nature* 350: 167
65. Jeng, T. W., Chiu, W. 1984. *J. Microsc.* 136: 35
66. Jeng, T. W., Chiu, W. 1987. *Ultramicroscopy* 23: 61
67. Jeng, T. W., Chiu, W., Zemlin, F., Zeitler, E. 1984. *J. Mol. Biol.* 175: 93
68. Jeng, T. W., Crowther, R. A., Stubbs, G., Chiu, W. 1989. *J. Mol. Biol.* 205: 251
69. Kabsch, W., Mannherz, H. G., Suck, D., Pai, E. F., Holmes, K. C. 1990. *Nature* 347: 37
70. Kabsch, W., Vandekerckhove, J. 1992. *Annu. Rev. Biophys. Biomol. Struct.* 21: 49
71. Klug, A., Crowther, R. A. 1972. *Nature* 238: 435
72. Kornberg, R. D., Darst, S. A. 1991. *Curr. Opin. Struct. Biol.* 1: 632
73. Koster, A. J., Chen, H., Sedat, J. W., Agard, D. A. 1992. *Ultramicroscopy*. 46: 207
74. Koster, A. J., DeRuijter, W. J. 1992. *Ultramicroscopy* 40: 89
75. Kühlbrandt, W. 1992. *Q. Rev. Biophys.* 25: 1
76. Kühlbrandt, W., Wang, D. N. 1991. *Nature* 350: 130
77. Lepault, J. 1987. *EMBO J.* 6: 1507
78. Lepault, J., Leonard, K. 1985. *J. Mol. Biol.* 182: 431
79. Liddington, R. C., Yan, Y., Moulai, J., Sahli, R., Benjamin, T. L., Harrison, S. C. 1991. *Nature* 354: 278
80. Makowski, L. 1984. In *Biological Macromolecules and Assemblies*, ed. F. A. Jurnak, A. McPherson, p. 203. New York: Wiley & Sons
81. Mandelkow, E. M., Mandelkow, E. 1985. *J. Mol. Biol.* 181: 123
82. McCrea, P. D., Engelman, D. M., Popot, J.-L. 1988. *TIBS* 13: 289
83. McCrea, P. D., Popot, J.-L., Engelman, D. M. 1987. *EMBO J.* 6: 3619
84. McPherson, A. 1982. *The Preparation and Analysis of Protein Crystals*. New York: Wiley and Sons
85. Michel, H. 1990. In *Crystallization of Membrane Proteins*, ed. H. Michel, p. 73. Boca Raton, FL: CRC Press
86. Michel, H., Oesterhelt, D. 1980. *Proc. Natl. Acad. Sci. USA* 77: 1283

87. Milligan, R. A., Flicker, P. F. 1987. *J. Cell Biol.* 105: 29
88. Milligan, R. A., Whittaker, M., Safer, D. 1990. *Nature* 348: 217
89. Moody, M. F. 1990. See Ref. 54a, p. 145
90. Morgan, D., DeRosier, D. 1992. *Ultramicroscopy.* 46: 263
91. Morgan, D. G., Grant, R. A., Chiu, W., Frank, J. 1992. *J. Struct. Biol.* 108: 245
92. Namba, K., Stubbs, G. 1986. *Science* 231: 1401
93. Nambudripad, R., Stark, W., Opella, S. J., Makowski, L. 1991. *Science* 252: 1305
94. Olson, N. H., Smith, T. J., Kolatkar, P. R., Oliveira, M. A., Rueckert, R. R., et al. 1992. In *Proc. Annu. Meet. Electron Microsc. Soc. Am., 50th,* 1: 524. Boston: San Francisco Press
95. Penczek, P., Radermacher, M., Frank, J. 1992. *Ultramicroscopy* 44: 33
96. Prasad, B. V. V., Burns, J. W., Marietta, E., Estes, M. K., Chiu, W. 1990. *Nature* 343: 476
97. Prasad, B. V. V., Degn, L. L., Jeng, T., Chiu, W. 1990. *Ultramicroscopy* 33: 281
98. Prasad, B. V. V., Prevelige, P. E., Marietta, E., Chen, R. O., Thomas, D., et al. 1993. *J. Mol. Biol.* In press
99. Radermacher, M. 1988. *J. Electron Microsc. Tech.* 9: 359
100. Rossmann, M. G., Arnold, E., Erickson, J. W., Frankenberger, E. A., Griffith, J. P., et al. 1985. *Nature* 317: 145
101. Saxton, W. O., Baumeister, W. 1982. *J. Microsc.* 127: 127
102. Schmid, M. F., DasGupta, B. R., Robinson, J. P. 1990. In *Proc. Int. Congr. Electron Microsc., 12th,* p. 496. Seattle, WA: San Francisco Press
103. Schmid, M. F., Jakana, J., Matsudaira, P., Chiu, W. 1992. In *Proc. Annu. Meet. Electron Microsc. Soc. Am., 50th,* 1: 512. Boston: San Francisco Press
104. Schmid, M. F., Matsudaira, P., Jeng, T. W., Jakana, J., Towns-Andrews, E., et al. 1991. *J. Mol. Biol.* 221: 711
105. Shaw, P. J., Hills, G. J. 1981. *Micron* 12: 279
106. Smith, M. F., Langmore, J. P. 1992. *J. Mol. Biol.* 226: 763
107. Spence, J. C. H. 1988. *Experimental High-Resolution Electron Microscopy.* New York: Oxford Univ. Press. 2nd ed.
108. Stewart, M. 1988. *J. Electron Microsc. Tech.* 9: 325
109. Stewart, P. L., Burnett, R. M., Cyrklaff, M., Fuller, S. D. 1991. *Cell* 67: 145
110. Toyoshima, C., Unwin, P. N. T. 1988. *Ultramicroscopy* 25: 279
111. Toyoshima, C., Unwin, P. N. T. 1988. *Nature* 336: 247
112. Trachtenberg, S., DeRosier, D. 1987. *J. Mol. Biol.* 195: 581
113. Unwin, P. N. T., Ennis, P. D. 1984. *Nature* 307: 609
114. Unwin, P. N. T., Henderson, R. 1975. *J. Mol. Biol.* 94: 425
115. Unwin, P. N. T., Klug, A. 1974. *J. Mol. Biol.* 87: 641
116. van Heel, M., Frank, J. 1981. *Ultramicrscopy* 6: 187
117. Wagenknecht, T., Grassucci, R., Frank, J., Saito, A., Inui, M., Fleischer, S. 1989. *Nature* 338: 167
118. Wagner, G., Hyberts, S. G., Havel, T. F. 1992. *Annu. Rev. Biophys. Biomol. Struct.* 21: 167
119. Wang, D. N., Fujiyoshi, Y., Kühlbrandt, W. 1992. In *Asia-Pacific Electron Microsc. Conf., 5th,* p. 138. Beijing: World Scientific
120. Wang, D. N., Kühlbrandt, W. 1992. *Biophys. J.* 61: 287
121. Wang, G., Porta, C., Chen, Z., Baker, T., Johnson, J. E. 1992. *Nature* 355: 275
122. Weinstein, S., Jahn, W., Hansen, H., Wittmann, H. G., Yonath, A. 1989. *J. Biol. Chem.* 264: 19138
123. Weiss, M. S., Wacker, T., Weckesser, J., Welte, W., Schulz, G. E. 1990. *FEBS Lett.* 267: 268
124. Whitby, F. G., Kent, H., Stewart, F., Stewart, M., Xie, X.-L., et al. 1992. *J. Mol. Biol.* 227: 441
125. Wright, P. 1989. *TIBS* 14: 255
126. Wüthrich, K. 1989. *Science* 243: 45
127. Yonath, A. 1992. *Annu. Rev. Biophys. Biomol. Struct.* 21: 77
128. Yu, X., Egelman, E. H. 1990. *Biophys. J.* 57: 555
129. Zeitler, E. 1990. See Ref. 54a, p. 289
130. Zemlin, F. 1992. *Ultramicroscopy.* 46: 25
131. Zhou, Z. H., Chiu, W. 1992. *Ultramicroscopy.* In press

# THE DESIGN OF METAL-BINDING SITES IN PROTEINS

## Lynne Regan

Department of Molecular Biophysics and Biochemistry, Yale University, New Haven, Connecticut 06511

KEY WORDS: protein design, metal-site design, optical absorption spectroscopy, zinc, cobalt, copper, calcium, carbonic anhydrase, zinc-finger peptide

CONTENTS

| | |
|---|---|
| PERSPECTIVES AND OVERVIEW............................................................................................... | 257 |
| METAL-BINDING SITES AS DISCRETE UNITS ........................................................................ | 259 |
|     The Square Planar Cu(II)-Binding Site of Human Albumin........................................... | 259 |
|     The Ca(II) Binding Loop of Thermolysin.............................................................................. | 259 |
| BINDING SITES FORMED BY TWO HISTIDINE RESIDUES.................................................... | 260 |
|     Affinity Purification by the Introduction of Two-Histidine Sites................................ | 261 |
|     Protein and Peptide Stabilization by Two-Histidine Sites ............................................. | 261 |
|     A Two-Histidine Design to Regulate Enzymatic Activity .............................................. | 262 |
| DESIGNED SITES WITH THREE PROTEIN LIGANDS .............................................................. | 263 |
|     Metalloantibodies....................................................................................................................... | 264 |
|     A (His)$_3$ Site in a Designed Four-Helix Bundle Protein................................................ | 266 |
|     A Modified Zinc-Finger Peptide ............................................................................................ | 268 |
|     Three-Coordinate Zn(II) Sites in a Designed Four-Helix Bundle Protein................. | 270 |
| DESIGNED SITES WITH FOUR PROTEIN LIGANDS............................................................... | 270 |
|     A Tetrahedral Binding Site in a Designed Four-Helix Bundle Protein..................... | 270 |
|     A Design for a Blue Copper Site........................................................................................... | 273 |
|     A Simplified (Cys)$_2$(His)$_2$ Zinc-Finger Peptide................................................................ | 275 |
| DISCUSSION AND PERSPECTIVES............................................................................................. | 277 |

## PERSPECTIVES AND OVERVIEW

Several recent reports have discussed the de novo design of proteins for the purpose of folding them into specified structures (13, 17, 21, 32). The first designs concentrated upon reproducing simple structural motifs that are found in natural proteins. More recent advances include designs in

1056–8700/93/0610–0257$02.00

which ligand-binding activities are introduced onto preexisting protein frameworks.

The introduction of a ligand-binding site into a new context allows one to attempt to delineate the factors responsible for high affinity ligand binding. Such studies are complementary to those that start with a natural site and seek to identify the important features with directed mutagenesis.

To date, most effort has focused upon the design of novel metal-binding sites, principally because metals can perform a vast array of different functions. Moreover, the geometries of metal-binding sites in natural metalloproteins and small compounds are well characterized from high resolution X-ray crystallographic studies.

Metal sites can be categorized as either structural or catalytic. Examples of naturally occurring structural sites include tetrahedrally coordinated Zn(II) sites in thermolysin and alcohol dehydrogenase, the various Zn(II) fingers and clusters of eukaryotic transcription factors, and the octahedrally coordinated Ca(II) sites that are found in proteins such as calmodulin and troponin C.

Catalytic activities that are associated with metal sites include redox activity, for example at the Cu site of plastocyanin and the Fe site of rubredoxin; oxygen transport, for example by iron in myohemerythrin and the globins; and electrophilic catalysis by Zn(II), for example in carbonic anhydrase and carboxypeptidase.

The brief survey presented in this review is not meant to be exhaustive, but rather to provide an overview of the types of sites for which designs have been attempted. For excellent reviews on natural metalloproteins, the reader is referred to the following articles (1, 4, 6, 12, 25, 30, 37, 39, 40).

The majority of designs concentrate on achieving correct positioning of the two, three, or four residues that will form the metal-binding site. This is an apparently simple route to generate proteins that have the potential to display an array of interesting properties. It will become evident, however, that the design of binding sites that behave as expected is in fact not as simple as one might first suppose.

A second, practical reason for focusing on the design of metal-binding sites is that detecting the binding of a metal ligand is experimentally straightforward. Moreover, a variety of spectroscopic techniques [typically optical absorption spectroscopy, nuclear magnetic resonance (NMR), and electron paramagnetic resonance (EPR)] can be used to determine the identity of the ligands and the geometry of the metal-binding site. Consequently, one can rapidly assess the success of a design before attempting a complete determination of the three-dimensional structure.

Specifically excluded from this discussion are peptides and proteins onto which metal-binding activity is introduced by the covalent attachment of metal-chelating organic groups (10, 34).

## METAL-BINDING SITES AS DISCRETE UNITS

In the majority of metalloproteins, the residues involved in metal binding come close together in the tertiary structure to form the binding site, but are dispersed along the amino acid sequence. Consequently, the metal site cannot generally be regarded as a discrete, movable unit. In this section two interesting exceptions are described.

### *The Square Planar Cu(II)-Binding Site of Human Albumin*

The Cu(II)-transport site of human albumin is located on a short linear sequence at the N terminus of the protein. The sequence is $NH_2$-Asp-Ala-His.... The square planar Cu(II) site is formed by the alpha-amino nitrogen, the two intervening peptide nitrogens, and the imidazole nitrogen of the histidine residue at the third position.

Sarcar and colleagues synthesized a short peptide, $NH_2$-Gly-Gly-His-N-methylamide, to mimic the site and to test the importance of the identity of the first two residues (23). Their peptide binds metal and displays an optical absorption spectrum that is identical to that of a human albumin–Cu(II) complex (see Figure 1). Equilibrium dialysis measurements using

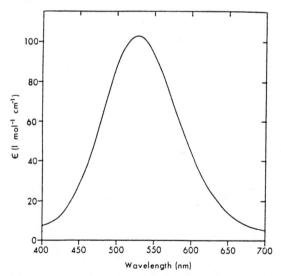

*Figure 1* The visible spectrum of the complex of Gly-Gly-His-methyl-amide with Cu(II) at pH 8. Reproduced with permission from Ref. 23 (copyright © 1976, American Chemical Society).

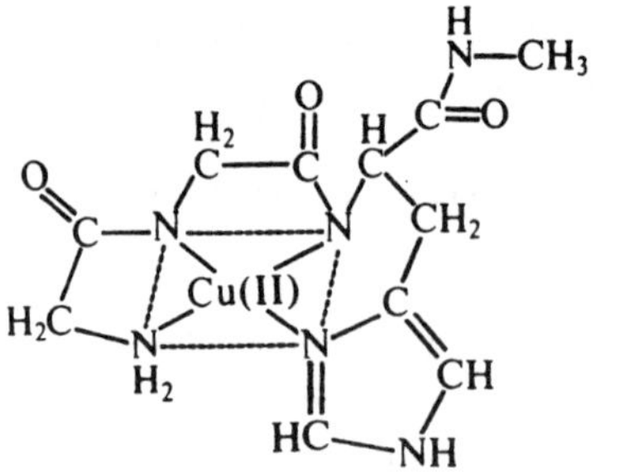

*Figure 2* Illustration of the proposed structure of the Cu(II)-Gly-Gly-His-methyl-amide complex. Reproduced with permission from Ref. 23 (copyright © 1976, American Chemical Society).

$^{67}$Cu(II) show that the peptide binds metal with remarkable affinity; the dissociation constant is $2.07 \times 10^{-17}$ M.

The crystal structure of the peptide-Cu(II) complex has been solved and confirms that the peptide binds metal with the expected geometry. A single atom of Cu(II) is tetradentately chelated by the amino-terminal nitrogen, the next two peptide nitrogens, and the histidyl nitrogen of a single peptide in a slightly distorted, square planar arrangement, as shown in Figure 2. As might be anticipated from the behavior of the isolated peptide, the same sequence may be added to the N terminus of an existing protein as a convenient means by which to introduce Cu(II)-binding activity.

## The Ca(II)-Binding Loop of Thermolysin

A related example of a discrete, transportable, metal-binding site is the Ca(II)-binding site of thermolysin. Thermolysin has a surface-exposed loop that includes all the residues involved in Ca(II) binding. The loop is called an Ω loop; the name is intended to describe its pinched-in shape (24). To determine if the loop can indeed be considered a discrete functional unit, Toma and colleagues (38) substituted it for an existing surface loop in neutral protease. The transplant was successful and the loop retains Ca(II)-binding activity in its new context.

# BINDING SITES FORMED BY TWO HISTIDINE RESIDUES

The simplest metal-binding sites that have been incorporated into proteins are those in which only two of the ligands are provided by the protein; the remaining coordination sites are filled by exogenous ligands. In this case, the goal has not been to mimic natural metalloproteins, but rather the metal-binding sites have been introduced with a view to applications in protein purification, stabilization, and the control of enzymatic activity. The designs realized to date are extremely simple and involve two histidine residues placed in appropriate positions, for example in an α-helix (His-$X_3$-His) or a β-sheet (His-X-His) (3).

## Affinity Purification by the Introduction of Two-Histidine Sites

The metal-binding affinity of the two-histidine sites varies between $2 \times 10^{-4}$ and $2 \times 10^{-6}$ M, depending on the molecular details of the site. The factors thought to increase the binding affinity include significant site rigidity and high solvent accessibility. Energetically unfavorable steric interactions between the liganding residues, or the metal itself, and the rest of the protein will result in lower-affinity sites (3).

For proteins with sites that bind metal with sufficient tightness, metal-affinity columns can be used as a single-step purification of the engineered protein from a crude lysate. Iminodiacetate (IDA) or nitrilotriacetic acid (NTA) attached to a solid support is commonly used, with Cu(II) as the metal. IDA provides three ligands to the metal ion, and the two histidine residues from the protein and solvent molecules complete the coordination shell. Protein purification in this fashion can also be accomplished without a "designed" site. A string of six histidine residues attached to the N or C terminus of a protein is often sufficient for the protein to be purified from crude extracts on metal-affinity columns, and several such systems are available commercially (29a) (see Figure 3).

## Protein and Peptide Stabilization by Two-Histidine Sites

Metal-binding and protein denaturation are coupled chemical equilibria. As a consequence, metal binding at the two-histidine sites stabilizes the protein towards denaturation, with the tightest binding sites having the greatest effect on protein stability (36). A possible practical application of the introduction of such sites is to create more resilient proteins for use in industrial or medicinal applications.

Histidine and cysteine residues appropriately positioned as the $i$ and $i+4$ residues on an α-helix can also stabilize helix formation in short peptides. Ghadiri and colleagues (9) have synthesized peptides with the sequence

Ac-Ala-Glu-Ala-Ala-Ala-Lys-Glu-Ala-Ala-Ala-Lys-$X_1$-Ala-

Ala-Ala-$X_2$-Ala-$CONH_2$

where either $X_1$ is cysteine and $X_2$ is histidine, or both $X_1$ and $X_2$ are

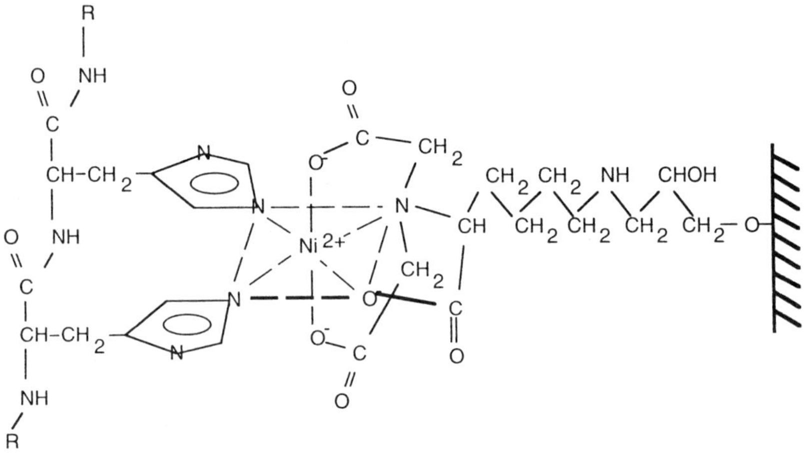

*Figure 3* Schematic illustration of the binding of a His-containing protein to a Ni(II)-NTA resin. Reproduced with permission from Ref. 29a.

histidine. In both examples, the addition of millimolar Cu(II), Zn(II), or Cd(II) results in an almost twofold increase in helicity of the peptide in aqueous solution at room temperature.

## A Two-Histidine Design To Regulate Enzymatic Activity

An inventive example of the use of a two-histidine metal-binding site may be found in the work of Craik and colleagues, who employed metal binding as a means to modulate the activity of trypsin (20). A histidine residue was introduced into trypsin such that it could form a two-ligand metal-binding site, in conjunction with the histidine residue of the active-site catalytic triad. A key feature of the design is that the side-chain of the catalytic-triad histidine must rotate out of its position in the active site if it is to bind a metal ion. It follows that the addition of metal ions to the solution should inhibit the enzyme. Figure 4 illustrates the expected movement of the histidine residue.

The properties of mutant trypsin are as follows: even in the absence of metal ions, this enzyme is somewhat less active than wild-type trypsin. The reason for this is not known, but the lower activity may derive from steric or electrostatic perturbations in the vicinity of the active site that result from the Arg-to-His substitution.

In the presence of Cu(II) or Zn(II) however, the activity of the modified enzyme is dramatically inhibited. The inhibition is noncompetitive with

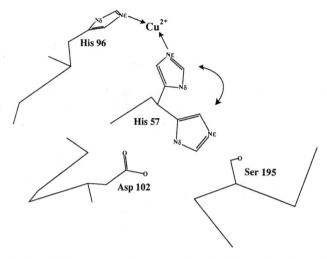

*Figure 4* A schematic illustration of the proposed metal-induced switch in trypsin-R96H. The $C_\alpha$ positions of the active-site region are shown along with the modeled coordination complex formed between a single Cu(II) ion, His96, and His57 of trypsin-R96H. Reproduced with permission from Ref. 20.

respect to substrate, with a $K_i$ for Cu(II) of $2.1 \times 10^{-5}$ M and for Zn(II) of $1.3 \times 10^{-4}$ M. These affinities are rather poor and suggest either that binding induces deformation of the protein's structure or that the metal may not be binding exactly as designed. Further structural and energetic analyses to address these points have not yet been presented.

## DESIGNED SITES WITH THREE PROTEIN LIGANDS

Several natural proteins contain three coordinated Zn(II) sites with tetrahedral or distorted tetrahedral geometry. The metal ions bound at such sites play an essential role in the enzymes' catalytic mechanisms, acting as electrophilic catalysts and stabilizing negative charges.

Carboxypeptidase and carbonic anhydrase are two interesting examples. In carboxypeptidase, Zn(II)-bound water is activated for nucleophilic attack on the carbonyl of the scissile peptide bond (7). Similarly, in carbonic anhydrase, metal-bound water provides a source of nucleophilic hydroxyl groups at neutral pH that can attack either the natural substrate $CO_2$ or hydrolyze exogenously added esters (35).

To date, the goal of the designs for three-ligand sites has been to achieve activation of metal-bound water for nucleophilic attack. For this aim, the designed sites must allow only a single exogenous ligand to bind. Otherwise, if two or three ligands bind, with trigonal bipyramid or octahedral geometry, the $pK_a$ of each will most likely not be reduced enough to achieve efficient catalysis at neutral pH (the $pK_a$ of water in $[Zn(H_2O)_6]^{2+}$ is about 10).

Model compounds give hope that such simple sites will display at least some level of enzymatic activity. In 1975, Woolley synthesized the macrocycle shown in Figure 5a (41). The objective was to determine whether a

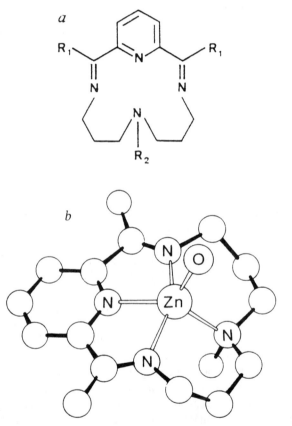

*Figure 5* (a) Structure of the free macrocyclic ligand. CR:$R_1$ = $CH_3$; $R_2$ = H:N-MeCR; $R_1$ = $R_2$ = $CH_3$; desdiMeCR:$R_1$ = $R_2$ = H. Reproduced with permission from Ref. 41. (b) The complex of the macrocycle with Zn(II). The four nitrogen atoms bind the metal ion, which also binds water. The skeleton of the $Zn(N-MeCR)(H_2O)^{2+}$ complex is shown (H is omitted and unlabeled atoms are C). Reproduced with permission from Ref. 41 (copyright © 1975, Macmillan Magazines Ltd.).

simple organic chelate of Zn(II) could mimic carbonic anhydrase activity by activating a molecule of water for nucleophilic attack, as illustrated in Figure 5b.

The catalytic activity of the Zn(II)-ligand complex was studied using two classical carbonic anhydrase assays: the hydration of acetaldehyde and the hydration of $CO_2$. The complex displays high activity in the aldehyde-hydration assay, comparable to that of carbonic anhydrase. In the carbon dioxide–hydration assay, the model compound has about 1/400 the activity of carbonic anhydrase, which is still remarkably high for such a simple enzyme model.

The compound {Tris[(4,5-dimethyl-2-imidazolyl)methyl]phosphene oxide}Zn(II) is also a mimetic of carbonic anhydrase and is a weak catalyst in both the $CO_2$-hydration and p-nitrophenyl acetate–hydrolysis assays (2). Similarly, model compounds that mimic carboxypeptidase and are capable of catalysis have been synthesized (14). No protein design has yet been as successful.

## Metalloantibodies

Tainer and colleagues have reported their attempts to introduce metal-binding activity onto an antibody framework (22, 33). The construction of such "metalloantibodies" is an attractive strategy because antibodies can potentially be raised against a desired substrate, or transition-state analogue, with the Zn(II)-activated water positioned close by for nucleophilic attack.

Underlying Tainer's design was the recognition that the antiparallel $\beta$ structure around the three histidine ligands of carbonic anhydrase B (see Figure 6) is similar to that around two of the complementarity-determining regions in antibody light chains (7). On the basis of this comparison, three histidine residues were introduced at appropriate positions in an antifluorescein antibody. The choice of this antibody allows fluorescence to be used as a convenient means to detect hapten binding in solution.

The interaction of protein with metal was monitored by detecting the quenching of the protein's intrinsic tryptophan fluorescence that occurred when Cu(II) was bound. When metal was added to an antibody-fluorescein complex, the fluorescence of both the protein's tryptophan residues, and of the hapten, was quenched, suggesting that the ternary complex of antibody, fluorescein, and metal is possible.

Quantitative fluorescence quenching studies were used to determine a binding constant for the Cu(II)-antibody complex of about $10^{-6}$ M. This value is several orders of magnitude poorer than that of a carbonic anhydrase–Cu(II) complex. It is closer to the values of dissociation constants

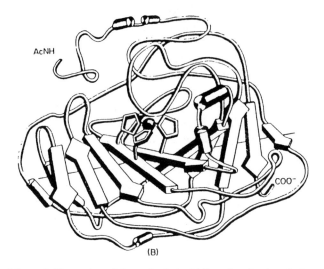

*Figure 6* Schematic illustration of the main-chain folding of carbonic anhydrase, with the Zn(II)-chelating His side chains shown. Cylinders represent α-helices and arrows β-sheets. Reproduced with permission from Ref. 6.

measured for two histidine-binding sites. This difference could potentially be caused by local structural perturbations or by incomplete histidine coordination.

A practical problem in using an antibody as a framework is the extreme difficulty that is encountered in obtaining significant quantities of the purified protein. This restriction precludes the use of informative spectroscopic techniques to determine if all three histidine residues are involved in binding and whether the binding site is tetrahedral. As an alternative approach towards answering the first of these questions, site directed mutagenesis is being used to mutate each of the proposed liganding residues to alanine. It will then be possible to compare the metal-binding properties of this family of mutant proteins and to determine the contribution of each desgined ligand.

## A $His_3$ Site in a Designed Four-Helix Bundle Protein

A three-histidine site has been introduced into a designed protein, $\alpha_4$. The $\alpha_4$ protein framework is an idealized four-helix bundle in which four identical helices are connected by three identical loops and there is minimal sequence complexity (21, 32). Three surface histidine residues were introduced into this protein, two on one helix as His-$X_3$-His (recall the two histidine sites with this arrangement) and the third on a neighboring helix (16). The protein contains no aromatic residues, which facilitates the assignment of the histidine C$\delta$2H and C$\epsilon$1H resonances in $^1$H NMR spectra. Spectra were recorded as increasing amounts of Zn(II) were added to the protein. The resonances of the C$\delta$2H and C$\epsilon$1H histidine ring protons of all three chelating residues were observed to shift until a 1:1 stoichiometry of metal to protein was reached. This result supports the conclusion that all three of the designed ligands are involved in interactions with the metal in a 1:1 protein:Zn(II) complex (see Figure 7).

Experiments to reveal the geometry of the site and its affinity for metal have not yet been reported.

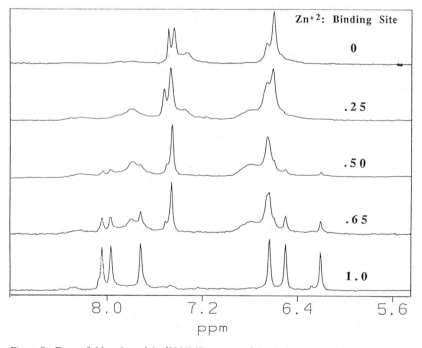

*Figure 7* Down-field region of the $^1$H NMR spectra of the designed Zn(II)-binding protein as a function of Zn(II) concentration. The peptide was in $^2$H$_2$O with 50 mM NaOAc and was titrated with zinc acetate. Spectra were recorded on a Bruker AM–600 spectrometer. Reproduced with permission from Ref. 16 (copyright © 1990, Americal Chemical Society).

## A Modified Zinc-Finger Peptide

Berg and colleagues reported an interesting example of a three-liganded site (26). The approach differed from those discussed above because in this case the site was created by modifying an existing 4-coordinate site in a 26-residue zinc-finger peptide. The peptide sequence is a consensus sequence for a prototypical TFIIIA-like zinc finger. The natural site is formed by two His residues on an α-helix and two Cys residues on an adjacent β-sheet. The sequence of the peptide is given below, with the residues involved in metal chelation underlined.

Pro-Tyr-Lys-Cys-Pro-Glu-Cys-Gly-Lys-Ser-Phe-Ser-Gln-Lys-Ser-

Asp-Leu-Val-Lys-His-Gln-Arg-Thr-His-Thr-Gly.

The first attempt to create a $Cys(His)_2$ coordination site by the removal of the first five amino acids of the peptide was unsuccessful, and the peptide did not bind Co(II). By contrast, deletion of the last four amino acids, to create a $(Cys)_2His$ site, resulted in a peptide that bound cobalt with an affinity within an order of magnitude of that of the parent peptide.

The optical absorption spectrum of the protein-Co(II) complex shows a long wavelength-absorption envelope with an intensity characteristic of tetrahedral coordination, which rules out higher coordination numbers (5) (see Figure 8a). This indicates that the new site successfully maintains the tetrahedral geometry of the original site, presumably with a molecule of water now bound as the fourth ligand. Moreover, titration of this complex with β-mercaptoethanol results in a series of spectra, with the final spectrum displaying wavelength maxima indicative of a three-thiolate, one-imidazole coordination sphere (see Figure 8b). This final spectrum represents the conditions in which the thiolate of β-mercaptoethanol coordinates Co(II) in the vacant ligand position.

Disappointingly, attempts to demonstrate carbonic anhydrase-like catalysis by peptide-Zn(II) complexes, in acetaldehyde hydration and p-

---

*Figure 8* (a) Absorption spectrum of the 1:1 complex of the modified zinc-finger peptide with Co(II) bound in 50 mM HEPES, 50 mM NaCl buffer, pH 7.0 under an atmosphere of 95% dinitrogen and 5% hydrogen. The inset shows a schematic of the proposed structure. Reproduced with permission from Ref. 26. (b) Titration of the peptide-Co(II) complex with β-mercaptoethanol. The offset is a derivative of the peptide with $(Cys)_3His$ ligands. This spectrum is similar to that at the end of the titration with mercaptoethanol. The inset shows the experimental data from spectral deconvolution and the fit using a β-mercaptoethanol dissociation constant of $2 \times 10^{-4}$ M. The measurements are made in 50 mM HEPES, 50 mM NaCl, pH 8 under an atmosphere of 95% dinitrogen and 5% hydrogen. Reproduced with permission from Ref. 26 (copyright © 1991, Americal Chemical Society).

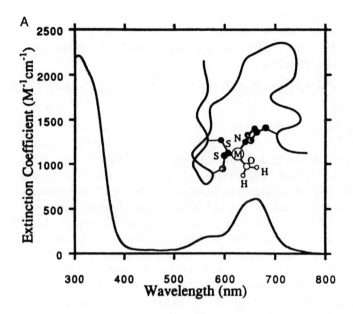

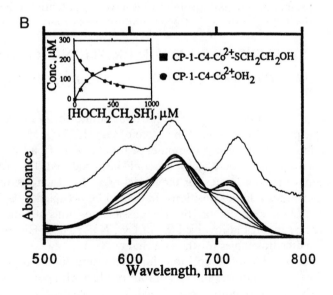

nitrophenyl acetate hydrolysis assays, have been unsuccessful. This is possibly because the $pK_a$ of the metal-bound water at this two-thiolate, one-imidazole site is estimated to be greater than 9, which compares unfavorably with the $pK_a$ of 7 for water bound at the three-imidazole site of carbonic anhydrase.

## Three-Coordinate Zn(II) Sites in a Designed Four-Helix Bundle Protein

$Z\alpha_4$, a four-helix bundle protein with a designed tetrahedral $(Cys)_2(His)_2$ Zn(II)-binding site is described in detail in the following section (31, 32). This protein was used as the starting point for the design of three-coordinate Zn(II)-binding derivatives. Each metal ligand was independently mutated to alanine, the small nonreactive side chain being chosen to enable a molecule of water to bind at the vacant coordination position.

Three of the four changes produce proteins that behave as anticipated. The proteins bind Zn(II) but with affinities that are reduced relative to $Z\alpha_4$ by up to an order of magnitude. The most important observation is that the optical absorption spectra of the protein-cobalt complexes suggest that a molecule of water binds at the vacant coordination position. Specifically, the spectra display absorption envelopes at long wavelengths in which the wavelength maxima are shifted as expected for sulfur-to-oxygen or nitrogen-to-oxygen substitutions of a single coordinating ligand. These proteins represent new metalloproteins that have the potential to be catalytically active (M. Klemba & L. Regan, unpublished observations).

# DESIGNED SITES WITH FOUR PROTEIN LIGANDS

## A Tetrahedral Binding Site in a Designed Four-Helix Bundle Protein

A tetrahedral Zn(II)-binding site was introduced into the designed four-helix bundle protein, $\alpha_4$, to create the protein $Z\alpha_4$ (31, 32).

The design resulted from a computer search to identify all potential positions in the protein where such a site could be introduced. The program takes as input the coordinates for the model of $\alpha_4$. For each residue, the program generates a list of positions that could be occupied by a Zn(II) ion, if that residue were Cys or His. The possible positions of a metal ion are dictated by the ligand involved ($S\gamma$ of Cys; $N\delta$ or $N\varepsilon$ of His) and the appropriate dihedral angles for the side chain (28).

The sets of possible metal positions for each residue in the protein are compared with one another, and sets of four residues with potential metal sites in fairly close proximity are analyzed further. At each site, the program iteratively attempts to find a consensus position for a metal, while allowing a more complete range of $\chi$ angles than was allowed in the initial screen.

Finally, the site is assessed for tetrahedrality. The best site in $\alpha_4$ was chosen and the geometry of the model for it improved by performing energy minimization and simulated annealing with the restraints of fixed main-chain atoms and the metal-site side chains. The final model differed little from the starting model; unfavorable steric contacts were absent, and the geometry of the site was good.

Optical absorption spectroscopy was used to obtain detailed information on the geometry of the binding site. Figure 9 shows the absorption spectrum of the Co(II)-substituted protein. The important features of the spectrum are the absorption envelope at long wavelength and the charge-transfer band at 300 nm.

The intensity of the absorption maximum at long wavelength is fully consistent with tetrahedral coordination of Co(II) and rules out higher coordination numbers. The charge-transfer band, with a maximum at 300 nm, is clearly indicative of cysteinate coordination. In addition, the wavelength maxima of the long wavelength transitions are consistent with Co(II) coordination in a (thiolate)$_2$(imidazole)$_2$ environment.

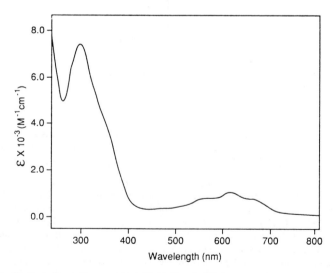

*Figure 9* Optical absorption spectrum of the Z$\alpha_4$-Co(II) complex in 20 mM HEPES at pH 7.5. The spectrum has been corrected by subtracting the spectrum of the fully reduced protein. Reproduced with permission from Ref. 31 (copyright © 1990, Americal Chemical Society).

Optical absorption spectroscopy was also used to directly measure the protein's affinity for Co(II) and to measure its affinity for Zn(II) by competition (see Figure 10). The dissociation constant for Co(II) is $1.6 \times 10^{-5}$ M and for Zn(II) is $2.5 \times 10^{-8}$ M. The lower affinity for Co(II) is again consistent with binding at a tetrahedral site.

As a final precaution, gel filtration chromatography with radioactive $Zn^{65}$ was used to demonstrate that the protein binds Zn(II) as a monomer. Confirmation of this point is important as there are precedents for metal-induced dimerization in insulin (His ligands) (5a); Tat, the transactivating protein of HIV (Cys ligands) (8a); and human growth hormone (His and Glu ligands) (8).

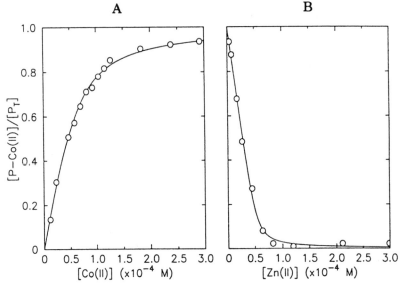

Figure 10 (a) Titration of $Z\alpha_4$ with Co(II) in 20 mM HEPES at pH 7.5. The concentration of the complex ([P-Co(II)]) divided by the total protein concentration ([PT]) is plotted against the total Co(II) concentration, [Co(II)]. The fit of the data, assuming noncooperative binding of one molecule of Co(II) to one molecule of protein, is shown by the solid line. (b) Titration of the protein-Co(II) complex with Zn(II). Aliquots of $ZnCl_2$ were added to a solution of $Z\alpha_4$ and $CoCl_2$, and the absorbance spectra were recorded. [P-Co(II)]/[Pt] is plotted against the total concentration of Zn(II). The solid line shows a fit of the data, assuming a simple competition between Co(II) and Zn(II) in binding to the protein, with a dissociation constant for the protein-Co(II) complex of $1.57 \times 10^{-5}$. Reproduced with permission from Ref. 31 (copyright © 1990, Americal Chemical Society).

METAL-SITE DESIGN 273

The results of the characterizations suggest that the design was successful and that protein binds Zn(II) in a 1:1 complex, with high affinity, at a site with tetrahedral coordination geometry.

## A Design for a Blue Copper Site

A similar approach to the design of a four-coordinate metal site was taken by Hellinga & Richards (19). Here, the target is a type I copper site, as found in blue copper proteins including azurin and plastacyanin. In these sites, copper is completely inaccessible to solvent and is chelated by by $S\gamma$ of Cys, two $N\delta s$ of His, and the $S\delta$ of methionine. The four coordinating atoms are arranged in a distorted tetrahedron (15) (see Figure 11). As their name suggests, blue copper–binding sites have a number of characteristic

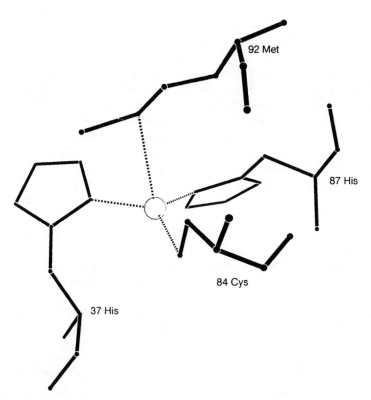

*Figure 11* The natural copper-binding site in plastocyanin, from the Brookhaven PDB entry 1PCY YY. Reproduced with permission from Ref. 19.

spectroscopic properties that reflect the detailed geometry of the copper coordination sphere.

Thioredoxin from *Escherichia coli*, a protein whose X-ray crystal structure is known, was chosen as the framework on which to introduce the new site (19). A computer program was used that takes the protein coordinates, keeping the main-chain atoms in their original positions, and substitutes each side chain with each of the four potential ligands in positions dictated by preferred rotamer dihedral libraries (29). The program then searches for a set of backbone positions that are arranged such that if appropriate side chains were attached they would form a Cu(II)-binding site.

The new model is subjected to energy minimization and checked for both steric clashes, which would result from the protein being overpacked, and solvent accessibility changes, which are taken as an indication that the protein is underpacked. If either problem exists, the protein is repacked around the site using a combinatorial repacking algorithm (29).

The first site required only four amino acid changes at the binding site (18). A complication occurred when characterization of the protein was initiated: even wild-type thioredoxin displays metal-binding activity and binds one atom of copper or two of mercury. The binding sites are on the surface of the protein and were first observed when heavy atom derivatives were obtained in crystallographic studies on the protein.

Changes in the protein's intrinsic tryptophan fluorescence induced by metal binding were used to determine the affinity and stoichiometry of the interaction of the redesigned thioredoxin with copper. The Scatchard plots of this data show classical two-site binding and indicate that an additional binding site for metal has been introduced that binds Cu(II) with an affinity of $4 \times 10^{-5}$ M compared with the surface site that binds Cu(II) with a dissociation constant of $4 \times 10^{-6}$ M (see Figure 12).

Optical absorption spectroscopy of protein-Cu(II) complexes was used to determine the coordination of the new metal-binding site; blue copper sites have a characteristic intense band at 600 nm that corresponds to charge-transfer interactions between copper and the cysteine thiolate and methionine thioether.

The absorption spectrum of the protein-Cu(II) complex does not have this band. It shows an unusual broad absorbance over the entire visible region (see Figure 13). This indicates that either the copper-thiolate and copper-thioether bonds are both missing or that the protein-metal complex has several different forms; the broad spectrum represents transitions with an overlapping range of energies. The spectrum does show strong bands at 240 and 300 nm that are characteristic of Cu-imidazole interactions.

EPR spectra of the protein-Cu(II) complexes suggest that the copper may be coordinated by two nitrogen and two oxygen ligands. The oxygen

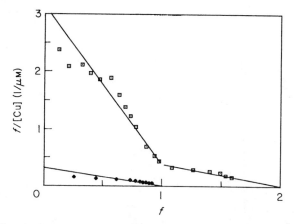

*Figure 12* Scatchard analysis of Cu(II) binding. The term $f$ is the mole fraction of bound metal, $f = n\Delta F/\Delta F_{max}$, where $\Delta F$ is the change in fluorescence at a given metal concentration and $\Delta F_{max}$ is the change in fluorescence at infinite concentration of metal, $n$ is the number of metal-binding sites, and [Cu(II)] is the concentration of Cu(II). The filled diamonds are data for wild-type thioredoxin; the open squares are data for thioredoxin with the additional designed site. Reproduced with permission from Ref. 18 (copyright © 1991, Academic Press, London, Ltd.).

ligands may come from nearby acidic groups or from solvent. Therefore, in an attempt to make the designed ligands the only potential chelators in the vicinity of the site, several neighboring acidic side chains have been removed. Absorption spectra of protein-cobalt complexes of this improved version of the design show charge transfer bands in the near UV that are indicative of thiol coordination. Therefore, it appears that the metal is now coordinated by at least three of the four designed ligands. Further characterizations and redesigns are progressing towards the aim of achieving a true type-I site (H. Hellinga, personal communication).

## A Simplified $(Cys)_2(His)_2$ Zinc-Finger Peptide

The following discussion of the modification of a natural zinc-finger peptide is included as a caution for protein design: Its behavior is quite unexpected.

A 26-residue peptide, MZF (minimalist zinc finger), based on the TFIIIA metal-binding sequence, was synthesized (27). It contains the conserved metal-binding and hydrophobic residues that are characteristic of this class of protein domain, with 16 of the remaining 19 residues being changed to alanine:

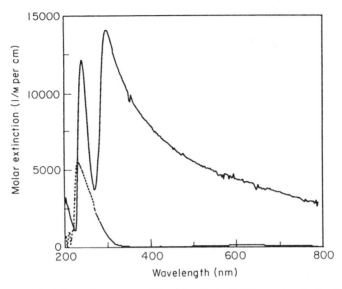

*Figure 13* Optical absorption spectra of the thioredoxin-Cu(II) complex (*broken line*) and the designed protein-Cu(II) complex (*solid line*). The contribution of the apoprotein has been subtracted from both spectra. Reproduced with permission from Ref. 18 (copyright © 1991, Academic Press, London, Ltd.).

Lys-**Tyr**-Ala-<u>Cys</u>-Ala-Ala-<u>Cys</u>-Ala-Ala-Ala-**Phe**-Ala-Ala-Lys-Ala-Ala-**Leu**-Ala-Ala-<u>His</u>-Ala-Ala-Ala-<u>His</u>-Ala-Lys.

The underlined residues are involved in metal chelation. The residues in bold are the conserved hydrophobic residues.

The unusual properties of this peptide were first revealed when the optical absorption spectrum was recorded during the titration of the protein with cobalt. The shape of the spectrum was observed to change during the course of the titration. A spectrum in which the ligand-field bands were relatively red-shifted was observed early in the titration. As increasing amounts of cobalt were added, the initial spectrum decayed and a new spectrum appeared and increased until saturation at a 1:1 protein:Co(II) ratio.

These unusual observations can be explained as follows. The spectrum at low peptide concentrations resembles the spectrum of Co(II) complexed by four cysteinate ligands. It is therefore likely that early in the titration, when the ratio of peptide:Co(II) is high, a dimer is formed in which cobalt is coordinated by two cysteine ligands from each of two peptides. Later in the titration, when the peptide:Co(II) ratio is low, the correct species is formed, with cobalt coordinated by two histidine and two cysteine residues from a single peptide. Figure 14a shows a superposition of the spectra of the two different species, and Figure 14b shows the amounts of each complex as a function of concentration. One- and two-dimensional NMR studies indicate that the peptide 1:1 complex has a structure that is very similar to that of an authentic TFIIIA zinc-finger peptide.

Additional mutations were made in MZF to disrupt its hydrophobic core by substituting the conserved hydrophobic amino acids (shown in bold, above) with alanine. These peptides form the 2:1 species to a much greater extent than the MZF peptide. Also, the absorption spectra of their 1:1 complexes with cobalt are much broader than that of MZF, which suggests that they may not form a unique structure, but rather an ensemble of structures.

These results are remarkable as they demonstrate how little structure is required to generate a peptide that can assemble and bind to a metal ion. The formation of the unexpected 2:1 complexes also demonstrates, as was observed in the design for a blue copper site, that if alternative liganding schemes are possible, the design must incorporate features to destabilize these structures relative to the intended structure.

## DISCUSSION AND PERSPECTIVES

Clearly, certain metal-site designs have been extremely successful, and the new proteins display the anticipated properties. A high resolution structure has not yet been determined for any of the designed proteins, so the exact details await confirmation.

The designed sites incorporate the minimum features required to obtain metal-binding. They are based primarily on geometric considerations, with additional checks to avoid unfavorable steric interactions. By comparison, natural metal sites display additional features, described below, that result in affinities for metal that are higher than those of the designed sites.

In many natural sites, the side-chains of the metal-binding residues are additionally hydrogen bonded to other side chains or to backbone carbonyl and amide groups to position them correctly (6, 12) (see Figure 15). Some natural metal-binding sites have evolved electrostatic surfaces with high negative potential in the region of the binding site. This is an additional feature that enhances the affinity of the site for positively charged metal ions (11; B. Honig, unpublished observations).

Finally, Eisenberg and colleagues have noted that metal ions appear to bind in regions of proteins that show "high hydrophobicity contrast" (42).

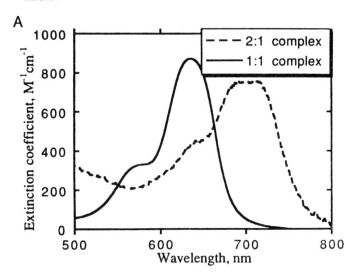

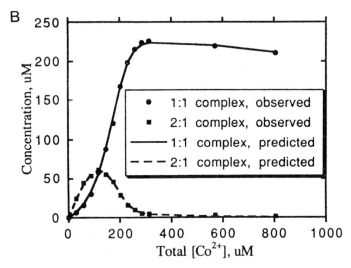

*Figure 14* (a) Derived spectra for the 2:1 and 1:1 complexes of the MZF peptide with Co(II). The spectra were calculated as linear combinations of the initial and final titration spectra. Reproduced with permission from Ref. 18. (b) Plot of the observed concentrations of the 1:1 peptide:Co(II) (*solid circle*) and 2:1 peptide:Co(II) (*solid square*) complexes as a function of total Co(II) concentration. The dashed and solid lines represent fits to the data with dissociation constants of $5 \times 10^{-7}$ M for the 1:1 species and $4 \times 10^{-11}$ M$^2$ for the 2:1 species. The 1:1 complex concentrations decrease at high Co(II) concentrations because of the increased volume of the peptide sample. Reproduced with permission from Ref. 27.

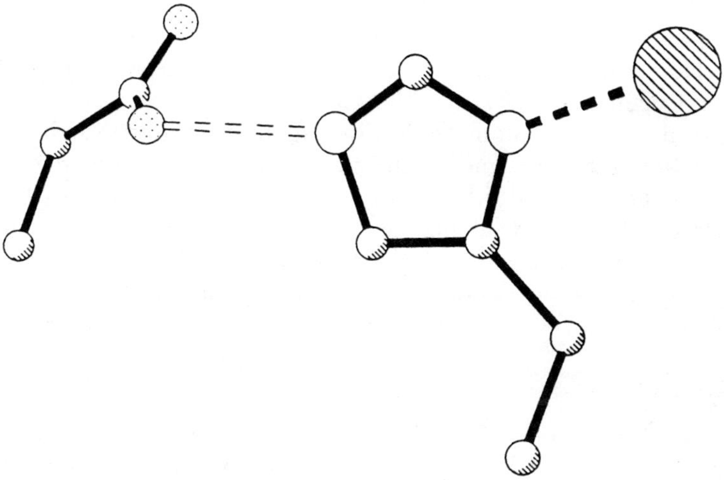

*Figure 15* The aspartate-histidine-zinc triad of the active site of carboxypeptidase that may contribute to enhanced zinc affinity at the metalloenzyme active site. Reproduced with permission from Ref. 27.

That is, the metal sites are centered in a shell of hydrophilic ligands that is surrounded by a shell of carbon-containing groups.

This observation could reflect the nature of the metal-ligand interaction: the metal-binding residues chelate through the electron pair donors oxygen, nitrogen and sulphur, which themselves are bound to the less polar carbon atoms of the side chain. Hence, there must necessarily be a large difference in hydrophobicity at this point. However, the nature of the second shell of residues around the site can serve to enhance that degree of hydrophobic contrast.

How could a hydrophobic shell around the site lead to enhanced metal-binding affinity? It is possible that the hydrophobic sphere restricts the flexibility of the site preorganizing it for metal binding. Alternatively, the low dielectric of the hydrophobic sphere could enhance electrostatic interactions between groups within it.

An important consideration in the design of high affinity sites is the potential for interaction of the metal with unanticipated additional ligands. These may either be groups on the protein or molecules of solvent. A close inspection and possible modification of potential liganding residues in the vicinity of the designed site would be a step towards addressing this problem. The participation of unwanted solvent ligands is a somewhat more difficult problem to address. One precaution is to ensure that interaction of the protein ligands with the metal does not require unfavorable side-chain dihedral angles or other energetically unfavorable distortions of the structure.

The above discussions make clear that additional features can be incorporated into future designs to generate sites that more closely reproduce the affinities of natural sites. However, it is most encouraging that there are already designed metal-binding proteins that can be the starting point for a systematic investigation of these factors. In addition to higher binding activities, future designs and redesigns should soon achieve some form of catalysis at the metal site.

ACKNOWLEDGMENTS

I thank my collaborators, Neil Clarke, Mike Klemba, Craig Ceol, and Mimi Shirasu for their important contributions to the metal-binding work. I thank Mike Klemba, Mary Munson, Athena Nagi, and Kate Smith for comments on the manuscript and Morris Gottlieb for his valuable assistance in manuscript submission. I also gratefully acknowledge E. I. du Pont de Nemours and Company and the British Ramsay Fellowship in Chemistry for early support and the Office of Naval Research (N00014-91-J-1578) for its ongoing support.

*Literature Cited*

1. Adman, E. T., Howard, J. B., Rees, D. G. 1991. *Adv. Protein Chem.* 42: 145
2. Armitage, I. M., Schoot-Uiterkamp, A. J. M., Chlebowski, J. F., Coleman, J. E. 1978. *J. Magn. Reson.* 29: 375
3. Arnold, F. H., Haymore, B. L. 1991. *Science* 252: 1796
4. Berg, J. M. 1990. *Annu. Rev. Biophys. Biophys. Chem.* 19: 405
5. Bertini, I., Luchinat, C. 1984. *Adv. Inorg. Chem.* 6: 71
5a. Blundell, T., Dodson, G., Hodgkin, D., Mercola, D. 1972. *Adv. Protein Chem.* 26: 279
6. Christianson, D. W. 1991. *Adv. Protein Chem.* 42: 224
7. Christianson, D. W., Lipscomb, W. N. 1989. *Acc. Chem. Res.* 22: 62
8. Cunningham, B. C., Mulkernin, M. G., Wells, J. A. 1991. *Science* 253: 545
8a. Frankel, A. D., Bredt, D. S. Pabo, C. O. 1988. *Science* 240: 70
9. Ghadiri, M. R., Choi, C. 1990. *J. Am. Chem. Soc.* 112: 1630
10. Ghadiri, M. R., Soares, C., Choi, C. 1992. *J. Am. Chem. Soc.* 114: 4000
11. Gilson, M. K., Honig, B. 1988. *Proteins* 4: 7
12. Glusker, J. P. 1991. *Adv. Protein Chem.* 42: 1
13. Goraj, K., Renard, A., Martial, J. A. 1990. *Protein Eng.* 3: 259
14. Groves, J. T., Olson, J. R. 1985. *Inorg. Chem.* 24: 2715
15. Guss, J. M., Freeman, H. C. 1983. *J. Mol. Biol.* 169: 521
16. Handel, T. M., DeGrado, W. F. 1990. *J. Am. Chem. Soc.* 112: 6710
17. Hecht, M. H., Richardson, J. S., Richardson, D. C., Ogden, R. C. 1990. *Science* 249: 884
18. Hellinga, H. W., Caragonna, J. P., Richards, F. M. 1991. *J. Mol. Biol.* 222: 787
19. Hellinga, H. W., Richards, F. M. 1991. *J. Mol. Biol.* 222: 763
20. Higaki, J., Fletterick, R. J., Craik, C. S. 1991. *TIBS* 17: 100
21. Ho, S. P., DeGrado, W. F. 1987. *J. Am. Chem. Soc.* 109: 6751
22. Iverson, B. L., Roberts, V. A., Iverson, S. A., Benkovic, S. J., Lerner, R. A., et al. 1986. *Science* 249: 659

23. Kruck, T. P. A., Lau, S.-J., Sarkar, B. 1976. *Can. J. Chem.* 8: 1300
24. Leszcznski, J. F., Rose, G. D. 1986. *Science* 234: 849
25. McPhalen, C. A., Strynadka, N. C. J., James, M. N. G. 1991. *Adv. Protein Chem.* 42: 77
26. Merkle, D. L., Schmidt, M. H., Berg, J. M. 1991. *J. Am. Chem. Soc.* 113: 5450
27. Michael, S. F., Kilfoil, V. J., Schmidt, M. H., Amann, B. T., Berg, J. M. 1992. *Proc. Natl. Acad. Sci. USA* 89: 4796
28. Pabo, C. O., Suchaneck, E. G. 1986. *Biochemistry* 25: 5987
29. Ponder, J. W., Richards, F. R. 1988. *J. Mol. Biol.* 193: 775
29a. Qiagen Inc. 1991. *The QIAexpressionist.* Chatsworth, CA: Qiagen. 2nd ed.
30. Regan, L. 1991. *Curr. Opin. Biotechnol.* 2: 544
31. Regan, L., Clarke, N. D. 1990. *Biochemistry* 29: 10879
32. Regan, L., DeGrado, W. F. 1988. *Science* 241: 976
33. Roberts, V. A., Iverson, B. L., Iverson, S. A., Benkovic, S. J., Lerner, R. A., et al. 1990. *Proc. Natl. Acad. Sci. USA* 87: 6654
34. Sasaki, T., Kaiser, E. T. 1989. *J. Am. Chem. Soc.* 111: 380
35. Silverman, D. N., Lingskog, S. 1988. *Acc. Chem. Res.* 21: 30
36. Suh, S.-S., Haymore, B. L., Arnold, F. H. 1991. *Protein Eng.* 4: 301
37. Tainer, J. A., Roberts, V. A., Getzoff, E. D. 1991. *Curr. Opin. Biotechnol.* 2: 582
38. Toma, S., Campagnoli, S., Margarit, I., Gianna, R., Grandi, G., et al. 1991. *Biochemistry* 30: 97
39. Vallee, B. L., Auld, D. S. 1990. *Proc. Natl. Acad. Sci. USA* 87: 220
40. Vallee, B. L., Coleman, J. E., Auld, D. S. 1991. *Proc. Natl. Acad. Sci. USA* 88: 999
41. Woolley, P. 1975. *Nature* 258: 677
42. Yamashita, M. M., Wesson, L., Eisenman, G., Eisenberg, D. 1990. *Proc. Natl. Acad. Sci. USA* 87: 5648

# ARTIFICIAL NEURAL NETWORKS FOR PATTERN RECOGNITION IN BIOCHEMICAL SEQUENCES

## S. R. Presnell and F. E. Cohen

Department of Pharmaceutical Chemistry, University of California, San Francisco, California 94143

KEY WORDS:  protein structure, gene determination, secondary structure prediction

CONTENTS

| | |
|---|---|
| PERSPECTIVES AND OVERVIEW | 283 |
| INTRODUCTION TO MACHINE LEARNING | 284 |
|     *Explanation-Based vs Similarity-Based Learning* | 284 |
|     *Artificial Neural Networks* | 285 |
| NETWORK TOPOLOGIES | 285 |
|     *Perceptrons* | 285 |
|     *Multilayer Networks* | 287 |
| COMPLEXITIES OF THE LEARNING SURFACE | 288 |
| MEMORIZATION VS GENERALIZATION | 288 |
| NETWORK-TRAINING ANALYSIS | 290 |
| APPLICATIONS | 291 |
|     *DNA Promotor Recognition* | 291 |
|     *mRNA Splicing Donor/Acceptor-Site Recognition* | 293 |
|     *Protein Coding-Region Recognition* | 293 |
|     *Protein–Secondary Structure Prediction* | 294 |
|     *Tertiary Structure–Class Prediction* | 296 |
| CONCLUSIONS | 296 |

## PERSPECTIVES AND OVERVIEW

Molecular biology has provided the tools to determine gene sequences and infer protein sequences. Given this wealth of basic biological information, how these sequences code for three-dimensional structures and biochemical functions are natural questions. The development of correlations between sequence and structure of nucleic and amino acid sequences continues to be a central challenge for biochemists. Some correlates are easy to recognize: the imino acid proline usually creates a change in the path of the polypeptide chain. This observation follows immediately from stereochemical aspects of the pyrrolidine ring structure and was confirmed by the first protein crystal structures. However, one may not recognize more subtle correlations between sequence and structure by simply perusing the data. The systematic determination of sequence structure correlates can offer new insights into biological function and provide new avenues for experimental work. Claims that these correlations are biologically relevant require statistical and experimental verification.

The field of machine learning should be well suited to the task of extracting information from large sequence data bases. When they are supplied with a hypothesis and supporting data, machine-learning techniques might be used to discern concepts hidden in the data. Unfortunately, the energetic stability of many biological systems is only marginal (1, 27). As a consequence, these systems may contain a significant amount of background noise. Furthermore, machine-learning tools are far from perfect in their ability to pinpoint previously unknown concepts. The result is a field in which both the data representation and the algorithmic tools are subject to revision and refinement.

In this review, we report on the application of one machine-learning technique, artificial neural networks (ANN), to the development of biologically relevant sequence structure correlates. We introduce the general theory of ANN computation, the importance of input representations, the utility of the methods, and the limitations of the resulting tools.

## INTRODUCTION TO MACHINE LEARNING

### Explanation-Based vs Similarity-Based Learning

Algorithms implementing explanation-based learning use preexisting knowledge of the field as a scaffold for additional learning. The existing knowledge is used to formulate plans for the attainment of specified goals. Execution of the plans is monitored for violated expectations, errors are then diagnosed, and the theory is modified to conform to the known data. This kind of technique does not require a large training set, but it does require the prior existence of applicable theories (11).

Similarity-based learning generates theories solely from the data at hand. It does not require preexisting knowledge of the field that applies to the task. On the other hand, the number of training instances required to resolve concepts in the data is substantially higher than that needed for explanation-based learning. Artificial neural networks are a form of similarity-based reasoning.

Implicit within ANNs are functional forms that are used to relate the input and output data. If a physical relationship that explains the data is best described by a functional form that is orthogonal to those available to the network, an approximate solution will be found for the data that may lead to erroneous extrapolations.

## Artificial Neural Networks

Artificial neural networks intend to mimic biological neural networks to some degree. In biological networks, many different neurons may report information to a synapse. The receiving neurons accept multiple inputs. If the information the neurons receive reaches an appropriate threshold, they fire. Neurotransmitter efficiency controls the effective strength or weight of the connection.

The nodes and connective arcs in ANNs are analogous to the neurons and synapses in biological neural networks. The most straightforward ANN network topology is an acyclic directed graph; this type of ANN network is referred to as feed-forward. However, this particular network configuration is not a requirement of the current ANN training algorithms; more complex network configurations may be used. McClelland & Rumelhart (18) provide a practical review of neural-network types and their merits.

Most of the neural-network models applied to problems in biochemistry use supervised learning algorithms. These were first developed by Rosenblatt (24). More recently, algorithms for multilayer networks were also developed (26). In the supervised learning mode, the strength or weight of the arcs between nodes begins at some initial (often random) value. Training examples are presented to the input nodes. The algorithm "learns" from the known answer. The weights are modified with each training iteration according to one of several different techniques in an attempt to minimize the difference between the desired output and the actual output. Having learned from the examples, the network with its fixed weights can be applied in a predictive mode to unknown testing examples.

# NETWORK TOPOLOGIES

## Perceptrons

The most straightforward topology utilized in neural networks is the perceptron (19, 25). In a perceptron network architecture, simple weights and biases directly connect inputs to outputs. In spite of the simplistic appearance, this system of nodes and arcs can represent an accurate input-

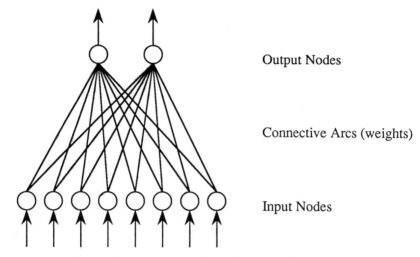

*Figure 1* A diagram of a single layer, feed-forward network. The input nodes are connected directly to the output nodes.

to-output mapping for many mathematical functions (Figure 1). However, only those problems that can be represented by linearly separable functions can be solved using perceptron networks; perceptrons cannot accurately represent some functions. Figure 2 shows three truth tables for a very simple perceptron network with two inputs and one output. Of the three binary logic operators, AND and OR can be described exactly by this network topology with known weights, but the exclusive OR (XOR) cannot be. A perceptron network with two inputs and one output has the power to separate two points on a plane with a line. The truth features of AND [(1,1)] and OR [(0,1), (1,0), (1,1)] can be separated easily as seen in

| Input Patterns | | Output Pattern | | |
|---|---|---|---|---|
| | | AND | OR | XOR |
| 00 | -> | 0 | 0 | 0 |
| 01 | -> | 0 | 1 | 1 |
| 10 | -> | 0 | 1 | 1 |
| 11 | -> | 1 | 1 | 0 |

*Figure 2* Truth tables for three logical operators, AND, OR, and XOR.

Figure 3a. By contrast, XOR requires two lines to segregate the truth features [(0,1), (1,0)] from the false counterparts [(0,0), (1,1)].

## Multilayer Networks

By adding an intermediate layer, one or more nodes between the input and output layers, we provide an additional dimension to the mathematical description of the XOR problem. Let $i$ and $j$ be the first and second input variables. Consider a cube with the logical value of $i$ as the $x$-coordinate, the logical value of $j$ as the $y$-coordinate, and the logical value of ($i$ AND $j$) as the $z$-coordinate. The $z$-coordinate acts as a hidden layer offering an intermediate level of input data processing that spreads the input into a three-dimensional object. It is then a simple matter to pick a plane that separates the truth values of the XOR [(0,1,0), (1,0,0)] from the false counterparts (Figure 3b). These intermediate layers are often called hidden

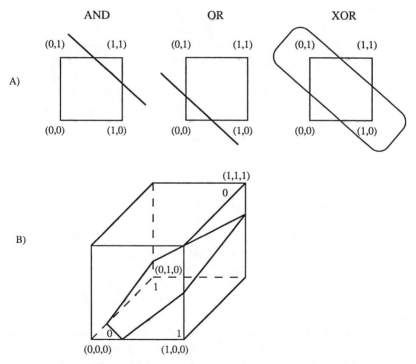

*Figure 3* (*a*) Geometric representations of the truth tables for AND, OR, and XOR. (*b*) The three-dimensional solution to the XOR network.

layers; the network configurations are referred to as multilayered. With this simple augmentation of ANNs, the XOR problem can be solved exactly.

## COMPLEXITIES OF THE LEARNING SURFACE

When the paradigm for the network architecture is a perceptron, and the concepts to be learned are linearly separable, the training solution for the network is unique. The learning surface has one global minimum if the architecture is that of a feed-forward network and back-propagation is used as the optimization algorithm (29). However, in multilayered networks, the learning surface exists in multiple dimensions and may contain multiple, suboptimal minima. Training of the network may become trapped in one of these minima. Most researchers treat this problem by running multiple training sessions, starting with randomly generated starting weights. The networks often converge on the same or a similar minimum, but there is no guarantee that this is the lowest minimum.

In the training of a simple feed-forward, hidden-layer network for the recognition of secondary structure in $\alpha/\beta$ proteins (15), the initial weights play a critical role in the development of predictive network weights. With random initial weights, a simple feed-forward network converges to a set of weights that do not predict any $\beta$ structure. However, if the initial connective weights between Phe, Ile, Leu, Met, Val, Trp, Tyr, and the $\beta$-output unit are set to the maximum value, and all other weights between the input layer and the $\beta$-output node are set to the minimum value, then the network weights converge to describe an energetic minimum that facilitated the prediction of $\beta$ structure. Learning algorithms and the variables used in controlling these algorithms remain a source of active research in the field.

## MEMORIZATION VS GENERALIZATION

One of the more troublesome issues in the construction and training of artificial neural networks is the determination of the state of mind of the network. If the number of independently variable weights is of the same order as the number of training examples, the network might "memorize" the specific features of the training examples rather than determine a generalized concept.

Some researchers have argued implicitly that recognition through memorization is indeed a desirable goal. In their application of ANNs to the prediction of secondary structure from primary sequence, Bohr et al (2) note that the homology between sequences in the test and training sets is reflected in the prediction efficiency of the network on the test sequence. These authors go on to suggest that neural networks might be valuable as a tool for measuring the similarity of two homologous structures. Unfortunately, this new metric provides an answer qualitatively similar to those techniques already in use. In general, neural networks of this type would only provide a basis for partial pattern recognition or playback of trained sequences.

Another goal in training neural networks is a representation of the data in an abstract, generalized form; one that would allow the prediction of structure in the absence of sequence homologues in the data used to train the network. To that end, some researchers have paid explicit attention to the degree of homology across and within the testing and training data bases used (15).

The work by Muskal et al (20) specifically addresses the issue of memorization vs generalization during network training in the design of the network architecture. In their system, the goal of the ANN was to predict the secondary structure content of a protein sequence. The input side of the network architecture contained 22 input nodes: 20 for amino acid composition, 1 for molecular weight and 1 for heme presence. Additionally, the network contained 5–10 hidden units. The two output units for this network were the percentage $\alpha$ and the percentage $\beta$ structures. In this particular case, the number of weights and the number of training examples were of the same order. This circumstance signals the potential for memorization in learning. The authors constructed a second network to monitor the activity of the hidden units in the first network. They defined a memory factor, $Mf$, that is proportional to the deviation of the activity of a hidden node, $j$, for a testing example from the average of the activities of the same hidden node over all training examples. The memory-factor values from each of the hidden nodes were used as input to the second network. The second network would then be trained and tested at regular intervals while training the first network (Figure 4). The two outputs of the second network would provide the extent of generalization for learning the percentage content of strand and helix in the first network.

Insofar as a predictive system is desired, individual network configurations must be evaluated for their training characteristics. In network training through back-propagation, determining the best point to terminate training is still another variable to consider. Historically, the best indicator of the state of the trained network and weights was considered the value returned by the chosen error function. As the change in error approaches some minimum value, the network training is considered complete. However, this criterion does not always result in the highest possible performance of the network. Indeed, under some circumstances the net-

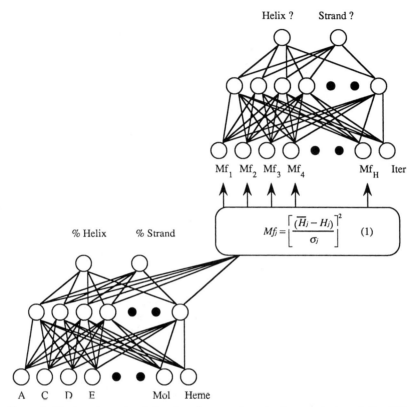

*Figure 4* The two-step network topology for secondary structure–content prediction (20). $\bar{H}_j$ represents the average activity of hidden node $j$ (training set). $H_j$ represents the activity of hidden node (testing example), and $\sigma_j$ is the standard deviation of the activities of the hidden node in the training set. $Mf_j$ is a memorization factor associated with $j$. In the first network there are 20 inputs for amino acid composition. The input Mol signals a measure of molecular weight, and the input Heme is used to signal heme presence or absence.

work reaches maximum performance against the test set substantially before the error function converges. To manage this situation, three data sets can be constructed: the training set for algorithm learning; the tuning set, a small but representative evaluation set for determining the efficiency of the network; and the testing set for gathering complete statistics on the ability of the network to predict results on new input data.

## NETWORK-TRAINING ANALYSIS

The results of ANN training can be evaluated in terms of the learned weights. This is often performed with the aid of the Hinton diagram (12)

(Figure 5). When evaluating perceptron networks, in which inputs are mapped directly to outputs, weights reflect the direct contribution of the various input units to the output units. Multilayer networks are more difficult to interpret. The weights of connections to the nodes in the hidden layers reflect the state of higher dimensionality in the network. This space tends to be difficult for researchers to examine and interpret. Networks may also be evaluated in an active sense by examining the dependence of the output on each weight for a given testing example.

## APPLICATIONS

Artificial neural networks have been applied to several problems in nucleic acid–sequence analysis: promotor recognition, intron/exon-boundary recognition, and gene recognition.

### DNA Promotor Recognition

In the first application of ANNs to promotor recognition, Lukashin et al (17) obtained excellent results using a three-layer, feed-forward network. The training set comprised 25 *Escherichia coli* RNA polymerase promotor sequences and 250 nonpromotor sequences. Testing was performed on 222 additional promotor sequences, the original 25 training sequences, and 2220 random sequences containing a uniform distribution of bases. The problem of input representation was split into recognizing the $-10$ and the $-35$ regions as two independent six-base blocks. Two bits of information represented each base. The network for each block contained 12 input nodes, a hidden layer of 4 nodes, and 1 output node. The sequence was recognized as a promotor when the output node of both networks was high. The tests indicated that the network could recognize 94–99% of actual promotor sequences with a 5% probability of a random pair of hexanucleotides being identified as a promotor (a false positive). When pBR322, a plasmid of 8726 nucleotides (both strands) was presented to the trained network, the 6 actual promotors in that sequence were found along with an additional 9–180 false-positive promotor sites, depending on the network parameters chosen.

Demeler & Zhou (8) obtained excellent results using a similar network architecture. In their work, the sequences were subdivided into three groupings: 20 bases around the $-10$ region, 20 bases around the $-35$ region, and an additional 44 bases containing the $-35$ region. In their experiments, the hidden layer varied from 4 to 10 nodes. Also, two-bit and four-bit data representations of the four actual nucleic acid bases were tried. The accuracy of the network fluctuated only by about 3% for the different multilayer network configurations. However, networks constructed and

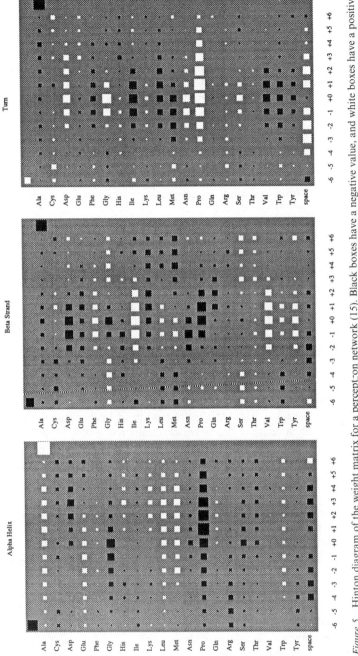

*Figure 5* Hinton diagram of the weight matrix for a perceptron network (15). Black boxes have a negative value, and white boxes have a positive value. Along the ordinate are the possible amino acid inputs, along the abscissa are the relative positions in the window of 13 residues. For a training or prediction of the network at sequence position $x$, residues evenly distributed from before and after $x$ are also used as input data. This is often referred to as the input window of residues. Input windows range widely, from 5 to 21 residues or more. The box at the upper right indicates the contribution of the periodic unit described in the text, and the box at the upper left indicates the bias on the output unit.

trained using the two-bit representation of bases generally had an accuracy near 80% while those networks utilizing a four-bit representation had an accuracy of 90%. The four-bit representation has the virtue of an identical Hamming distance for all possible input vectors (21).

Recently, more advanced methods have been applied to promotor detection. Shavlik et al (28) have developed a technique called knowledge-based artificial neural networks (KBANN) and applied it to the recognition of promoter sequences in *E. coli*. In this technique, knowledge of the field (domain knowledge) is outlined as a formal specification, or inference rules. This set of rules is then translated into a network topology and a partial set of initial weights; the network is then trained and evaluated. Once training is complete, the network topology and weights are translated back into a set of rules to be examined by the implementor. In practice, this technique compares favorably with the other techniques for promotor detection: it has a 97% rate of accuracy with a 1.2% rate of false positives. This work represents a change in the conceptual development of neural-network architectures. In previous efforts, decisions about network connection topologies were made in a haphazard fashion. In translating appropriate domain-theory information into network connectivity, Shavlik et al have begun to provide a coherent approach to the selection of an ANN network topology.

## *mRNA Splicing Donor/Acceptor-Site Recognition*

In their most recent work, Brunak et al (4) studied the prediction of mRNA donor and acceptor sites in human DNA. These experiments were analogous to the promotor-recognition studies—separate networks were constructed to recognize donor and acceptor sites in the DNA. While the results were somewhat better than the weight-matrix methods of Staden (30), the false-positive rate of the donor sites was about five times the number of true positives. For acceptor sites, the ratio of false positives to true positives was eight to one. To compensate for these over-predictions, Brunak et al provided an additional network that was trained to recognize coding regions. With this additional network, the false positives for both donors and acceptors were reduced by a factor of 2–30. In the course of the work, they identified approximately 16 typographical and reading frame–assignment errors in the European Molecular Biology Laboratory (EMBL) and GenBank Sequence data base (3).

## *Protein Coding-Region Recognition*

Artificial neural networks have also been applied to the determination of protein-coding regions in eukaryotic DNA sequences. Lapedes et al (16) performed the initial efforts in this area. In this work, the data representation consisted of isolated codon information. Markov dependence of the codons was not taken into consideration in the analysis of the coding regions. The average accuracy of intron and exon sensitivities was 98.4%, but with a significant rate of false positives (the data were not fully analyzed in the paper). In their subsequent work (9), the use of dicodon frequencies raised the accuracy of prediction to 99.4%. Unfortunately, this report also does not fully describe the specificity of the trained ANNs. High specificity as well as sensitivity would be critical to the expedient determination of coding regions in large anonymous fragments of genomic DNA.

Uberbacher & Mural (31) used a multisensor ANN approach to determine the location of protein-coding regions in human DNA sequences. Several different algorithms evaluate the sequence using different underlying principles. The generated metrics for a given region of DNA sequence are fed as training input to a feed-forward, multilayer ANN. This method locates 90% of the coding exons, with a false-positive rate of 19%. When they are evaluated on the basis of individual nucleic acid bases, 92% of the bases are assigned correctly with 8% false positives. With respect to false-positive rates, the method of Uberbacher & Mural compares favorably to that of Lapedes et al. However, even current rates of false positives would be intolerable for complete genomic analysis.

## Protein–Secondary Structure Prediction

The work of Qian & Sejnowski (23) has been the impetus for much of the additional experimentation with ANNs in the prediction of secondary structure in proteins. Buoyed by their success with neural networks for the generation of speech from text, they applied similar techniques to the development of an ANN for the prediction of secondary structure from sequence. A standard feed-forward network architecture was used. They also experimented with a hidden layer, and the number of nodes therein. The window width used for examining the sequence and the representation used for amino acids determined the number of input units. Several input representations were tried, but none worked better than one in which one bit represents a single amino acid, although representations using biophysical properties scored equally well. Their results improved the fraction of correctly predicted residues ($Q_3$) in a three-class prediction scheme ($\alpha$, $\beta$, coil) by 20% over the classic methods of Chou & Fasman (5) and Garnier et al (10). Qian & Sejnowski found that a window width of 13 performed the best, and that hidden layers provided a very small addition to the predictive accuracy of the trained networks.

Holley & Karplus (13) performed similar experiments and found that a window of 17 residues and a hidden layer of multiple nodes gave the best accuracy on their test set of sequences ($Q_3 = 63\%$). These results were comparable to those of Qian & Sejnowski. Holley & Karplus also noted that the strength of the network outputs correlates with the accuracy of the prediction.

Kneller et al (15) explored some novel data representations for the application of ANNs to protein–secondary structure prediction while continuing to use the feed-forward network architecture favored by many researchers. Regular secondary structure in globular proteins is periodic. Helices have 3.6 residues per turn, and $\beta$-strands have two residues per pleat. Helices are often amphipathic in nature; one face of the helix has a stronger hydrophobic character than the opposite face. Strands can also have different characters for each face. Although an appropriately constructed neural network can recognize periodic functions, these functions may be in a region of the learning surface that is difficult to reach. By explicitly determining the hydrophobic moments with a period appropriate for helices, Kneller et al were able to obtain a modest increase in the accuracy of the trained network. They were concerned that the addition of an input node simply increased the capacity for memorization in the network and tested this hypothesis by adding one more unit to the hidden layer instead. This change of the network architecture did not increase the testing-set accuracy, which suggests that the addition of a periodic node resulted in an increase of general information encoded by the network, rather than an increase in the capacity for memorization.

Characteristics of secondary structure differ with the tertiary-structure class of the protein under consideration (6, 7). With this in mind, Kneller et al partitioned training and testing sequence data based on tertiary structural class. Of the 105 proteins, 22 were all-$\alpha$, 24 were all-$\beta$, 20 were $\alpha/\beta$, and 40 were unclassified. The all-$\alpha$ protein class was augmented by a hydrophobic-moment calculation that used a period appropriate for helices. The increase in prediction accuracy obtained by separating the all-$\alpha$ proteins into their own class was 16%: from 63% to 79%. After an adjustment for the gain of moving from a three-state to a two-state prediction, the actual gain in accuracy was 4%. For all-$\beta$ structures, the increase in prediction accuracy was 6%, only half that expected for the change in the number-prediction states. The prediction accuracy in the $\alpha/\beta$ class was unchanged.

Alternative data representations for amino acids have been tried. L. Hunter (personal communication) has developed a representation for amino acids based on the atoms, orbitals, and hydrogens in each amino acid (AOH). To uniquely define the standard 20 amino acids, 48 bits of information were used. Training experiments demonstrated that the AOH representation could be used to generalize a wide variety of chemical properties. Hence, it might be a useful addition to the possible representations to consider when training a network. However, our results (S. R. Presnell & F. E. Cohen, unpublished work) show that this representation does no better at providing accurate secondary-structure prediction than does a simple 20-bit vector for representing amino acids.

## Tertiary Structure–Class Prediction

Information on the tertiary-structure class of a protein can be used to improve the quality of secondary-structure prediction (10, 15, 22). Muskal & Kim (20) adapted ANNs to predict the content of $\alpha$-helix and $\beta$-sheet in a protein from its amino acid composition and molecular weight. To begin, a consistent strategy for determining $\alpha$-helix and $\beta$-sheet content in a protein was required. Muskal & Kim used the computer program DSSP (Define Secondary Structure in Proteins) (14) to assign secondary-structure content in 80 of the 104 structures used as the network-training data base and compared this to the author-assigned secondary-structure features (and therefore amino acid content). Values for the implicit uncertainty and deviation in the assignment of secondary-structure content were estimated from this data: 5.2% $\pm$ 5.4 for $\alpha$-helix and 4.4% $\pm$ 5.5 for $\beta$-strand. The method for prediction of secondary structure content proposed by Muskal & Kim performs to this level of error: 4.1% $\pm$ 4.5 for $\alpha$-helix, and 4.1% $\pm$ 3.4 for $\beta$-strand. Other approaches to tertiary structure–class prediction, such as a two-layer ANN and the indirect determination of secondary-structure content from the ANN for secondary-structure prediction of Qian & Sejnowski (23), have approximately twice the rate of error. Muskal & Kim provide only a heuristic basis for the determination of trained network generalization. Furthermore, training examples of their second network clearly demonstrate the existence of multiple minima in the learning surface. However, within these limitations, these networks have reached the theoretical minimum error in the prediction of secondary-structure content. As high-resolution structures become available, it remains to be seen whether additional tests of this technique will bear out its general applicability.

## CONCLUSIONS

Initially, computational scientists explored only simple network connectivities. Now, network architecture is an experimental battlefield of artificial neural networks. Randomly placed hidden nodes provide no additional accuracy in the predictive mode of a network, but logically constructed topologies may provide some advancement.

Several different data representations have been attempted. Preprocessing the input data has a noticeable affect in some systems. Input vectors with the same Hamming distance between them perform consistently well.

Perceptron networks can only resolve linearly separable functions. However, multilayer networks may contain many suboptimal minima. Few well-defined mechanisms are available for dealing with this problem.

Rather than acquiring general concepts, ANNs may simply memorize the training set. Methods to detect and avoid this limitation are now in the early stages of development.

The weights representing trained perceptron networks can be successfully interpreted for the contribution of individual input nodes. However, intermediate layers of multilayer networks are much more difficult to interpret.

Even with these drawbacks, some current ANN applications represent a significant advancement in the recognition of certain correlations between sequence and structure in biochemistry.

ACKNOWLEDGMENTS

We thank Nomi Harris for her helpful discussions. F. E. Cohen is supported by a grant from the National Institutes of Health, GM #39900.

*Literature Cited*

1. Benner, S. A. 1988. In *Redesigning the Molecules of Life*, ed. S. A. Benner, pp. 115–75. Heidelberg: Springer-Verlag
2. Bohr, H., Bohr, J., Brunak, S., Cotterill, R. M. J., Lautrup, B., et al. 1988. *FEBS Lett.* 241: 223–28
3. Brunak, S., Engelbrecht, J., Knudsen, S. 1990. *Nucleic Acids Res.* 18: 4797–4801
4. Brunak, S., Engelbrecht, J., Knudsen, S. 1991. *J. Mol. Biol.* 220: 49–65
5. Chou, P. Y., Fasman, G. D. 1978. *Annu. Rev. Biochem.* 47: 251–76
6. Cohen, F. E., Abarbanel, R. M., Kuntz, I. D., Fletterick, R. J. 1986. *Biochemistry* 25: 266–75
7. Deleage, G., Roux, B. 1987. *Protein Eng.* 1: 289–94
8. Demeler, B., Zhou, G. 1991. *Nucleic Acids Res.* 19: 1593–99
9. Farber, R., Lapedes, A., Sirotkin, K. 1992. *J. Mol. Biol.* 226: 471–79
10. Garnier, J. R., Osguthorpe, D. J., Robson, B. 1978. *J. Mol. Biol.* 120: 97–120
11. Hayes-Roth, F. 1983. In *Machine Learning: An Artificial Intelligence Approach*, ed. R. S. Michalski, J. G. Carbonell, T. M. Mitchell, pp. 221–40. Palo Alto, CA: Tioga
12. Hinton, G. E., Sejnowski, T. J. 1986. See Ref. 26a, pp. 282–317
13. Holley, L. H., Karplus, M. 1989. *Proc. Natl. Acad. Sci. USA* 86: 152–56
14. Kabsch, W., Sander, C. 1983. *Biopolymers* 22: 2577–2637
15. Kneller, D. G., Cohen, F. E., Langridge, R. 1990. *J. Mol. Biol.* 214: 171–82
16. Lapedes, A., Barnes, C., Burks, C., Farber, R., Sirotkin, K. 1989. In *Computers and DNA, SFI Studies in the Sciences of Complexity*, ed. G. Bell, T. Marr, pp. 157–82. Reading, MA: Addison-Wesley
17. Lukashin, A. V., Anshelevich, V. V., Amirikyan, B. R., Gragerov, A. I., Frank-Kamenetskii, M. D. 1989. *J. Biomol. Struct. Dyn.* 6: 1123–33
18. McClelland, J. L., Rumelhart, D. E. 1988. *Explorations in Parallel Distributed Processing*. Cambridge, MA: MIT Press
19. Minsky, M., Papert, S. 1969. *Perceptrons: An Introduction to Computational Geometry*. Cambridge, MA: MIT Press
20. Muskal, S. M., Kim, S.-H. 1992. *J. Mol. Biol.* 225: 713–27
21. Pao, Y.-H. 1989. *Adaptive Pattern*

22. Presnell, S. R., Cohen, B. I., Cohen, F. E. 1992. *Biochemistry* 31: 983–93
23. Qian, N., Sejnowski, T. J. 1988. *J. Mol. Biol.* 202: 865–84
24. Rosenblatt, F. 1958. *Psychol. Rev.* 62: 559
25. Rosenblatt, F. 1959. *Mechanisation of Thought*. London: Her Majesty's Stationary Office
26. Rumelhart, D. E., Hinton, G. E., Williams, R. J. 1986. See Ref. 26a, pp. 318–62
26a. Rumelhart, D. E., McClelland, J. L., eds. 1986. *Parallel Distributed Processing*. Cambridge, MA: MIT Press
27. Serrano, L., Fersht, A. R. 1989. *Nature* 342: 296–99
28. Shavlik, J. W., Towell, G. G., Noordewier, M. O. 1992. *Int. J. Genome Res.* In press
29. Sontag, E. D., Sussman, H. J. 1989. *Proc. IEEE* 1: 639–42
30. Staden, R. 1984. *Nucleic Acids Res.* 12: 551–67
31. Uberbacher, E. C., Mural, R. J. 1991. *Proc. Natl. Acad. Sci. USA* 88: 11261–65

*Recognition and Neural Networks*. Reading, MA: Addison-Wesley

# THE STRUCTURE OF THE FOUR-WAY JUNCTION IN DNA

David M. J. Lilley

CRC Nucleic Acid Structure Group, Department of Biochemistry, The University, Dundee DD1 4HN, United Kingdom

Robert M. Clegg

Abteilung Molekulare Biologie, Max-Planck-Institut für biophysikalische chemie, Am Fassburg, D–3400 Göttingen, Germany

KEY WORDS: DNA structure, recombination, fluorescence, thermodynamics, Holliday junction

CONTENTS

| | |
|---|---|
| FOUR-WAY JUNCTIONS AND RECOMBINATION | 300 |
| THE FOLDED STRUCTURE OF THE JUNCTION | 300 |
|     The Stacked X-Structure | 300 |
|     Gel Electrophoresis | 301 |
|     Fluorescence Resonance Energy Transfer | 305 |
|     Probing Experiments | 305 |
|     Other Physical Methods | 307 |
|     All Available Data Are Consistent with the Stacked X-Structure | 308 |
| STEREOCHEMISTRY OF THE FOUR-WAY JUNCTION | 310 |
| THE ROLE OF METAL IONS IN THE STRUCTURE OF THE FOUR-WAY JUNCTION | 313 |
|     Structure of the Junction in the Absence of Metal Ions | 313 |
|     The Manner of Ion Binding, and the Folding of the Four-Way DNA Junction by Cations | 314 |
| THERMODYNAMICS OF THE STRUCTURE OF THE FOUR-WAY JUNCTION | 316 |
|     Coaxial Stacking of Helical Arms | 317 |
|     Free Energy Difference Between the Two Stacking Isomers | 317 |
|     Electrostatic Repulsions Between Negative Charges at the Point of Strand Exchange | 318 |
|     Stability of the Junction Compared with the Stability of the Duplex Arms | 319 |
|     The Relative Stability of the Parallel and Antiparallel Conformation | 320 |
| DYNAMIC ASPECTS OF THE STRUCTURE OF THE JUNCTION | 321 |
| INTERACTIONS WITH PROTEINS | 322 |
| IN CONCLUSION | 324 |

1056–8700/93/0610–0299$02.00

# FOUR-WAY JUNCTIONS AND RECOMBINATION

Branched DNA species are commonly postulated as intermediates in DNA rearrangements. The four-way (Holliday) junction was proposed to be the central intermediate in homologous genetic recombination (6, 39, 81, 88, 89, 92, 93) (Figure 1), and there is good evidence for a role in the integrase class of site-specific recombination events (38, 44, 53, 80). Branched DNA can also arise in other ways, including the replication of DNA as exemplified by bacteriophage T4 replication. In each case, enzymes must interact specifically with these structures to bring about a resolution or repair event. These proteins recognize their DNA substrates at the level of tertiary structure—challenging us to understand how this may be achieved.

Despite the potential importance of DNA junctions, their structure has remained poorly understood until relatively recently. The following review describes our current view of the structure of the four-way DNA junction, and the way this is recognized by proteins.

# THE FOLDED STRUCTURE OF THE JUNCTION
## The Stacked X-Structure

Early attempts to model the structure of a four-way junction in DNA (7, 92, 93) involved pairwise stacking of helical arms by means of coaxial stacking (akin to the stacking of the arms of tRNA). The majority of models placed the resulting quasicontinuous helices side-by-side, with a parallel alignment of continuous strands. Experimental results (below)

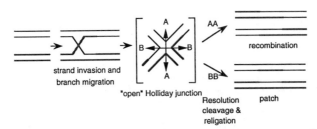

*Figure 1* The simplest model for homologous genetic recombination (39, 70). Homologous DNA molecules undergo strand invasion to generate a four-way (Holliday) junction. This species may undergo branch migration by exchange of base pairing between the homologous sequences. The junction is ultimately cleaved by a resolving enzyme, followed by religation, to regenerate the unjoined DNA molecules. The junction may be cleaved (resolved) in two ways. Cleavage across the axis BB can lead to local exchange of genetic information, but no long-range exchange of markers, while cleavage across AA leads to true recombination.

confirm the propensity for coaxial stacking of arms but indicate that the most stable arrangement in solution is neither side-by-side nor parallel.

The first four-way DNA junctions to be studied were cruciform structures formed by inverted repeat sequences in supercoiled DNA (30, 58, 82). Examination of cruciform structures with chemical probes indicated that the junction could be formed without gross distortion of local base pairing, suggesting that complete pairing could be preserved. The free energy of cruciform formation was significantly positive [between 13 and 18 kcal mol$^{-1}$ for typical inverted repeats (18, 34, 60, 72)], but the contribution of the junction to this was unknown. Structural studies were difficult because cruciform structures are only stable when contained within topologically constrained circular DNA. Therefore, progress was greater when junctions were constructed from smaller pieces of DNA. Junctions were constructed from cloned (2, 32, 41) or synthesized (47) DNA by hybridizing appropriate sequences incapable of extensive branch migration.

Early studies indicated that stable four-way junctions could be formed and that full base pairing was preserved. Chemical-probing experiments could detect no single-stranded character in any of the bases at the point of strand exchange in the junction (29, 33), and NMR studies revealed resonances for all imino protons (103). Other early studies attempted to model the structure of a four-way junction in DNA (7, 92, 93). These all led to the conclusion that the helical arms would undergo coaxial stacking, akin to the stacking of the arms of tRNA. The majority of models placed the quasicontinuous stacked helices side-by-side, with a parallel alignment of continuous strands.

Our current view of the structure adopted by the four-way DNA junction is based on coaxial pairwise helical stacking, with a rotation to form an overall X-shape. But instead of the commonly depicted parallel alignment, the energetically preferred arrangement is an approximately antiparallel conformation. The evidence for this stacked X-structure has come from a variety of techniques; such higher-order DNA structure is not easily studied by the now-classical methods of crystallography and NMR, and one must adopt a multidisciplinary approach employing several different techniques, including electrophoresis, probing, and physical methods.

## Gel Electrophoresis

Gel electrophoresis has been an immensely powerful method for studying the structures of nucleic acids. It has been employed to study a variety of structural problems from the topology of supercoiled circular DNA (90, 102) to the curvature of linear DNA fragments (21, 37, 55, 68, 107). This simple technique has completely replaced older methods such as analytical ultracentrifugation, and yet the physical basis of the method is still relatively poorly understood. Despite this limitation, one can nevertheless perform experiments that give valuable insight into a structure. Gel electrophoresis turned out to provide the first real indication of the overall geometry of the structure of the junction, i.e. the configuration of the arms in space.

Several years ago, we showed that the introduction of a four-way junction into the center of a DNA fragment retarded the mobility of that DNA in polyacrylamide (32) and that the relative retardation increased with the percentage of polyacrylamide. This observation was rather similar to the behavior of kinetoplast DNA (68) and was consistent with the presence of a pronounced bend or kink at the position of the junction. The electrophoretic mobility depended on the concentration and type of cation present (20), suggesting a role for ion interactions in the structure. Cooper & Hagerman (15) developed a technique based on the study of changes in electrophoretic mobility of a four-way junction following extension of pairs of arms. They concluded that the symmetry of the junction was lower than tetrahedral, and that two of the strands were more severely bent than proposed in the model of Sigal & Alberts (92).

We employed a closely related gel electrophoretic technique of comparing isomeric junctions with two long arms and two short arms generated by restriction cleavage (25). We compared the electrophoretic mobility of the six species (i.e. the permutations of two short arms from four), assuming that the speed of migration would reflect the angle subtended between the longer arms. When metal cations (e.g. $\geqslant 100$ $\mu$M $Mg^{2+}$) were present in the electrophoresis buffer, the pattern of migration obtained for a variety of sequences showed that the electrophoretic mobilities were grouped into three comigrating pairs of bands—the 2:2:2 pattern. Figure 2 illustrates an example. These experiments suggested that the junction adopts an X-shape in the presence of added cations, because there are two ways by which the long arms may be related to each other across an acute angle, two ways they may be related by an obtuse angle, and two possibilities that they may be colinear.

An X-shaped junction is formed from a four-way junction by the pairwise coaxial stacking of helical arms, followed by a rotation that resembles opening a pair of scissors. This arrangement generates a favorable increase in base-pair stacking interactions while reducing steric and electrostatic interaction between the stacked pairs of arms. The reduction to two-fold symmetry divides the four strands of the junction into two classes; two strands (continuous) have effectively continuous helical axes, whereas the other two strands (exchanging) pass between the two helical stacks at the point of strand exchange, i.e. the junction itself.

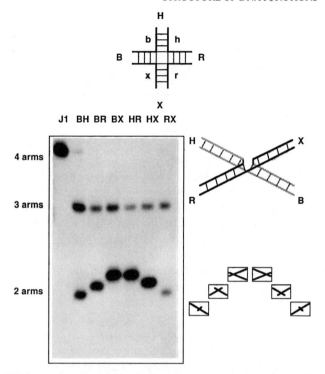

*Figure 2* Gel electrophoretic analysis of a four-way junction in the presence of metal ions. A junction was constructed from four 80-base oligonucleotides (strands b, h, r, and x) (25). Each arm (B, H, R, or X) contained a unique site for a restriction enzyme and could be selectively shortened by enzyme cleavage. The electrophoretic mobility of the six possible double digests of the junction were compared in a 5% polyacrylamide gel in 90 mM Tris.borate (pH 8.3) containing 100 $\mu$M $MgCl_2$. The leftmost track (J1) contains uncleaved junction. The following six tracks contain the junctions after restriction digestion. The digested species are named according to the uncleaved arms (e.g. BR is the junction that has been cleaved into the H and X arms). The digests are incomplete, and the slower species in each track (3 long arms) have undergone cleavage in a single arm. Those that have been cleaved in two arms (2 long arms) exhibit three types of mobility, consistent with a two-fold symmetrical X-shape. Assuming that the mobility is inversely proportional to the distance between the ends of the long arms, the antiparallel X-shape results, as shown. This junction contains a coaxial stacking of the B and H arms and of the R and X arms. By changing the base sequence at the junction, we can alter the stacking preference so that there is coaxial stacking of the B and X arms.

Upon alteration of the sequence at the point of strand exchange, we observed that the electrophoretic pattern of our long/short arm junctions changed, with an exchange of mobilities for the fast and slow species (25). This indicates that the stacking partners depend upon the local sequence at the junction, and indeed two isomers of the stacked X-structure are possible, differing in the choice of helical stacking partners. This isomerization changes the nature of each individual strand; exchanging strands become continuous strands and vice versa. The identity of the most stable isomer will be governed by the thermodynamics of the interactions at the point of strand exchange, e.g. by the stacking interactions. The choice of stacking isomers could be made on the basis of relatively small differences in free energy, discussed below.

Assuming that electrophoretic mobility is determined by end-to-end distance, we then assign the slower species to those in which the longer arms are related by the smallest angle of the X-structure. This means that the structure must be approximately antiparallel, i.e. the two continuous strands run in opposite directions. This model contrasts with the normal depiction of Holliday junctions and with the model of Sigal & Alberts (92), but this conclusion has been confirmed by many other studies (see below).

In the absence of added ions, a quite different gel pattern emerged (25), comprising four slow and two fast species (4:2 pattern). Figure 3 shows an example. Junctions of every sequence that we have examined reveal this pattern of mobilities in the absence of added ions, indicating a different symmetry; it is consistent with an extended structure in which the four arms are directed towards the corners of a square but is inconsistent with a tetrahedral disposition of arms. The same pattern was obtained from certain four-way junctions in which particular non-Watson-Crick mis-

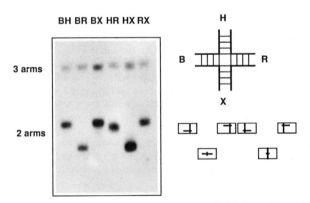

*Figure 3* Structure of a four-way junction in the absence of added metal ions. Gel electrophoretic analysis of a junction in 90 mM Tris.borate (pH 8.3), 0.1 mM EDTA (25). Under these conditions, the two-fold symmetry of the earlier pattern of fragments (Figure 2) is replaced by one containing four slow and two fast species. This is consistent with a square configuration of helical arms—yielding four species in which the long arms subtend approximately 90° (slow species), and two with angles of approximately 180° (faster species).

matches were placed at the point of strand exchange (23), even in the presence of metal ions. The role of ions in folding the junction is discussed below.

## Fluorescence Resonance Energy Transfer

The electrophoresis experiment compares relative angles subtended between the arms of the junction. An alternative approach would be to measure end-to-end distances in a junction with arms of equal length. This requires the determination of distances in solution in the range of 20–80 Å, and we exploited fluorescence resonance energy transfer (FRET) for this purpose (11, 77). In this experiment, different fluorophores are conjugated to the 5' termini of two of the four different strands, thereby positioning the dyes at specific locations in the structure. Excitation of the donor may lead to energy transfer to the acceptor dye by nonradiative dipole-dipole coupling between transition moments (28), the efficiency of which depends on the inverse sixth power of the distance between them (for nonoriented donor-acceptor pairs). We applied this method to the study of a series of DNA duplexes varying in length between 8 and 20 bp and obtained the expected reduction in FRET efficiency as the end-to-end distance increased (12). We observed the cylindrical geometry of the DNA as a sinusoidal modulation of the FRET efficiency, and we obtained excellent agreement between the experimental data and the calculated values based on dipolar energy transfer and a knowledge of the geometry of double-stranded DNA.

We have made an extensive study of four-way junctions using FRET (11, 77). Figure 4 shows an example of the data. In the absence of added metal ions, all six end-to-end vectors give very low efficiencies of transfer, consistent with an extended structure. Upon addition of magnesium, two of the vectors exhibited greatly increased energy transfer, indicating that these pairs of ends approach each other. In fact, these are the ends that are predicted to be closest in the antiparallel stacked X-structure, and thus the agreement is excellent between the spectroscopic and gel electrophoretic methods. We have determined the relative FRET efficiencies in several different ways, by measuring the normalized enhancement of acceptor fluorescence, the reduction in donor fluorescent quantum yield, the reduction in the fluorescent nanosecond lifetime of the donor, and the reduction in the measured anisotropy of the acceptor fluorescence (11). All these techniques lead to the same conclusion about the antiparallel stacked X-structure of the junction.

## Probing Experiments

Enzyme and chemical probes have provided important information on the conformation of the DNA junction. Probing of synthetic DNA junctions

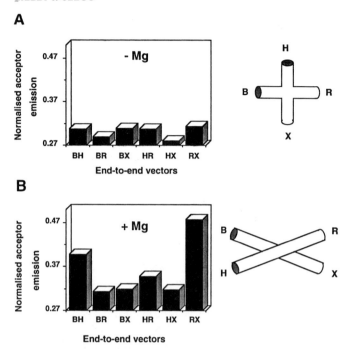

*Figure 4* Analysis of the structure of the four-way junction by fluorescence resonance energy transfer (11, 77). Analysis of junction 3 that folds by coaxial B upon X-stacking according to gel electrophoretic experiments (25). Junctions were assembled from two strands conjugated at their 5′ termini with fluorescein and rhodamine, respectively, and two unlabeled strands, generating a junction carrying the fluorophores at the termini of two arms. The efficiency of FRET between the fluorophores is inversely proportional to the distance between them. This is performed for each of the six possible end-to-end vectors. The data are presented as normalized acceptor emission values (11), which contain both fluorescent emission from rhodamine resulting from direct excitation and FRET. For these experiments, the numerical value is 0.27 in the absence of fluorescein (direct excitation of rhodamine only), i.e. under these experimental conditions a value greater than 0.27 indicates acceptor emission due to FRET. (*A*) No added magnesium ions. Under these conditions the value of normalized acceptor emission is close to 0.27 for all six end-to-end vectors, indicating an extended structure. (*B*) With added magnesium ions. Two end-to-end vectors exhibit significant increases in FRET efficiency, i.e. B to H and R to X. This is completely consistent with the antiparallel stacked X structure in which B-on-X coaxial stacking occurs.

with hydroxyl radicals provided an early indication that the structure of the four-way junction was two-fold symmetric (10), as two of the four strands were protected against radical attack at the point of strand exchange. Moreover, when the sequence at the junction was altered, the pattern changed in a manner consistent with an exchange of stacking partners (8). These observations are fully consistent with the stacked X-structure deduced from the electrophoresis and FRET experiments.

Kallenbach and coworkers (35, 36) have examined the binding of a series of intercalating agents to four-way junctions. On the basis of footprinting studies, these authors concluded that the junction creates a point of selective binding for a series of intercalating drugs, including porphyrins (64) and cyanines (66). Similar sites of high affinity have been observed in the stacked base pairs next to base mismatches and bulges (79). This seems to be related to the decreased rigidity of the last two stacked base pairs next to the perturbed site that aids the intercalation site. Thus, the base pairs at the center of the junction may be perturbed from the normal conformation. Other studies also indicate that the conformation of the folded junction generates ligand-binding sites of elevated affinity. Kirshenbaum et al (52) have observed enhanced cleavage in the vicinity of a cruciform junction by the compound $Rh(DIP)_3(III)$, and we recently observed a region of hypersensitivity to uranyl ion in a four-way junction at the point of strand exchange (N.-E. Møllegaard, A. I. H. Murchie, P. Nielsen & D. M. J. Lilley, in preparation).

Enzymes have also been used to probe the structure of the four-way junction. To test the colinear stacking of arms in the junction, we employed the enzyme *Mbo*II. This is a member of the class of restriction enzymes in which the recognition site is a significant distance from the site of cleavage. We constructed a junction containing an *Mbo*II recognition site positioned such that cleavage would be required in a DNA arm coaxially stacked on the arm containing the binding site (78). We observed efficient cleavage at the expected nucleotides, implying good stacking at the junction, without a large degree of bending or twisting deformation. The nuclease DNase I has been applied to the study of junction structure (65, 76), and once again two-fold symmetry was evident from the patterns of cleavage. Significant sections of each of the four strands were protected around the point of strand exchange, but protection was more extensive on the exchanging strands. The continuous strands were protected for about five nucleotides to the 3' side of the point of strand exchange (76).

## Other Physical Methods

Several physical studies have examined the shape of four-way DNA junctions, mainly based on hydrodynamic approaches. Cooper & Hagerman studied the geometry of a set of junctions with permuted pairs of long and short arms (i.e. analogous to the molecules used for gel electrophoretic experiments discussed above) using transient electric birefringence (16). In this method, the rate of decay of birefringence of the solution is measured following orientation of the DNA by a strong electric field pulse, and rotational decay times may be calculated. Shorter decay times are expected to be associated with smaller interarm angles. Cooper & Hagerman found that in the presence of 1 mM Mg two of the six angles were characterized by significantly shorter decay times, consistent with an X-shaped structure. In the absence of added cations, all six times were approximately equal, suggesting a more symmetrical structure.

We have also employed magnetic birefringence to examine the shape of the four-way junction (J. Torbet, A. I. H. Murchie & D. M. J. Lilley, unpublished data). In this technique, the optical birefringence of a solution of the DNA is measured while it is under orientation by a magnetic field. The magnitude and sign of the effect were again consistent with the stacked X-structure. In an attempt to distinguish experimentally between parallel and antiparallel structures, we used neutron scattering to study a junction comprising two arms of 50 bp and two of 15 bp (J. Torbet, A. I. H. Murchie & D. M. J. Lilley, unpublished data). From the measured radius of gyration, we concluded that the structure was antiparallel, with a small angle of $67 \pm 20°$. Diffusion constants measured using dynamic light scattering (J. Langowski, W. Cramer, A. I. H. Murchie & D. M. J. Lilley, unpublished data) were also consistent with the stacked X-structure.

## *All Available Data Are Consistent with the Stacked X-Structure*

The right-handed, antiparallel stacked X-structure provides a consistent explanation for virtually all the available data obtained in the presence of cations, and this can be fairly described as the consensus view even if it may lack complete unanimity. Although we so far have discussed the data from our own laboratories in most detail, an examination of the work of other investigators, who use DNA junctions of sequences unrelated the ones we use, is instructive. One sequence in particular, shown in Figure 5, has been extensively studied in the laboratories of Seeman & Kallenbach and Hagerman. In their gel electrophoretic analysis, Cooper & Hagerman (15) presented an approximately 2:2:2 pattern, in which the junctions with arms 1 and 4 extended exhibited slow mobility, as did junctions with arms 2 and 3 extended. According to the stacked X-model, these results would indicate pairwise stacking of arms 1 with 2 and 3 with 4, as illustrated. This structure would be completely consistent with all other data for junctions based on this sequence. Churchill et al (10) observed protection against hydroxyl-radical cleavage in strands a and c (our nomenclature, Figure 5); these become the exchanging strands when folded into the stacked X-isomer with 1-on-2 stacking. Moreover, the structure is consistent with the observations of shortest rotational decay times (smallest angle) from transient electric birefringence for the angles subtended

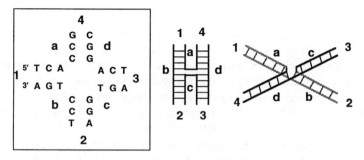

*Figure 5* Data for another four-way junction are completely consistent with the stacked X-structure. The sequence shown is that of the center of the junction studied initially by Churchill et al (10). The arms are numbered using the scheme followed by Cooper & Hagerman (15–17), and we have labeled the strands using the letters shown. This junction is folded into the stacked X-structure by coaxial stacking of arms 1 and 2 and of arms 3 and 4 as shown, together with a rotation to form the antiparallel X-shape. This model is consistent with the available experimental data, discussed in the text.

between arms 1 and 4 and between arms 2 and 3 (16) and with the measurement of greatest FRET efficiency across the end-to-end vector of arms 1 and 4 and across that of arms 2 and 3 (17). Thus, there is complete agreement between all these techniques, and the antiparallel stacked X-structure provides a consistent explanation for all the data.

Are other structures consistent with these data? Alternative explanations for the data would have to be based on an approximately two-fold symmetrical X-shaped structure, but pairwise helical stacking is not a required conclusion. However, this feature is included in the model for several reasons. Base stacking is a common feature of folded nucleic acids [tRNA being an obvious example (43, 50, 73, 96, 106)], which clearly indicates a free-energy advantage unless outweighed by other factors. All model-building exercises apparently confirm this predisposition. Moreover, it is very hard to explain the well-defined isomerization that junctions exhibit upon alteration of the sequence at the point of strand exchange (8, 25) in the absence of helix-helix stacking in this region. In the absence of added cations, the junction appears to adopt an extended structure (discussed below); under these conditions the unstacked base pairs at the ends of the helices at the point of strand exchange are reactive to addition by osmium tetroxide. However, upon folding by addition of metal ions, these bases become fully unreactive, suggesting that they have become stacked into a quasicontinuous helix (25, 26). Finally, coaxial stacking of arms provides the best explanation for the observation of *Mbo*II cleavage in one arm when the enzyme-recognition site is located in a different arm (78).

How perfect is the two-fold symmetry indicated by the stacked X-structure of the four-way junction? Although this structure accounts in a general way for all the data, one should not assume that the symmetry is perfect. Indeed, close examination of the data suggests deviations from full symmetry. For example, the fluorescence data for our junction 3 (Figure 4) show that two end-to-end vectors exhibit significantly greater efficiency of FRET than the remaining four vectors; yet the two short vectors do not have equal FRET efficiencies (11). Similarly, the two deduced small angles in the junction studied with transient electric birefringence do not give exactly equal rotational decay times (16). However, this should not come as a great surprise. The sequences of the four arms of the junctions studied in vitro are unequal, and so the junctions are only pseudosymmetric at best. Moreover, the expected symmetry of the antiparallel stacked X-structure is lower than that of a planar X; the distinction between major and minor groove sides leaves only two dyad axes (ignoring sequence differences) in the structure, and thus the two species in which long arms are related by the obtuse angle are not equivalent.

There is good agreement that the structure of the junction leads to an antiparallel alignment of helices. However, unless the angle between the axes of the two helical stacks is exactly 0 or 180°, alternative projections of any given structure that show either parallel or antiparallel alignment are possible, and therefore an X-shape is neither fully parallel nor antiparallel. We therefore distinguish operationally between parallel and antiparallel structures by measuring the angle subtended between the helical axes of the two ends of the exchanging strands; if this angle is less than 90°, we define the structure as antiparallel. In this type of structure, the chemical polarities of the continuous strands run in approximately opposite directions. All methods including gel electrophoresis, FRET, transient electric birefringence, and neutron scattering indicate that the structure is indeed antiparallel according to these definitions. This idea contrasts with the usual depiction of Holliday junctions drawn in textbooks on genetics. Of course, all the structural experiments on junctions have been performed on DNA in isolation, and therefore, the conformations deduced concern the structural properties of the DNA alone. DNA-protein interactions could alter the conformation of the DNA.

## STEREOCHEMISTRY OF THE FOUR-WAY JUNCTION

Although a relative wealth of data attest to the global geometry of the four-way junction as an antiparallel stacked X-structure, many fewer experimental data provide insights about the local stereochemistry at the point of strand exchange. However, modeling at different levels of sophistication has been very helpful in outlining stereochemical possibilities and refining ideas. The X-shaped conformation of the junction is a rather compact, complex higher-order folding of DNA, with many mutually repulsive charged phosphate groups; hence, avoidance of clash between backbones is probably a major determinant of the folded structure. Very simple modeling suggests that the most effective way to avoid stereochemical clash is by means of a right-handed, antiparallel structure, as observed experimentally, shown in Figure 6 (77). In this conformation, a favorable juxtaposition of strands and grooves minimizes steric clash between backbones; this is most effective when the small angle is 60°. Similar packing has been observed in the crystal structures of three different double-stranded oligonucleotides (99; R. E. Dickerson, personal communication; U. Heinemann, personal communication); this is clearly a natural way for DNA molecules to achieve close association. Experimental determinations of the small angles in different four-way junctions by

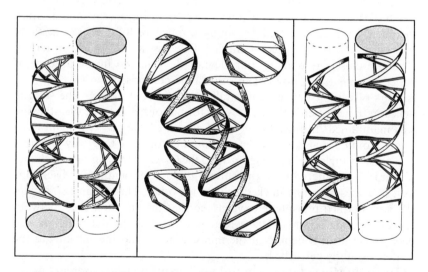

**major groove side**  **face**  **minor groove side**

*Figure 6* Three views of the right-handed, antiparallel stacked X-structure of the four-way DNA junction. (*Center*) Face view, showing the X-shape of the folded junction. The ribbon model shows the accommodation of the continuous strands in the major grooves of the opposing helix; this is optimal if the small angle is 60°. The two sides of the structure are not equivalent. On one side (*right*) the four base pairs at the point of strand exchange all present minor groove edges, while on the other side (*left*), the corresponding major groove edges are presented.

FRET, neutron scattering, and electric and magnetic birefringence are all close to the theoretical expectation of 60°. The location of the continuous strands in the major grooves 3' to the point of strand exchange is fully consistent with the localized protection against cleavage by DNase I (65, 76). The right-handedness of the X-structure was confirmed by FRET experiments in which fluorophores were systematically moved around the arms of a junction in order to map out the faces of closest approach (77).

The antiparallel stacked X-junction presents two dissimilar sides, as seen in Figure 6. The four base pairs that define the point of strand exchange are oriented in the same direction, leading to chemically different sides with major and minor groove characteristics. On the major-groove side, a continuous major groove passes down one arm, through the point of strand exchange, and continues on an arm of the opposite helical stack. An analogous continuous minor groove is on the other side. This inherent asymmetry in the structure of the junction has important consequences for interaction with ligands.

An extensive molecular mechanical modeling exercise was performed for the structure of a four-way junction (100). Using energy minimization methods designed to avoid becoming trapped in local energy minima, two kinds of structure were identified. One was an extended, unstacked structure closely resembling our view of the junction in the absence of added cations (see below). The other type comprised the two stacking isomers of a right-handed antiparallel stacked X-structure. The stacked helices retained essentially B-DNA conformations, and favorable stacking was maintained through the point of strand exchange. Two main deviations from normal B-DNA were required to form the four-way junction. One torsion angle ($\varepsilon$—the angle around the $C_{3'}$-$O_{3'}$ bond) on each exchanging strand was altered from the normal $t$ to $g^-$, reorienting the backbone as it traverses from one stacked helix to the other. This action also rotates the nonesterified phosphate oxygens, so that they avoid a clash between the exchanging strands. The exchange of strands in the junction also generates a local widening of the minor groove, and the stacking at the center is slightly perturbed.

Timsit & Moras (98) performed a different modeling exercise, extrapolating from their crystal structure of the *Nar* I dodecamer (99). This structure contains a trigonal packing of double-helical DNA molecules, locked together at each triangular vertex by an interpenetration of deoxyribophosphate backbones in grooves. The symmetry of the packing generates an angle between helical axes of 60°, giving an optimal alignment of grooves and backbones as noted above. They used this structure as a starting point from which to build a variety of junctions by a simple reconnection of strands. In this way, three different junction stereochemistries were explored, one parallel and two antiparallel structures.

Ultimately one would hope to have a full description of the stereochemistry of the four-way junction from conventional high-resolution techniques, notably NMR and crystallography, but progress to date has been limited. Such junctions are a relatively large problem for NMR, the size presenting potential problems in both linewidth and spectral complexity. Wemmer et al (103) studied the imino NMR region of a junction comprising four arms, each 8 bp in length, from which they concluded that base pairs at the point of strand exchange were unbroken. Chazin and coworkers (9) have assigned the majority of nonexchangeable protons in two four-way junctions, and this will hopefully lead to detailed structural information in due course. Crystallographic progress naturally requires crystals of acceptable quality, and so far these have not been obtained despite the best efforts of several laboratories. In the long run, complexes with binding proteins may provide the best means of obtaining crystals and solving the structure of the junction as part of a complex.

## THE ROLE OF METAL IONS IN THE STRUCTURE OF THE FOUR-WAY JUNCTION

### Structure of the Junction in the Absence of Metal Ions

Metal ions have an important role in the structure of the four-way DNA junction. In the absence of added cations, the junction cannot fold into the stacked X-structure and remains extended with unstacked arms. A considerable body of evidence points to this proposition. Gel electrophoretic experiments (Figure 3) indicate that the junction adopts a structure with approximately square symmetry in the absence of metal ions (25, 26), which FRET experiments confirm (R. M. Clegg, A. I. H. Murchie & D. M. J. Lilley, in preparation). Transient electric birefringence experiments in the absence of added ions gave results that were significantly different from those in the presence of 1 mM $Mg^{2+}$ (16), with more uniform rotational decay times. Taken together, these results suggest that the folding of the junction into the stacked X-structure brings negatively charged phosphate groups into close proximity, which is destabilizing unless electrostatic repulsion is reduced by ion screening. The four-fold symmetry of this unfolded form indicates that interarm stacking cannot occur under these conditions, which is supported by the reactivity to osmium tetroxide of thymine bases present at the point of strand exchange (25). The unstacked end of the helix is susceptible to out-of-plane electrophilic addition at the 5,6 bond of thymine, which is not normally possible when an AT base pair is sandwiched between neighboring base pairs in B-DNA. A variety of ions can bring about the folding, with differing efficiencies (26). Group II metals such as magnesium and calcium fold the junction at concentrations greater than about 80 $\mu$M. Complex ions and polyamines are more efficient; 2 $\mu$M [Co(NH$_3$)$_6$](III) or 25 $\mu$M spermine fold the junction, for example. Group I metal ions, such as sodium or potassium, bring about at least a partial folding of the junction, but very high concentrations are required.

The exact configuration of arms in the junction in the absence of added metal ions is more difficult to describe with certainty than the folded structure in the presence of magnesium or other ions. A priori, one could imagine two kinds of extended structure, based on square or tetrahedral symmetries. Although four charges at a fixed radius from a center will adopt a tetrahedral distribution as the configuration with minimum repulsion, this is not necessarily the case for four charges linked by fixed lengths. In our hands, the gel electrophoresis results argue unequivocally for a square structure (25, 26), whatever the sequence of the junction. This result was supported by sensitive FRET measurements (R. M. Clegg, A. I. H. Murchie & D. M. J. Lilley, in preparation), where FRET efficiencies for selected end-to-end vectors were compared using a procedure that involved titrating the molar fraction of donor fluorophore while observing the enhanced fluorescent emission of the acceptor. This experiment showed that the distance across the diagonal of the square was significantly greater than that across the sides. The transient electric birefringence results are not completely in agreement with a square structure (while supporting an extended structure of some kind), but this may reflect the anticipated flexibility of the unstacked structure, which could be deformed in electric fields. Although the equilibrium structure seems well described by a square configuration of arms, the structure is not necessarily planar; in fact a planar structure is rather unlikely. The two sides of the extended structure are chemically different—one comprises all major groove edges of base pairs, while the other has minor groove edges (exactly equivalent to the two sides of the stacked X-structure). This asymmetry makes a pyramidal structure of some degree most probable.

## The Manner of Ion Binding, and the Folding of the Four-Way DNA Junction by Cations

The association of ions in solution, including the interaction of small ions with polyelectrolytes, can be roughly divided into two major categories:

1. A dynamic association of small ions with the extended high charge density of polyelectrolytes (ionic atmosphere).
2. More stable, static, and localized associations of ions at distinct molecular sites, or pockets.

The second class of binding usually means that at least one of the closest coordinated water molecules is replaced by a direct attachment to a ligand atom, forming an inner sphere complex. In helical DNA, the alkali metal ions of the group are usually considered to interact predominantly with the phosphates according to the first mechanism, and fast kinetic data indicate that both $Mg^{2+}$ and $Ca^{2+}$ ions do not form inner sphere complexes to the polynucleotides poly(A) and poly(C) (86, 87).

FRET experiments indicate that the junction folds into very similar stacked X-structures in the presence of sufficient $Na^+$, $Mg^{2+}$, and $Ca^{2+}$ ion concentrations (11); however, in the presence of only $Na^+$, but not $Mg^{2+}$ or $Ca^{2+}$, the thymine bases directly at the exchange point remain sensitive to attack by osmium tetroxide (26). The apparent stability of the stacked structure at the point of strand exchange is greater in the presence of the alkaline earth cations than in the alkali metal ions. The fact that the folded structure has the same FRET efficiency in the presence of all the cations is a strong indication that the global X-structure is very similar, but apparently there are differences in the local region of the strand exchange point.

Do the ion titrations provide evidence to indicate which mode of ion association enables the junction to fold into the stacked X-form? We do not observe directly a complex containing a bound ion, but rather a conformational change that is mediated by ion interactions with the macromolecule. The close proximity of the helical arms, especially at the point of strand exchange, makes the application of the common linear polyelectrolyte theories inapplicable, but we can rationalize the results in terms of the simplest notions of ion-polyelectrolyte interactions. Although $Na^+$ does not normally bind to DNA in the form of a molecular complex, our ion titrations show that the junction folds into an X-form with $Na^+$. The ultimate extent of energy transfer in the fully folded junction (from FRET experiments) is the same for all cations that have been titrated (11). This seems to exclude a requirement that an immobilized site-binding cation occupy a negatively charged phosphate pocket to permit folding. However, a negative charge density of phosphates most certainly accumulates at the exchange point in the folded form, as indicated by the resistance to folding at lower ionic strengths and molecular modeling calculations of the folded junction (100). A local density of highly mobile positive charge will lower the repulsion between phosphates, allowing them to approach each other more closely. Thus, even without site binding, it is understandable how cations can act as an ion switch and assist the junction to fold in an ion concentration–dependent manner.

The effects of the singly and doubly charged ions on the junction folding differ in other important ways. The ion titrations can be interpreted in terms of an apparent affinity of the ions for the DNA, by fitting a titration function to the progress curve of the FRET efficiency as the ion concentration is increased (11). This is an apparent affinity because such a titration curve can be produced without requiring site binding, as defined above (57). The apparent ion-dissociation constants from the FRET ion titrations corresponding to the folding of the junction from an extended structure differ greatly (a factor of $\sim 400$) between the alkali and alkaline earth ions (11); the relative difference is greater than that found when comparing the effect of different cations on the melting temperatures of linear DNA. Taken together, the facts that the nucleotides at the exchange point of junctions in the presence of only $Na^+$ are still chemically reactive, and that the junctions completely folded globally, indicate that doubly charged cations have specific interactions beyond those considered by general ion-polyelectrolyte theories, which do not account for ion interactions at specific localized sites on the DNA molecule. Apparently the $Mg^{2+}$ and $Ca^{2+}$ ions can lock the nucleotides at the point of strand exchange into a conformation that is inaccessible even to small reactive chemicals. At the present time, we cannot determine whether this protection results primarily from equilibrium or dynamic differences in the presence of different ions. In light of this difference between $Na^+$ and the $Mg^{2+}$ or $Ca^{2+}$ ions, it is interesting that, in a solution containing only $Na^+$, the junction can nevertheless fold into the antiparallel X-structure, still discriminating between the two possible stacking isomers of the arms. Thus, the interactions of ions with DNA junctions are complex, and apparently more than one mode of interaction leads to clearly differentiable observable effects. Further investigations are necessary to better define the details of the ion interactions.

## THERMODYNAMICS OF THE STRUCTURE OF THE FOUR-WAY JUNCTION

The major physical interactions determining the molecular structure of the four-way junction are those contributing to all DNA structures, i.e. base stacking; hydrogen bonding; electrostatic repulsions; the chemical structures of the bases, sugars, and phosphates; van der Waals forces; and hydration interactions. Electrostatic repulsion of the phosphate charges, together with van der Waals excluded volumes of the phosphate ridges of the helix and the major groove, probably determine the global stereochemistry of the folded junction molecule that retains maximum stacking of the bases even at the exchange point. It is convenient to break down the free energy of the four-way junction folding into several separate components, although this division does not constitute a unique independent set of free energy contributions. We discuss the major characteristics and consequences of the free-energy contributions within the context of the four-way junction structure.

## Coaxial Stacking of Helical Arms

There is considerable evidence that pairs of neighboring helical arms of the four-way junction stack upon each other (see above), and it is reasonable to assume that this stacking is a major stabilizing force. The actual free energy of stacking depends on the local sequence, the temperature, and the solvent composition. At 25°C these stacking interactions are on the order of 1.6 kcal mol$^{-1}$ nucleotide base pair (that is $\sim 2$–$3 \cdot RT$) in duplex DNA.

## Free Energy Difference Between the Two Stacking Isomers

The variation in the stacking free energy in duplex DNA between the different nearest-neighbor sequences (at 25°C) is reported to be relatively small compared to the total stacking free energy. The authors of an extensive recent study (22) estimated the maximum difference in the base-pair stacking free energy of normal duplexes to be $\sim 1/3 \cdot RT$; this estimate applies to the ionic strengths used in most four-way junction studies. A difficulty in interpreting the selection of the stacking isomer in terms of known stacking interactions is that the corresponding sequence variation in the stacking free energy is not really well understood even for simple duplex DNA. If the difference in the stacking free energies between the two possible stacking isomers of the junction were derived solely from these interactions, their population ratio would be no more than 6:4. For the majority of sequences studied, the population bias is much greater (11, 25, 77, 78). Apparently, the sequence discrimination operative for normal duplex base stacking does not suffice to explain the experimentally determined stacking selectivity at the center of the four-way junction. Perhaps geometric constraints are imposed on the spatial configurations of the center base pairs of the folded four-way DNA junction that augment sequence specificity. Stacking partners may be influenced by small variations in the relative positions of the phosphates in the vicinity of the point of strand exchange, leading to a sequence-dependent electrostatic potential (see below). Sequence specificity of stacking can be overridden by the neutralization of specific charges. Substitution of phosphate groups at the position of strand exchange with electrically neutral methyl phosphonates has a specific effect on the stacking choice; the junction folds so as to present the neutralized phosphonates on the exchanging strands (26).

The hydrogen-bonding contributions at the point of strand exchange probably are used to align the correct complementary base, as in normal duplex DNA. Hydrogen bonds may not contribute much to the stability of the complex formation as long as the correct complementary base is paired; however, we only know a little about the requirements of the stacked structure at the exchange point, and alternative hydrogen-bonding structures, or the participation of water, may be involved. We have studied the structure of junctions containing single-base mismatches at the point of strand exchange (23). Inclusion of some mismatches had a very destabilizing effect on the folded structure. In general, the effects of these mismatches depended both on the nature of the mismatch [there was an approximate correlation between the effect on the junction and the destabilization of short duplex species (1, 104)] and its context.

In spite of our ignorance concerning the molecular and physical contributions to the thermodynamic stability of junction structure with different sequences, the data are clear: the identity of the nucleotides directly facing the exchange point influences which of the two possible stacking isomers forms when sufficient salt is present; the nucleotides farther into the helical arms do not appear to affect the stacking process to a first approximation (25). This sequence dependence may influence the outcome of a recombination event.

## Electrostatic Repulsions Between Negative Charges at the Point of Strand Exchange

We have proposed that the screening effect associated with the mobile ion atmosphere is sufficient to cause the junction to fold into the stacked conformation as the concentration of ions increases (11). As soon as the unfavorable electrostatic repulsion between the phosphates at the strand-exchange point is reduced below the favorable free energy of the stacking interactions, the junction can fold in the manner of an ion switch.

Ion titrations, monitored by FRET, indicate that the four-way junction can fold without ion binding to a specific site on the DNA, and this implies that the folding process can be interpreted in the context of general polyelectrolyte theories. One of the most detailed treatments of ion-polyelectrolyte effects is the potential of mean force (PMF) model (94). This model has been applied to the structure of the four-way junction, and the general features of the ion titrations are reproduced (R. Klement, D. M. Soumpasis, E. von Kitzing, R. M. Clegg & D. M. J. Lilley, in preparation). The calculated ion distribution around the junction molecule (including the center and the distal helices) indicates an elevated electrostatic potential at the point of strand exchange, with a concomitant elevated density of counterions at this location compared with the purely helical regions. The difference between the calculated relative electrostatic free energy of the extended and folded forms of the junction is a function of the surrounding ion concentration, and this agrees with the general aspects of the salt-dependent isomerization found in the FRET ion titrations (see above).

## Stability of the Junction Compared with the Stability of the Duplex Arms

Markey et al (69) compared the thermal stability of the junction with that of four duplexes with sequences identical to the arms of the junction by using calorimetry and UV melting curves. They concluded that the calorimetric enthalpy of melting equals the sum of the enthalpies of melting the four duplex molecules, in agreement with the view that the helices of the arms are apparently intact in the folded conformation. The spectroscopic melting curves were interpreted in terms of a two-state dissociation reaction. The enthalpies from the spectroscopic experiments differed from those determined from calorimetry, and the values of the apparent $\Delta H$ depended upon the method of analysis (concentration variance or shape analysis). These differences probably arise from the failure of the two-state model; there may be intermediates within the melting transition. The complementary sequences of the arms would probably tend to interact almost independently, leading to a population distribution of intermediate structures, until four strands finally combine to form a tetrameric complex. The closure of the last arm of the junction would probably progress differently. Even though the overall reaction can be formally written as a single strand–to–tetramer reaction, the significant contribution of such intermediates to the measured signal within the thermal transition region would invalidate any analysis that requires the whole association reaction to proceed in one effective (tetramolecular) step. An analysis of similar spectroscopic melting curves in these laboratories (F. Walter, R. M. Clegg, A. I. H. Murchie & D. M. J. Lilley, submitted) supports this latter view. However, these experiments provide little information about the molecular details of the stability of the folded junction, or about the folding process, other than that the arm helices are apparently fully intact below the melting transition.

In a recent study, Lu et al (63) applied electrophoretic and titration calorimetric methods to estimate the difference in free energy between the formation of two junction molecules with complementary sequences, and the corresponding duplex molecules formed from the same sequences. Although the $\Delta G$ (18°C) for the reaction of two 16-bp duplexes forming a junction was found to be relatively small ($\sim$ 2 kcal mol$^{-1}$) by all methods, the $\Delta H$ was endothermic, large, and temperature-dependent (with a considerable heat capacity). They interpreted their results to indicate a loss of favorable stacking and pairing interactions and an entropic compensation by ionic and solvent interactions. An advantage of the electrophoresis experiments is that using large concentrations and remaining far away from the thermal melting transition may allow the experimenter to avoid potential partially completed junction molecules as intermediates. If the stacking contributions in the difference between duplexes and the point of strand exchange in the junction are different, we might expect to observe significant sequence-dependence in these experiments.

## The Relative Stability of the Parallel and Antiparallel Conformation

Kimball et al (51) constructed four-way junctions from two normal DNA strands and one long strand (effectively two strands connected by a $T_9$ linker), such that the ends of two arms were forced into proximity by the relatively short length of the $T_9$ tether. By placing the tether between different strands, the same junction sequence could be constrained into being either parallel or antiparallel. The method can also be used to force the selection of stacking isomer. Using a competition gel assay and thermal melting, Lu et al (67) studied the difference between the free energies of the constrained parallel and antiparallel orientations and concluded that the antiparallel orientation was thermodynamically favored by only 2 or $3 \cdot RT$. This free-energy difference is sufficient to render the population of the parallel orientation negligible and agrees with the observation that only the antiparallel orientation is observed (we note again that this energy difference is on the order of a stacking interaction). On the other hand, the energy difference between the two forms is low enough so that an interaction with proteins may make the parallel configuration possible. These authors (62, 67) have suggested that this equilibrium result implies a low activation barrier for the interconversion between the parallel and antiparallel conformation, which may be important for certain processes (e.g. branch migration) during recombination. However, the free-energy difference between two equilibrium states is not related to a kinetic activation barrier. Interconversion between the parallel and antiparallel conformations must involve large rotations of the DNA helices, and the activation energy of such a process is probably quite large, even if the free-energy difference between the equilibrium states is much smaller. In vivo this observation is more likely because the strands exchange in a crowded environment, and the arms are part of very long DNA molecules.

Lu et al (67) defined $\Delta\Delta G$ to be the difference in the free energy of melting between the constrained antiparallel (JA) and parallel (JP) molecules at 25°C, where both junctions are stable. If tethering does not contribute to $\Delta\Delta G$ (which may be questionable because the geometry of the tether could be different in JP and JA), $\Delta\Delta G$ would correspond to the free-energy difference in the parallel-antiparallel orientations. The free energies of melting of the two isomers, $\Delta G$ (JA) and $\Delta G$ (JP), were determined from spectrophotometric melting curves, assuming two-state processes. Different urea concentrations were used to reduce the melting temperature of higher melting species, and then the $\Delta G$s were extrapolated to 25°C using

$$\Delta G = \Delta H(1 - T/T_m).$$

However, the two tethered junctions are not structurally equivalent because one junction molecule (JP) includes a hairpin structure (of 39 base pairs) in the complete three-stranded junction complex. The authors performed melting experiments on the isolated hairpin and calculated $\Delta G$ (hairpin), the total free energy of melting the hairpin at 25°C. They calculated the required free energy difference as $\Delta\Delta G = \Delta G$ (JA) $- \Delta G$ (JP) $- \Delta G$ (hairpin). However, the $\Delta G$ (JP) is determined from experiments of complete thermal denaturation that are analyzed as two-state processes, so within the assumptions of this model, $\Delta G$ (JP) is the difference in the free energies of the fully folded and fully denatured states, and $\Delta G$ (hairpin) should not simply be added to $\Delta G$ (JP) to calculate the overall stability of JP. More likely, the very large hairpin loop in JP may destabilize the JP junction because of a statistical contribution of a significant loop weighting factor. This situation would require a different analysis of the melting data. Because the values of $\Delta G$ (JA), $\Delta G$ (JP), and $\Delta G$ (hairpin) are much larger than $\Delta\Delta G$, this procedure could introduce significant errors in the estimate of $\Delta\Delta G$; without correcting for the hairpin, $\Delta\Delta G$ is apparently quite large. Its apparent size would lead to the conclusion that the antiparallel configuration is more stable than the parallel by a much larger factor than the value reported. The tethers may also introduce significant differences, as may the urea conditions in the different experiments. Thus, the thermal analysis is not conclusive, although similar conclusions are drawn about the relative stabilities of JA and JP from the competition assay. In addition, if the two-state model does not apply the statistical error in estimating, $\Delta\Delta G$ may be larger than the value itself.

## DYNAMIC ASPECTS OF THE STRUCTURE OF THE JUNCTION

In the cell, genetic recombination will occur between between homologous DNA molecules, and four-way Holliday junctions created will therefore be capable of branch migration in principle. This is an important dynamic aspect of the structure of the junction, whereby the point of strand exchange moves along the strands by a mutual exchange of base-pairing partners. Unfortunately, we know very little about the mechanism of branch migration at the moment, although measured rates have been found to be slow (R. Fishel, personal communication). Two broad kinds of mechanisms may be envisaged. The junction may be required to open up into an unstacked conformation (perhaps similar to the extended low-salt structure) during branch migration, requiring large relative movements of the helical arms. Alternatively, the junction may remain in a conformation more closely related to the stacked X-structure, in which case the branch migration could proceed by a simple rotation of the two helical stacks around their respective helical axes and by a translation. Migration of the point of strand exchange could occur by exchange of two base pairs per step without changing the stacking choice among the four arms; in this case, the structure would pass through intermediate stages at which the stacking isomer of lower stability is present, although the energy difference between isomers is never likely to be large (see above). The presence of a very unfavorable stacking sequence at one particular location might have significant genetic consequences.

## INTERACTIONS WITH PROTEINS

Isolated four-way DNA junctions serve as excellent substrates for a class of structure-specific nucleases, or resolving enzymes (reviewed in greater detail in 24). Such activities have been isolated from many sources from bacteriophage to human cells (13, 14, 19, 42, 45, 46, 54, 95, 97, 101, 105), and are probably ubiquitous enzymes for the manipulation and/or repair of branched DNA. Also, several proteins selectively bind DNA junctions without nucleolysis (5, 27, 84).

The best-characterized junction-resolving enzyme is probably endonuclease VII of bacteriophage T4 (48). Phage T7 possesses a resolving activity similar to that of T4, called endonuclease I (19). T4 endonuclease VII is the product of gene 49, which is required for the maturation of T4 DNA prior to packaging (49). The enzyme cleaves isolated four-way junctions of various sequences (25, 74), as well as supercoil-stabilized cruciform structures (61, 71). It acts with considerable specificity, cleaving the exchanging strands of the junction, two or three bases 3' to the point of strand exchange.

When the positions of cleavage by T4 endonuclease VII are placed onto the stacked X-structure, one can see that the exchanging strands are cleaved on the minor-groove side of the junction. Figure 7 shows the cleavage positions for this and two other resolving enzymes. Clearly, each enzyme cuts two of the four strands, consistent with the two-fold structural symmetry of the junction. However, the three enzymes cleave the junction in different ways, and the cleavage sites appear to have little in common at first sight, apart from the shared symmetry element. Yet when these

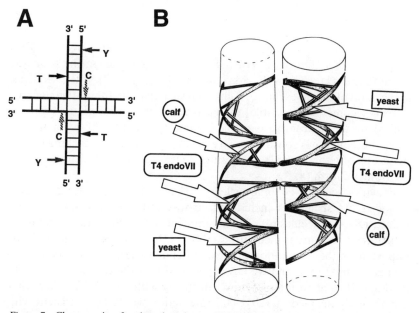

*Figure 7* Cleavage sites for three junction-resolving enzymes on the minor groove side of the four-way DNA junction. (*A*) Schematic showing the cleavage positions for the different enzymes on the four-way junction. The arrows indicate the phosphodiester bonds cleaved by T4 endonuclease VII (T), yeast (Y), and calf thymus (C) resolving enzymes. The three enzymes appear to cleave the junction in very different ways; T4 endonuclease VII cleaves the exchanging strands, while the yeast and calf enzymes cleave the continuous strands, 5' and 3', respectively, to the point of strand exchange. (*B*) Positions of cleavage on the minor-groove side of the stacked X-structure. Arrows indicate the positions of cleavage by the three resolving enzymes. Despite the differences in cleavage sites apparent in *A*, all three enzymes cleave on the minor-groove side of the junction, introducing two cleavages related by the central dyad axis.

sites are placed on the stacked X-structure (Figure 7), a clear relationship becomes apparent. All three sets of cleavages are clustered on one side of the junction—the minor-groove side—suggesting that these enzymes all bind to this face of the DNA junction. This is consistent with hydroxyl radical footprinting (4, 83) and with experiments showing a requirement for about one helical turn of each of the cleaved arms (75). Thus, the three enzymes probably have rather similar mechanisms for binding their substrates, despite their wide evolutionary separation. However, because the cleavages are introduced into different positions on the minor-groove side, the binding and catalytic functions of these proteins could be separable, perhaps as different domains. Of course, we are not suggesting that all junction-resolving enzymes will bind in the same way to four-way DNA junctions. Several structurally distinct faces to the stacked X-structure might be recognized by different proteins.

Enzymes of the T4 endonuclease VII class select primarily for DNA structure rather than base sequence. The cleavage sites in the four-way junction are related by a dyad symmetry axis (Figure 7); if the sequence at the point of strand exchange is changed such that the alternative stacking isomer becomes more stable, the cleavage pattern changes so as to be related by the alternative two-fold symmetry axis (25). If a junction of constant sequence is constrained to exist in one or another stacking isomer by means of tethering (51), the resulting cleavage sites are determined by the isomeric form (4).

Enzymes such as T4 endonuclease VII and T7 endonuclease I cleave various DNA structures, including four-way and three-way junctions and bulged DNA molecules (4). The common structural feature among these substrates is a mutual inclination of two DNA helices, which could be important in the recognition process. T4 endonuclease VII exists in solution as a dimer (56), and we have suggested that the two subunits could be oriented so as to interact most effectively with two DNA helices that are mutually inclined at 120°. This interaction would be consistent with the observation that T4 endonuclease VII cleaves bulged molecules fastest when they have particular numbers of bulged bases (4); bulges kink the axis of the helix in a way that depends on the number and type of bases (3, 40, 91), and thus, bulges with an axial kinking that is optimal for interaction with the enzyme should be cleaved most efficiently. Such structure-specific recognition of axial bending or kinking could be significant for other proteins, such as general DNA-repair enzymes and proteins that are required to manipulate DNA structure.

One junction-binding protein of particular interest is the relatively abundant eukaryotic nuclear protein HMG1 (31). Bianchi (5) has shown that this binds DNA junctions in a structure-specific manner, and we have suggested that the true role of this protein might be to bind to inclined DNA helices, and that binding to junctions is a direct consequence of this (reviewed in 59).

## IN CONCLUSION

The evidence for the stacked X-structure of the four-way DNA junction is now very strong. Left to its own devices, DNA assembles as a right-handed, antiparallel X-shaped structure in the presence of sufficient concentration of metal ions. The folding principles of this proposed structure may be summarized as follows:

1. The helical arms form two quasicontinuous helices by coaxial stacking in pairs. Two isomers of the structure are possible, depending upon the relative stabilities of alternative stacking partners.
2. The helical stacks are rotated into a right-handed, antiparallel cross, determined by the mutual interaction of the continuous strand from one helix with the major groove of the other. This interaction is optimal for a small angle of 60°.
3. The structure behaves as an ion-dependent conformational switch. In the absence of sufficient cation concentrations, the arms of the junction remain fully extended and unstacked.

DNA-protein interactions may play an important role in recombination processes; a junction formed as part of a multistranded filament of DNA generated by a pairing protein such as RecA might be rather different from a junction in naked DNA. However, the isolated four-way junction folds in such a precise and well-behaved way that the structure is probably preserved despite mediation by proteins. Isolated DNA junctions are recognized by a whole series of structure-specific proteins (see above), and the structure of the junction appears to remain unaltered by interaction with these enzymes. It is sometimes implied that the antiparallel structure of the junction would require an antiparallel alignment of entire chromosomes. Clearly this is not the case; the junction would confer a local alignment but nothing more.

The structure adopted by the four-way DNA junction is quite unlike most other biologically functional DNA geometries, which for the most part are variants of standard B-DNA. It is the best example of an ion-dependent, folded higher-order structure in DNA. In this regard, the structure is perhaps more similar to the kinds of folding found in RNA structures, and so perhaps we can learn something from it. A four-way RNA junction would not be expected to adopt the same right-handed, antiparallel structure because the folding principle of strand-groove alignment is quite different for an A-form double helix. Preliminary experiments confirm this. But other folding principles are likely to be more general, such as the tendency for coaxial helical stacking [the RNA pseudoknot would be a good example (85, 108)], and the important role of metal ions.

We may also learn from the methods and approaches used in the study of the DNA junctions; these could be valuable in other comparable problems, notably in RNA structure. In the course of these investigations, new approaches have been applied to determine folded nucleic acid structure. For example, FRET has been shown to a very powerful method for the comparison of relatively long distances and can be applied to nucleic acids in a straightforward manner. We look forward to the further application of these techniques to new biological problems, and to a deeper understanding of the structure and dynamics of DNA junctions and their interactions with proteins.

ACKNOWLEDGMENTS

We thank all our colleagues who have participated in studies of four-way DNA junctions, including D. R. Duckett, A. I. H. Murchie, A. Bhattacharyya, J. Portugal, W. A. Carter, A. Zechel, F. Walter, G. Vámosi, C. Gohlke, E. von Kitzing, and S. Diekmann, and the MRC, SERC, CRC, DFG, and NATO for continued financial support.

*Literature Cited*

1. Aboul-ela, F., Koh, D., Tinoco, I. 1985. *Nucleic Acids Res.* 13: 4811–24
2. Bell, L. R., Byers, B. 1979. *Proc. Natl. Acad. Sci. USA* 76: 3445–49
3. Bhattacharyya, A., Lilley, D. M. J. 1989. *Nucleic Acids Res.* 17: 6821–40
4. Bhattacharyya, A., Murchie, A. I. H., von Kitzing, E., Diekmann, S., Kemper, B., Lilley, D. M. J. 1991. *J. Mol. Biol.* 221: 1191–1207
5. Bianchi, M. E., Beltrame, M., Paonessa, G. 1989. *Science* 243: 1056–59
6. Broker, T. R., Lehman, I. R. 1971. *J. Mol. Biol.* 60: 131–49
7. Calascibetta, F. G., de Santis, P., Morosetti, S., Palleschi, A., Savino, M. 1984. *Gazz. Chim. Ital.* 114: 437–41
8. Chen, J.-H., Churchill, M. E. A., Tullius, T. D., Kallenbach, N. R., Seeman, N. C. 1988. *Biochemistry* 27: 6032–38
9. Chen, S., Heffron, F., Leupin, W., Chazin, W. J. 1991. *Biochemistry* 30: 766–71
10. Churchill, M. E., Tullius, T. D., Kallenbach, N. R., Seeman, N. C. 1988. *Proc. Natl. Acad. Sci. USA* 85: 4653–56
11. Clegg, R. M., Murchie, A. I. H., Zechel, A., Carlberg, C., Diekmann, S., Lilley, D. M. J. 1992. *Biochemistry* 31: 4846–56
12. Clegg, R. M., Murchie, A. I. H., Zechel, A., Lilley, D. M. J. 1993. *Proc. Natl. Acad. Sci. USA*. In press
13. Connolly, B., Parsons, C. A., Benson, F. E., Dunderdale, H. J., Sharples, G. J., et al. 1991. *Proc. Natl. Acad. Sci. USA* 88: 6063–67
14. Connolly, B., West, S. C. 1990. *Proc. Natl. Acad. Sci. USA* 87: 8476–80
15. Cooper, J. P., Hagerman, P. J. 1987. *J. Mol. Biol.* 198: 711–19
16. Cooper, J. P., Hagerman, P. J. 1989. *Proc. Natl. Acad. Sci. USA* 86: 7336–40
17. Cooper, J. P., Hagerman, P. J. 1990. *Biochemistry* 29: 9261–68
18. Courey, A. J., Wang, J. C. 1983. *Cell* 33: 817–29
19. de Massey, B., Studier, F. W., Dorgai, L., Appelbaum, F., Weisberg, R. A. 1984. *Cold Spring Harbor Symp. Quant. Biol.* 49: 715–26
20. Diekmann, S., Lilley, D. M. J. 1987. *Nucleic Acids Res.* 14: 5765–74
21. Diekmann, S., Wang, J. C. 1985. *J. Mol. Biol.* 186: 1–11
22. Doktycz, M. J., Goldstein, R. F., Paner, T. M., Gallo, F. J., Benight, A. S. 1992. *Biopolymers* 32: 849–64
23. Duckett, D. R., Lilley, D. M. J. 1991. *J. Mol. Biol.* 221: 147–61
24. Duckett, D. R., Murchie, A. I. H., Bhattacharyya, A., Clegg, R. M., Diekmann, S., et al. 1992. *Eur. J. Biochem.* 207: 285–95
25. Duckett, D. R., Murchie, A. I. H., Diekmann, S., von Kitzing, E., Kemper, B., Lilley, D. M. J. 1988. *Cell* 55: 79–89
26. Duckett, D. R., Murchie, A. I. H., Lilley, D. M. J. 1990. *EMBO J.* 9: 583–90
27. Elborough, K., West, S. 1988. *Nucleic Acids Res.* 16: 3603–14
28. Förster, T. 1948. *Ann. Phys.* 2: 55–75
29. Furlong, J. C., Lilley, D. M. J. 1986. *Nucleic Acids Res.* 14: 3995–4007
30. Gellert, M., Mizuuchi, K., O'Dea, M. H., Ohmori, H., Tomizawa, J. 1979. *Cold Spring Harbor Symp. Quant. Biol.* 43: 35–40

31. Goodwin, G. H., Sanders, C., Johns, E. W. 1973. *Eur. J. Biochem.* 38: 14–19
32. Gough, G. W., Lilley, D. M. J. 1985. *Nature* 313: 154–56
33. Gough, G. W., Sullivan, K. M., Lilley, D. M. J. 1986. *EMBO J.* 5: 191–96
34. Greaves, D. R., Patient, R. K., Lilley, D. M. J. 1985. *J. Mol. Biol.* 185: 461–78
35. Guo, Q., Lu, M., Seeman, N. C., Kallenbach, N. R. 1990. *Biochemistry* 29: 570–78
36. Guo, Q., Seeman, N. C., Kallenbach, N. R. 1989. *Biochemistry* 28: 2355–59
37. Hagerman, P. J. 1985. *Biochemistry* 24: 7033–37
38. Hoess, R., Wierzbicki, A., Abremski, K. 1987. *Proc. Natl. Acad. Sci. USA* 84: 6840–44
39. Holliday, R. 1964. *Genet. Res.* 5: 282–304
40. Hsieh, C.-H., Griffith, J. D. 1989. *Proc. Natl. Acad. Sci. USA* 86: 4833–37
41. Hsu, P. L., Landy, A. 1984. *Nature* 311: 721–26
42. Iwasaki, H., Takahagi, M., Shiba, T., Nakata, A., Shinagawa, H. 1991. *EMBO J.* 10: 4381–89
43. Jack, A., Ladner, J. E., Klug, A. 1976. *J. Mol. Biol.* 108: 619–49
44. Jayaram, M., Crain, K. L., Parsons, R. L., Harshey, R. M. 1988. *Proc. Natl. Acad. Sci. USA* 85: 7902–6
45. Jensch, F., Kosak, H., Seeman, N. C., Kemper, B. 1989. *EMBO J.* 8: 4325–34
46. Jeyaseelan, R., Shanmugam, G. 1988. *Biochem. Biophys. Res. Commun.* 156: 1054–60
47. Kallenbach, N. R., Ma, R.-I., Seeman, N. C. 1983. *Nature* 305: 829–31
48. Kemper, B., Garabett, M. 1981. *Eur. J. Biochem.* 115: 123–31
49. Kemper, B., Janz, E. 1976. *J. Virol.* 18: 992–99
50. Kim, S.-H., Quigley, G. J., Suddath, F. L., McPherson, A., Sneden, D., et al. 1973. *Science* 179: 285–88
51. Kimball, A., Guo, Q., Lu, M., Cunningham, R. P., Kallenbach, N. R., et al. 1990. *J. Biol. Chem.* 265: 6544–47
52. Kirshenbaum, M. R., Tribolet, R., Barton, J. K. 1988. *Nucleic Acids Res.* 16: 7943–60
53. Kitts, P. A., Nash, H. A. 1987. *Nature* 329: 346–48
54. Kleff, S., Kemper, B. 1988. *EMBO J.* 7: 1527–35
55. Koo, H.-S., Wu, H.-M., Crothers, D. M. 1986. *Nature* 320: 501–6
56. Kosak, H. G., Kemper, B. W. 1990. *Eur. J. Biochem.* 194: 779–84
57. Kotin, L., Nagasawa, M. 1962. *J. Chem. Phys.* 16: 873–79
58. Lilley, D. M. J. 1980. *Proc. Natl. Acad. Sci. USA* 77: 6468–72
59. Lilley, D. M. J. 1992. *Nature* 357: 282–83
60. Lilley, D. M. J., Hallam, L. R. 1984. *J. Mol. Biol.* 180: 179–200
61. Lilley, D. M. J., Kemper, B. 1984. *Cell* 36: 413–22
62. Lu, M., Guo, Q., Kallenbach, N. R. 1992. *Crit. Rev. Biochem. Mol. Biol.* 27: 157–90
63. Lu, M., Guo, Q., Marky, L. A., Seeman, N. C., Kallenbach, N. R. 1992. *J. Mol. Biol.* 223: 781–89
64. Lu, M., Guo, Q., Pasternack, R. F., Wink, D. J., Seeman, N. C., Kallenbach, N. R. 1990. *Biochemistry* 29: 1614–24
65. Lu, M., Guo, Q., Seeman, N. C., Kallenbach, N. R. 1989. *J. Biol. Chem.* 264: 20851–54
66. Lu, M., Guo, Q., Seeman, N. C., Kallenbach, N. R. 1990. *Biochemistry* 29: 3407–12
67. Lu, M., Guo, Q., Seeman, N. C., Kallenbach, N. R. 1991. *J. Mol. Biol.* 221: 1419–32
68. Marini, J. C., Levene, S. D., Crothers, D. M., Englund, P. T. 1982. *Proc. Natl. Acad. Sci. USA* 79: 7664–68
69. Markey, L. A., Kallenbach, N. R., McDonough, K. A., Seeman, N. C., Breslauer, K. J. 1987. *Biopolymers* 26: 1621–34
70. Meselson, M. S., Radding, C. M. 1975. *Proc. Natl. Acad. Sci. USA* 72: 358–61
71. Mizuuchi, K., Kemper, B., Hays, J., Weisberg, R. A. 1982. *Cell* 29: 357–65
72. Mizuuchi, K., Mizuuchi, M., Gellert, M. 1982. *J. Mol. Biol.* 156: 229–43
73. Moras, D., Comarmond, M. B., Fischer, J., Weiss, R., Thierry, J. C., et al. 1980. *Nature* 288: 669–74
74. Mueller, J. E., Kemper, B., Cunningham, R. P., Kallenbach, N. R., Seeman, N. C. 1988. *Proc. Natl. Acad. Sci. USA* 85: 9441–45
75. Mueller, J. E., Newton, C. J., Jensch, F., Kemper, B., Cunningham, R. P., et al. 1990. *J. Biol. Chem.* 265: 13918–24
76. Murchie, A. I. H., Carter, W. A., Portugal, J., Lilley, D. M. J. 1990. *Nucleic Acids Res.* 18: 2599–2606
77. Murchie, A. I. H., Clegg, R. M., von Kitzing, E., Duckett, D. R., Diekmann, S., Lilley, D. M. J. 1989. *Nature* 341: 763–66
78. Murchie, A. I. H., Portugal, J., Lilley, D. M. J. 1991. *EMBO J.* 10: 713–18
79. Nelson, J. W., Tinoco, I. 1985. *Biochemistry* 24: 6416–21
80. Nunes-Düby, S. E., Matsomoto, L., Landy, A. 1987. *Cell* 50: 779–88

81. Orr-Weaver, T. L., Szostak, J. W., Rothstein, R. J. 1981. *Proc. Natl. Acad. Sci. USA* 78: 6354–58
82. Panayotatos, N., Wells, R. D. 1981. *Nature* 289: 466–70
83. Parsons, C. A., Kemper, B., West, S. C. 1990. *J. Biol. Chem.* 265: 9285–89
84. Parsons, C. A., Tsaneva, I., Lloyd, R. G., West, S. C. 1992. *Proc. Natl. Acad. Sci. USA* 89: 5452–56
85. Pleij, C. W. A., Rietveld, K., Bosch, L. 1985. *Nucleic Acids Res.* 13: 1717–31
86. Pörschke, D. 1976. *Biophys. Chem.* 4: 383–94
87. Pörschke, D. 1978. *Nucleic Acids Res.* 6: 883–98
88. Potter, H., Dressler, D. 1976. *Proc. Natl. Acad. Sci. USA* 73: 3000–4
89. Potter, H., Dressler, D. 1978. *Proc. Natl. Acad. Sci. USA* 75: 3698–3702
90. Pulleyblank, D. E., Shure, M., Tang, D., Vinograd, J., Vosberg, H.-P. 1975. *Proc. Natl. Acad. Sci. USA* 72: 4280–84
91. Rice, J. A., Crothers, D. M. 1989. *Biochemistry* 28: 4512–16
92. Sigal, N., Alberts, B. 1972. *J. Mol. Biol.* 71: 789–93
93. Sobell, H. M. 1972. *Proc. Natl. Acad. Sci. USA* 69: 2483–87
94. Soumpasis, D. M., Jovin, T. M. 1987. *Nucleic Acids and Molecular Biology*, ed. F. Eckstein, D. M. J. Lilley, 1: 85–111. Heidelberg: Springer-Verlag
95. Stuart, D., Ellison, K., Graham, K., McFadden, G. 1992. *J. Virol.* 66: 1551–63
96. Sussman, J. L., Holbrook, S. R., Wade Warrant, R., Church, G. M., Kim, S.-H. 1978. *J. Mol. Biol.* 123: 607–30
97. Symington, L., Kolodner, R. 1985. *Proc. Natl. Acad. Sci. USA* 82: 7247–51
98. Timsit, Y., Moras, D. 1991. *J. Mol. Biol.* 221: 919–40
99. Timsit, Y., Westhof, E., Fuchs, R. P. P., Moras, D. 1989. *Nature* 341: 459–62
100. von Kitzing, E., Lilley, D. M. J., Diekmann, S. 1990. *Nucleic Acids Res.* 18: 2671–83
101. Waldman, A. S., Liskay, R. M. 1988. *Nucleic Acids Res.* 16: 10249–66
102. Wang, J. C., Peck, L. J., Becherer, K. 1983. *Cold Spring Harbor Symp. Quant. Biol.* 47: 85–91
103. Wemmer, D. E., Wand, A. J., Seeman, N. C., Kallenbach, N. R. 1985. *Biochemistry* 24: 5745–49
104. Werntges, H., Steger, G., Riesner, D., Fritz, H.-J. 1986. *Nucleic Acids Res.* 14: 3773–90
105. West, S. C., Korner, A. 1985. *Proc. Natl. Acad. Sci. USA* 82: 6445–49
106. Woo, N. H., Roe, B. A., Rich, A. 1980. *Nature* 286: 346–51
107. Wu, H.-M., Crothers, D. M. 1984. *Nature* 308: 509–13
108. Wyatt, J. R., Puglisi, J. D., Tinoco I. Jr. 1990. *J. Mol. Biol* 214: 455–70

# STRUCTURE AND FUNCTION OF HUMAN GROWTH HORMONE: Implications for the Hematopoietins

## James A. Wells and Abraham M. de Vos

Department of Protein Engineering, Genentech, Inc., 460 Point San Bruno Boulevard, South San Francisco, California 94080

KEY WORDS: hormone-receptor complex, mutagenesis, crystal structure, receptor activation

CONTENTS

| | |
|---|---|
| PERSPECTIVE AND OVERVIEW | 329 |
| LIGANDS FOR HEMATOPOIETIC RECEPTORS ARE FOUR-HELIX BUNDLES | 331 |
| THE EXTRACELLULAR DOMAINS OF THE HEMATOPOIETIC RECEPTORS | 334 |
| ONE hGH BINDS TWO RECEPTORS | 337 |
|     Mutational Analysis | 337 |
|     Crystallographic Analysis | 341 |
| FUNCTIONAL AND STRUCTURAL EPITOPES ON hGH OVERLAP | 342 |
| IMPLICATIONS FOR OTHER HEMATOPOIETINS | 344 |
|     Hematopoietic Hormones | 344 |
|     Extracellular Receptor Domains | 345 |
| MECHANISM(S) FOR ACTIVATION OF THE GROWTH-HORMONE RECEPTOR | 346 |
| FUTURE APPLICATIONS | 348 |

## PERSPECTIVE AND OVERVIEW

Molecular recognition events represent essential steps in the proper functioning of many critical biological processes. For instance, antibodies must bind antigens that are foreign to an organism, but discriminate against many very similar molecules derived from the host itself. Specific inhibitors and activators need to interact only with the proteins to which they are directed and not with close homologues. Membrane-spanning receptors must recognize their own specific ligand(s) and not the ligand(s) of other, sometimes related receptors. In each of these examples, highly selective molecular recognition mechanisms operate to differentiate between the proper targets and several structurally similar molecules. A detailed understanding of this specificity requires analysis at the atomic level of the relevant molecular contact surfaces, employing both structural and functional approaches to reveal the individual interactions involved and to estimate the relative importance of each of these interactions. Here, we review the progress in this area, emphasizing a specific class of receptor-ligand interactions, and suggest where the findings may be extrapolated to related receptors.

The pituitary hormones from the growth hormone (GH) family are involved in the regulation of many physiological processes, including growth and differentiation of muscle, bone, and cartilage cells (for reviews, see 15, 42). Members of this family include GH, prolactin (PRL), and placental lactogen (PL), proteins that are likely to have the same overall three-dimensional fold. Growth and differentiation of hematopoietic cells is regulated by a large, nonpituitary family of hormones that includes interleukins 1–7 (IL–1 to IL–7), granulocyte–macrophage colony stimulating factors (GM-CSF and G-CSF), and erythropoietin (EPO) (for reviews, see 54, 56). Even though these proteins share little if any sequence homology, both among themselves and with the pituitary hormones, most have been shown to have the same three-dimensional fold as the pituitary hormones. Because of the similarities in function as well as structure, we refer to pituitary and hematopoietic hormones as hematopoietins.

The receptors for the hematopoietins were recently recognized as belonging to the same superfamily (8, 10, 26, 36–38, 41, 43, 61). All are single-pass transmembrane receptors, with a three-domain architecture: an extracellular domain that binds the activating ligand, a short transmembrane segment, and a domain residing in the cytoplasm. The extracellular domains of these receptors have low but significant homology within their ligand-binding region of about 200 amino acids. This observation led to the proposal that these ligand-binding regions share a common three-dimensional fold (8, 10, 26, 36–38, 41, 43, 61). Interestingly, the intracellular domains of the hematopoietic receptors have little apparent sequence similarity to each other or to other known proteins, including the tyrosine kinase receptors.

Here, we review in detail the structural and functional evidence that is available in the human growth hormone (hGH) example. We evaluate the binding sites between hGH and its receptor, and discuss the match of epitopes defined by high resolution mutational and structural analysis. We also review the evidence showing that activation of the GH receptor occurs via a receptor dimerization mechanism. Finally, we discuss how a detailed understanding of structure, function, and mechanism can be used to produce receptor-specific analogues and to rationally design receptor agonists and antagonists. Throughout the discussion, these results are generalized to other members of the hematopoietic superfamily. This generalization is based on the observation that the hematopoietic ligands and receptors are linked by similar protein structures and biological functions, and that therefore the hormone-receptor binding interfaces may be topologically similar, and the signal-transduction mechanisms analogous, for all members of this family.

## LIGANDS FOR HEMATOPOIETIC RECEPTORS ARE FOUR-HELIX BUNDLES

The first crystal structure reported for a member of the hematopoietic family of hormones was for porcine growth hormone (pGH), whose 2.8-Å crystal structure was determined in 1987 by Abdel-Meguid and coworkers (1). The molecule contains four helices 21–30 residues long that are arranged in a left-handed bundle. The topology is unusual in that the first two helices are parallel to each other and antiparallel to the last two helices (Figure 1a). To achieve this arrangement, long crossover connections link the two sets of parallel helices, and a short segment connects helix 2 to helix 3. There is a distinctive bend in helix 3 at P86. Disulfides connect C53 in the first crossover connection to C164 in helix 4, and C181 in helix 4 to C189 near the C terminus. For several years, pGH remained the only example of this type of fold, and only recently did the elucidation of similar structures confirm its generality (11, 28, 29, 62, 66, 70).

The structure of the complex between human growth hormone (hGH) and the extracellular domain of its receptor (28) demonstrated that the bound hGH has virtually the same fold as pGH (Figure 1b). However, the local structures of the connecting loops differ significantly. For example, two short helical segments (residues K38-N47 and R64-K70, respectively) in the first long crossover connection were not found in pGH. Because residues in both these helices are involved in hormone-receptor contacts (see below), the difference from pGH may represent conformational changes in the hormone that are induced by receptor binding. Second, residues R94-S100 in the short linker between helix 2 and 3 also have a helical conformation, in contrast to the omega-loop conformation described for pGH (1). Because this region is not involved in receptor contacts and the local sequences are virtually identical between hGH and pGH, the cause of this difference is unclear. It may reflect the preliminary state of the refinement of the reported pGH structure ($R = 33\%$).

The core of hGH is almost exclusively made up of hydrophobic side

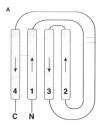
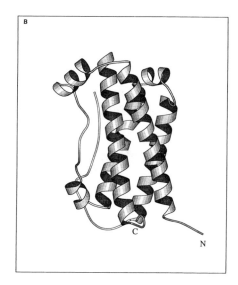
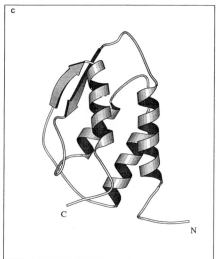

*Figure 1* Hematopoietic ligands. (*a*) Topology diagram showing the four-helix bundle core with the characteristic connectivity. (*b*) Schematic drawing of the structure of hGH (coordinates from 28). (*c*) Schematic drawing of the structure of GM-CSF [coordinates (29) taken from the Protein Data Bank (12)]. Cartoon drawings made with the program MOLSCRIPT (45).

chains. Residues contributing to the interior of the four-helix bundle include F10, A13, L20, and A24 of helix 1; L76, S79, I83, W86, and V90 of helix 2; V110, L114, L117, I121, and L124 of helix 3; and F166, D169, M170, V173, L177, and V180 of helix 4. The only two hydrophilic side chains in this list are S79, whose OG hydrogen bonds back to the carbonyl oxygen of L75, and D169, whose OD1 is hydrogen-bonded to W86. The OD2 of D169 juts out from the core and forms a long salt bridge with the side chain of K172. Smaller hydrophobic clusters hold the connection segments to the four-helix bundle. For example I36, F44, C53, F54, and I58 in the segment between helix 1 and 2 pack together with L75 and I78 of helix 2 and with L157, Y160, Y164, C165, and F176 in helix 4; similarly, L93, V96, and F97 in the short connection between helices 2 and 3 interact with F31 of helix 1 and L162 and L163 of helix 4 (28).

The structures of several other hematopoietic ligands were determined recently. Granulocyte–macrophage colony stimulating factor and interleukin 4 are about 60 residues shorter than growth hormone. Both the crystal structure of GM-CSF (29, 70) and the NMR structure of IL–4 (62, 66) reveal the same topology as GH, but with an additional structural motif not seen before—a short segment of $\beta$-ribbon formed by residues in the long crossover connections (Figure 1c). As pointed out by Diederichs et al (29), the relative disposition of the crossover connections is different from that seen in GH, resulting in the first segment being located on the opposite side of helix 4. The $\alpha$-helices are significantly shorter than those of GH (10–16 residues for GM-CSF and 18–22 residues for IL–4) and the twist angles between the helices are larger than those seen in GH. Like GH, GM-CSF has a kink in helix 3 (at P76), whereas all helices in IL–4 appear to be regular. GM-CSF has two disulfide bridges: C54 at the beginning of helix 2 is linked to C98, just before the second $\beta$-strand, and C88 immediately following helix 3 is connected to C121 in the C-terminal tail. The three disulfide bonds in IL–4 connect the N and C termini (C7 to C131), the first crossover connection and the short segment between helices 2 and 3 (C28 to C69), and helix 2 and the second crossover connection (C50 to C103).

Recently, IL–2 was also shown to have a GH-like topology, based on a reevaluation of the available mutational data in the light of the structures of GM-CSF and IL–4 (11). This GH-like fold has been confirmed by an independent crystal structure determination at high resolution (M. Hatada, personal communication). Finally, the crystal structures of G-CSF (C. Hill, T. Osslund & D. Eisenberg, personal communication) and macrophage colony stimulating factor (60) again show the GH fold. This last structure is of special interest, because the ligand-binding domain of the receptor for this cytokine belongs to the immunoglobulin-like superfamily. Therefore, the occurrence of the hematopoietin fold is not limited to ligands of hematopoietic receptors.

## THE EXTRACELLULAR DOMAINS OF THE HEMATOPOIETIC RECEPTORS

A few years ago, investigators recognized that the extracellular portions of many receptors share certain characteristics, suggesting a similar overall structure (for review, see 9). Among these shared features are four cysteine residues near the N terminus, a short segment near the C terminus, consisting of the sequence WSXWS (X can be any amino acid), and some common patterns of hydrophobic and hydrophilic residues, including a few conserved residues. Based on analysis of these features in numerous sequences of many receptor types and for the same receptor in different species, the extracellular part of each receptor was predicted to consist of two domains containing seven $\beta$-strands each, with a topology similar to that of immunoglobulin constant domains (10). The family includes receptors belonging to the hematopoietic superfamily (growth hormone; prolactin; interleukins 2–4, 6, and 7; granulocyte and granulocyte-macrophage colony stimulating factors; erythropoietin; interferon-$\alpha$; interferon-$\gamma$; and tissue factor) (Figure 2) as well as the receptor for ciliary neutrophic factor.

At present the only three-dimensional structure for a member of this family is for the extracellular domain of the hGH receptor (hGHbp) in complex with hGH (28). This structure shows that the receptor does indeed consist of two distinct immunoglobulin-like domains of the type described above (Figures 3a,b). In contrast to the topology of immunoglobulin constant domains, however, sheet switching takes place, similar to the structure of the second domain of chaperone protein PapD (40), domain D2 of CD4 (64, 71) and fibronectin type III domains (5, 46). The two hGHbp domains are linked by a short, four-residue segment with approximately helical conformation. This confers a relative orientation between the domains that has not been observed before (Figure 3b). Because there appear to be only few interdomain interactions, it is unclear whether this orientation is maintained in the free receptor. The finding that residues from both domains contribute to the binding interface with the hormone (see below) makes it tempting to speculate that the observed orientation is induced by ligand binding.

The N-terminal domain of the hGHbp contains six cysteine residues. These cysteines are all linked to form three disulfide bridges, with connectivity C38-C48, C83-C94, and C108-C122 (34). C38, C48, C83, and C94 are conserved in the superfamily and are buried in the core of the N-

STRUCTURE AND FUNCTION OF hGH 335

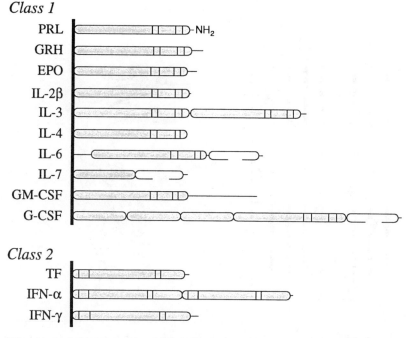

*Figure 2* Pictorial representation of the extracellular domains of receptors belonging to the hematopoietic superfamily. Homologous domains are drawn with the same symbol. The domain characteristic of the superfamily is shown as a long, shaded bar; short lines indicate the positions of the conserved cysteines (after 9). Short shaded bars represent fibronectin type III–homologous domains; open bars represent immunoglobin-homologous domains (after 9).

terminal $\beta$ sandwich. The first of these disulfides connects neighboring strands A and B of the three-stranded sheet, whereas the second cross-links the two sheets by linking strand C to strand E. The third disulfide is only found in the hGHbp and bridges strands F and G located on top of the sandwich. Both the N-terminal and the C-terminal domain of the extracellular domains in the superfamily contain a strictly conserved tryptophan, W50 and W157, respectively. Other hydrophobic residues that are less strictly conserved complete the interior of the $\beta$ sandwiches, for example F35, F46, I64, L66, F96, I109, L111, F123, and I128 in the N-terminal domain and P131, P134, W139, I153, L172, V199, V212, and V233 in the C-terminal sandwich.

The characteristic WSXWS box present near the C terminus of the

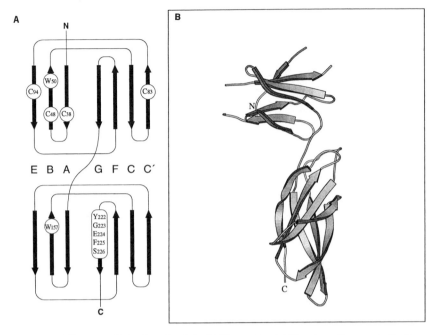

*Figure 3* The extracellular domain of the hGH receptor. (*a*) Topology diagram, indicating the positions of conserved residues and of the residues analogous to the WSXWS-box. (*b*) Schematic drawing of the structure of the hGHbp. Made with the program MOLSCRIPT (45).

extracellular domains is replaced by YGEFS beginning at residue 222 in the hGHbp. The crystal structure shows that this sequence is part of a stretch of irregular extended structure just before the beginning of strand G of the C-terminal domain. A bulge at residues G223–E224 positions both aromatic side chains on the solvent-accessible face of the $\beta$ sandwich, while the side chain of S226 is hydrogen bonded to the neighboring strand F (Figure 3). The two aromatic side chains are part of an interesting pattern generated on the exposed face of the four-stranded sheet of the sandwich. This pattern is composed of a series of pairs of charged or hydrophilic residues separated by the hydrophobic side chains Y222, F225, and W186. This surface pattern consists of E173 and K215, Y222, E175 and R213, F225, Q177 and R211, W186, and K179 and E209 and is conserved in most other members of the superfamily (28). Since this surface is located away from all binding interfaces observed in the crystal structure of the complex (below), it is probably not directly involved in ligand binding or receptor dimerization.

Considering the similarities in the extracellular domains of the receptors in the hematopoietic superfamily, the structure of the hGHbp can probably be used as a model for these related receptors. Thus, throughout the superfamily, the positions of the $\beta$-strands are expected to be similar, the conserved amino acids will occupy equivalent positions in the hydrophobic cores, and the residues seen to interact with the hormone in the hGH-hGHbp complex are possible candidates for ligand-binding determinants in the other receptors. These considerations suggest that residues in the WSXWS box probably do not contact the ligands. Based on the structure of the hGH-hGHbp complex, we can only speculate that these residues are used to interact with a putative accessory protein required for the formation of the active signal-transduction complex. Nonetheless, mutations in the WSXWS box decreased ligand binding in the cases of IL-2 (55), erythropoietin (18), and prolactin (63) receptors. These mutations may have disrupted the local structure of the receptor, and so indirectly affected ligand binding. More evidence is needed to elucidate the biological function of the WSXWS box.

## ONE hGH BINDS TWO RECEPTORS

### Mutational Analysis

Mutational mapping of ligand-receptor binding sites and three-dimensional structures of ligand-receptor complexes provides the complementary information required for a full understanding of ligand-receptor interactions. In these terms, the hGH-receptor complex is the most thoroughly characterized member of the hematopoietic superfamily, and thus we focus on it before discussing the other family members.

Prior to 1987, the functional analysis of GH-receptor interactions had been limited to studies with peptide fragments and natural or chemically modified hormone variants (for reviews, see 3, 48). Peptide fragments bound very weakly if at all to GH receptors. Natural variants provided for interesting sequence comparisons (for reviews, see 17, 57), but these hormones were not available in sufficient quantities, nor did their sequences vary systematically to provide a thorough means of dissecting the residues important for hormone binding.

Recent advancements have dramatically changed this situation. First, an efficient secretion system was developed in *Escherichia coli* for production of recombinant hGH (14), which greatly facilitated the production of hGH mutants (21). Second, a structural model of pGH was reported (1) that was crucial in verifying the proximity of functionally important residues in the folded hormone. Next, the hGH receptor was cloned (47, 67), enabling researchers to obtain it free of other receptors that bind hGH, such as the prolactin receptor. Finally, high yields of the extracellular domain of the receptor, known as the hGH binding protein (hGHbp), which is found naturally in serum and binds hGH tightly (7, 39), were expressed in *E. coli* in a manner similar to that used for hGH (34). Importantly, the *E. coli*-derived hGHbp bound hGH with the same affinity as its glycosylated natural counterpart and provided a rich source of the highly purified hGHbp for binding, biophysical, and crystallographic studies. All these developments facilitated the high resolution functional analysis that followed.

A strategy called homologue-scanning mutagenesis was used to obtain a general idea of where the hGHbp bound to hGH (21; for a review, see 73). In this approach, segments 7–30 residues long were substituted into hGH from hGH homologues that do not bind tightly to the hGH receptor [human placental lactogen (hPL), pGH, and human prolactin (hPRL)]. Based on a model of pGH, segment substitutions were confined to a given unit of secondary structure (such as a helix or loop) and did not cross disulfide boundaries. Moreover, buried hydrophobic residues in the amphipathic helices were not altered, because mutations at buried positions are more likely to destabilize the protein and cause problems in protein expression (2, 59). A set of 17 segment-substituted hGH mutants was created that collectively altered about 50% of the residues in the protein.

Mutants were analyzed for binding to the hGHbp (21) by competitive displacement of (I–125)hGH and immunoprecipitation using Mab5 (4), which binds to the hGHbp and precipitates a 1:1 hGH·hGHbp complex (67). Six of the homologue-scan variants caused a 10-fold or greater reduction in binding (21). These segment substitutions were from three discontinuous regions of the molecule but together formed a patch when mapped upon a model of the folded hormone. This technique was also used to map the binding sites (epitopes) for eight monoclonal antibodies for hGH. Virtually all these epitopes were shown to be discontinuous. The binding sites for many of these conformation-sensitive antibodies did not overlap the hGHbp site and were not affected by segment substitutions that disrupted binding to the hGHbp. These observations suggested that each segment substitution did not cause global misfolding of the molecule because otherwise none of these conformation-sensitive probes should have bound.

A higher resolution functional analysis called alanine scanning was then applied (24; for a review, see 73). In this approach, all side chains larger than alanine that were within or closely flanking the homologue-scan epitope were converted individually to alanine. From some 62 single alanine replacements, about 20 side chains were identified in which alanine replacements caused a twofold or greater decrease in binding affinity. These residues were confined to the N-terminal portion of helix 1, the loop before helix 2 extending from residues 54–68, and the C-terminal portion of helix 4 (Figure 4). Generally these side chains extend from the same side of hGH and form a highly discontinuous epitope, which explains why peptide fragments are not hormone mimics. The epitope consists of both hydrophobic and hydrophilic side-chains, of which basic groups dominate. This latter observation complements an analysis on the hGHbp showing that acidic side chains from the receptor are functionally important in binding (6).

Homologue and alanine scanning of the binding site on hGH for the extracellular domain of the prolactin receptor (hPRLbp) showed that while the two sites overlapped they were not identical (25), thus indicating that hGH bound somewhat differently to the hGHbp compared with the hPRLbp. In one striking example, the hGH mutant E174A improved binding to the hGHbp four- to fivefold while causing a 350-fold decrease

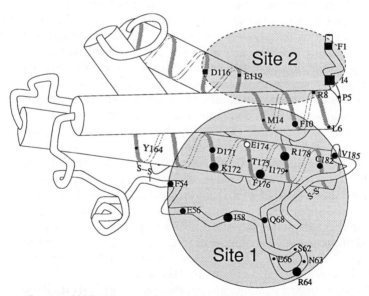

*Figure 4* Receptor binding sites on hGH, as mapped by homologue (21) and alanine (23, 24) scanning. Binding determinants are marked with circles (site 1) or squares (site 2). The size of the symbol is proportional to the decrease in receptor binding observed when the residue was changed to alanine, larger symbols corresponding to greater reductions. The open circle indicates an increase in receptor binding affinity for the E174A mutant.

in binding to the hPRLbp. Also shown was that Glu174 along with His18 and His21 coordinate one $Zn^{2+}$ when hGH forms a tight complex with the prolactin receptor (19). In addition, hGH is believed to be stored in the pituitary as a weak dimeric complex, $(Zn^{2+} \cdot hGH)_2$, and this dimerization requires these same zinc ligands (20). Thus, some of the residues in hGH are involved in binding to both receptors while others are specific for one and can even be detrimental to binding of the other. This analysis allowed the production of receptor-selective analogues (25) that are discussed below.

In addition to providing information about residues in the hormone that modulate binding affinity in these receptors, the high resolution functional information was used to recruit weak or nonbinding homologues of hGH, namely hPL (53) and hPRL (19), respectively, to bind tightly to the hGHbp. Surprisingly, these tight-binding analogues were inactive in rat weight-gain assays for hGH (R. Clark, D. Mortensen, M. Cronin, B. Cunningham, H. Lowman & J. Wells, unpublished results). This suggested that receptor activation required more than simple binding at this site.

Hormone-induced receptor oligomerization has been suggested as a general mechanism for activation of tyrosine kinase single-transmembrane receptors (for reviews, see 13, 68). While hGH does enhance chemical cross-linking of the hGHbp (G. Fuh & J. Wells, unpublished results), establishing the stoichiometry of the complex is difficult because of incomplete yields, multiple cross-linked products, and the possibility that cross-linking is dominated by the chemical reactivity of groups rather than the actual concentration of oligomers in solution. Subsequently, a series of crystallization and solution biophysical experiments established that hGH forms a 1:2 complex with the hGHbp. For example, when crystals of the complex were dissociated and components separated by high-performance liquid chromatography (HPLC), there were two molecules of hGHbp per hGH (69). Gel filtration studies of various mixtures of hGH and the hGHbp showed that hGH forms a 1:2 complex with the hGHbp (23). This result was supported by titration calorimetry experiments that showed that the heat of binding saturates at half an equivalent of hGH per hGHbp. In addition, recruited analogues of hPL and hPRL were biologically inactive and only formed 1:1 complexes with the hGHbp. This observation linked receptor dimerization and hGH activity.

The experiments demonstrating formation of the $hGH \cdot (hGHbp)_2$ complex were in apparent conflict with the 1:1 stoichiometry assessed by Scatchard analysis from competitive displacement and immunoprecipitation of the complex. This contradiction was resolved by the determination that the Mab5 antibody used in the immunoprecipitation studies blocked binding of a second hGHbp to hGH (23). Additional Mabs to the hGHbp were produced subsequently that allowed two molecules of the hGHbp to bind to hGH. However, the mutational analysis conducted with the Mab5 immunoprecipitation assay (21) identified only one of two functional sites on hGH for the hGHbp. This finding was fortuitous because dissecting two sites at once by mutational analysis would have been a much more difficult task.

A fluorescence assay was developed (23) to follow the dimerization of the hGHbp by hGH in solution. The hGHbp was specifically labeled with fluorescein-iodoacetamide on an engineered free thiol. Addition of hGH caused homoquenching of the fluorescein as the hGHbp dimerized (23), thus allowing sensitive detection of the hGH·(hGHbp)$_2$ complex without the potential artifacts associated with antibody precipitation. Using this assay, a series of homologue- and alanine-scanning experiments identified functional determinants for a second receptor-binding site on hGH (Figure 4). The site was adjacent to site 1, and was located in the amino-terminal region and the central portion of helix 3 (Figure 4).

## Crystallographic Analysis

After the mutagenesis experiments, the crystal structure of the hGH·(hGHbp)$_2$ complex was determined (28) (Figure 5). As was shown earlier (23, 69), the hormone binds two copies of the hGHbp. In addition, the structure revealed a significant contact interface between the two hGHbps. The bound hGHbps in the complex are related by approximate two-fold symmetry, therefore they contribute essentially the same regions to the interfaces with the hormone. Moreover, the interface between the two receptors involves the same part of their structure, namely the three-stranded sheet of their C-terminal domain. Note that, in the hormone binding sites, the symmetry holds up even at the individual amino acid level. Many of the residues contributed by the receptors to the interfaces with the hormone are present in both binding interfaces, even though the hormone itself has no internal symmetry or sequence duplication (28). For example, solvent accessibility calculations showed that in each receptor W104 has considerably more contact surface with the hormone than any other receptor residue, consistent with the observation that mutation of this residue to alanine virtually abolishes binding to the hormone (6). Other residues from the N-terminal domain that have a large contact area with the hormone include R43, P106, and E127, while each receptor also uses K167, W169, and N218 from the C-terminal domain (28). Whereas the receptors use only a few amino acids for interaction with hGH, the number of residues making up the binding sites on the hormone is relatively large (Figures 4, 6). Site 1 consists of residues on helix 1, the short segments of helix in the first long crossover connection, and a large part of one face

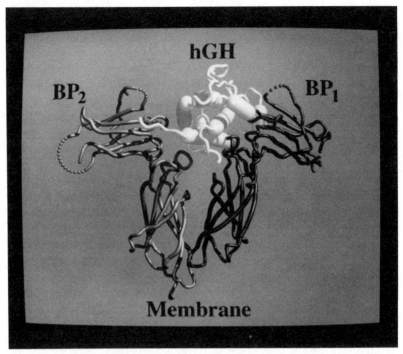

*Figure 5* The structure of the hGH·(hGHbp)$_2$ complex (28). The hormone is in white, the receptor binding to hGH site 1 is labeled BP1, and the receptor bound to hGH site 2 is labeled BP2.

of helix 4 (about 24 residues total). The second binding site is formed by the N terminus, the beginning of helix 1, and residues on helix 3 (for a total of 13 amino acids). The surfaces of these binding sites are characterized by pockets and clefts in which the receptor residues fit tightly.

## FUNCTIONAL AND STRUCTURAL EPITOPES ON hGH OVERLAP

For the purpose of discussion, we refer to the residues at the contact interface as a structural epitope, and to those residues substituted with alanine that cause greater than twofold reduction in binding affinity as a functional epitope. We find that for both site 1 and site 2 the functional and structural epitopes overlap but do not superimpose (Figure 6). The structural epitope for site 1 includes both the functional epitope for the hGHbp as well as that for the prolactin receptor (25). Generally, for both sites 1 and 2, the functional epitopes for the hGHbps are smaller than the

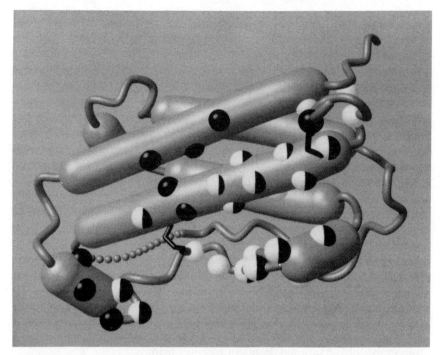

*Figure 6* Structural and functional site 1 epitope on hGH. Black dots depict residues with significant (>20 Å$^2$) contact area with a receptor, whereas white dots denote residues that result in twofold decreased receptor binding when changed to alanine. The N terminus is at the top right.

corresponding structural epitopes and reveal that some of the contacts are functionally silent or even deleterious. Almost all of the silent or deleterious contacts are hydrophilic or charged side chains. In site 1, these include residues in helix 1 (H18 and 21) and helix 4 (R167, K168, E174); for site 2 these include residues in helix 1 (N12, R16, and R19) and helix 3 (N109). Functionally silent hydrophilic contacts may result from the fact that desolvation of charged and hydrogen-bonding groups on the hormone is energetically costly and may not be offset by more favorable complementary interactions at the hormone-receptor interface. These hydrophilic side chains may nevertheless be preserved to ensure binding specificity, because for other binding sites such interactions may be much more deleterious than for the hGHbp. Note that we refer to a functional interaction as impacting overall binding affinity; it may however be possible to detect these silent contacts through other thermodynamic and/or kinetic measurements.

Most of the side chains that compose the functional epitopes on hGH for the hGHbp are seen from the crystal structure of the complex to be in direct van der Waals contact with the receptor; however, some are partially or completely buried (notably F10, M14, F54, I58, and F176). Alanine substitutions at any of these large hydrophobic side chains do not abolish binding of most Mabs to hGH, suggesting the variants are not grossly misfolded (21, 24, 44). Nonetheless, removal of these buried side chains must cause local structural changes that indirectly affect contact residues. This observation demonstrates that binding affinity can be changed by altering residues that indirectly affect the binding interface.

The concept that structural epitopes are considerably larger than functional or energetic epitopes has been suggested based upon theoretical considerations of antibody-antigen complexes (58). In fact, alanine-scanning of epitopes on hGH for binding of 21 different anti-hGH Mabs shows that on average only about 3 side chains dominate the binding interaction, and only 4 more contribute factors ranging from 2 to 20 to the binding affinity (44). This compares with an average of about 15 different side chains from the protein antigen that make van der Waals contacts in typical antibody-antigen complexes (for a review, see 27). These studies emphasize the need for both structural and functional analyses if we are to fully understand the chemistry of the noncovalent interactions so important for the rational design of receptor ligands.

# IMPLICATIONS FOR OTHER HEMATOPOIETINS

## Hematopoietic Hormones

With respect to other hormones in the superfamily, the recent determination of the three-dimensional structures of GM-CSF and IL-4 has made possible a structural interpretation of previous mutational findings that is consistent with the evidence from the hGH example. For IL-4, it was proposed that the molecule contains two receptor-binding sites, one involving the surface formed by helices 1 and 3; the other formed by the first $\beta$-strand, the end of the crossover connection between helices 1 and 2, and part of the surface of helix 4 (62). For GM-CSF, some of the residues implicated in receptor binding form a patch on helices 1 and 3, which may therefore represent a receptor-binding site (29). However, analysis of the mutational data allow two more regions to be identified, with the resulting surface encircling the molecule at one end (70). It is tempting to speculate that GM-CSF also contains two receptor binding sites. Here again, one site may consist of the exposed face of helices 1 and 3, the other of the accessible surface of helix 4 together with parts of the first crossover connection. A similar story now appears to be true for IL-

2: the binding site for the low affinity p55 α-chain is formed by residues at one end of the long crossover connections, while residues interacting with the β- and γ-chains are clustered on adjacent helices 1 and 4 (11). Finally for IL-3, even though a three-dimensional structure is not yet available, the mutational evidence is again consistent with the postulation that two binding sites reside on opposing sides of a four-helix bundle (49–51).

## Extracellular Receptor Domains

The evidence thus far available indicates that two topologically conserved receptor-binding sites are a common theme throughout the hematopoietins. Whereas hGH uses these two sites to bind two copies of the same receptor, in many other cases such as IL-2, IL-3, GM-CSF, and others, the equivalent segments may form binding interfaces for two different receptor subunits. Structure determinations of more ligand-receptor complexes are needed to shed further light on the similarities and differences in the superfamily.

Mapping of regions of hematopoietic receptors that are involved in ligand binding include studies of the G-CSF receptor and the prolactin (PRL) receptor. The extracellular domain of the G-CSF receptor consists of an immunoglobulin-like domain, a domain homologous to the hematopoietic superfamily, and three fibronectin type III domains. Consistent with the results for the growth hormone case, mutational analyses of this receptor (35) indicate that only the hematopoietic domain is essential for ligand binding. Furthermore, deletion studies on this domain showed that its C-terminal half could be removed without abolishing ligand binding completely, whereas the N-terminal half was indispensable (35). The prolactin receptor binds not only prolactin, but in the presence of zinc, also binds hGH. Mutational analysis of the hPRL receptor shows that one of the zinc ligands in the hGH-hPRL receptor complex is His188 of the hPRL receptor (19). His188 aligns with Asn218 of the hGH receptor, which is a contact residue for binding hGH via either site 1 or site 2 as seen in the crystal structure of the complex (28). This alignment strongly suggests that hGH binds to an analogous region in the prolactin receptor. Furthermore, an assessment of the importance of residues 12–68 of the extracellular domain of the PRL receptor indicated that R13, D16, and E18 of the first disulfide loop, and more generally the region between the two disulfide bonds, are key determinants for PRL binding specificity (63). This is in agreement with the structural (28) and mutational (6) results for the hGH-hGHbp complex, where functional and structural determinants are located in the equivalent disulfide loop. However, the most important binding determinants of the hGHbp were found farther from the N terminus (6, 28).

## MECHANISM(S) FOR ACTIVATION OF THE GROWTH-HORMONE RECEPTOR

Clustering of different receptor subunits is required for signal transduction of several of the hematopoietins. However, in cases such as growth hormone, prolactin, IL–4, and erythropoietin, only a single receptor type may exist. In these cases, hormone-induced receptor homodimerization may cause signal transduction. Addition of excess hGH to the hGHbp was shown to induce dissociation of the dimeric hGH·(hGHbp)$_2$ complex into a monomeric hGH·hGHbp complex, suggesting that the binding sites on the receptor for hGH overlapped each other (23). Indeed, the structure of the complex revealed the striking fact that virtually all the residues on the first receptor for binding site 1 on hGH were identical to the set used by the second receptor to bind site 2, even though sites 1 and 2 on hGH are very different (28).

Further studies suggested that binding of hGH to the hGHbp occurred sequentially; here hGH bound a first molecule of the hGHbp via site 1, and then a second molecule of hGHbp via site 2 (Figure 7). For example, for hGH to induce dissociation of the hGH·(hGHbp)$_2$ complex, only a functional site 1 is required (23). Moreover, mutants of hGH with a nonfunctional site 2 could still form 1:1 complexes with the hGHbp, whereas those containing a nonfunctional site 1 could not (23, 33). The structure of the hGH·(hGHbp)$_2$ complex provided additional support for this mechanism (28) because the structural epitope at site 1 covers $\sim 1230$ Å$^2$, whereas that for site 2 covers only $\sim 800$ Å$^2$. The smaller area of burial for site 2 is compensated for by an additional $\sim 500$ Å$^2$ of buried surface area where the receptors meet. Thus, site 2 for the second receptor is fully created only after the first receptor has bound to site 1 on hGH.

Additional support for the sequential dimerization mechanism of hGH binding was provided by cell-based assays of hGH mutants (33). Nagata and coworkers (35) created a cell line containing the cloned murine G-CSF receptor that is induced to proliferate in response to G-CSF, a member of the hematopoietic superfamily. The G-CSF binding portion of this receptor was replaced with the hGHbp, thus creating a cell line that proliferated in the presence of hGH but not G-CSF (33). The dimerization mechanism for hGH (Figure 7) predicted that excess hGH would actually antagonize cell growth, and this was in fact observed for the first time. More recently, others have also found that hGH can self-antagonize various cell-based assays (65; P. DeMeytes, personal communication). Moreover, many antibodies to the hGHbp were potent agonists, whereas their corresponding Fab fragments were not. Finally, mutants of hGH that are nonfunctional in site 1 are neither good agonists nor antagonists. In

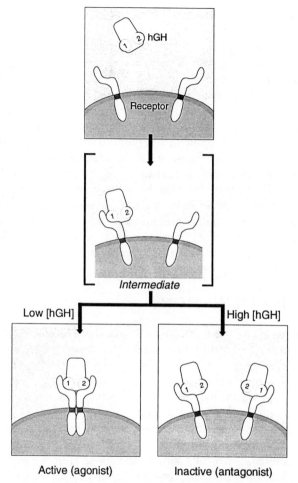

*Figure 7* Mechanism of activation of the human growth hormone receptor. Binding occurs via an intermediate 1:1 complex, in which a receptor is bound to hGH site 1. An active signal transduction complex is formed when a second receptor binds to hGH site 2. At high hormone concentrations, hGH can antagonize by taking up all receptors with its site 1.

contrast, variants that are nonfunctional in site 2 are poor agonists but potent hGH antagonists in the hybrid receptor cell–based assay (33) and in transgenic mice (16).

A similar dimerization mechanism likely operates for the prolactin receptor. For example, some bivalent rat prolactin-receptor antibodies can weakly stimulate rat Nb2-cell proliferation, whereas their monovalent Fab fragments cannot (31). Recent experiments on Nb2 cells showed that hGH acts on this prolactin receptor in a fashion very similar to the way it acts on the hGH receptor, that is by stimulating proliferation at high concentrations (74). Moreover, the proliferation activity is modulated by the presence of $Zn^{2+}$, consistent with the role of $Zn^{2+}$ for binding of hGH to the prolactin receptor. As already pointed out, the epitopes for hGH on the PRL receptor and on the GH receptor overlap. Finally, there is circumstantial evidence for receptor dimerization as the mechanism of activation of the erythropoietin receptor. A constitutively active mutant of the EPO receptor was found to have a single point mutation, in which R129 was changed to cysteine, resulting in disulfide cross-linked dimers (72). This result is consistent with the structure of the hGH-hGHbp complex: the analogous position in hGHbp for residue 129 of the erythropoietin receptor immediately precedes a series of residues involved in the receptor-receptor interface. Taken together, these results suggest that the mechanisms for activation of these receptors are similar to that of the growth hormone receptor.

## FUTURE APPLICATIONS

Our understanding of binding and activation of the hematopoietic receptors has increased tremendously over the past few years. How can this new information be put to use to design hematopoietins with altered and hopefully improved pharmaceutical properties? Many of these hormones have pleiotropic effects that result from binding to more than one receptor. For example, hGH exhibits a variety of pharmacologic effects (for reviews, see 15, 42) and is known to bind to the hGH and hPRL receptors and possibly others. When the binding epitopes in site 1 were mapped using alanine scanning, the results showed that these epitopes overlapped, but were not identical (25). Receptor-selective analogues were produced that bound tightly to either or neither receptor so that they would have receptor-selective binding profiles. Such analogues were used to assign specific receptor-binding epitopes with a given pharmacologic effect. For example, hGH binds and activates macrophages (30). Using the receptor-selective analogues, Fu et al (32) determined that the principal functional epitope for this effect results from the prolactin-binding epitope on hGH. Thus, it should be possible to make hematopoietin variants with narrower pharmacological profiles in cases where multiple receptors are used with nonidentical binding sites.

Conversely, broadening the specificity of hematopoietins should be possible. For example, by incorporating functionally important site 1 residues from hGH into hPRL (19) or hPL (53), it was possible to engineer variants that bound tightly to the hGH receptor and maintained their ability to bind their parent receptors. These recruited analogues lacked residues important for binding to site 2 and were therefore very important in separating binding to site 1 from receptor dimerization resulting from binding to site 2. However, it is possible that they could be rendered functionally active by making the appropriate changes in site 2 determinants.

The mechanistic information, that the hGH receptor is activated by sequential dimerization, is useful for the design of hormone antagonists. For example, engineering variants of hGH that are blocked in their ability to bind to site 2 but capable of binding tightly at site 1 produced potent antagonists to the hGH receptor (33). Further improvements in binding to site 1 could be made by monovalent phage display (52; H. Lowman & J. Wells, unpublished results), which would make these analogues even better antagonists. Such analogues may be useful in the treatment of acromegaly, a disease caused by excess secretion of hGH. An analogous strategy may be applied to the other hematopoietin family members when producing antagonists of their action would be desirable.

Finally, the high resolution structural and functional data base provides a necessary foundation for rational small-molecule design. There is good news and bad news from this story. The bad news is the binding epitopes for sites 1 and 2 are very large, and to build a small molecule that can mimic these is difficult to imagine, much less achieve. Moreover, the fact that some structural contacts are functionally silent needs to be better understood. On the other hand, the functional epitopes are smaller than the structural epitopes, suggesting that only a few key contacts are necessary for tight binding.

The hematopoietic hormone family is of great scientific and commercial interest. (In fact three of the members of this family, hGH, G-CSF, and EPO, are important protein pharmaceutics). Clearly, detailed structural and functional analyses will continue to elaborate the fundamental and perhaps common mechanisms for binding and activation of the receptors in this superfamily.

ACKNOWLEDGMENTS

The authors acknowledge Dr. T. Hynes for making Figures 5 and 6, Mr. W. Anstine for Figures 1a, 2, 3a, and 7, and Brian Cunningham, Germaine Fuh, Mark Ultsch, Henry Lowman, and Tony Kossiakoff for many stimulating discussions and their support.

## Literature Cited

1. Abdel-Meguid, S. S., Shieh, H.-S., Smith, W. W., Dayringer, H. E., Violand, B. N., Bentle, L. A. 1987. *Proc. Natl. Acad. Sci. USA* 84: 6434–37
2. Alber, T., Sun, D.-P., Nye, J. A., Muchmore, D. C., Matthews, B. W. 1987. *Biochemistry* 26: 3754–58
3. Aubert, M. L., Bewley, T. A., Grumbach, M. M., Kaplan, S. L., Li, C. H. 1986. *Int. J. Peptide Protein Res.* 28: 45–57
4. Barnard, R., Bundesen, P. G., Rylatt, D. B., Waters, M. J. 1984. *Endocrinology* 115: 1805–13
5. Baron, M., Main, A. L., Driscoll, P. C., Mardon, H. J., Boyd, J., Campbell, I. D. 1992. *Biochemistry* 31: 2068–73
6. Bass, S. H., Mulkerrin, M. G., Wells, J. A. 1991. *Proc. Natl. Acad. Sci. USA* 88: 4498–4502
7. Baumann, G., Stolar, M. W., Amburn, K., Barsano, C. P., DeVries, B. C. 1986. *J. Clin. Endocrinol. Metab.* 62: 134–41
8. Bazan, J. F. 1989. *Biochem. Biophys. Res. Commun.* 164: 788–95
9. Bazan, J. F. 1990. *Immunol. Today* 11: 350–54
10. Bazan, J. F. 1990. *Proc. Natl. Acad. Sci. USA* 87: 6934–38
11. Bazan, J. F., McKay, D. B. 1992. *Science* 257: 410–13
12. Bernstein, F., C., Koetzle, T. F., Williams, G. J. B., Meyer, E. F., Brice, M. D., et al. 1977. *J. Mol. Biol.* 112: 535–42
13. Bormann, B. J., Engelman, D. M. 1992. *Annu. Rev. Biophys. Biomol. Struct.* 21: 223–66
14. Chang, C. N., Rey, M., Bochner, B., Heyneker, H., Gray, G. 1987. *Gene* 55: 189–96
15. Chawla, R. K., Parks, J. S., Rudman, D. 1983. *Annu. Rev. Med.* 34: 519–47
16. Chen, W. Y., Wright, D. C., Mehta, B. V., Wagner, T. E., Kopchick, J. J. 1991. *Mol. Endocrinol.* 1845–51
17. Chene, N., Martal, J., De la Llosa, P., Charrier, J. 1989. *Reprod. Nutr. Dev.* 29: 1–25
18. Chiba, T., Amanuma, H., Todokoro, K. 1992. *Biochem. Biophys. Res. Commun.* 184: 485–90
19. Cunningham, B. C., Bass, S., Fuh, G., Wells, J. A. 1990. *Science* 250: 1709–12
20. Cunningham, B. C., Henner, D. J., Wells, J. A. 1990. *Science* 247: 1461–65
21. Cunningham, B. C., Jhurani, P., Ng, P., Wells, J. A. 1989. *Science* 243: 1330–36
22. Cunningham, B. C., Mulkerrin, M. G., Wells, J. A. 1991. *Science* 253: 545–48
23. Cunningham, B. C., Ultsch, M., de Vos, A. M., Mulkerrin, M. G., Clauser, K. R., Wells, J. A. 1991. *Science* 254: 821–25
24. Cunningham, B. C., Wells, J. A. 1989. *Science* 244: 1081–85
25. Cunningham, B. C., Wells, J. A. 1991. *Proc. Natl. Acad. Sci. USA* 88: 3407–11
26. D'Andrea, A. D., Fasman, G. D., Lodish, H. F. 1989. *Cell* 58: 1023–24
27. Davies, D. R., Sheriff, S., Padlan, E. A. 1990. *Annu. Rev. Biochem.* 59: 439–73
28. de Vos, A. M., Ultsch, M., Kossiakoff, A. A. 1992. *Science* 255: 306–12
29. Diederichs, K., Boone, T., Karplus, P. A. 1991. *Science* 254: 1779–82
30. Edwards, C. K., Ghiasuddin, S. M., Schepper, J. M., Yunger, L. M., Kelley, K. W. 1988. *Science* 239: 769–71
31. Elberg, G., Kelly, P. A., Djiane, J., Binder, L., Gertler, A. 1990. *J. Biol. Chem.* 265: 14770–76
32. Fu, Y.-K., Arkins, S., Fuh, G., Cunningham, B. C., Wells, J. A., et al. 1992. *J. Clin. Invest.* 89: 451–57
33. Fuh, G., Cunningham, B. C., Fukunaga, R., Nagata, S., Goeddel, D. V., Wells, J. A. 1992. *Science* 256: 1677–80
34. Fuh, G., Mulkerrin, M. G., Bass, S., McFarland, N., Brochier, M., et al. 1990. *J. Biol. Chem.* 265: 3111–15
35. Fukunaga, R., Ishizaka-Ikeda, E., Pan, C.-X., Seto, Y., Nagata, S. 1991. *EMBO J.* 10: 2855–64
36. Fukunaga, R., Ishizaka-Ikeda, E., Seto, Y., Nagata, S. 1990. *Cell* 61: 341–50
37. Gearing, D. P., King, J. A., Gough, N. M., Nicola, N. A. 1989. *EMBO J.* 8: 3667–76
38. Goodwin, R. G., Griend, D., Ziegler, S. F., Jerzy, R., Falk, B. A., et al. 1990. *Cell* 60: 941–51
39. Herrington, A. C., Ymer, S. I., Stevenson, J. L. 1986. *J. Clin. Invest.* 77: 1817–23
40. Holmgren, A., Brändén, C.-I. 1989. *Nature* 342: 248–51
41. Idzerda, R. L., March, C. J., Mosley, B., Lyman, S. D., Vanden Bos, T., et al. 1990. *J. Exp. Med.* 171: 861–73
42. Isaksson, O., Eclen, S., Jansson, J. O. 1985. *Annu. Rev. Physiol.* 47: 483–99
43. Itoh, N., Yonehara, S., Schreurs, J., Gorman, P. M., Maruyama, K., et al. 1990. *Science* 247: 324–27
44. Jin, L., Fendly, B. M., Wells, J. A. 1992. *J. Mol. Biol.* 226: 851–65
45. Kraulis, P. J. 1991. *J. Appl. Cryst.* 24: 946–50
46. Leahy, D. J., Hendrickson, W. A., Aukhil, I., Erickson, H. P. 1992. *Science* 257: 987–91
47. Leung, D. W., Spencer, S. A.,

Cachianes, G., Hammonds, G., Colins, C., et al. 1987. *Nature* 330: 537–43
48. Lewis, U. J. 1984. *Annu. Rev. Physiol.* 46: 33–42
49. Lokker, N. A., Movva, N. R., Strittmatter, U., Fagg, B., Zenke, G. 1991. *J. Biol. Chem.* 266: 10624–31
50. Lokker, N. A., Strittmatter, U., Steiner, C., Fagg, B., Graff, P., et al. 1991. *J. Immunol.* 146: 893–98
51. Lokker, N. A., Zenke, G., Strittmatter, U., Fagg, B., Movva, N. R. 1991. *EMBO J.* 10: 2125–31
52. Lowman, H. B., Bass, S. H., Simpson, N., Wells, J. A. 1991. *Biochemistry* 30: 10832–38
53. Lowman, H. B., Cunningham, B. C., Wells, J. A. 1991. *J. Biol. Chem.* 266: 10982–88
54. Metcalf, D. 1989. *Nature* 339: 27–30
55. Miyazaki, T., Maruyama, M., Yamada, G., Hatakeyama, M., Taniguchi, T. 1991. *EMBO J.* 10: 3191–97
56. Nicola, N. A. 1989. *Annu. Rev. Biochem.* 58: 45–77
57. Nicoll, C. S., Mayer, G. L., Russell, S. M. 1986. *Endocr. Rev.* 7: 169–203
58. Novotny, J., Bruccoleri, R. E., Saul, F. A. 1989. *Biochemistry* 28: 4735–49
59. Pakula, A. A., Young, V. B., Sauer, R. T. 1986. *Proc. Natl. Acad. Sci. USA* 83: 8829–33
60. Pandit, J., Bohm, A., Jancarik, J., Halenbeck, R., Koths, K., Kim, S.-H. 1992. *Science* 257: 1358–62
61. Patthy, L. 1990. *Cell* 61: 13–14
62. Powers, R., Garrett, D. S., March, C. J., Frieden, E. A., Gronenborn, A. M., Clore, G. M. 1992. *Science* 256: 1673–77
63. Rozakis-Adcock, M., Kelly, P. A. 1992. *J. Biol. Chem.* 267: 7428–33
64. Ryu, S.-E., Kwong, P. D., Truneh, A., Porter, T. G., Arthos, J., et al. 1990. *Nature* 348: 419–26
65. Silva, C. M., Weber, M. J., Thorner, M. O. 1992. *Endocrinology*. In press
66. Smith, L. J., Redfield, C., Boyd, J., Lawrence, G. M. P., Edwards, R. G., et al. 1992. *J. Mol. Biol.* 224: 899–904
67. Spencer, S. A., Hammonds, R. G., Henzel, W. J., Rodriguez, H., Waters, M. J., Wood, W. I. 1988. *J. Biol. Chem.* 263: 7862–67
68. Ullrich, A., Schlessinger, J. 1990. *Cell* 61: 203–12
69. Ultsch, M., de Vos, A. M., Kossiakoff, A. A. 1991. *J. Mol. Biol.* 222: 865–68
70. Walter, M. R., Cook, W. J., Ealick, S. E., Nagabhushan, T. L., Trotta, P. P., Bugg, C. E. 1992. *J. Mol. Biol.* 224: 1075–85
71. Wang, J., Yan, Y., Garrett, T. P., Liu, J., Rodgers, D. W., et al. 1990. *Nature* 348: 411–18
72. Watowich, S. S., Yoshimura, A., Longmore, G. D., Hilton, D. J., Yoshimura, Y., Lodisch, H. F. 1992. *Proc. Natl. Acad. Sci. USA* 89: 2140–44
73. Wells, J. A. 1991. *Methods Enzymol.* 202: 390–411
74. Fuh, G., Colosi, P., Wood, W. I., Wells, I. A. 1993. *J. Biol. Chem.* 268: In press

# REALISTIC SIMULATIONS OF NATIVE-PROTEIN DYNAMICS IN SOLUTION AND BEYOND

## V. Daggett[1] and M. Levitt

Beckman Laboratories for Structural Biology, Department of Cell Biology, Stanford University Medical School, Stanford, California 94305–5400

KEY WORDS: molecular dynamics, macromolecular solution simulations, water-protein interactions, protein unfolding

## CONTENTS

| | |
|---|---|
| PERSPECTIVES AND OVERVIEW | 353 |
| HOW ARE SIMULATIONS PERFORMED? | 355 |
|     The Potential Function | 355 |
|     Energy Minimization | 357 |
|     Molecular Dynamics | 357 |
|     Special Procedures and Problems | 358 |
| WATER MODELS | 362 |
| NATIVE PROTEIN–WATER SIMULATIONS | 364 |
|     Water Behavior | 364 |
|     Protein Behavior | 365 |
|     Protein-Water Interactions | 373 |
| NON-NATIVE PROTEIN–WATER SIMULATIONS | 375 |
|     High Temperature | 375 |
|     High Pressure | 378 |
| CONCLUSIONS | 378 |

## PERSPECTIVES AND OVERVIEW

Motion is clearly important to biological function. Proteins are not frozen, static structures but instead exhibit a variety of complex motions even in

[1] Current address: Department of Medicinal Chemistry BG–20, University of Washington, Seattle, Washington 98195

1056–8700/93/0610–0353$02.00

their resting states. The magnitude of the motion experienced by the protein can increase dramatically during binding, catalytic, and signal-transduction events. In fact, considerable motion can occur even within a protein crystal; substrates and other ligands can be diffused into binding sites without disrupting the crystal lattice. In some cases, a direct route for ligand penetration of the protein does not exist, and investigators have postulated that residues may move many Ångstroms to accommodate the ligand [at least 3.5 Å in the case of T4 lysozyme (26)]. A recent study also suggests that normal room-temperature movements of proteins in solution can span ~15 Å (16). Whether this observation can be generalized to other proteins remains to be seen. However, motion is an incontrovertible consequence of a room-temperature existence, in which the kinetic energy per atom is almost 1 kcal/mol and the thermal velocities can be several Ångstroms per picosecond.

Crystal structures represent average structures, but not all residues have single preferred conformations even when constrained in a low-solvent crystal environment. This phenomenon is even more evident in solution. Two-dimensional nuclear magnetic resonance (NMR) experiments have been instrumental in providing a view of proteins in solution, and in most cases, the solution and crystal structures are similar. Equally important is the absence of interactions as measured by the nuclear Overhauser effect (NOE) with NMR. A lack of NOEs in an area of the protein where interactions are expected based on the crystal structure indicates that the residues have adopted different conformations in the two environments and/or that they are very mobile, and actually many different conformations are sampled. Studies have shown that proteins (or portions of a protein) can adopt different structures in solution and in the crystalline environment (72).

Experimental studies generally only yield limited amounts of information on the actual molecular details of the motion experienced by a protein. Therefore, theoretical studies are needed to better elucidate these events. To be most effective, experimental and theoretical studies should be linked. In this way, the experimental results are used to test the validity of the simulations, and when simulations accurately mimic reality, they can be used predictively.

Because many atoms are involved in the protein and solvent, in the past investigators generally chose to perform simulations in vacuo. Although corrections can be made to such simulations to screen out the exaggerated electrostatic interactions, these simulations should be viewed cautiously, and every attempt should be made to compare to experiment because of the possible artifacts introduced by neglecting solvent. This review focuses on molecular-dynamics simulations of proteins that aimed for realism by including a fairly large number of water molecules. Even with the inclusion of explicit solvent molecules, only some simulations appear to succeed in realistically modeling protein dynamics. We have attempted to outline the criteria for judging simulations: e.g. what makes a simulation good? After establishing that native proteins at room temperature can be modeled realistically, we look at studies of a protein's response to perturbing conditions such as high temperature and pressure and present recent simulations in this area. Our focus here is narrow—numerous reviews have examined the general area of molecular dynamics and its relation to experiment. In this regard, the reader is referred to recent reviews by van Gunsteren & Mark (76) and Teeter (70) for a discussion of protein-water interactions.

## HOW ARE SIMULATIONS PERFORMED?

### The Potential Function

In order to perform molecular-dynamics simulations, one needs a mathematical function, or force field, to describe the system. The force fields are empirically derived and describe the potential energy of the system as a function of the positions of the atoms. The parameters for these equations are derived primarily from the results of ab initio quantum mechanics, spectroscopic data, and crystallographic data. The force field is calibrated by fine tuning these parameters to reproduce structures and energy trends for relevant model compounds. The goal is to derive a force field that is transferable to a variety of systems, for example small organic molecules, proteins, carbohydrates, lipids, RNA, and DNA. One can often achieve good agreement with experiment because the functions are parameterized to reproduce the experimental data and therefore include the effects of any omitted terms that would make the approach more theoretically rigorous.

An empirical potential function like the one shown below is generally employed:

$$U = \sum_{\substack{\text{Bond} \\ \text{lengths}}} K_{bi}(b_i - b_{0i})^2 + \sum_{\substack{\text{Bond} \\ \text{angles}}} K_{\theta i}(\theta_i - \theta_{0i})^2 +$$

$$\sum_{\substack{\text{Torsion} \\ \text{angles}}} K_{\varphi i} \{1 - \cos[n_i(\varphi_i - \varphi_{0i})]\} +$$

$$\sum_{\substack{\text{Nonbonded pairs } i,j \\ \text{closer than cut-off}}} e_{0ij}(r_{0ij}/r_{ij})^{12} - 2e_{0ij}(r_{0ij}/r_{ij})^6 + \sum_{\substack{\text{Partial charges} \\ \text{closer than cut-off}}} q_i q_j / r_{ij}.$$

The first three terms in the potential-energy function describe the bonded interaction acting between atoms that are separated by one, two, or three bonds. Bond-length stretching is generally represented by a quadratic potential in which the energy is minimal when the $i$th bond length, $b_i$, has the equilibrium value $b_{0i}$, and the stretching-force constant is $K_{bi}$. The second term describes bond-angle bending and takes the same form as the first. The third term describes dihedral- (or torsion-) angle twisting and is represented by one or more cosine functions, to give a barrier height $K_{\varphi i}$, a periodicity $n_i$ and a minimum value at $\varphi_{0i}$.

The fourth term in the potential energy function describes the van der Waals interactions acting between atoms that are separated by three or more bonds, while the fifth term describes the electrostatic interaction also acting between these same pairs of atoms. Because proteins have thousands of atoms, these two nonbonded terms are much more important than the three bonded terms. The van der Waals interactions are represented by the Lennard-Jones 12–6 potential, which has a short-range term, $e_{oij}(r_{oij}/r_{ij})^{12}$, for repulsion when atoms overlap, and a longer range term, $-2e_{oij}(r_{oij}/r_{ij})^6$, for the weak attraction that exists between all atoms ($r_{ij}$ is the separation of the atoms). The electrostatic terms are represented by the Coulomb potential, $q_i q_j/r_{ij}$, which can be both attractive or repulsive depending on the signs of the partial charge parameters, $q_i$, assigned to all atoms. In simulations that include explicit solvent, there is no need for any dielectric screening of the electrostatic terms, which is equivalent to setting $\varepsilon = 1$ in a screened Coulomb potential, $q_i q_j/\varepsilon r_{ij}$. This is convenient because it avoids uncertainties associated with the value and distance dependence of the dielectric screening. The range of the nonbonded terms is important in simulations of macromolecules in solution and is discussed in more detail below.

The variables in these potential energy terms are the bond lengths, $b_i$, the bond angles, $\theta_i$, the torsion angles, $\varphi_i$, and the interatomic distances, $r_{ij}$, which in turn depend on the Cartesian coordinates of the atoms. The other quantities ($K_{bi}$, $b_{0i}$, $K_{\theta i}$, $\theta_{0i}$, etc) are fixed energy parameters with values that depend on the type of interaction (e.g. a C-C bond, an N-C-O angle). The values of the energy parameters are determined from experimental data on small molecules. Parameterization of bond lengths and bond angles is generally provided by X-ray crystallography for the equilibrium values and by infrared spectroscopy for the force constants. These sources also provide the parameters for torsion angles, but NMR can provide additional information on barrier heights. Parameters for van der Waals interactions are best determined from crystal-packing geometries (for $r_{oij}$) and sublimation energies (for $e_{oij}$). Atomic partial charges can also be determined from crystals but they are often determined from quantum mechanical calculations of electron distributions in small molecules.

The results of any simulation depend critically on the values of the energy parameters, and efforts are normally made to ensure that all values used have been calibrated against experimental data or quantum calculation. One should realize that the use of an empirical potential energy function is always an approximation to reality that can be expected to break down under certain conditions. Careful analysis of results and comparison with experimental properties where available is essential at all stages of simulation. More thorough discussions of the details of parametrization of force fields can be found elsewhere (30, 31, 45, 77).

## Energy Minimization

Energy minimization is accomplished by adjusting the positions of the atoms until the gradient of the potential energy is zero (e.g. $\partial U/\partial x_i = 0$). Several different methods are available for performing minimization, and the reader is referred to Allinger (6) and Levitt (44) for further information. Most of these methods require the gradient or first derivative of the potential in analytical form. For realistic simulations of proteins, minimization is no longer used, other than for relieving bad contacts in preparing a system for molecular dynamics. A minimized molecule is equivalent, at best, to a structure at 0 K, and as mentioned above, motion is critical to the function of proteins.

## Molecular Dynamics

Molecular dynamics calculates the motion of a molecule by generating the changes of the atomic coordinates as a function of time. These sets of related coordinates define a trajectory in conformational space. A static structure is used as the starting point for these calculations, usually the minimized crystal structure. The velocities of the atoms are slowly increased from zero to values corresponding to a specified temperature in a process known as temperature equilibration, which is part of a start-up procedure needed to ensure that the initial forces are reasonable. The atomic velocities, $v_i$, are related to the absolute temperature, $T$, through the total kinetic energy with $1/2 \Sigma m_i v_i^2 = 3/2 N k_b T$, where $m_i$ is mass of the atom, $k_b$ is the Boltzmann constant, and the sum is over $N$ atoms. Note that if the temperature doubles (say from 200 K to 400 K, which is $-73°C$ to 127°C), the atomic velocities only increase by a factor $2^{1/2}$ or 40%.

New positions and velocities of all the atoms are determined by solving the equations of motion using the old positions, the old velocities, and the accelerations. Accelerations are related to forces via Newton's second law ($F = ma$), and the forces are related to the gradients or first derivatives of the potential energy needed for energy minimization ($F = -\partial U/\partial x_i$) so that $a_i = -(1/m_i)\partial U/\partial x_i$. The new atomic positions, $x_i(t+\Delta t)$, and velocities, $v_i(t+\Delta t)$, are derived from elementary mechanics as $x_i(t+\Delta t) = x_i(t)$

$+v_i(t)\Delta t + 1/2 a_i(t)\Delta t^2$ and $v_i(t+\Delta t) = v_i(t) + a_i(t)\Delta t$. For this simple scheme to work, the time step, $\Delta t$, must be so small that positions (and forces) change very slightly with each step (less than 1/100 Å); typical values of $\Delta t$ range from 0.5 to 2 femtoseconds ($10^{-15}$ seconds!). In practice, the equations used to integrate equations of motions in protein simulations are slightly more complicated than those given above and use accelerations at earlier times (45).

Simulating for any reasonable time period requires many iterations of these equations: for example, simulating motion for 1 ns requires 1 million iterations with a 1-fs time step. Numerical stability becomes an issue in this iterative scheme in which a new value is calculated from an old value. After a million time steps, errors can build up and spoil the simulation. Fortunately, elementary mechanics provides a check of numerical stability in that the total energy of the system, potential plus kinetic, should remain constant. This conservation law applies whether or not the protein conformation is native, the system is well equilibrated, or the energy parameters are appropriate: it serves as a test of the accuracy with which equations of motion are solved. Given the importance of energy conservation as an independent check, it is a pity that many simulations do not attempt to conserve energy and instead adjust atomic velocities to artificially keep the total energy constant by using thermal-bath coupling (12) (see many entries with dynamics protocol B in Table 4, below). One reason for this manipulation may be that conserving energy places very stringent demands on the accuracy with which computer programs calculate the potential energy and its derivatives (the forces).

Molecular dynamic simulations provide a series of conformations that should represent the way the macromolecule moves at the specified temperature. The amount of data generated is enormous (a small protein in water produces half a million sets of Cartesian coordinates in a nanosecond, each describing the positions of about 10,000 atoms), and considerable skill is needed for data analysis.

## Special Procedures and Problems

So far, this review has dealt with the general nature of the potential and methods for changing conformation. There are a number of additional points that may seem to be minor details but can make the difference between a realistic and an unrealistic simulation.

INCLUDING HYDROGEN ATOMS Which atoms should be included? This question may seem surprising as the obvious answer is all the atoms of the protein, the water, the counterions, etc. Simulations often omit some or all of the hydrogen atoms on the protein. Instead "united atoms" are used where the properties of, say, a carbon in a methyl group are altered to account for the three hydrogen atoms (6, 45, 77). This approach saves significant amounts of computer time when doing calculations in vacuo. In solution, the saving is much smaller. All water models explicitly include the hydrogen atoms on the water, and omitting the much smaller number of hydrogen atoms on nonpolar protein groups saves very little. A more fundamental reason for including hydrogen atoms on protein molecules is that treatment of hydrogen bonds, which once used special directional potentials, now follows the work of Hagler et al (30) and utilizes the balance of van der Waals and electrostatic interactions with no additional terms. This method requires the explicit treatment of hydrogen atoms bonded to polar atoms. Nonpolar hydrogen atoms connected to carbon atoms may be just as important, as these atoms are best parametrized as carrying small positive partial charges balanced by compensating charges on the carbon atom (48). This distribution of charges gives rise to the aromatic-ring hydrogen bond that is of importance in protein/drug interactions (56). Many simulations restrict the motional freedom of the hydrogen atoms by restraining bond lengths by using the SHAKE algorithm introduced by Ryckaert & Bellemans (61) and popularized by van Gunsteren & Berendsen (74). However, the hydrogen atoms do not need to be restrained as energy conservation can be obtained with a large time step (2 fs) and fully flexible hydrogens (46, 49).

SIMULATING BULK SOLVENT   Bulk water consists of many water molecules; a single cell contains billions of water molecules. How can bulk water be simulated with the much fewer water molecules possible in a simulation? The answer is surprising: with suitable boundary conditions, as few as a hundred water molecules can reproduce the properties measured for bulk water. The pioneering work of Rahman & Stillinger (60) on water and the previous study of Rahman on liquid argon (59) used periodic boundaries with great success. In this method, a cubic (or rectangular) box of water molecules is treated as a periodic system to ensure that waters at the edges of the box are surrounded by as many neighboring waters as waters at the center of the box; a water leaving the box immediately enters again from the opposite face. Other periodic geometries are possible, the most interesting being a truncated octahedron (1, 75), but the periodic rectangular box has been used most frequently.

Computing under periodic boundary conditions can be tricky, especially if one tries to make the calculation as efficient as possible. Programs should be carefully tested. Fortunately, one can run a 10-ps simulation of 216 water molecules in a periodic box in less than an hour on a desktop workstation. Rahman & Stillinger used this system composed of 216 waters in 1971, but they could only accumulate a 5-ps simulation using the most powerful machines of that day. Even these short simulations can give reliable estimates of water properties known from experiment, which is more important for calibrating and error checking. Such experimental verification is an essential part of any study of protein dynamics in solution.

Using periodic boundary conditions raises a question: are nonbonded interactions to be calculated between all pairs of atoms in all the repeating boxes of the periodic structure? One can use two approaches in liquid simulations. One approach follows the original Rahman & Stillinger (60) scheme and truncates the nonbonded potential (both van der Waals and electrostatic) at a range less than half the size of the box. (With this range, no water molecule can interact with more than one copy of another water molecule; this phenomenon is known as minimum image.) The other approach deals with the system as a periodic crystal and includes all interactions [this can be done relatively efficiently using the Ewald summation (27)]. Currently, most work on proteins uses the former scheme. We discuss nonbonded truncation next.

TRUNCATING INTERACTIONS Truncating nonbonded interactions has advantages and disadvantages in simulations of proteins in solution. The two advantages are a dramatic increase in speed of computing and the elimination of spurious effects of dealing with a protein as a period crystal (albeit a very wet crystal) rather than one in solution. The two disadvantages are that numerical discontinuities caused by truncation can spoil accurate integration of the equations of motion and that electrostatic interactions between net charges are significant over very long distances.

The increase in computational speed is especially important in simulations of proteins in water that involve at least 5000 atoms. Without truncation, the number of nonbonded pairs for $N$ atoms is approximately $N^2/2$; with truncation, the number of pairs depends on the range but varies as $N$ not $N^2$. The desire to treat the protein-water system with the same minimum-image conditions used for most liquid simulations would limit the range to half the shortest side of the box. For a typical simulation of a small protein, this would give a range of 20 Å. With this range, many water molecules would interact with two copies of the protein, and a shorter cut-off is needed to ensure that only one copy of any water or protein molecule ever interacts. In practice, the cut-off is determined by the need to speed the simulation and varies from 6 to 12 Å.

Numerical discontinuities arise if the nonbonded energy is simply neglected when the pair of atoms are farther apart than the cut-off range. Investigators have long appreciated this, and many schemes are available for making the energy and forces go to zero smoothly as the separation gets close to the cut off (5, 14, 39, 46, 50, 58, 60).

Ideally, the smoothed, truncated potential should not generate any spurious forces near the cut-off range and should depend only on the separation of the particular pair of atoms. This is not easily achieved. For example, in the F3C potential used by Levitt & Sharon (49) and Levitt (46), the truncation was done by smoothing the total interaction of a neutral group of atoms (e.g. a water molecule or $C=O$ group). This method resulted in good energy conservation, but because the group truncation depends on the separation of the group of atoms, not just the particular pair of atoms, spurious forces can arise. A new version of the potential function that uses smooth-atom truncation was needed for simulations of nucleic acids (35) and is now used for all of our simulations. This form of smoothing is most like that used by van Belle et al (73) and that used in earlier simulations of pure liquids (5).

If the nonbonded potential is truncated properly, integration of the equations of motion will be accurate and the total energy will be conserved. Once the system has been equilibrated and reaches the target temperature, no further adjustment of the energy is needed even for simulations that last hundreds of picoseconds. Without proper truncation, the spurious forces at the cut-off distance lead to a slow temperature increase that must be corrected by removing energy from the system. Energy can be removed by simply rescaling the atomic velocities periodically to ensure that the total energy remains close to its target value (45). A better scheme is the coupling to a thermal bath introduced by Berendsen et al (12), in which the velocities are continually adjusted to keep the total energy fixed. Both methods are dangerous in that they provide symptomatic treatment of the problem caused by inaccurate truncation. While the energy is artificially conserved, other properties may be affected. For example, in simulations of liquid water, smooth group truncation and energy conservation gave diffusion constants that were half those obtained using the same potential with sharp group truncation and velocity rescaling (46). In some simulations, the instabilities caused by inappropriate truncation are reduced by increasing the masses of the hydrogen atoms by a factor of three (33). Such changes in mass should not affect thermodynamic properties (such as energy or structure) but will dramatically affect all dynamic properties.

At present, there is only limited information for estimating the errors associated with the neglect of electrostatic interactions that may be appreciable at long range. Nevertheless, Hirshberg's simulations of highly charged systems like the DNA double helix are very stable with short-range truncation of electrostatic interactions (35). As speed of computation continues to grow, computation of large solvated systems with different cut-off ranges will be possible. It is also worth pointing out that several methods significantly speed computation of long-range electrostatic interactions for large systems (11, 29, 62).

## WATER MODELS

Because the quality of any simulation of a solvated protein will depend on the quality of the water model in addition to the factors discussed above, a comparison of the different water models used routinely in protein simulations, TIP3P (36, 37), SPC (13), and F3C (46, 49; V. Daggett, M. Hirshberg, R. Sharon & M. Levitt, in preparation), is worthwhile. These water models are similar in terms of geometries and the nonbonded parameters (Table 1). One should also note that whereas the F3C model is flexible, the other two models are rigid, and in fact, the TIP3P model contains a third nonphysical bond between the two hydrogens of the water molecule. Even though the differences in the entries in Table 1 may appear minor, they yield very different dynamical and structural results (Table 2).

The potential energy per water molecule in the various simulations is in fairly good agreement with experiment (within 0.5 kcal/mol; Table 2).

**Table 1** Geometrical properties and nonbonded parameters for the water models commonly used in protein simulations[a]

| Model | $d$(OH) (Å) | $\theta$(HOH) (Degrees) | $q$(H) (Electrons) | $q$(O) (Electrons) | $r_0$ (Å) | $e_0$ (kcal/mol) | Reference |
|---|---|---|---|---|---|---|---|
| TIP3P | 0.957 | 104.520 | 0.417 | −0.834 | 3.1506 | 0.1521 | 36 |
| SPC | 1.000 | 109.467 | 0.410 | −0.820 | 3.1656 | 0.1554 | 13 |
| F3C | 1.000 | 109.470 | 0.410 | −0.820 | 3.4515 | 0.2200 | 46 |

[a] The equilibrium bond length is $d$(OH); the equilibrium bond angle is $\theta$(HOH); the partial charge on the hydrogen is $q$(H) and on the oxygen is $q$(O); the minimum separation in the van der Waals potential is at $r_0$ and has a depth of $e_0$.

**Table 2** Dynamical and structural properties exhibited by the water molecules during molecular dynamics[a]

| Model | $T$ (K) | $U_{pot}$ (kcal/mol) | $D$ (Å$^2$/ps) | $r1$ (Å) | $h1$ | $r2$ (Å) | $h2$ | $r3$ (Å) | $h3$ | Reference |
|---|---|---|---|---|---|---|---|---|---|---|
| TIP3P | 298 | −10.45 | 0.40 | 2.80 | 2.90 | — | — | — | — | 36 |
| SPC | 298 | −10.18 | 0.36 | 2.80 | 2.80 | 3.30 | 0.90 | 4.50 | 1.00 | 13 |
| F3C | 298 | −9.57 | 0.24 | 2.70 | 3.20 | 3.30 | 0.81 | 4.40 | 1.10 | In prep.[b] |
| Experiment | 298 | −9.90 | 0.23 | 2.90 | 3.00 | 3.30 | 0.73 | 4.40 | 1.10 | 46 |

[a] The headings $r1$, $h1$, $r2$, $h2$, $r3$, and $h3$ are positions and areas from the radial distribution function as depicted in Figure 1: $r$ represents the distances, and the heights of the peaks and valleys, $h$ represents the number of water molecules at that distance. The experimental values of the radial distribution function are taken from the more recent values of Soper & Phillips (65) rather than the original values of Narten & Levy (55).
[b] V. Daggett, M. Hirshberg, R. Sharon & M. Levitt, in preparation.

Considering the theoretical and experimental uncertainties, the SPC and F3C models best reproduce experimental results. The only water model to reproduce the experimental diffusion constant is the F3C model. This is a very important parameter because it is one of the few time-dependent properties that can be measured directly both experimentally and in simulations. In fact, how reasonable protein motions are in a particular simulation is unclear when the water motion is 60–75% too fast.

Not only is correctly modeling the water dynamics important, but the experimentally determined inherent water structure must also be reproduced. Here, reproduction of the radial distribution function determined from X-ray diffraction experiments is the aim. Figure 1 shows such a distribution of the interoxygen distances for the F3C water model. The first peak at 2.8 Å in any direction from a central water molecule ($r1$) corresponds to two hydrogen-bonded water molecules. All of the models give too many water molecules in this first shell, although F3C is the worst. The second peak ($r3$) relates to the tetrahedral structure of near-neighbors and is the distance between two water molecules hydrogen bonded to a third molecule. Local ordering of this sort is not observed in simpler liquids and is responsible for water's special properties at room temperature.

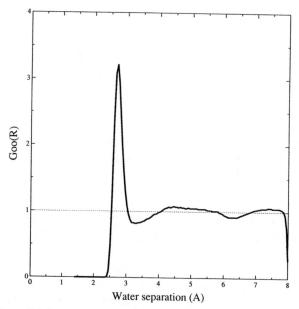

*Figure 1* The radial distribution function is given for a 40-ps simulation of a periodic box of 216 waters at 298 K using the F3C model.

Another peak appears around 7 Å (Figure 1), and in fact, evidence supports structure in liquid water to about 8 Å at room temperature (55). Therefore, for realistic simulations of the properties of water alone and when solvating proteins, it seems particularly important that a water model displays this behavior. The SPC and F3C models are in good agreement with experiment regarding water structure, while TIP3P does not show structure beyond approximately 3 Å.

Currently, much effort is devoted to better, more complicated water models incorporating polarizability (15, 66). These new models give improved results in some model systems, but whether they will do the same in protein simulations and whether the increased computational requirements are worth it remains to be seen. However, the F3C model is a good alternative and does not force the investigator to compromise on the dynamics and structural properties of water. The water self-diffusion constant is far too high with the SPC and TIP3P models. In addition, the TIP3P model does not reproduce the structural properties of water. One must remember that unrealistic dynamics behavior can be caused by a variety of effects, including not only the water model but the computational procedures outlined above. Thus, a test of the different water models using the same dynamics protocols and procedures under conditions of strict energy conservation would be interesting.

## NATIVE PROTEIN–WATER SIMULATIONS

Protein simulations in solution under normal native conditions have as their primary aim the realistic calculation of protein dynamics and structure in solution. Many such simulations have now been performed (summarized in Tables 3 and 4). First, we present those simulations for which data are available on the behavior of the bulk solvent. Then those same simulations, in addition to further studies, are discussed from the point of view of protein behavior. Finally, the details of the water-protein interactions are presented.

### Water Behavior

All of the simulations presented in Table 3, with the exception of trypsin (TRP) and ferrocytochrome $c$ (FCYT), were performed for a fairly long period of time and contain many water molecules, especially because all of the simulations made use of periodic boundary conditions. They should, then, provide a fair appraisal of the properties of bulk water in protein simulations. As mentioned above, the self-diffusion of water is a sensitive gauge of the accuracy of the degree of motion experienced in molecular dynamics. The experimental values for pure water at room temperature

**Table 3** Water motion in protein simulations[a]

| Protein | T (K) | Time (ps) | Water model | No. waters | Bulk diffusion constant (Å²/ps) | Reference |
|---|---|---|---|---|---|---|
| TRP | 300 | 28.8 | SPC | 4785 | 0.50 | 79 |
| PARV | 303 | 106 | SPC | 2327 | 0.61 | 2 |
| BPTI | 298 | 210 | F3C | 2607 | 0.24 | 49 |
| FCYT | 300 | 20 | SPC | 3128 | 0.27 | 80 |
| CAL | 302 | 124 | SPC | 2248 | 0.67 | 4 |
| OVO | 298 | 100 | TIP3P | 1721 | 0.45 | 71 |
| CTF | 298 | 200 | F3C | 2455 | 0.23 | 22 |
| BPTI | 298 | 550 | F3C | 2509 | 0.22 | 23 |
| ALP1 | 298 | 250 | F3C | 4535 | 0.24 | In prep.[b] |
| ALP2 | 298 | 100 | F3C | 4402 | 0.24 | In prep.[b] |
| UB | 298 | 125 | F3C | 2797 | 0.23 | In prep.[b] |

[a] Abbreviations used: TRP, trypsin containing a benzamidine inhibitor; PARV, parvalbumin; BPTI, bovine pancreatic trypsin inhibitor; FCYT, ferrocytochrome $c$; CAL, calbindin $D_{9k}$; OVO, third domain of ovomucoid; CTF, C-terminal domain of the L7/L12 ribosomal protein; ALP1, α-lytic protease with a bound inhibitor; ALP2, apo form of ALP1; UB, ubiquitin.
[b] V. Daggett, in preparation.

are between 0.21 and 0.27 Å²/ps and the diffusion decreases somewhat in protein solutions (18).

Only the simulations using the F3C model agree with the experimental data. This is not surprising since the SPC and TIP3P simulations of pure water also exhibited exaggerated water motion. Although some of the simulations described below and given in Table 4 employed improved versions of these water models and more careful simulation protocols, they do not provide data on the water motion and therefore it is not known if the new models perform any better in protein simulations. Again, the water models themselves may not be the only source of the errors observed. All of the simulations using the SPC and TIP3P models utilized temperature-coupling schemes. That the water motion is exaggerated even when the motion is constantly being attenuated is most disturbing.

## Protein Behavior

Table 4 contains a chronologically arranged list of free molecular-dynamics simulations of proteins containing a sufficient amount of solvent to approximate a dilute solution. Simulations with thin shells of water, partial solvation, or those that allow only a subset of the protein atoms to move freely have not been included. Thus, these studies aim to investigate solution structures of proteins. Recently, interesting studies involving partial

**Table 4** Results of protein simulations in solution[a]

| Protein[b] | T (K) | Time (ps) | C$^\alpha$ rmsd (Å) end | C$^\alpha$ rmsd (Å) coord | C$^\alpha$ FLUC (Å) MD | C$^\alpha$ FLUC (Å) Expt | No. of waters | Protocol[c] | Reference |
|---|---|---|---|---|---|---|---|---|---|
| BPTI1   | 300 | 20    | 1.50 | 1.10 | NA   | —    | 1467 | SUPB | 75 |
| APP     | 300 | 15    | 1.40 | 1.05 | 0.53 | 0.62 | 1022 | SUPB | 41 |
| TRP     | 300 | 28.8  | NA   | 1.72 | 0.52 | 0.73 | 4785 | SUPB | 79 |
| CYT1    | 300 | 140   | 1.20 | NA   | NA   | —    | 1222 | TASB | 78 |
| PARV    | 303 | 106   | 2.72 | NA   | NA   | —    | 2327 | SUPB | 3  |
| BPTI2   | 298 | 210   | 1.43 | 0.74 | 0.42 | 0.63 | 2607 | FAPC | 49 |
| BPTI3   | 298 | 180   | NA   | 1.04 | 0.41 | 0.63 | 2601 | FAPC | 46 |
| SOD     | 300 | 23    | 1.49 | NA   | NA   | —    | 1761 | SUSB | 63 |
| LAC     | 300 | 55    | 1.30 | NA   | 0.60 | NA   | 3950 | SUPB | 25 |
| FCYT    | 300 | 20    | NA   | NA   | 0.58 | —    | 3128 | SUPB | 80, 81 |
| BAR     | 304 | 40    | NA   | 1.04 | NA   | —    | 2359 | TAPB | 73 |
| CAL     | 302 | 163   | NA   | NA   | 0.67 | 0.66 | 2248 | SUPB | 4  |
| OVO     | 298 | 100   | 1.43 | 1.28 | 1.32 | 0.78 | 1721 | TUPB | 38, 71 |
| HIV     | 300 | 96    | 1.50 | NA   | NA   | —    | 6990 | SUPC | 34, 69 |
| CTF1    | 300 | 55    | NA   | 0.67 | 0.59 | 0.68 | 2352 | SUPB | 8  |
| CTF2    | 298 | 200   | 1.24 | 0.80 | 0.64 | 0.68 | 2455 | FAPC | 22 |
| INSM1   | 300 | 100   | NA   | 1.60 | 1.10 | 0.75 | 1708 | SUPB | 52 |
| INSM2   | 300 | 100   | NA   | 1.58 | 0.90 | 0.75 | 1604 | SUPB | 52 |
| INSD-1  | 300 | 100   | NA   | 1.23 | 1.10 | 0.75 | 2849 | SUPB | 52 |
| INSD-2  | 300 | 100   | NA   | 2.20 | 1.10 | 0.75 | 2849 | SUPB | 52 |
| CYT2    | 300 | 100   | 2.80 | NA   | NA   | —    | 1001 | TUSB | 68 |
| MBN     | 300 | 150   | NA   | 1.07 | 0.58 | 0.66 | 3832 | TASC | 67 |
| BPTI4   | 298 | 550   | 1.95 | 1.42 | 0.63 | 0.63 | 2509 | FAPC | 23, 24 |
| LYZ1    | 300 | 50    | 1.40 | NA   | NA   | —    | 4689 | SASB | 62 |
| IL1     | 300 | 500   | 1.60 | 1.25 | 0.75 | 1.00 | 3783 | TUSC | 17 |
| PRT     | 300 | 116   | 2.62 | 2.52 | 0.60 | 0.63 | 8500 | TAPB | 33 |
| BPTI5   | 298 | 100   | NA   | 1.85 | 0.73 | 0.63 | 2943 | SAPB | 40 |
| UB      | 298 | 125   | 2.05 | 1.31 | 0.59 | 0.60 | 2797 | FAPC | In prep.[d] |
| ALP1    | 298 | 200   | 2.15 | 1.60 | 0.63 | 0.48 | 4535 | FAPC | In prep.[d] |
| ALP2    | 298 | 100   | 2.44 | 1.65 | 0.65 | NA   | 4402 | FAPC | In prep.[d] |
| LYZ2    | 300 | 550   | NA   | 2.20 | NA   | —    | 5345 | SUPB | 53 |

[a] Header abbreviations used: $T$, the simulation temperature; C$^\alpha$ RMSD, the rms deviation of the α-carbons from the crystal structure at the end of the simulation (End) or the coordinate averaged value (Coord)—in some cases the authors report backbone deviations; C$^\alpha$ FLUC, the fluctuations of the α-carbons about the mean structure during the simulation (MD) or as estimated from the experimental B factors (Expt, see text). Some of the entries in this table are approximate, as they were estimated from complicated, small figures; wherever possible the benefit of the doubt is given.

[b] Abbreviations used for the proteins; BPTI, bovine pancreatic trypsin inhibitor; APP, avian pancreatic polypeptide; TRP, trypsin with a benzamidine inhibitor; CYT1, cytochrome c–cytochrome $b_5$ complex; PARV, parvalbumin; SOD, superoxide dismutase; LAC, lac repressor headpiece; FCYT, ferrocytochrome c; BAR, barnase; CAL, calbindin D$_{9k}$; OVO, ovomucoid; HIV, HIV protease; CTF, C-terminal fragment of the L7/L12 ribosomal protein; INSM, insulin—1 refers to one monomer and 2 refers to the other when run as monomers as well as when run as the dimer, INSD; CYT2, cytochrome c–rubredoxin complex; MBN, carboxy-myoglobin; LYZ, lysozyme; IL1, interleukin 1β; PRT, prothrombin fragment 1; UB, ubiquitin; ALP1, α-lytic protease with a bound boron inhibitor; ALP2, the apo form of α-lytic protease.

[c] Abbreviations used for the protocols: water model (T, TIP3P; S, SPC; F, F3C); atoms included (U, united atom model; A, all atom model); water system (S, shell of water molecules; P, box of waters using periodic boundary conditions); dynamic scheme (B, coupling to a thermal bath; C, classical dynamics with energy conservation; NA, not available.

[d] V. Daggett, in preparation.

hydration of proteins with extensive comparison to experiment (51, 67) and crystalline simulations with limited amounts of solvent (9) were reported, but these studies fall outside of the scope of this review.

One of the main concerns with protein simulations is the length of the simulation. Any molecular-dynamics simulation is short on the biological time-scale. Therefore, running a simulation as long as possible is all the more important to at least ensure that the system is equilibrated. Many simulations have been performed for very short periods of time (tens of picoseconds), and elaborate analysis has been performed on artifactual structures. Thus, one of the most important parameters for a molecular-dynamics simulation is the root-mean-square positional deviation of the generated structures from the crystal structure as a function of time. Figure 2 gives an example of such a plot for the α-carbons of bovine pancreatic trypsin inhibitor (BPTI) at 298 K from our own work. As can be seen, large deviations from the crystal structure occur early in the simulation, drifting further with time. This simulation was the first to be performed for such a long period (>0.5 ns) and points out that any solution simulation under 50-ps duration is probably not sufficiently equilibrated for one to draw any conclusions about the behavior of proteins.

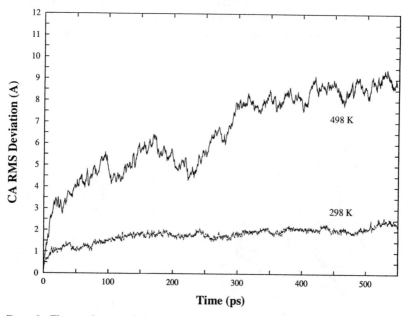

*Figure 2* The α-carbon rms deviation as a function of time for BPTI at room temperature (298 K) and high temperature (498 K) (23, 24).

Even after 200 ps, the model drifts slightly from the crystal structure. This drift could result from a variety of factors. The potential energy parameters may need to be adjusted slightly; there may be instabilities in the molecular-dynamics algorithms and programs; or the drift may indicate that the simulations need to be run longer to equilibrate.

The magnitude of the deviation from the crystal structure also deserves mention. It is not clear what the rms deviation should be to accurately model the solution structure of a protein. Different crystal structures of a particular protein can give main-chain rms deviations of up to 0.7 Å (47). Solution structures determined by two-dimensional NMR are generally within 1 Å of the crystal structure. As mentioned above, for a solution simulation, one should compare an ensemble of structures consistent with the NMR data. Recent work suggests that α-lactalbumin in solution may adopt a different structure than that obtained in crystals (72). Whether this proves to be more general and bears up under scrutiny of other experimental methods remains to be seen. However, these various issues point out that rms deviations of up to ~1.5 Å are probably acceptable. Figure 3 illustrates what such a deviation looks like, with the average structure following equilibration (50–550 ps, Figure 2) superimposed on the crystal structure. The rms deviation of the structures is 1.42 Å. As can be seen, the structures are remarkably similar, with particular regions deviating more than others. The portions of the structure that deviate most from the crystal structure are in the turns and loops. Such behavior may

*Figure 3* The main-chain atoms of the BPTI crystal structure and the average structure from the molecular-dynamics simulation at 298 K shown in Figure 2.

be perfectly reasonable in solution. And in fact, crystallographers often report a single conformation even when disordering is present (64). Furthermore, the average errors in atomic positions from NMR are at least 1 Å (43).

One ultimately performs molecular-dynamics simulations of proteins to investigate the role of movement in biologically relevant processes. Therefore, we believe that one must accurately represent the protein-solvent interactions and dynamics to have confidence in contentions about the role of motion in determining structure and function. So, as minimum criteria for good simulations, we require simulation times of over 50 ps, preferably longer than 100 ps, and rms deviations from the crystal or NMR structures of no greater than approximately 1.5 Å. Deviations of greater magnitude may be acceptable if specific experimental data suggest larger changes. Even in such cases, however, a control calculation is desirable to establish that the program, potential function, and protocols are capable of realistically modeling proteins.

The only other concept to be discussed prior to comparing the simulations listed in Table 4 is the rms atomic fluctuation about the mean structure following equilibration. This fluctuation is related to the crystallographic B-factor, $\langle \Delta r^2 \rangle^{1/2} = (3B/8\pi^2)^{1/2}$. In addition to internal mobility, B-factors also represent static disorder and rigid body motion. This last term is generally not considered in molecular-dynamics simulations, and in fact, B-factors can only be used as approximate guides to the degree of motion one would expect given the different environments, e.g. solution vs crystal and the fact that conformations about different average structures are being sampled. If the molecular dynamics–generated fluctuations are less than or equal to the internal motion indicated by the B-factor, the degree of motion is probably reasonable. For example, the rms fluctuation of the $\alpha$-carbons about the mean structure shown in Figure 3 is 0.63 Å, and the value from the B-factors is also 0.63 Å. Given that the component of the B-factor representing static disorder is not included, the result from molecular dynamics is either too high or indicates that internal motion is greater in solution than in the constrained crystalline environment, which is perfectly reasonable (43).

The first simulation of a protein in water was reported in 1984 by van Gunsteren & Berendsen (75). The protein simulated was bovine pancreatic trypsin inhibitor (BPTI1, Table 4), which has been the focus of many other simulations as well. The simulation was only 20 ps long but no doubt took a considerable amount of computer time. Fortunately, computers have become much faster and more affordable. As a result, the length of simulations has increased steadily over the years except for a few exceptions in 1989 (Table 4). As mentioned above, the problem with short simulations is that one does not know if a rms deviation of 1.5 Å after 20 ps would be representative of the value after another 500 ps. Instead, the 1.5 Å could increase more with time such that analysis at 20 ps would be artifactual.

The first simulation of sufficient duration was reported in 1987 by Wendoloski and coworkers (78) and involved studies of a complex of cytochrome $c$ and cytochrome $b_5$ (CYT1, Table 4). The authors used a relatively small water shell and temperature coupling, which was probably necessary because this system is large. In any case, the deviations from the crystal structure are quite reasonable. Other properties of the simulation necessary to evaluate it were not given.

The next long simulation to be reported was of parvalbumin by Ahlström and coworkers (3) (PARV, Table 4). Structural deviations were not put into the terms used here; however, the protein almost certainly displayed too much internal motion: many distances between phenylalanine residues in the core of the molecule changed by 3 Å when compared with the crystal structure (3). This simulation also exhibited unexpected hydration properties, which are discussed further below.

The next two simulations listed in Table 4 were of BPTI. The first remained very near the crystal structure (0.74 Å) and was a good test of the method as it was the longest, most accurate solution simulation performed up to that time (49) (BPTI2, Table 4). The authors employed classical molecular dynamics, periodic boundary conditions, and an all-atom model. The rms fluctuations of the $\alpha$-carbons were reasonable. The self-diffusion of water was also in good agreement with experiment (Table 3). Unfortunately, the arginine residues were inadvertently given zero net charge. This matter was corrected, and neutralizing counterions were added for the next BPTI run (46) (BPTI3, Table 4). In this case, the $\alpha$-carbon rms deviation increased by 0.3 Å but remained reasonable. The rms fluctuation and water-diffusion constant remained the same.

Several short but interesting simulations followed the BPTI3 simulation: SOD, LAC, FCYT, and BAR. The next long simulation was of calbindin D9k reported by Ahlström et al (4) (CAL, Table 4). The rms fluctuation was reasonable but the diffusion constant was high (Table 3). The deviations from the crystal structure were not given, making further evaluation of this simulation difficult. Nevertheless, the authors do compare their results to many other experimental results and provide an interesting discussion of the comparison of simulated reorientation processes with NMR and fluorescent studies. The simulations also provide insights into calcium binding.

Jorgensen & Tirado-Rives (38, 71) reported a 100-ps simulation of the third domain of ovomucoid (OVO, Table 4). Both the coordinate and final rms deviations from the crystal structure are good. However, even after 100 ps, the simulation does not appear to have equilibrated; a plot of the backbone rms deviation with time shows that the protein may still be drifting away from the crystal structure (Figure 2 of 71). The rms fluctuation and water diffusion are disturbingly high (Tables 3 and 4).

The following simulation of the HIV protease (34, 69) (HIV, Table 4) remained close to the crystal structure, although no other data were presented to allow evaluation of the simulation. The authors do, however, present an interesting analysis of the motion observed and discuss the potential importance of particular motions to inhibitor binding and domain interactions that may be of catalytic importance.

The next simulation listed is of the C-terminal fragment from the L7/L12 ribosomal protein by Åqvist & Tapia (8) (CTF1, Table 4). This simulation was short (55 ps) but the protein was very well behaved by all measures given in Table 4. This simulation gave the lowest $\alpha$-carbon coordinate deviation observed to date. The rms fluctuation is also quite good. Because this protein was so well-behaved, we also began to study it and performed a longer simulation to test the time dependence of the results and to develop a control for unfolding studies discussed below (22) (CTF2, Table 4). In our simulation, the protein remained very close to the crystal structure even after 200 ps and the rms fluctuation remained reasonable (Table 4).

Recently simulations of two different insulin conformations were reported, both as monomers and as a dimer (INS, Table 4). Overall, the rms deviations are high, especially for the coordinate-averaged deviations. Also, the rms fluctuations are very large; however, the values are only approximate as they were estimated from plots of the rms fluctuations as a function of sequence (Figure 2 of 52). The authors argue that these deviations may be real and can explain several experimentally observed phenomena. Unfortunately, we cannot really know whether or not this is true given that, by the authors' own admission, the systems do not equilibrate in the simulation time and because the protocols and program used have never been shown to accurately reflect protein dynamics in long simulations. This work is encouraging, however, in that we can now study biologically interesting properties such as protein-protein interactions and compare them with the monomeric forms, although such simulations will need to be longer and more accurate to provide trustworthy insights.

The next simulation is of a cytochrome $c$–rubredoxin complex (CYT2, Table 4). The rms deviation given in this case is for all atoms. Unfortunately, a plot of this deviation with time reveals that the system has not equilibrated (Figure 5 of 68). Nonetheless, the authors compare their solution simulation of this complex to one in vacuo and show how the presence of water may participate in interactions at the interface of the two proteins.

Steinbach and coworkers (67) reported a fairly long simulation of carboxy-myoglobin (MBN, Table 4) with a large water shell. The average structure generated during molecular dynamics following equilibration is in good agreement with the crystal structure as judged by the $\alpha$-carbon rms deviation and radius of gyration (67). Details of this simulation are not really discussed by the authors since it was not the main subject of their study. Instead the simulation was provided as a control for investigations of the effects of temperature and different degrees of partial hydration on the protein's structure and dynamics.

We have performed even longer simulations of BPTI (23) (BPTI4, Table 4). This is the longest reported simulation of a protein in solution. This simulation exhibited higher deviations from the crystal structure than the previous simulations from this laboratory, mostly because of the increased simulation time. The rms deviation of 1.4 Å at 210 ps reported by Levitt & Sharon (49) also occurs in this longer simulation at approximately 200 ps (Figure 2). The inclusion of charged arginines and counterions also contributes to the greater deviation from the crystal structure in this later simulation. Such an inclusion increased the deviation before (46). Also, reports of the two previous BPTI simulations from this laboratory show deviations for residues 1–56 and neglect the last two residues. When we followed the same procedure for BPTI4, the coordinate averaged rms deviation dropped to 1.3 Å. As before, the rms fluctuations and water diffusion are in good agreement with experiment, as is the radius of gyration.

Saito (62) recently reported a 50-ps simulation of lysozyme in a very large water shell (LYZ1, Table 4). This simulation was short, but the aim of the work was to test different dynamics protocols, not to perform an in-depth study of the protein motions.

Recently, Chandrasekhar et al (17) performed a long simulation (500 ps) of interleukin $1\beta$ (IL1, Table 4). This simulation was performed with a large shell of solvating waters. The protein remained close to the crystal structure (1.2 Å) and exhibited reasonable fluctuations. The authors extensively compare their structural results with NMR relaxation data as well as with the crystal structure and the more relevant reference state for such simulations, the NMR-derived family of structures (1.2-Å rms deviation, where the deviation between the crystal and NMR structures is 0.9 Å and the deviation between the three available crystal structures is 0.4 Å). In any event, this simulation further confirms the use of these methods for studying macromolecules, especially because the authors used a different program, potential function, and dynamics procedures.

A 116-ps simulation of prothrombin fragment 1 has been reported (33) (PRT, Table 4). The system contained a huge number of water molecules, and the backbone rms deviations were high in this simulation (2.5 Å). The protein is made up of different domains, and a hinge region about which the domains move was identified. When the rms deviation of particular regions of the protein were calculated, the deviations became more reasonable (0.9–1.7 Å) (33). This particular motion and others identified in the simulation may be of biological importance.

Another simulation of BPTI by Kitchen et al (40) (BPTI5, Table 4) has just been described. This simulation shows a coordinate-average deviation from the crystal structure (1.85 Å) that is high, especially because only residues 3–56 (58 residues total) were considered. The rms fluctuations are also a bit high.

Mark & van Gunsteren (53) have performed another lyzozyme simulation (LYZ2, Table 4). The simulation was performed for a long period of time with numerous water molecules. However, the coordinate averaged rms deviation from the crystal structure is high, 2.2 Å.

Even the long simulations of $\geq 0.5$ ns need to be repeated and extended to test these approaches, which is in progress. However, we recently directed our efforts to studying other proteins, because we feel confident that properties we observe on the 200-ps time frame will be stable with time provided the proteins are well behaved following a 50-ps equilibration period and because computer resources are always limited. Therefore, we have been studying the native dynamics of other proteins, primarily as controls for unfolding studies (discussed below), and we have reported preliminary results of the dynamics of ubiquitin (22) (UB, Table 4). The simulation has only completed 126 ps thus far, but the coordinate rms deviation and fluctuations are reasonable. Further studies have been undertaken of $\alpha$-lytic protease both with (ALP1) and without (ALP2) a bound inhibitor molecule (Table 4).

## Protein-Water Interactions

Some of the details of water-protein interactions are discussed below for those simulations in Table 4 that provide such analyses: TRP, PARV, BPTI, FCYT, OVO, and CTF. For all of these simulations, the self-diffusion constant of water is reported as a function of distance from the protein surface. The diffusion constant increased with increasing distance in all of the protein simulations except that of parvalbumin.

Experimentally, the protein affects two layers of water, thus manifesting lower diffusion constants (32). NMR relaxation experiments measuring local reorientational motion indicate a 5- to 10-fold decrease in translational diffusion within 10 Å of the protein (57). Other studies have indicated that the drag resulting from close proximity to the protein only leads to a decrease of a factor of two (18). The simulations that give reliable diffusion constants for the bulk solvent display a decrease in diffusion of the hydration waters of approximately a factor of two or less (all BPTI simulations from this laboratory and CTF), as does the OVO simulation that displays excessive water motion. The TRP simulation gives a factor of 5 difference. So it is encouraging that, although the water motion is exaggerated in many of the simulations, the proper qualitative behavior of the water's interactions with the protein are reproduced. The basis for the lower diffusion constants near the protein is the ordering of water molecules by electrostriction, whereby polar groups pull water toward the protein, and in the case of nonpolar atoms, the mobility of the nearby waters is upset because of the inability of the waters to make hydrogen bonds with the hydrophobic groups, which limits the conformations that the molecules can adopt because they attempt to maintain hydrogen bonding with neighboring waters.

Another curious feature worth mentioning is that a couple of the simulations (FCYT, PARV) exhibit diffusion constants at intermediate distances that are roughly twice the bulk values. McCammon and coworkers suggested that the competing structure-determining forces of the protein and bulk solvent cause this strange behavior (80, 81), but there is no experimental basis for this assertion.

Levitt & Sharon (49) have presented the most extensive analysis of hydration. They observed some clustering of the waters close to the surface of BPTI, which appears mostly to result from electrostriction. The closest waters have slightly strained bonds and angles and less favorable interactions with neighboring waters. The binding energy of these waters is on average 1.9 kcal/mol more favorable than for bulk water. The lowered binding energy is offset by a decrease in the entropy of the waters because of restricted rotation and translation. However, these hydration waters were still very mobile with mean residence times of 4 ps, and after 100 ps only 34% of the waters initially solvating the polar groups are still present. Only 12% of the waters around the nonpolar atoms remain. Very similar results have been observed in the other BPTI (23, 46) and CTF (22) simulations from this laboratory. Ahlström et al (2) observed similar results in their parvalbumin simulation, although the actual numbers are different, in part because of the different classification of water shells.

Quantification of the hydration layer can also be made both in simulations and experimentally. One of the simplest, most direct experimental approaches is to measure the amount of unfreezable water in a protein solution at $-35°C$ with NMR, which yields a measure of the amount of rotationally restricted water at the protein-water interface (42). These values range from 0.31–0.45 g water/g protein. In the BPTI2 simulation (49), a value of 0.42 g water/g protein was reported. The later, longer simulation of BPTI4 (23) yielded a value of 0.31 g water/g protein, which is derived from the number of waters in contact with the protein ($\leq 3$ Å). The values decreased to 0.17 if only waters forming hydrogen bonds with the protein were considered. The CTF2 simulation gave 0.52 g water/g protein (22). If only the water forming hydrogen bonds is considered, the value decreases to 0.29 g water/g protein. These results suggest that the hydration layer feels the field of the protein and experiences restricted motion even in the absence of specific hydrogen bonds. It is not clear why there is such a difference between the BPTI and CTF simulations, but the degree of hydration in all cases appears to be reasonable.

# NON-NATIVE PROTEIN–WATER SIMULATIONS

## High Temperature

Simulations are proving to be very important in elucidating the properties of proteins under perturbing, nonnative conditions because of both the confidence one can now have in molecular-dynamics simulations of native proteins and the limited experimental structural information one can garner under these conditions. The number of such simulations is small (Table 5). For any simulations under nonnative conditions, it is crucial that the potential function, software, and protocols realistically model the native

**Table 5** Protein simulations under nonnative conditions

| Protein | $T$ (K) | Time (ps) | $C^\alpha$ rmsd (Å) (end) | $C^\alpha$ fluctuations (Å) | Reference |
|---|---|---|---|---|---|
| High temperature | | | | | |
| BPTI | 423 | 550 | 5.1 | 2.1 | 23, 24 |
| | 498 | 550 | 8.1 | 2.5 | |
| CTF | 498 | 166 | 3.8 | 2.1 | 19 |
| ALP | 498 | 300 | 8.1 | 2.5 | In prep.[a] |
| LYZ | 500/320 | 190 | 4.3 | NA | 53 |
| | 500 | 180 | 7.0 | NA | |
| High pressure | | | | | |
| BPTI | 1000[b] | 100 | 1.5 | 0.59 | 40 |

[a] V. Daggett, in preparation.
[b] This value is of pressure (bar) rather than temperature.

state. Without such a control, one can have little faith that the properties observed under nonnative conditions are physically relevant. Again, we only present the results of free molecular-dynamics simulations in solution; there was, however, an interesting report of the unfolding of α-lactalbumin, but portions of the molecule were restrained to maintain the structure (28).

We have chosen to use high temperature as a means to disrupt the native structure. The advantage of using temperature is that processes involving traversing energy barriers (such as dihedral transitions) can be greatly accelerated without drastically changing the velocities of the atoms. For example, if the activation enthalpy is 20 kcal/mol and the process is normally seen on the millisecond time scale at room temperature, increasing the temperature to 500 K moves it into the nanosecond regime, which is accessible. At such high temperatures, the density of the water is decreased to relieve excess pressure and assure that the solvent remains a liquid.

In the first example of protein unfolding in solution, reduced BPTI at 423 and 498 K was examined (23) (Table 5). Controls with the oxidized protein at high temperature and with the reduced protein at low temperature were also performed. These simulations were run for over 0.5 ns, and there was clearly considerable movement away from the crystal structure at 498 K (Figure 2, Table 5). The α-carbon fluctuations were large, indicating that many different conformations were sampled. Figure 4 shows some representative snapshots from the simulation at 498 K, which illustrate the wide range of conformations sampled. Between 100 and 300 ps, the protein stabilized at an ensemble of conformations that were expanded, compared to the native state, but still remained intact (Figure 2). These conformations also contained native-like amounts of secondary structure. In addition to these properties, the partially unfolded forms of BPTI reproduce all of the known properties of the molten-globule state, and BPTI does appear to adopt such a conformation under some conditions (7). Because the resulting denatured state was physically reasonable, we could analyze the unfolding pathway itself with some confidence and there was further unfolding after 300 ps (24).

This same approach was used to study the unfolding of CTF (Table 5). This protein behaved in a manner very similar to BPTI and also appeared to adopt a molten-globule conformation (19). Work is in progress to ascertain whether CTF does indeed form a molten globule (P. E. Evans, personal communication).

We have also investigated a larger protein that adopts a molten-globule conformation, α-lytic protease (V. Daggett, in preparation) (ALP, Table 5). The simulation reproduces the experimental properties: an increase in the radius of gyration, increased exposure of the tryptophan residues, and

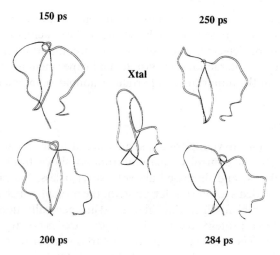

*Figure 4* Snapshots from the high-temperature simulation of BPTI at 498 K (as shown in Figure 2 between 100 and 300 ps) illustrating the conformational heterogeneity of the molten-globule state.

near native-like amounts of secondary structure (10). These studies are on-going, as simulations, as experimental studies of the solution structure of the molten globule, and as hydrogen-exchange experiments that seek to determine the steps in folding/unfolding (D. Agard & V. Basus, personal communication).

Mark & van Gunsteren (53) recently reported a high-temperature dynamics simulation of lysozyme employing conditions much like those used by us (described above) (LYZ, Table 5). In contrast to the other unfolding work, the putative molten-globule state reached in their simulation was not stable; however, it became more stable when the temperature was lowered. Also, the control simulation at room temperature exhibited a large deviation from the crystal structure (Table 4). Further discussion of this simulation must await presentation of their results.

Simulations pertaining to protein folding are very important because experimental approaches will probably never be able to yield detailed structural information about the folding pathway and folding intermediates—even those like molten globules, which are equilibrium intermediates. This inability results mainly from the lack of dominant structure in the molecules, because the molecules are so mobile, portions of the protein are unstructured, and the species tend to be only transiently populated. Some headway is being made in this area, and limited amounts of structural information are becoming available; however, because of the many limitations imposed due to the nature of the system, only native-like structure can be detected. Thus, simulations are important in this regard as well, since all interactions are visible, and experimentally testable predictions about the importance of particular nonnative interactions can be made.

## High Pressure

Kitchen et al (40) have reported a simulation at high pressure (Table 5). Unfortunately, the control simulation probably shows too much motion (BPTI5, Table 4, and discussed above). Nevertheless, the high-pressure simulation yielded a lower rms deviation for the protein, which is consistent with experimental findings that high pressure generally does not cause large changes in protein structure below pressures necessary to denature the molecule. The fluctuation of the $\alpha$-carbons is also lower at high pressure, which is a manifestation of the small free volume available to the protein. Interestingly, the simulation showed large changes in the hydration layer. The compressibility of the waters around nonpolar groups increased while it decreased around polar atoms. This study has shown the great potential in using simulations to study the effect of pressure on protein structure and the details of water-protein interactions that are difficult to elucidate experimentally.

# CONCLUSIONS

We have attempted to briefly discuss all free molecular-dynamics simulations in solution that have been performed to date. In addition, we have outlined how molecular-dynamics simulations are performed and highlighted special concerns when the goal is to accurately simulate protein motion in solution.

Various realistic simulations of native proteins have been performed and provide interesting insights into the details of the dynamics experienced at room temperature and of the interactions with water. The next logical step is to perform simulations of proteins under conditions in which experimental characterization is difficult. These simulations are necessarily more speculative; however, if realistic control simulations are performed, one can have some confidence in the simulation of unknown properties and attempt to predict experimentally testable behavior. Furthermore, BPTI has been a prototype for simulation studies. BPTI was the focus of the first molecular-dynamics simulation in vacuo (54) and in solution (75), the first truly realistic solution simulation (49), and the first simulations at high temperature in solution (23, 24) and at high pressure in solution (40).

Because the validity of such simulations has now been established, even though further work on the potential functions and protocols is still necessary and on-going, we should move away from simulations of BPTI and attempt to elucidate the importance of dynamical effects on the structure and function of other more biologically relevant proteins. Available studies have been discussed and work in this area is in progress. Finally, the increase in processing power and the lower cost of computers further enables the simulation of interesting biological properties in solution.

ACKNOWLEDGMENTS

We are grateful to the NIH (GM-41455 to M. L.) and the Jane Coffin Child's Fund for Medical Research (to V. D.) for support.

*Literature Cited*

1. Adams, D. J. 1979. *Chem. Phys. Lett.* 62: 329–32
2. Ahlström, P., Teleman, O., Jönsson, B. 1988. *J. Am. Chem. Soc.* 110: 4198–4203
3. Ahlström, P., Teleman, O., Jönsson, B., Forsèn, S. 1987. *J. Am. Chem. Soc.* 109: 1541–51
4. Ahlström, P., Teleman, O., Kördel, J., Jönsson, B. 1989. *Biochemistry* 28: 3205–11
5. Allen, M. P., Tildesley, D. J. 1987. *Computer Simulation of Liquids*, Oxford: Clarendon
6. Allinger, N. L. 1976. *Adv. Phys. Org. Chem.* 13: 1–82
7. Amir, D., Haas, E. 1988. *Biochemistry* 27: 8889–93
8. Åqvist, J., Tapia, O. 1990. *Biopolymers* 30: 205–9
9. Avbelj, F., Moult, J., Kitson, D. H., James, M. N. G., Hagler, A. T. 1990. *Biochemistry* 29: 8658–76
10. Baker, D., Sohl, J. L., Agard, D. A. 1992. *Nature* 356: 263–65
11. Barnes, J., Hut, P. 1986. *Nature* 324: 446
12. Berendsen, H. J. C., Postma, J. P. M., van Gunsteren, W. F., DiNola, A., Haak, J. R. 1984. *J. Chem Phys.* 81: 3684–90
13. Berendsen, H. J. C., Postma, J. P. M., van Gunsteren, W. F., Hermans, J. 1981. In *Intermolecular Forces*, ed. B. Pullman, pp. 331–42. Dordrecht: Reidel
14. Brooks, B. R., Bruccoleri, R. E., Olafson, B. D., States, D. J., Swaminathan, S., Karplus, M. 1983. *J. Comp. Chem.* 4: 187–217
15. Caldwell, J., Dang, L. X., Kollman, P. A. 1990. *J. Am. Chem. Soc.* 112: 9144–47
16. Careaga, C. L., Falke, J. J. 1992. *J. Mol. Biol.* 226: 1219–35
17. Chandrasekhar, I., Clore, G. M., Szabo, A., Gronenborn, A. M., Brooks, B. R. 1992. *J. Mol. Biol.* 226: 239–50
18. Cooke, R., Kuntz, I. D. 1974. *Annu. Rev. Biophys. Bioeng.* 3: 95–126
19. Daggett, V. 1993. In *Techniques in Protein Chemistry IV*, ed. R. H. Angeletti. New York: Academic. In press
20. Deleted in proof
21. Deleted in proof
22. Daggett, V., Levitt, M. 1991. *Chem. Phys.* 158: 501–12
23. Daggett, V., Levitt, M. 1992. *Proc. Natl. Acad. Sci. USA* 89: 5142–46
24. Daggett, V., Levitt, M. 1993. *J. Mol. Biol.* In press
25. de Vlieg, J., Berendsen, H. J. C., van Gunsteren, W. F. 1989. *Proteins Struct. Funct. Genet.* 6: 104–27
26. Eriksson, A. E., Baase, W. A., Wozniak, J. A., Matthews, B. W. 1992. *Nature* 355: 371–73
27. Ewald, P. P. 1921. *Ann. Phys.* 64: 253
28. Fan, P., Kominos, D., Kitchen, D. B., Levy, R. M., Baum, J. 1991. *Chem. Phys.* 158: 295–301
29. Greengard, L., Rokhlin, V. 1987. *J. Comp. Phys.* 73: 325–48
30. Hagler, A. T., Huler, E., Lifson, S. 1974. *J. Am. Chem. Soc.* 96: 5319–35
31. Hall, D., Pavitt, N. 1984. *J. Comp. Chem.* 5: 441–50
32. Halle, B., Andersson, T., Forsén, S., Lindeman, B. 1981. *J. Am. Chem. Soc.* 103: 500–8
33. Hamaguchi, N., Charifson, P., Darden, T., Xiao, L., Padmanabhan, K., et al. 1992. *Biochemistry* 31: 8840–48

34. Harte, W. E. Jr., Swaminathan, S., Mansuri, M. M., Martin, J. C., Rosenberg, I. E., Beveridge, D. L. 1990. *Proc. Natl. Acad. Sci. USA* 87: 8864–68
35. Hirshberg, M. 1990. *Simulations of the static and dynamic properties of the DNA double helix*. PhD thesis, Weizmann Inst. Science, Rehovot, Israel
36. Jorgensen, W. L. 1981. *J. Am. Chem. Soc.* 103: 335–45
37. Jorgensen, W. L., Chandrasekhar, J., Madura, J. D., Impey, R. W., Klein, M. L. 1983. *J. Chem. Phys.* 79: 926–35
38. Jorgensen, W. L., Tirado-Rives, J. 1989. *Chem. Scr.* 29A: 191–96
39. Kitchen, D. B., Hirata, F., Westbrook, J. D., Levy, R. 1990. *J. Comp. Chem.* 11: 1169–80
40. Kitchen, D. B., Reed, L. H., Levy, R. M. 1992. *Biochemistry* 31: 10083–93
41. Krüger, P., Strassburger, W., Wollmer, A., van Gunsteren, W. F. 1985. *Eur. Biophys. J.* 13: 77–88
42. Kuntz, I. D. 1969. *Science* 163: 1329–31
43. Kuntz, I. D. 1987. *Protein Eng.* 1: 147–48
44. Levitt, M. 1982. *Annu. Rev. Biophys. Bioeng.* 11: 251–71
45. Levitt, M. 1983. *J. Mol. Biol.* 168: 595–620
46. Levitt, M. 1989. *Chem. Scr.* 29A: 197–203
47. Levitt, M. 1992. *J. Mol. Biol.* 226: 507–33
48. Levitt, M., Perutz, M. F. 1988. *J. Mol. Biol.* 201: 751–54
49. Levitt, M., Sharon, R. 1988. *Proc. Natl. Acad. Sci. USA* 85: 7557–61
50. Loncharich, R. J., Brooks, B. R. 1989. *Proteins Struct. Funct. Genet.* 6: 32–45
51. Loncharich, R. J., Brooks, B. R. 1990. *J. Mol. Biol.* 215: 439–55
52. Mark, A. E., Berendsen, H. J. C., van Gunsteren, W. F. 1991. *Biochemistry* 30: 10866–72
53. Mark, A. E., van Gunsteren, W. F. 1992. *Biochemistry* 31: 7746–48
54. McCammon, J. A., Gelin, B. R., Karplus, M. 1977. *Nature* 267: 585–89
55. Narten, A. H., Levy, H. A. 1969. *Science* 165: 447–54
56. Perutz, M. F., Fermi, G., Abraham, D. J. Poyart, C, Bursuax, E. 1986. *J. Am. Chem. Soc.* 108: 1064–78
57. Polnaszek, C. F., Bryant, R. G. 1984. *J. Chem. Phys.* 81: 4038–46
58. Prévost, M., van Belle, D., Lippens, G., Wodak, S. 1990. *Mol. Phys.* 71: 587–603
59. Rahman, A. 1971. *Phys. Rev.* 136A: 405
60. Rahman, A., Stillinger, F. H. 1971. *J. Chem. Phys.* 55: 3336–59
61. Ryckaert, J. P., Bellemans, A. 1975. *Chem. Phys. Lett.* 30: 123–27
62. Saito, M. 1992. *Mol. Simul.* 8: 321–33
63. Shen, J., Subramaniam, S., Wong, C. F., McCammon, J. A. 1989. *Biopolymers* 28: 2085–96
64. Smith, J. L., Hendrickson, W. A., Honzatko, R. B., Sheriff, S. 1986. *Biochemistry* 25: 5018–5127
65. Soper, A. K., Phillips, M. G. 1986. *J. Chem. Phys.* 107: 47–60
66. Sprik, M., Klein, M. L. 1988. *J. Chem. Phys.* 89: 7556–60
67. Steinbach, P. J., Loncharich, R. J., Brooks, B. R. 1991. *Chem. Phys.* 158: 383–94
68. Stewart, D. E., Wampler, J. E. 1991. *Proteins Struct. Funct. Genet.* 11: 142–52
69. Swaminathan, S., Harte, W. E. Jr., Beveridge, D. L. 1991. *J. Am. Chem. Soc.* 113: 2717–21
70. Teeter, M. M. 1991. *Annu. Rev. Biophys. Biophys. Chem.* 20: 577–600
71. Tirado-Rives, J., Jorgensen, W. L. 1990. *J. Am. Chem. Soc.* 112: 2773–81
72. Urbanova, M., Dukor, R. K., Pancoska, P., Gupta, V. P., Keiderling, T. A. 1991. *Biochemistry* 30: 10479–85
73. van Belle, D., Prévost, M., Wodak, S. J. 1989. *Chem. Scr.* 29A: 181–89
74. van Gunsteren, W. F., Berendsen, H. J. C. 1977. *Mol. Phys.* 34: 1311–27
75. van Gunsteren, W. F., Berendsen, H. J. C. 1984. *J. Mol. Biol.* 176: 559–64
76. van Gunsteren, W. F., Mark, A. E. 1992. *Eur. J. Biochem.* 204: 947–61
77. Weiner, S. J., Kollman, P. A., Case, D. A., Singh, U. C., Ghio, C., et al. 1984. *J. Am. Chem. Soc.* 106: 765–84
78. Wendoloski, J. J., Matthew, J. B., Weber, P. C., Salemme, F. R. 1987. *Science* 238: 794–97
79. Wong, C. F., McCammon, J. A. 1986. *Israel J. Chem.* 27: 211–15
80. Wong, C. F., Shen, J., Zheng, C., Subramaniam, S., McCammon, J. A. 1989. *J. Mol. Liq.* 41: 193–206
81. Wong, C. F., Zheng, C., McCammon, J. A. 1989. *Chem. Phys. Lett.* 154: 151–54

# HYDROGEN BONDING, HYDROPHOBICITY, PACKING, AND PROTEIN FOLDING

## George D. Rose

Department of Biochemistry and Molecular Biophysics, Washington University School of Medicine, Box 8231, 660 South Euclid Avenue, St. Louis, Missouri 63110

## Richard Wolfenden

Department of Biochemistry and Biophysics, CB # 7260 Faculty Laboratory Office Building, University of North Carolina, Chapel Hill, North Carolina 27599-7260

KEY WORDS: secondary structure, protein engineering, solvent entropy, solute entropy, stereochemical code

---

CONTENTS

| | |
|---|---|
| PERSPECTIVES AND OVERVIEW | 382 |
|    The Protein-Folding Problem | 382 |
|    Stability of the Native Structure—the Evolving View | 382 |
|    Scope and Purpose of this Review | 384 |
| H-BONDS | 384 |
|    How Strong Are H-Bonds Between Polar Groups in Proteins? | 385 |
|    Does the Hydrogen Bond Play a Directing Role in Protein Folding? | 390 |
| HYDROPHOBIC EFFECT | 396 |
|    Scales of Hydrophobicity | 398 |
|    Generalizations for Protein Folding | 400 |
| PACKING | 403 |
|    Internal Packing and Protein Folding | 404 |
| A MODEL FOR PROTEIN FOLDING | 407 |
|    A Stereochemical Code for Protein Folding | 409 |

## PERSPECTIVES AND OVERVIEW

The literature abounds with arguments about whether a protein hydrogen bond is stabilizing or destabilizing. Controversy also surrounds the hydrophobic effect. Long thought to be the principal stabilizing force in folded proteins, it recently was reexamined and found to oppose folding; meanwhile other studies reached an opposite conclusion, finding that hydrophobicity is even more stabilizing than previously believed. In the case of packing, the ubiquitous jigsaw puzzle–like fit of side chains in the protein interior suggests structural restriction and close tolerances, whereas recent mutational studies foster a contrasting view of structural malleability and forgiving tolerances.

The dialectic is often heated, as familiar premises are questioned and alternative ones proposed. The purpose of this review is to delineate these topical issues and perhaps contribute to their resolution. First, the protein-folding problem is defined and placed in historical perspective.

### The Protein-Folding Problem

A protein molecule adopts its native three-dimensional structure spontaneously under normal physiological conditions (2). Despite intense research, a general mechanistic understanding of the folding transition remains obscure; this important problem is called the protein-folding problem.

The protein-folding problem was recognized clearly at least half a century ago in work of Anson & Mirsky, who observed that denaturation is a reversible process (4), and Mirsky & Pauling (108), who hypothesized that native proteins have a characteristic structure that is abolished upon denaturation. Such observations culminated ultimately in the experiments of Anfinsen and coworkers (3), who showed that reduced, denatured ribonuclease will renature spontaneously in vitro, with full restoration of enzymatic activity and return of its four native disulfide bridges. Such work has led to the contemporary view that protein tertiary structure is dictated by the amino acid sequence, although molecular chaperones may influence the folding kinetics in some cases (58).

A solution to the folding problem remains one of the principal challenges of Twentieth Century chemistry and biology. Many recent (1, 27, 34, 36, 49, 60, 77, 78, 111, 138, 152) and earlier (74, 76, 137) reviews have been written on these topics.

### Stability of the Native Structure—the Evolving View

Current perspectives in protein folding have been conditioned by ideas developed during the preceding several decades of research. The driving force for folding was initially thought to be intramolecular hydrogen bonding. This was Pauling's view, and it led him to the crucial role ascribed to hydrogen bonding in model structures for $\alpha$-helix and $\beta$-sheet (121, 122). Such models, which were constrained by the known geometry of the peptide unit (101), sought to optimize both the number and the geometry of $\rangle$N-H$\cdots$O=C$\langle$ hydrogen bonds. The existence of these predicted structures was soon confirmed in ongoing X-ray studies of proteins (124), bolstering the assumption that hydrogen bonds play the formative role in folding and stability.

Less than a decade later, this point of view was to change dramatically. In a seminal review, Kauzmann, reasoning from model compounds, showed that burial of apolar groups must be a significant source of the stabilization energy in proteins (74). It had long been observed that liquid water dissolves polar substances readily but apolar substances only sparingly (168). Upon mixing, water squeezes out hydrophobic molecules, resulting in a segregation into polar and nonpolar phases. The spontaneous separation of oil and water after mixing is a familiar example of this phenomenon. Thus, that hydrophobicity plays a key role in organizing the self-assembly of protein molecules is entirely plausible because some amino acid residues are abundantly water soluble while others are only sparingly so (167, 180).

Kauzmann's observations have evolved into the contemporary textbook view that the hydrophobic effect serves as the driving force for protein folding. According to this popular idea (referred to as the oil-drop model), the protein interior is enriched in apolar (oily) residues that are expelled from water and engender, in effect, a separate organic phase (101a, 175). However, protein folding cannot be simply a matter of burying apolar residues while exposing polar ones. Even residues with hydrophobic side chains have pronounced hydrogen-bonding capacity because of the presence of backbone $\rangle$N-H and $\rangle$C=O groups. Were such residues unable to realize hydrogen bonds within the molecular interior, then hydrogen bonding would favor denaturation, since presumably these same groups could hydrogen bond readily to water in the unfolded state. For this reason, hydrogen bonding came to be regarded as energetically neutral, or even unfavorable (79, 166), with respect to folding, with intramolecular hydrogen bonds in the native structure supplanted by intermolecular hydrogen bonds in the unfolded state. A neutral or unfavorable hydrogen bonding balance sheet leaves the hydrophobic effect as the presumed driving force toward the folded state.

The hydrophobic effect alone seems insufficient to account for the existence of a unique equilibrium structure in proteins because aggregates of oil, which form spontaneously in water, lack specific internal architecture.

Specific packing interactions within the molecular interior are believed to be a major source of structural specificity in proteins (136). Richards (161) showed that globular proteins are packed, on average, as well as crystals of small organic molecules, with packing densities that are more reminiscent of solids than of oil. The inside of a typical protein contains side chains that fit together with striking complementarity, like pieces of a three-dimensional jigsaw puzzle.

The high packing densities seen in globular proteins have been interpreted to mean that protein conformation is linked tightly to internal packing. Thus, for example, lysozyme does not have the same folded conformation as ribonuclease, although both proteins have approximately the same size and composition, because the lysozyme sequence cannot achieve efficient internal packing when organized into a ribonuclease fold. This interpretation of packing is consistent with classical studies of protein evolution in which the most conserved residues are found in the buried interior (153).

In summary, the hydrophobic effect came to be viewed as the principal force that drives the protein toward globular collapse and engenders a solvent-shielded molecular interior. Within this interior, specific packing interactions are thought to be the primary determinant of structural specificity.

## Scope and Purpose of this Review

Many recent developments have appeared in each of these areas: hydrogen bonding, the hydrophobic effect, and packing. One purpose of this review is to assess whether and how our understanding of these issues has changed during the 15 years since Richards' influential review in this series (137). Also, the final section introduces a novel model for protein folding that is consistent with data from earlier sections.

## H-BONDS

The importance of the hydrogen bond was made apparent in Pauling's text (120) and in the early treatise of Pimentel & McClellan (125). Of more recent interest is the monograph by Jeffrey & Saenger (71), which emphasizes the role of hydrogen bonding in biological systems.

In proteins, the patterns and principles of hydrogen bonding have been analyzed by surveying X-ray-elucidated molecules to identify recurrent themes. Baker & Hubbard used such an approach in an influential review (8) that laid the groundwork for many later studies. Hydrogen-bond geometry in proteins has been analyzed in both individual molecules (7) and systematic surveys (8, 10, 66, 71, 160, 163). Small-molecule crystal structures have also provided a wealth of information about H-bond geometry (23, 70, 110, 169, 170, 174).

## How Strong Are H-Bonds Between Polar Groups in Proteins?

The thermodynamics of hydrogen bonding in proteins have been controversial and remain so. Since early work of Schellman (151), Susi et al (166), and Klotz & Franzen (79), studies have found that the stability of hydrogen bonds between polar groups in water is marginal, at best. Thus, it came as a surprise when Privalov & Gill, in a recent review (130), concluded that hydrogen bonding is the principal source of stabilization energy in folded proteins. Earlier studies by Fersht and coworkers had reported favorable H-bond energies in protein engineering experiments, with values ranging from $-0.5$ to $-1.8$ kcal/mol for polar partners and $-3.5$ to $-4.5$ kcal/mol when one of the partners bears a charge (52). More recently, Williams reckoned the amide-amide H-bond to be favorable by $-1$ to $-4$ kcal/mol (38) by analyzing the binding between antibiotics (e.g. vancomycin and ristocetin A) and suitable peptides. Also, Pace and coworkers (158), using protein engineering, found that, on average, an H-bond contributes $-1.3$ kcal/mol to the stability of the protein ribonuclease T1.

The lack of suitable model compounds can often confound thermodynamic analysis. In particular, the dimerization of two independent, freely diffusing molecules A and B to AB is not entropically analogous to formation of an intramolecular H-bond within a protein. According to Stahl & Jencks (162, p. 4201):

> This does not mean that hydrogen bonding is not an important driving force for the maintenance of the native structure of proteins and nucleic acids. Most of the hydrogen bonds in proteins and other macromolecules are formed in intramolecular reactions, so that the equilibrium constants for hydrogen-bond formation can be much more favorable than for bimolecular reactions because of the smaller loss of translational and rotational entropy in intramolecular reactions.

How strong are H-bonds between polar groups? In proteins, the inherent strengths of individual intramolecular H-bonds are modulated by direct competition with water molecules that supply alternative binding partners and by other polar groups in the surroundings. For estimates of the intrinsic strengths of individual H-bonds that are unperturbed by these additional effects, the only uncomplicated reference state is the dilute vapor phase, in which intermolecular interactions are absent. In the vapor phase, mass spectroscopic methods show that groups carrying a single positive or negative charge form single H-bonds to water molecules with the release of 16–23 kcal/mol of enthalpy (105). Second virial coefficients indicate that between molecules in which no net charge is present, formation of single

H-bonds is accompanied by the release of 3–6 kcal/mol of enthalpy (84). Figure 1 presents a scale of vapor phase values, showing second virial coefficients from cluster ion experiments.

In principle, such high free energies of formation of H-bonds can be realized not only in the vapor phase, but also in condensed phases if

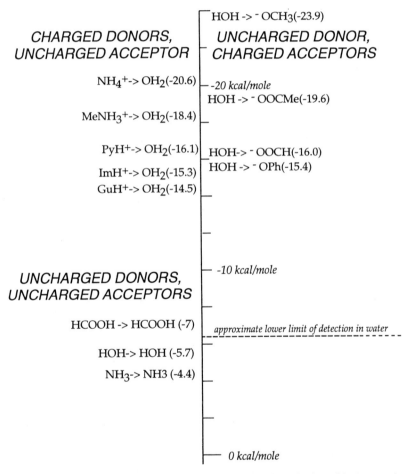

*Figure 1* Scale of vapor phase H-bond values. Groups carrying a single positive or negative charge form single H-bonds to water molecules with the release of 16–23 kcal/mol of enthalpy (105). When no net charge is present, formation of single H-bonds is accompanied by the release of 3–6 kcal/mol of enthalpy (84).

elimination of a group from one of the H-bonding partners leaves a gap (in their complex) that cannot be filled by solvent water. Thus, high levels of equilibrium-binding discrimination, favoring OH-substituted compounds over their H-containing counterparts by factors as large as $10^7$, have been observed in inhibitory complexes formed by adenosine and cytidine deaminases (183). After correction for differences in free energies of solvation between H- and OH-containing ligands, the desolvated OH-containing ligand is bound roughly $10^{12}$ times more favorably than the desolvated H-containing ligand. If the site itself must lose solvent to bind either ligand, then the difference in binding affinities could in principle approach the value that would be observed for H-bonding in the vapor phase. In fact, the observed difference in enthalpies is approximately $-17$ kcal/mol, approaching the value observed in the vapor phase for H-bonding between a water molecule and a partner bearing a single positive or negative charge (Figure 1). In the actual structures, a charged group is present but additional bonds are also present.

In practice, these extremes are moderated by multiple factors, including competition from solvent, molecular flexibility, cooperativity, and any constraints imposed by excluded volume. When solvent water has free access to both binding partners, as presumably tends to be the case in a fully unfolded protein, the effective strengths of their H-bonding interactions are expected to be greatly attenuated. Nevertheless, substantial free energies of interaction have been observed in enzyme-ligand interactions, even in the presence of water. For example, in Fersht's experiments involving mutations in tyrosyl-tRNA synthetase, which compare ligand affinities as reflected by $k_{cat}/K_m$, H-bonds between neutral partners typically seem to have free energies of formation of $-0.5$ to $-1.8$ kcal, and H-bonds involving one charged partner attain effective values of $-3.5$ to $-4.5$ kcal/mol (51). These values may not correspond to true equilibrium-binding constants, because they are based on kinetic constants that may describe reactions having transition states at different stages of advancement along the reaction coordinate.

Recent efforts to analyze the strengths of H-bonds in water compared true binding affinities of small model peptides, using the antibiotics ristocetin and vancomycin (38). Reassuringly, after correction for losses in rotational and translational entropy, the resulting negative free energies of H-bond formation appear to be comparable to those inferred from earlier mutagenesis experiments (179). However, Williams' analysis underscores the degree to which the apparent intrinsic enthalpy of H-bond formation depends upon underlying assumptions about differences in configurational entropy between free monomers and the H-bonded dimer. Related experimental observations concerning effects of mutations on protein stabilities (see above/below) suggest that individual H-bonds are not usually very directional in their preferences, so that bond rotors may not in fact be fully frozen in these complexes. Figure 2, a "blurogram" that depicts the ensemble of conformations over which a glutamate side chain can maintain an H-bond with a given backbone $\rangle$N-H group, illustrates this point pictorially.

Another interesting question raised by Williams' observations is the degree to which variations in the binding contributions of individual H-bonds are entropic, rather than enthalpic, in origin. The suggestion has been made that H-bond formation between amides is accompanied by a major gain in entropy resulting from the release of bound water molecules, which had been immobilized when they were H-bonded to the amides (38). Vapor-to-water transfer equilibria of simple solutes indicate, however, that for molecules of similar size, differences in polarity appear to be reflected almost entirely in enthalpies of solvation (Table 1). Thus, any entropy that is gained probably requires some other explanation.

Several investigators have used protein engineering to toggle selected hydrogen bonds off and on by replacing a native hydrogen-bonding side chain with a steric homologue that lacks a polar moiety (e.g. Asn → Val), followed by determination of the $\Delta\Delta G$ between native and mutant structures (1, 13, 17, 22, 155, 158, 159). Such experiments, while facile, cannot distinguish cleanly between energy differences resulting from H-bonds and from complicating side effects of comparable magnitude due to attendant changes in configurational entropy, hydrophobic burying, or van der Waals interactions.

In molecules capable of forming several H-bonds, one must also consider the possibility of competition from other groups that may furnish alternative binding partners. This effect is clearly illustrated by the solvation of ethylene glycol, which is much weaker than expected (16). Microwave spectroscopy shows that this diminution results from intramolecular H-bonding in the vapor phase. Alternatively, formation of one H-bond can affect the strength of another by electronic effects transmitted through the compound in question, and such effects may work in either direction. The existence of cooperativity has been demonstrated in the solvation of molecules such as p-nitrophenol by water, and anticooperativity is evident in the solvation of the p-nitrophenolate ion (182a). The structures of protein complexes with carbohydrates suggest that secondary -OH groups of sugars are typically involved in cooperative H-bonds with the protein (134). Finally, volumes of ionization of carboxylic acids (75), and of covalent hydration of aliphatic aldehydes (93), show that the solvation requirements of different parts of the same solute can be in conflict, leading to marked departures from additivity. The systematic approach of

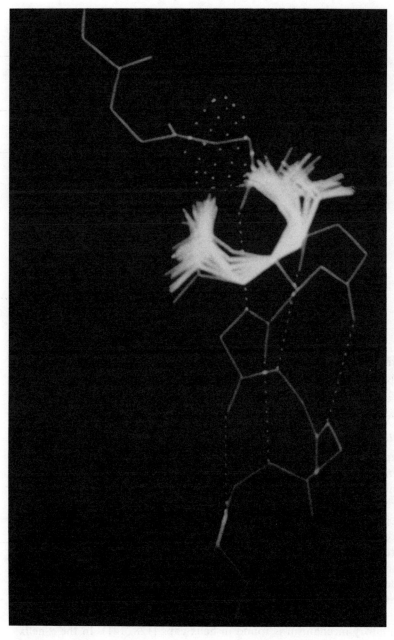

*Figure 2* Hydrogen-bond rotors need not be fully frozen. The blurogram shows a glutamate side chain at the N3 position of a helix hydrogen bonded to the backbone ⟩N-H at the N-cap position (128). Stippled surface represents the van der Waals radius around the ⟩N-H donor; solid lines mark acceptable conformations (163) for an oxygen acceptor.

**Table 1** Equilibria of transfer from the vapor phase to water (25°C)[a]

|  | $K_{hyd}$ vapor → water | $\Delta G$ kcal/mol | $\Delta H$ kcal/mol | $T\Delta S$ kcal/mol |
|---|---|---|---|---|
| $H_2$ | $1.7 \times 10^{-2}$ | +2.4 | −1.3 | −3.7 |
| He | $8.5 \times 10^{-3}$ | +2.7 | −0.8 | −3.5 |
| Ne | $1.0 \times 10^{-2}$ | +2.7 | −1.9 | −4.6 |
| Ar | $3.1 \times 10^{-2}$ | +2.1 | −2.7 | −4.8 |
| $N_2$ | $1.4 \times 10^{-2}$ | +2.5 | −2.1 | −4.6 |
| $O_2$ | $2.9 \times 10^{-2}$ | +2.1 | −3.0 | −5.1 |
| $Cl_2$ | 2.0 | −0.4 | −5.0 | −4.6 |
| $CO_2$ | 0.8 | +0.2 | −4.7 | −4.9 |
| $H_2O$ | $5.1 \times 10^4$ | −6.4 | −11.9 | −5.4 |
| $CH_4$ | $3.8 \times 10^{-2}$ | +1.9 | −3.2 | −5.1 |
| $CH_3$-$CH_3$ | $4.6 \times 10^{-2}$ | +1.8 | −4.4 | −6.2 |
| $CH_3$-OH | $5.3 \times 10^3$ | −5.1 | −11.2 | −6.1 |
| $CH_3$-$CH_2$-OH | $4.7 \times 10^3$ | −5.0 | −12.9 | −7.9 |
| $CH_3$-$CH_2$-$NH_2$ | $2.4 \times 10^3$ | −4.6 | −12.9 | −8.3 |
| $CH_3$-CO-$CH_3$ | $7.7 \times 10^3$ | −3.8 | −10.1 | −6.3 |
| $CH_3$-CO-OH | $1.0 \times 10^5$ | −6.8 | −12.6 | −5.8 |
| $CH_3$-CO-$OCH_3$ | $7.7 \times 10^2$ | −3.9 | −10.1 | −6.2 |
| $CH_3$-CO-$OC_2H_5$ | $1.8 \times 10^2$ | −3.1 | −11.7 | −8.6 |

[a] Data from Butler & Ramchandani (20a) and Frank & Evans (55). For an extensive collection of data for organic compounds, see Cabani et al (21a).

Gellman and coworkers (57), using NMR to study solution equilibria of model amides, has been designed to recognize and correct for these complicating factors.

## Does the Hydrogen Bond Play a Directing Role in Protein Folding?

Complementary base-pairing in DNA represents a discriminatory mechanism based upon hydrogen bonding (176), tantamount to a stereochemical code. Does a corresponding code exist in proteins, and, if so, where is it found?

Regular secondary structure—α-helix (122) and β-sheet (121)—does not represent such a code because discrimination is not at issue. Indeed, one reason why these structural motifs can recur with such high frequency in proteins is that the hydrogen bonding in question is between backbone amides ($\rangle$N-H···O=C$\langle$), and all residues (except proline) are indiscriminate in this respect.

A candidate stereochemical code with hydrogen bonding as a discriminatory mechanism is found in helix caps (128, 141). In the α-helix, the initial four N-H groups and final four $\rangle$C=O groups necessarily lack intrahelical hydrogen-bonding partners. These eight polar groups account for half of the total backbone hydrogen bonds in a protein helix of average length ($\sim 12$ residues), with intrahelical Pauling-Corey-Branson H-bonds (122) accounting for the remaining half. It has been hypothesized (128) that hydrogen bond partners (i.e. caps) for these initial four amide hydrogens and final four carbonyl oxygens are provided characteristically by side chains of polar residues that flank the helix termini—Asp, Glu, Ser, Thr, Asn, Gln, and neutral His at the N terminus and Lys, Arg, Ser, Thr, Asn, Gln and His(+) at the C terminus.

To what extent do the data support this hypothesis? By exhaustive stereochemical modeling, Presta & Rose (128) showed that the helices in 13 proteins of known structure are flanked either by residues that could provide complete capping (i.e. all four backbone groups at either end of the helix) or by combinations that include Gly and/or Pro. Do these potential capping interactions actually occur? In X-ray-elucidated proteins, approximately one-half of the initial four amide hydrogens and one-third of the final four carbonyl oxygens are indeed capped by side-chain partners of residues nearby in sequence. An additional 5% of the amide hydrogens and 9% of the carbonyl oxygens are satisfied by partners that are distant in sequence. In some of the remaining instances, capping is provided by lattice neighbors, and in these cases, the X-ray structure does not model an isolated molecule in solution. Any ostensibly uncapped groups, which are presumed to be solvated, may be capped during the folding process and subsequently liberated, once the nascent helix is fixed within the tertiary fold, although an alternative explanation is raised by the discussion that follows.

In their analysis of X-ray-elucidated proteins, Richardson & Richardson (141) found sharply differentiated residue preferences at positions that flank helix termini. For example, Asn in the N-cap position occurs 3.5 times more frequently than expected by chance. Their statistics as well as the stereochemical analysis of Presta & Rose (128) were truncated at two residues beyond the helix proper, within the adjacent peptide-chain turn. Figure 3 extends these statistical data in a series of histograms derived from helices in proteins of known structure. Each histogram represents the normalized frequency of a single residue type at the six positions on either side of both termini. The position-dependent frequencies in Figure 3 are similar, though not identical, to those of Richardson & Richardson (141). Differences probably resulted because Richardson & Richardson and Presta & Rose used different criteria to identify helix termini from X-ray coordinates.

It is often argued that intramolecular hydrogen bonding of this type is unlikely because solvent water is so concentrated, and so efficient in forming H-bonds to solutes, that solute-solute interactions cannot compete. That is, for a donor (D) and an acceptor (A) at the surface of the protein, the solvation reaction

D·A + solvent → D·solvent + A·solvent

is always favored. Whereas the preceding section cites examples of favorable intramolecular H-bonds, it might still be argued that these are typically buried within the protein where they are effectively shielded from solvent water.

Extending the argument, the attraction between any H-bonding donor or acceptor and solvent water is substantial. The vapor-to-water distribution coefficient of methanol, for example, is about $10^5$-fold more favorable than that of methane, indicating the existence of an attraction between the hydroxyl group of methanol and solvent water, with a free energy of $-7$ kcal/mol.

However, solvent water does not appear to be extremely efficient at forming hydrogen bonds with solutes. Experimentally determined enthalpies of transfer, from vapor to water, are for methanol $-11.24$, ethanol $-12.88$, propanol $-14.42$, butanol $-15.94$, amyl alcohol $-17.50$, or -CH$_2$ $-1.56$. Extrapolation gives an enthalpy of $-9.7$ kcal for transfer of a hydroxyl group from vapor to water. The hydroxyl group of an alcohol should in principle be able to form three hydrogen bonds to solvent water, and the enthalpies of formation of those bonds, based on calculations by Kollman et al (177), should be worth $\sim -12.3$ kcal. In the actual case, the shortfall is about 20%, suggesting that the average hydroxyl group makes roughly 2.4 H-bonds to solvent water. The preexisting structure of bulk water is expected to be incompatible, at least to some extent, with the

---

*Figure 3* Histograms derived from helices in Presta & Rose (128). Each histogram represents the normalized frequency of a single residue type at the six positions on either side of both termini. Nomenclature is as follows: $N^{-6}$-$N^{-5}$-$N^{-4}$-$N^{-3}$-$N^{-2}$-$N^{-1}$-N-cap-$N^1$-$N^2$-$N^3$-$N^4$-$N^5$-$C^{-5}$-$C^{-4}$-$C^{-3}$-$C^{-2}$-$C^{-1}$-C-cap-$C^1$-$C^2$-$C^3$-$C^4$-$C^5$-$C^6$, with the helix proper extending from N-cap to C-cap. To normalize a frequency, the number of instances of residue X at position $i$ was divided by the total number of residues at position $i$, yielding the fraction of times that residue X was found at position $i$. This fraction, converted to a percentage, was then divided by the percentage of residue X in the total data base. For example, at the $N^{-6}$ position: among the 73 residues *in toto*, 6 are Ala, representing 8.22%. There are 294 Ala in the data base of 3423 residues, so in general, the percent composition of Ala is 8.59%. Thus, the normalized frequency of occurrence for Ala at $N^{-6}$ is 8.11/8.59 = 0.94. A normalized value equal to 1 means that the frequency of occurrence of residue X at position $i$ is the same as the frequency of occurrence of residue X at large. A value of 2, for example, would then mean that the residue occurs at this position twice as often as would be expected from the empirically determined percentage composition in the data base.

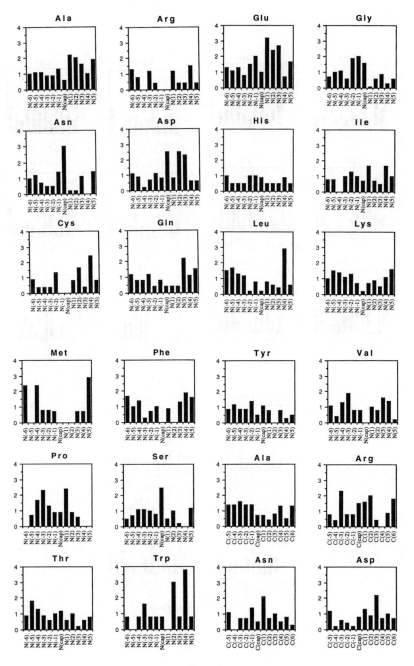

*Figure 3*

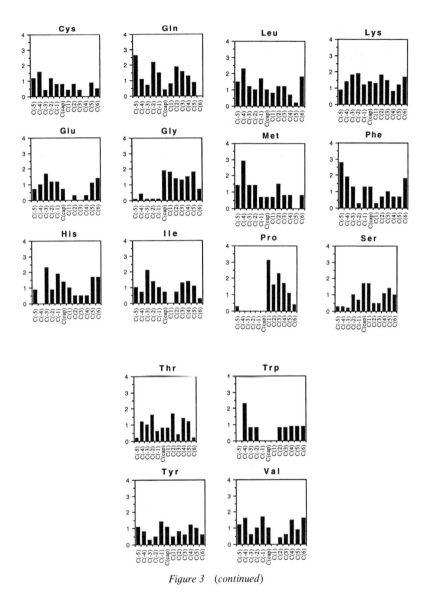

*Figure 3* (*continued*)

introduction into that bulk water of almost any solute, and in simple organic compounds, a -$CH_2$- increment is typically observed to make an unfavorable entropic contribution of $\sim 1.6$ kcal to the free energy of transfer from vapor to water. As in other groups of similar size, a foreign hydroxyl group contributes $+1.0$ to $+1.6$ kcal to the observed free energy of transfer (21).

This rationale for favorable intramolecular H-bonding, even at the solvent-accessible molecular surface, is fortified by experimental results from several groups. In a mutational analysis of the protein barnase, Serrano & Fersht (154) found that a side chain to backbone H-bond at the N-terminus of either helix stabilizes the protein by up to $\sim 2.5$ kcal/mol relative to a non-hydrogen-bonding residue in the corresponding position. (Of course, the conclusion is subject to caveats already mentioned.) Bruch et al (20) synthesized a peptide with the sequence of a carboxypeptidase helix and its flanking polar groups, and compared the helicity of the native-like sequence with variants in which alanine substituted for the flanking polar groups. Using circular dichroism to assess helicity, Bruch et al found that substitution of Ala for the three residues at either end reduced $\theta_{222}$ to two-thirds the value measured for the native peptide. A telling example is provided by the finding that a probable pH-dependent switch in which protonation of an H-bonded histidine situated at the N-cap position of a helix in apo-cytochrome $b_5$ lowers the helix$\rightleftharpoons$coil equilibrium constant by more than an order of magnitude (86a). On the other hand, Fairman et al (48) replaced the amide-blocking group of a model peptide (which can potentially H-bond to one of the four terminal main-chain $\rangle$C=O groups) with its methyl-ester (in which no H-bond is possible), with no observed effect on stability. In this case, either the intramolecular H-bond is unfavorable or, more likely, the C terminus of the peptide helix is frayed.

Kallenbach and coworkers conducted a more extensive test of helix capping using a reference peptide together with a series of derivatives (99). In the reference peptide, each position bears the residue found to occur with highest frequency at the corresponding position among protein helices (141). Derivatives are devised in matched pairs; each pair quantifies the effect of a single substitution, made either at the N-cap or at a central position. For example, Ala, the most helix-stabilizing residue in a central position (98, 106, 115, 119), diminishes helicity by a factor of two when at the N-cap position. Conversely, Ser, a statistically preferred residue at the N-cap position, diminishes helicity by a factor of two when in a central position. The study shows that statistical preferences seen in X-ray studies (5, 141) correspond to structural differences in a related series of peptides.

Taken in sum, these experimental findings suggest widespread helix capping in natural proteins. This conclusion is supported by a recent study of Stickle et al (163), who conducted a global census of hydrogen bonding in high resolution, X-ray-elucidated proteins. In their study, the set of all side chain–to-backbone hydrogen bonds is subdivided into (*a*) backbone donors with side-chain acceptors and (*b*) side-chain donors with backbone acceptors. Histograms of these two groups show that pronounced peaks are situated at helix-capping loci, against an otherwise undifferentiated, low-level background. In view of these histograms, a directed analysis of existing X-ray structures may disclose the existence of further, previously overlooked, interactions of this type.

## HYDROPHOBIC EFFECT

The hydrophobic effect has been a topic of chemical interest for more than a century (172). Early on, Edsall noted that the transfer of an apolar compound from an organic medium to water is accompanied by a large positive change in heat capacity (41). Proteins exhibit similar, striking differences in heat capacity between their folded and unfolded states (129). Prompted by this similarity, Kauzmann proposed that the transfer of small apolar solutes from water to liquid hydrocarbon could model the process in which apolar side chains in a protein are sequestered from solvent upon folding (74). However, as X-ray-elucidated structures became available, Richards surveyed protein interiors and found them to be remarkably close-packed in a manner more reminiscent of a solid than of liquid hydrocarbon, and he questioned the suitability of the liquid hydrocarbon model (137). Since that time, the field has been mired in controversy, with disagreement over the meaning (63), sign (37, 109, 131), and magnitude (156) of the hydrophobic contribution to folding.

Despite ongoing controversy, there can be no disagreement about the tendency of saturated hydrocarbons to leave water and enter other solvents. Dispute arises over whether solute molecules tend to leave water and enter less polar solvents primarily because they are repelled by water, or because they are attracted to the less-polar solvent. This question can be analyzed by using some absolute standard of reference such as the vapor phase that neither attracts nor repels solutes. Framing the question in these terms, one finds that methane exhibits an appreciable water-leaving tendency, with an equilibrium distribution between water and the vapor phase of 27 in favor of the vapor phase. This tendency, expressed as an equilibrium constant for transfer from dilute aqueous solution to the dilute vapor phase, increases in the normal alkanes as follows: methane 27, ethane 20, propane 29, butane 38, pentane 51, hexane 74, heptane 83. Thus, for each methylene increment ($-CH_2-$), transfer to the vapor phase is enhanced very gradually by an average factor of 1.3, equivalent to 0.14

kcal in free energy. This increment is about the same in the normal series of hydrocarbons, acetic acid alkyl esters, primary amines, and primary alcohols. In contrast, each methylene increment increases the distribution coefficient for transfer from water to a nonpolar solvent by a factor of 4.0, equivalent to 0.8 kcal in free energy (32, 72, 182).

Summarizing, saturated hydrocarbon molecules have an appreciable tendency to leave water, and are thus hydrophobic in any sense of the word. As the size of the hydrocarbon increases (addition of a methylene increment), this tendency is enhanced only gradually, while the corresponding tendency to enter a nonpolar solvent, such as a hydrocarbon, is considerably stronger.

With this background in mind, the definition of hydrophobicity used here is the operational process in which an apolar group is transferred from a polar or neutral phase to an apolar phase. The effect is pertinent to protein folding to the degree that residues with apolar side chains are expelled from water and engender a solvent-shielded molecular interior.

Many investigators have studied the transfer process (36, 74, 167, 180). The effect of a methylene increment on the free energy associated with simple removal of a solute from water is small at room temperature, but major compensating changes occur in other thermodynamic parameters. Removal of nonpolar molecules or groups from water is accompanied by increases in entropy (and volume) and by a compensating uptake of heat from the surroundings. Thus, hydrophobic bonds are distinguished by a tendency to become stronger with increasing temperature (74, 112). Because the properties of liquid water, and of water of solvation in particular, are still not fully understood, an ongoing discussion has examined the likely origins of the entropy increases that accompany the formation of hydrophobic bonds, on the removal of hydrocarbons from water to the vapor phase. Conditions have been found (9, 132, 184) under which the major changes in entropy and heat capacity disappear.

The self-cohesive properties of water, although not unique (47), are unusual (42). A reasonable assumption is that changes in the properties of water in the immediate neighborhood of the solute could account for the loss of entropy that accompanies introduction of a nonpolar molecule or methylene increment into water from the vapor phase or from a nonpolar solvent. Frank & Evans (55) suggested that the observed changes in entropy and heat capacity might imply the formation around solutes of a kind of clathrate or iceberg structure, in which water molecules were more ordered than in the bulk solute, but did not resemble Ice I in any literal sense.

More recent evidence suggests that entropic effects associated with introducing nonpolar solutes into water may arise, not only from restrictions on the mobility of water molecules, but also from restrictions on the mobility of solutes when they are introduced into the aqueous environment (6). Solutes experience a three- to fivefold enhancement in $^{13}$C spin-lattice relaxation times when they are transferred to water from nonhydroxylic solvents of similar viscosity, so that water appears to be unusual in the restrictions that it imposes on the motion of dissolved solutes (65). In normal aliphatic compounds of increasing size, solubility might (according to this view) be reduced by progressive restrictions on internal rotation. Compounds with internal rotations already restricted would not be affected to the same extent. It is therefore of interest that steroids and cycloalkanes display considerably lower activity coefficients in water than nonrigid compounds, with reference both to nonpolar solvents and the vapor phase (116). For similar reasons, the hydrophilic character of proline is greater than that of acyclic amino acids of similar size; this effect has been shown to be entirely entropic in origin (59). Such disparate solutes as argon, $CO_2$, and water itself show almost the same entropy of solution, despite major differences in their affinities for solvent water and the restrictions that their polar interactions might have been expected to impose on the solvent.

## Scales of Hydrophobicity

The free energy required to transfer amino acid solutes from an aqueous to an organic phase, $\Delta G^0_{\text{water}\rightarrow\text{organic}}$, is related to their relative solubility in either phase. As shown by Cohn & Edsall (30), the free energy of transfer, $\Delta G^0_{p\rightarrow p'}$, between phases p and p' is related to the partition coefficient between the phases, $K_{p\rightarrow p'}$, by the equation $\Delta G^0_{p\rightarrow p'} = -RT \ln K_{p\rightarrow p'}$. Cohn & Edsall provided the solubilities for many amino acids in both water and ethanol.

Beginning with an early compendium based on the relative solubilities of some amino acids in water compared with dioxan or ethanol (113), several experimental scales have been developed for comparing the affinities of amino acid side chains for solvent water (31, 147). The simplest of these uses the gas phase, devoid of attractive or repulsive forces, as a reference phase in water-to-vapor distribution measurements at infinite dilution (181). More complex scales use a second solvent, whose solvation properties must be considered. Of these, one of the least complicated is cyclohexane, with a dielectric constant of 2 (135). Water-to-vapor and water-to-cyclohexane distributions (Figure 4) are related to each other by a third scale (not shown) of cyclohexane-to-vapor distribution coefficients. Positions of amino acids on the latter scale are closely related to surface area, consistent with the view that cyclohexane can participate in van der Waals' interactions with solutes and not much else. Other solvents, such as 1-octanol (50), dissolve substantial quantities of water and are capable

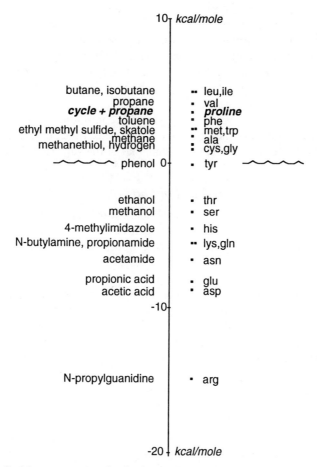

*Figure 4* Cyclohexane-to-water distribution coefficients. The scale includes a recent value for proline (*bold*) that was generated by a cycle of transformations of the side chain of norvaline, as discussed in the text.

of specific polar interactions with solutes such as indole, resulting in the misleading impression that tryptophan, for example, is extremely nonpolar.

The scale shown in Figure 4 includes a recent value for proline, which does not have a side chain in the ordinary sense. The value for proline was generated by a cycle of transformations of the side chain of norvaline, for which exact experimental information could be obtained. The results indicate that proline is more hydrophilic than would have been expected from simple considerations of surface area. Proline exemplifies a general tendency of ring compounds, mentioned previously, to enter water more readily than their acyclic counterparts. The exceptional hydrophilic character of proline is entropic rather than enthalpic in origin. Ring compounds have less freedom of rotation than open-chain compounds, and may experience fewer constraints upon entering the relatively structured environment of solvent water. However, ring compounds have slightly less surface area than their acyclic homologues; e.g. the standard-state area of the prolyl side chain is 111.0 $\text{Å}^2$ while that of the valyl side chain is 128.4 $\text{Å}^2$ (90).

## Generalizations for Protein Folding

How are these scales related to protein structure? The above distribution coefficients of amino acid side chains can be related to various measures of the tendencies of these side chains to appear on the surfaces or in the interiors of proteins (107, 146). However, a note of caution remains in order. Despite the strengths of these correlations, the tendencies of residues of each kind to be exposed to the solvent in real proteins vary as an extremely shallow function of their water-to-vapor or water-to-cyclohexane distribution coefficients, i.e. the former values, expressed as equilibrium constants, vary over a numerical range of 15 orders of magnitude, whereas the latter values vary over a range of only 2 orders of magnitude. This insensitivity is hardly surprising: the interiors of proteins tend to be more polar than cyclohexane, with evident hydrogen-bond partners for virtually every buried polar group (8, 163), and most surface locations may be less polar than bulk water.

Nevertheless, it has been possible to arrive at generalizations that are valuable for understanding the effects of mutations and for predicting protein structure. After correction for accessibility, the mean change in stability (i.e. $\Delta\Delta G$) for a large collection of hydrophobic mutants was found to scale linearly with the free energy of transfer from water to $n$-octanol (118). In a mutational study of myoglobin, where Ala was substituted at 16 selected sites, approximately 80% of the measured change in stability ($\Delta\Delta G$) can be accounted for by the difference in surface area between the native residue and the alanine substitution (126). Table 2 shows energy differences, expressed as $\Delta\Delta G$ values, between a set of mutations of large nonpolar residues and those of smaller residues upon denaturation of bacterial nucleases by guanidine. These are seen to be numerically equivalent, within experimental error, to values based on the cyclohexane-to-water distribution coefficients. This near-identity suggests

Table 2  Effects of mutations on the stability of staphylococcal nuclease[a]

| Mutation | Denaturation equilibrium, GU+ $\Delta\Delta G$ (kcal) | Cyclohexane → water $\Delta\Delta G$ (kcal) | Octanol → water $\Delta\Delta G$ (kcal) |
|---|---|---|---|
| Leu → Ala (11 mutants) | $-3.10 \pm 1.5$ | $-3.11$ | $-1.90$ |
| Val → Ala (9 mutants) | $-2.40 \pm 1.5$ | $-2.23$ | $-1.24$ |
| Ile → Ala (5 mutants) | $-3.60 \pm 1.0$ | $-3.11$ | $-2.04$ |
| Ala → Gly (5 mutants) | $-0.95 \pm 0.2$ | $-0.87$ | $-0.42$ |

[a] Data from Shortle et al (159), Radzicka & Wolfenden (135), Eisenberg & McLachlan (43), and Fauchère & Pliska (50).

that when staphylococcal nuclease is denatured, these nonpolar residues pass from a cyclohexane-like to a water-like environment. Evidently, these nonpolar residues can find environments in the native protein with dielectric constants of approximately 2, and denaturation involves complete exposure of their side chains to solvent water. This relationship fails for the more-polar amino acids, in such a way as to suggest that they find more-polar environments in the native protein, as might be expected from the observation that buried polar groups are usually hydrogen bonded.

The distribution of hydrophobic groups along the linear sequence of a polypeptide chain is often used in a predictive manner. A plot of the average hydrophobicity per residue against the sequence number reveals loci of minima and maxima in hydrophobicity. Such a plot, termed a hydrophobicity profile, has been applied to proteins of known sequence but unknown structure to predict (*a*) the probable location of peptide-chain turns (144), (*b*) segmentation of the molecule into interior/exterior regions (83, 148), (*c*) likely antigenic sites (64), and (*d*) the location of membrane-spanning segments (44, 83). Any profile of this type is based upon some scale of hydrophobicity for the naturally occurring amino acid residues, of which there are many (31, 43, 147).

Perhaps the most widely applied generalization is that made by Chothia, who noted that free energies of transfer of side chains of nonpolar amino acids to water, from a diverse collection of organic solvents, show a slope of 25 cal of free energy per squared Ångstrom of buried surface area (24). However, this value was called into question recently. From consideration of the effects of the relative molecular volumes of the solute and water, Honig and coworkers (157) concluded that concave regions such as protein-binding sites (or sites in a protein from which a residue is removed by mutation) can be more hydrophobic than had been realized, resulting in a slope of 46 cal of free energy per squared Ångstrom of buried surface area. Debate about this topic is ongoing (165).

On consideration, one would not expect that any single solvent or physical environment could, except by an occasional coincidence, represent the variety of environments experienced by residues within a protein. Thus, correlations between various thermodynamic variables, e.g. heat capacity or free energy of transfer from water to nonaqueous solvent, may seem surprising and, by inference, informative. Again, a note of caution is in order. Some time ago, Bigelow showed that the ratio of hydrophilic to hydrophobic groups in globular proteins is confined within a narrow range of values (15). Later, Janin (67) and Teller (171) demonstrated that the surface area of a protein monomer is a simple function of its molecular weight, i.e. upon folding, proteins bury a constant fraction of their available surface. Chothia (25) found that this fraction can be subdivided into characteristic contributions from polar and apolar residues. Extending these observations to an atomic level, Lesser & Rose (90), using the algorithm of Lee & Richards (87), calculated the mean area buried upon folding for every atom in a data base of X-ray-elucidated proteins and found that, on average, each atom type buries a constant fraction of its standard-state area (Figure 5). Surprisingly, the mean area buried by most (though not all) residues can be closely approximated by summing contributions from three characteristic parameters corresponding to three generic atom types: (a) carbon or sulfur, which are 86% buried, on average; (b) neutral oxygen or nitrogen, which are 40% buried, on average; and (c) charged oxygen or nitrogen, which are 32% buried, on average.

Although such characteristic behavior seems to confer welcome predictability on the folding process, its existence leads inescapably to a complex of confounding functional interrelationships. For example, if proteins bury a well-behaved and constant fraction of their apolar surface and also have a constant specific heat capacity, then any strong correlation between these properties need not be causal, as noted by Record and coworkers (96).

Indeed, to the extent that the mean behavior of individual hydrophobic groups can be predicted by a single line, they are indistinguishable from aliphatic compounds, which exhibit analogous linear behavior in both liquid and condensed phases (161). Although aliphatic compounds do have characteristic thermodynamic and geometric properties, they lack the unique molecular architecture that is the very hallmark of globular proteins. Thus, the regularities described in the preceding paragraphs suggest that the hydrophobic effect may push the chain to collapse to a compact, water-excluding state as an effective means to sequester apolar

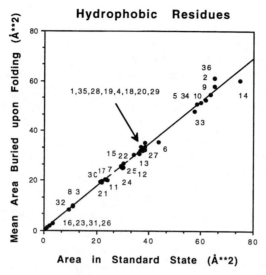

*Figure 5* Plot of the mean area buried upon folding vs the standard-state area for 36 atom types from 8 apolar residues in a data base of 61 proteins of known structure (90). Residue types included are Val, Ile, Leu, Met, Phe, Trp, Cys, and 1/2-Cys. The identity of individual groups (which can be obtained from the original paper) is of less significance than the fact that all are well-described by a single straight line, as shown. Groups 2 and 36, the worst outliers, are -SH and -S- atoms, respectively.

groups from solvent access. However, the reason why the chain adopts a specific, unique, equilibrium fold instead of an ensemble of micelle-like structures is unlikely to be the hydrophobic effect per se, although Lau & Dill (86) have argued to the contrary.

## PACKING

In the 1970s, after the first several protein structures had been elucidated by X-ray crystallography, Richards made a fundamental observation (137). Upon calculating the packing density—the ratio of the summed atom volumes to the molecular volume—he determined that the interiors of proteins are packed as well, on average, as crystals of small organic molecules with few cavities of atomic dimension, more reminiscent of a solid than of liquid hydrocarbon. A typical cross-section through a protein molecule reveals lumpy, irregular-residue side chains that fit together with exquisite complementarity, almost like pieces of a three-dimensional jigsaw puzzle. This surprising observation prompted a general reassessment and altered prevailing notions about protein structure. Richards' seminal review appeared earlier in this series (137).

Richards' work led to the appealing idea that interior packing is the property that distinguishes a protein from an oil drop. According to this view, lysozyme, for example, adopts the lysozyme fold and not the ribonuclease fold because efficient internal packing of the lysozyme sequence can be achieved only in the former case. Like a jigsaw puzzle, the side chains fail to fit together unless the conformation is "right." This interpretation motivated many subsequent attempts to predict and/or characterize conformation based on internal packing constraints (88, 127, 188, 189; J. S. Fetrow & S. Bryant, submitted).

## Internal Packing and Protein Folding

Richards' observations are consistent with a restricted architecture in which constituent parts fit together with snug tolerances. However, that picture must be incomplete because further evidence supports a somewhat different view. Both NMR (185) measurements and molecular-dynamics calculations (104) indicate that buried aromatic rings experience a highly mobile microenvironment, tight packing notwithstanding. Furthermore, the ensemble of methyl rotors, which is coextensive with the molecular interior, is mobile and preferentially populates staggered (i.e. relaxed) configurations (80), suggestive of more fluid-like surroundings. Indeed, side chains in X-ray-elucidated protein structures populate staggered conformations (69, 100, 127), which is again consistent with internal relaxation.

Compounding the paradox, recent mutational studies of proteins have demonstrated that the buried interior can tolerate a broad diversity of residue substitutions, usually with only minor effects on structure, stability, or function (1, 46, 95, 102, 103). Thus, although static measurements of packing suggest a tight particle with constrained tolerances, both dynamics and mutagenesis studies portray the molecule as malleable with forgiving tolerances.

This conundrum prompted Behe et al (12) to conduct a simple test of whether protein conformation is determined primarily by packing interactions. In this test, they analyzed proteins of known structure for the presence of preferred interactions either between residue pairs or among larger composites, reasoning that if side-chain complementarity is an important source of conformational discrimination, then sets of residues that interact favorably should be readily apparent. Also, if no interactions are especially favorable, then efficient packing—an undeniable experimental fact—is achieved without severe limitation of the individually allowed side-chain orientations. The analysis of packing was conducted in two parts. First, all residue pairs, X-Y, were assayed for the existence of particularly favorable interactions between X and Y (i.e. binary interactions). None were found. However, preferred higher-level packing arrangements (i.e. tertiary, quaternary, etc) may exist despite an absence of preferred binary interactions. To address this remaining possibility, the interaction of all X-Y pairs with the rest of the protein was assessed. Again, no preferred interactions were found. This analysis leads to the unexpected conclusion that high packing densities so characteristic of globular proteins are readily attainable among clusters of the naturally occurring residues.

These two ostensibly conflicting views of protein architecture—tight versus malleable—would be reconciled if tight packing could be achieved primarily by local adjustment of the structure, without global rearrangement. In this case, a perturbation event—either kinetic (e.g. a ring flip) or structural (e.g. a mutation)—would be followed by local relaxation of the structure, with minimal impact beyond the immediate microenvironment (56). Mounting evidence supports the view that the ordering imposed by packing requirements does not propagate beyond a highly local neighborhood. For example, a single mutation in a protein typically results in many small structural changes, but all are confined within a several-Ångstrom sphere centered about the site of mutation (1). Such mutations do measurably affect packing (173), usually reducing stability (150), but the changes are generally quite small. However, even in extreme cases where a mutation results in a substantial packing fault (45), the overall conformation of the molecule does not undergo global rearrangement to eliminate the fault.

Because a single mutation is most often destabilizing, one might think the combined effect of several such mutations, each disruptive of packing, could cause gross conformational rearrangement. However, the cumulative effect on stability is unlikely to be realized in this fashion for reasons described below:

According to Lattman & Rose (85, p. 439):

> The folding reactions of small, globular proteins typically exhibit two-state kinetics, in which the folded and unfolded states interconvert readily without observable intermediates (76).[1] The free energy difference, $\Delta G$, between the native and denatured states of such proteins is quite small, lying in the range of $\sim -5$ to $-15$ kcal/mol (117). In the usual thermodynamic interpretation, if a single mutation destabilizes a protein by $\sim 1$–2 kcal/mol (a typical change), and if the difference in free energy between native and denatured states is $\sim 8$ kcal/mol (a typical $\Delta G$), then a few such mutations might be expected to destabilize the molecule altogether (i.e. $\Delta G > 0$).
>
> While mutations may well result in structural changes, they are not expected to affect two-state behavior materially. Logically, the persistence of two-state behavior implies that a population of native-like molecules will exist, even in the presence of the most

---

[1] Citation numbers correspond to those used in this review.

destabilizing mutations. To emphasize this point, suppose that several destabilizing mutations were introduced, shifting the equilibrium constant such that $\Delta G = +5.5$ kcal/mol. Assuming two-state behavior at physiological $RT$, in a population of 10,000 molecules, $\sim 9999$ would be unfolded, but the remaining molecule would still adopt the native-like conformation.

Thus, jigsaw-puzzle packing is unlikely to be the main factor that governs conformational specificity, i.e. the stereochemical code that discriminates the native conformation from other conceivable chain folds. The two arguments in summary are:

1. X-ray studies indicate that mutations do alter packing slightly, but the changes are accommodated locally, without gross rearrangement of the conformation.
2. From folding studies, one can infer that packing faults push the folding equilibrium toward the unfolded state, not toward a novel conformation.

The conclusion that packing is not a principal source of conformational specificity prompts two obvious questions: given the close-packed nature of the protein core, how can perturbations that affect packing be accommodated without global disruption, and if packing is not the source of specificity, what is? We turn now to the former question; the next section examines the latter one.

Generalizing from current studies, a structural perturbation, such as a mutation, can be accommodated without global rearrangement. This finding appears enigmatic, in view of the close-packed nature of the protein core; were a tennis ball added to the middle of a stack of close-packed tennis balls, the perturbation would be propagated throughout the stack. How can the perturbing effect of adding a carbon atom to the close-packed protein core be absorbed with only local rearrangement?

We propose an explanation. The interior of a globular protein is comprised largely of assemblages of repetitive secondary structure (92)—helix, sheet, and their superstructures—with turns and loops relegated to the exterior (81, 91, 144). At least in part, the high packing densities observed in proteins are a consequence of the fact that segments of repetitive secondary structure pack together tightly: helix against helix (28), strand-of-sheet against strand-of-sheet (189), and helix against strands-of-sheet (68). Even turns and loops, though typically situated on the molecular exterior, are known to be intrinsically well packed (91).

In general, perturbation of a residue (e.g. by mutation) will involve some segment of secondary structure that includes the affected residue, together with other impinging segments. In the case of interacting helices, the side chain–into-groove packing should be somewhat insensitive to the precise details of the perturbation. The globins, being predominantly helical, are a useful model for accommodation of this kind. As shown by Lesk & Chothia (89), natural mutations do not promote global change. Instead, the axis of the involved helix realigns slightly, with adjacent turns acting as flexible joints. Sheet is also highly flexible within its allowed modes of deformation, as demonstrated by Salemme (149) and Chothia (26).

In summary, the high packing densities of globular proteins are an undeniable experimental fact (137). Given the lumpy, idiosyncratic shapes of residue side chains, this fact begs rationalization, and an explanation has been proposed. However, regardless of the path by which high packing densities are realized, the molecular interior of a folded protein should be well packed. Otherwise, dispersion forces would favor denaturation, because presumably efficient packing can be achieved between protein and solvent in the denatured state. High packing densities have been interpreted to mean that internal packing is a principal source of conformational specificity, but recent structural and folding studies imply that such a conclusion may be unwarranted.

## A MODEL FOR PROTEIN FOLDING

We propose a hierarchic framework model of protein folding, an expanded hybrid of two earlier models (76, 145), in which nascent elements of secondary structure associate in step-wise fashion, leading to larger modules and further association. Proteins are known to exhibit hierarchic architecture (35, 145), and numerous examples of independently folding protein domains and subdomains have been documented, both in natural (73, 178) and synthetic (114) systems. Hierarchic organization is a familiar and simplifying pattern in the macroscopic world, so the existence of hierarchic architecture and autonomous domains in proteins has been a suggestive finding. The following paragraphs describe major features of the model.

HYDROPHOBIC COLLAPSE  Akin to a critical micelle concentration (168), under physiological conditions, a polypeptide with both a sufficient fraction of hydrophobic residues and sufficient chain length to enclose a suitable volume will undergo spontaneous collapse, because constituent hydrophobic groups are a better self-solvent than bulk water.

MOLTEN GLOBULE  Under conditions that promote the molten-globule state (39, 62, 82), which includes some extraneous water and precedes fixed tertiary interactions, the usual cohesive properties of water would be reduced or eliminated because of the high local concentration of protein. For this reason, the hydrophobic effect is expected to be markedly diminished once the molten globule has formed, and further folding will arise from the drive to return interior waters to the bulk phase.

PACKING   Because the molecule traverses paths leading to the folded state, from either an unfolded state or the molten globule, the search for close-packed interactions would not be rate-limiting. This critical feature arises from the fact that protein structure is reducible, almost entirely, to just four secondary-structure formats: $\alpha$-helix, $\beta$-sheet, turns, and loops (54, 139). The core is comprised primarily of the two repetitive structures—$\alpha$ and $\beta$—with connecting elements of nonrepetitive structure turns and loops situated typically on the outside. Both $\alpha$ and $\beta$ have extensible, open-ended structural formats, like tile patterns (123). These elements can pack together tightly in any pairwise combination (92)—$\alpha$ with $\alpha$, $\beta$ with $\beta$, and $\alpha$ with $\beta$—resulting in the high packing densities that are characteristic of globular proteins.

HYDROGEN BONDS   Because the core is comprised largely of $\alpha$-helix and $\beta$-sheet, many polar groups will necessarily be buried. Hydrogen-bonding requirements for these polar groups must be satisfied within the molecular interior, or hydrogen bonding would favor denaturation. An effective way to satisfy such requirements is to establish H-bonds prior to or concomitant with formation of the core. The finding that H-bonds are predominantly local (8, 163) is consistent with this order of events.

AUTONOMY OF SECONDARY STRUCTURE   Several years ago, Baldwin and coworkers (14), expanding upon work of Brown & Klee (19), demonstrated the existence of stable, short, isolated helices in water, under near-physiological conditions. Though not as well studied, stable, isolated turns (40) and loops (61) have also been reported. Studies of $\beta$-sheet in isolation have lagged behind other categories, because sheet, by its very nature, tends to aggregate in isolation. Unlike earlier studies of homopolymers (164), contemporary studies utilize sequences drawn from natural proteins (98, 106, 115, 119); it is plausible that factors that promote secondary structure in isolation will also do so in proteins.

HIERARCHY   Once secondary structure is fixed, only a few assembly patterns will both satisfy steric constraints and bury exposed apolar surfaces effectively (29, 133, 142). The additional imposition of hierarchic constraints (35, 145) would further winnow this set, possibly to uniqueness. In such a folding process, segments of secondary structure associate, spawning supersecondary structure. In turn, these small hierarchic units coalesce into larger modules, in stepwise progression, leading to the hierarchic architecture that is observed in globular proteins.

COOPERATIVITY   In globular proteins, folding is accompanied by a hierarchy of energetically favored and mutually reinforcing conformational restrictions. Dipeptides preferentially populate discrete regions of a Ramachandran ($\phi,\psi$) plot (18). Larger structures, e.g. helix and sheet, are found within these preferred regions (140), although their characteristic hydrogen bonding is not a factor in determining the preferred conformation of a dipeptide. Supersecondary structure involves favorable packing between segments of helix and sheet (28); again at this level, the burial of hydrophobic surface promotes segment association but is not a factor in segment formation. Thus, at each successive level in the accretion of structure—dipeptide, secondary structure, and supersecondary structure—new quasi-independent, stabilizing factors come into play that serve to enhance those of earlier steps, with no energetic compromises required at any level.

These levels of structure emerge regardless of their order of formation. Early steps in the process, with individual stabilities approximating physiological $RT$ (18), should be readily reversible. However, as secondary structure is nucleated and structures of persisting stability form, the folding reaction proceeds spontaneously, leading to the observed cooperativity of protein folding under physiological conditions.

Central to any model is the question of why proteins have a unique equilibrium structure. In the features described above, the set of allowed conformations would be restricted to a small and possibly unique number, once segmentation of the polypeptide chain into elements of secondary structure has been fixed. As such, the formation of secondary structure plays a pivotal role in our model. In light of recent work, the secondary-structure format may be governed by an underlying stereochemical code.

## A Stereochemical Code for Protein Folding

To a close approximation, protein structure can be dissected into just four categories of secondary structure: helix, sheet, turns, and loops (54, 138). If so, then factors governing secondary structure formation need only be sufficient to discriminate among these four formats. What factors determine which of the four available structural formats will be adopted by an arbitrary chain segment?

Selection of a secondary structure format is likely mediated by a redundant stereochemical code that arises from the interplay between the shape and polarity of residue side chains and secondary-structure conformation. Initial evidence for such a code is now emerging in the case of helices. Capping of helix termini involves residue polarity (128, 141), while residue preferences at central positions of the helix are governed largely by side-chain shape (33, 186, 187), as described next.

The key feature of the Pauling-Corey-Branson $\alpha$-helix is a repeating pattern of main-chain hydrogen bonds between each amide hydrogen and the carbonyl oxygen located four residues upstream (122). However, this feature fails to discriminate between helix and other categories of secondary structure because all residues (except proline) have identical backbones.

A likely source of structural discrimination is found in helix capping (128, 141). In the helix backbone, the first four amide hydrogens and last four carbonyl oxygens necessarily lack intrahelical hydrogen-bond partners. Since the mean length of a helix in a globular protein is 12 residues, these 8 unsatisfied groups represent 50% of main-chain hydrogen bonding, on average. It has been hypothesized that hydrogen bonds are essential for these otherwise unsatisfied backbone amides and, typically, are provided by side chains of polar residues that flank the helix termini. Experimental evidence in favor of this hypothesis has been accumulating, as described in the section on hydrogen bonding. Recent NMR studies of peptides (97) have been particularly convincing in this regard.

In addition to capping at helix termini, side-chain shape appears to play a discriminatory role in central residues of the helix (33, 186, 187). For example, valine can populate all three side-chain conformers (*gauche*+, *gauche*−, *trans*) in the unfolded state, but this $\beta$-branched side chain is restricted to essentially one conformer (*trans*) in a helix, as shown in Figure 6. The corresponding energy loss ($T\Delta S$), due solely to the reduction in

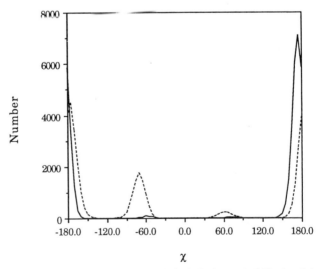

*Figure 6* Rotamer distributions for Val in both the helical state (*solid line*) and the standard state (*dashed line*) (33). See text for further details.

side-chain configurational entropy, is $\sim RT \ln 3$, on the order of physiological $RT$. Such energy terms can cause residues to favor one type of secondary structure over others, and, although modest individually, they can be substantial in the aggregate.

In a stereochemical code, such as the one described here for helices, each residue makes an energetically small contribution to conformation: e.g. a single H-bond for a polar residue at a helix cap or an (unfavorable) entropic contribution of $\sim RT \ln 3$ for a Val in the center of the helix. It is the aggregate effect of these individual terms, summed over the entire molecule, that determines (and probably overdetermines) conformation. One prediction from this type of distributed code is a scarcity of conformational tender spots wherein single mutations result in major disruption of the overall fold. Currently, the prediction is in good agreement with experiment (1, 46, 95, 102, 103). Even in extreme cases, the protein apparently can tolerate a major packing fault without precipitating global rearrangement (45).

This postulated stereochemical code, with the protein fold determined by many individually small terms, represents a robust folding mechanism. One might think that the existence of conserved residues in proteins argues against these ideas, but on further consideration, the opposite appears to be true. Of approximately 450 globins of known sequence, only one residue (i.e. the proximal histidine) is absolutely conserved (11), ostensibly for functional, not conformational, reasons. Typically, the mutational experiments of nature, as well as those in the laboratory, do not dead-end in conformational catastrophe, thereby allowing evolution to proceed apace.

ACKNOWLEDGMENTS

We thank Eaton Lattman for helpful discussion, Don Engelman for forbearance, and Robert Baldwin and Neville Kallenbach for preprints of papers. This review is dedicated to Kensal E. Van Holde on the occasion of his sixty-fifth birthday. Support from grants NIH (GM 29459 to G. R.) and NSF (DMB 8705085 to R. W.) is acknowledged.

*Literature Cited*

1. Alber, T. 1989. *Annu. Rev. Biochem.* 58: 765–98
2. Anfinsen, C. B. 1973. *Science* 181: 223–30
3. Anfinsen, C. B., Haber, M., Sela, M., White, F. H. Jr. 1961. *Proc. Natl. Acad. Sci. USA* 47: 1309–14
4. Anson, M. L., Mirsky, A. E. 1925. *J. Gen. Physiol.* 9: 169–79
5. Argos, P., Palau, J. 1982. *Int. J. Pept. Protein Res.* 19: 380–93
6. Aronow, R. H., Witten, L. 1960. *J. Phys. Chem.* 64: 1643–48
7. Artymiuk, P. J., Blake, C. C. F. 1981. *J. Mol. Biol.* 152: 737–62
8. Baker, E. N., Hubbard, R. E. 1984. *Prog. Biophys. Mol. Biol.* 44: 97–179
9. Baldwin, R. L. 1986. *Proc. Natl. Acad. Sci. USA* 83: 8069–72
10. Barlow, D. J., Thornton, J. M. 1988. *J. Mol. Biol.* 201: 601–19
11. Bashford, D., Chothia, C., Lesk, A. M.

1987. *J. Mol. Biol.* 196: 199–216
12. Behe, M. J., Lattman, E. E., Rose, G. D. 1991. *Proc. Natl. Acad. Sci. USA* 88: 4195–99
13. Bell, J. A., Becktel, W. J., Sauer, U., Baase, W. A., Matthews, B. W. 1992. *Biochemistry* 31: 3590–96
14. Bierzynski, A., Kim, P. S., Baldwin, R. L. 1982. *Proc. Natl. Acad. Sci. USA* 79: 2470–74
15. Bigelow, C. C. 1967. *J. Theor. Biol.* 16: 187–211
16. Bone, R., Cullis, P., Wolfenden, R. 1983. *J. Am. Chem. Soc.* 105: 1339–43
17. Bowie, J. U., Reidhaar-Olson, J. F., Lim, W. A., Sauer, R. T. 1990. *Science* 247: 1306–10
18. Brant, D. A., Miller, W. G., Flory, P. J. 1967. *J. Mol. Biol.* 23: 47–65
19. Brown, J. E., Klee, W. A. 1971. *Biochemistry* 10: 470–76
20. Bruch, M. D., Dhingra, M. M., Gierasch, L. M. 1991. *Proteins* 10: 130–39
20a. Butler, J. A. V., Ramchandani, C. N. 1935. *J. Chem. Soc.* pp. 952–55
21. Cabani, S., Gianni, P. 1979. *J. Chem. Soc. Faraday Trans. 1* 75: 1184–95
21a. Cabani, S., Gianni, P., Mollica, L. 1981. *J. Sol. Chem.* 10: 563
22. Carter, P., Wells, J. A. 1987. *Science* 237: 394–99
23. Ceccarelli, C., Jeffrey, G. A., Taylor, R. 1981. *J. Mol. Struct.* 70: 255–71
24. Chothia, C. 1974. *Nature* 248: 338–39
25. Chothia, C. 1976. *J. Mol. Biol.* 105: 1–14
26. Chothia, C. 1983. *J. Mol. Biol.* 163: 107–17
27. Chothia, C., Finkelstein, A. V. 1990. *Annu. Rev. Biochem.* 59: 1007–39
28. Chothia, C., Levitt, M., Richardson, D. 1977. *Proc. Natl. Acad. Sci. USA* 74: 4130–34
29. Cohen, F. E., Kuntz, I. D. 1989. See Ref. 49, pp. 647–705
30. Cohn, E. J., Edsall, J. T. 1943. *Proteins, Amino Acids, and Peptides as Ions and Dipolar Ions.* Princeton, NJ: Van Nostrand–Reinhold
31. Cornette, J. L., Cease, K. B., Margalit, H., Spouge, J. L., Berzofsky, J. A., DeLisi, C. 1987. *J. Mol. Biol.* 195: 659–85
32. Cramer, R. D. 1977. *J. Am. Chem. Soc.* 99: 5408–12
33. Creamer, T. P., Rose, G. D. 1992. *Proc. Natl. Acad. Sci. USA* 89: 5937–41
34. Creighton, T. E. 1990. *Biochem J.* 270: 1–16
35. Crippen, G. M. 1978. *J. Mol. Biol.* 126: 315–32
36. Dill, K. A. 1990. *Biochemistry* 29: 7133–55
37. Dill, K. A. 1990. *Science* 250: 297
38. Doig, A. J., Williams, D. H. 1992. *J. Am. Chem. Soc.* 114: 338–43
39. Dolgikh, D. A., Gilmanshin, R. I., Brazhnikov, E. V., Bychkova, V. E., Semisotnov, G. V., et al. 1981. *FEBS Lett.* 136: 311–15
40. Dyson, H. J., Cross, K. J., Houghten, R. A., Wilson, I. A., Wright, P. E., Lerner, R. A. 1985. *Nature* 318: 480–83
41. Edsall, J. T. 1935. *J. Am. Chem. Soc.* 57: 1506–7
42. Edsall, J. T., McKenzie, H. A. 1978. *Adv. Biophys.* 10: 137–207
43. Eisenberg, D., McLachlan, A. D. 1986. *Nature* 319: 199–203
44. Engelman, D. M., Steitz, T. A., Goldman, A. 1986. *Annu. Rev. Biophys. Biophys. Chem.* 15: 321–53
45. Eriksson, A. E., Baase, W. A., Wozniak, J. A., Matthews, B. W. 1992. *Nature* 355: 371–73
46. Estell, D. A., Graycar, T. P., Miller, J. V., Powers, D. B., Burnier, D. B., et al. 1986. *Science* 233: 659–63
47. Evans, D. F., Chen, S., Schriver, G. W., Arnett, E. M. 1981. *J. Am. Chem. Soc.* 103: 481–82
48. Fairman, R., Shoemaker, K. R., York, E. J., Steward, J. M., Baldwin, R. L. 1989. *Proteins* 5: 1–7
49. Fasman, G. D., ed. 1989. *Prediction of Protein Structure and the Principles of Protein Conformation.* New York: Plenum
50. Fauchère, J.-L., Pliska, V. E. 1983. *Eur. J. Med. Chem. Chim. Ther.* 18: 369–75
51. Fersht, A. R. 1987. *Trends Biochem. Sci.* 12: 301–4
52. Fersht, A. R., Shi, J.-P., Jack, K.-J., Lowe, D. M., Wilkinson, A. J., et al. 1985. *Nature* 314: 235–38
53. Deleted in proof
54. Fetrow, J. S., Zehfus, M. H., Rose, G. D. 1988. *Bio/Technology* 6: 167–71
55. Frank, H. S., Evans, M. W. 1945. *J. Chem. Phys.* 13: 507–32
56. Frauenfelder, H., Sligar, S. G., Wolynes, P. G. 1991. *Science* 254: 1598–1603
57. Gellman, S. H., Adams, B. R., Dado, G. P. 1990. *J. Am. Chem. Soc.* 112: 460–61
58. Gething, M.-J., Sambrook, J. 1992. *Nature* 355: 33–45
59. Gibbs, P. R., Radizick, A., Wolfenden, R. 1991. *J. Am. Chem. Soc.* 113: 4714–15
60. Gierasch, L. M., King, J., eds. 1990. *Protein Folding.* Washington, DC: Am. Assoc. Adv. Sci.

61. Gooley, P. R., Carter, S. A., Fagerness, P. E., MacKenzie, N. E. 1988. *Proteins* 4: 48–55
62. Goto, Y., Fink, A. L. 1989. *Biochemistry* 28: 945–52
63. Hildebrand, J. H. 1978. *Proc. Natl. Acad. Sci. USA* 76: 1–194
64. Hopp, T. P., Woods, K. R. 1981. *Proc. Natl. Acad. Sci. USA* 78: 3824–28
65. Howarth, O. W. 1975. *J. Chem. Soc. Faraday Trans.* 1 71: 2303–9
66. Ippolito, J. A., Alexander, R. S., Christianson, D. W. 1990. *J. Mol. Biol.* 215: 457–71
67. Janin, J. 1976. *J. Mol. Biol.* 105: 13–14
68. Janin, J., Chothia, C. 1980. *J. Mol. Biol.* 143: 95–128
69. Janin, S., Wodak, S., Levitt, M., Maigret, B. 1978. *J. Mol. Biol.* 125: 357–86
70. Jeffrey, G. A., Maluszynska, H. 1982. *Int. J. Biol. Macromol.* 4: 173–85
71. Jeffrey, G. A., Saenger, W. 1991. *Hydrogen Bonding in Biological Structures*. New York: Springer-Verlag
72. Jencks, W. P. 1966. In *Current Aspects of Biochemical Energetics*, ed. N. O. Kaplan, E. P. Kennedy, p. 273. New York: Academic
73. Jeng, M.-F., Englander, S. W. 1991. *J. Mol. Biol.* 221: 1045–61
74. Kauzmann, W. 1959. *Adv. Protein Chem.* 16: 1–64
75. Kauzmann, W., Bodanszky, A., Rasper, J. 1962. *J. Am. Chem. Soc.* 84: 1777–88
76. Kim, P. S., Baldwin, R. L. 1982. *Annu. Rev. Biochem.* 51: 459–89
77. Kim, P. S., Baldwin, R. L. 1990. *Annu. Rev. Biochem.* 59: 631–60
78. King, J. 1989. *Chem. Eng. News* 67: 32–54
79. Klotz, I. M., Franzen, J. S. 1962. *J. Am. Chem. Soc.* 84: 3461–66
80. Kossiakoff, A. A., Shteyn, S. 1984. *Nature* 311: 582–83
81. Kuntz, I. D. 1972. *J. Am. Chem. Soc.* 94: 8568–72
82. Kuwajima, J. 1989. *Proteins* 6: 87–103
83. Kyte, J., Doolittle, R. F. 1983. *J. Mol. Biol.* 157: 105–32
84. Lambert, J. D. 1953. *Discuss. Faraday Soc.* 15: 226–36
85. Lattman, E. E., Rose, G. D. 1993. *Proc. Natl. Acad. Sci. USA* 90: 439–41
86. Lau, K. F., Dill, K. A. 1989. *Macromolecules* 22: 3986–97
86a. Lecomte, J. T. J., Moore, C. D. 1991. *J. Am. Chem. Soc.* 113: 9663–65
87. Lee, B. K., Richards, F. M. 1971. *J. Mol. Biol.* 55: 379–400
88. Lee, C., Subbiah, S. 1991. *J. Mol. Biol.* 217: 373–88
89. Lesk, A. M., Chothia, C. 1980. *J. Mol. Biol.* 136: 225–70
90. Lesser, G. J., Rose, G. D. 1990. *Proteins* 8: 6–13
91. Leszczynski, J. F., Rose, G. D. 1986. *Science* 234: 849–55
92. Levitt, M., Chothia, C. 1976. *Nature* 261: 552–58
93. Lewis, C. A. Jr., Wolfenden, R. 1973. *J. Am. Chem. Soc.* 95: 6685–88
94. Deleted in proof
95. Lim, W. A., Sauer, R. T. 1991. *J. Mol. Biol.* 219: 359–76
96. Livingstone, J. R., Spolar, R. S., Record, M. T. Jr. 1991. *Biochemistry* 30: 4237–44
97. Lyu, P., Wemmer, D. E., Zhou, H. X., Pinker, R. J., Kallenbach, N. R. 1992. *Biochemistry*. In press
98. Lyu, P. C., Liff, M. I., Marky, L. A., Kallenbach, N. R. 1990. *Science* 250: 669–73
99. Lyu, P. C., Zhou, H. X., Jelveh, N., Wemmer, D. E., Kallenbach, N. R. 1992. *J. Am. Chem. Soc.* 114: 6560–62
100. MacGregor, M. J., Islam, S. A., Sternberg, M. J. E. 1987. *J. Mol. Biol.* 198: 295–310
101. Marsh, R. E., Donohue, J. 1967. *Adv. Protein Chem.* 22: 235–56
101a. Mathews, C. K., van Holde, K. E., 1990. *Biochemistry*. Redwood City, CA: Benjamin/Cummings
102. Matouschek, A., Kellis, J. T., Serrano, L., Fersht, A. R. 1989. *Nature* 340: 122–26
103. Matthews, B. W. 1987. *Biochemistry* 26: 6885–88
104. McCammon, J. A., Harvey, S. C. 1987. *Dynamics of Proteins and Nucleic Acids*. Cambridge: Cambridge Univ. Press
105. Meot-ner, M., Sieck, L. W. 1986. *J. Am. Chem. Soc.* 108: 7525–29
106. Merutka, G., Lipton, W., Shalongo, W., Park, S.-H., Stellwagen, E. 1990. *Biochemistry* 29: 7511–15
107. Miller, S., Janin, J., Lesk, A. M., Chothia, C. 1987. *J. Mol. Biol.* 196: 641–56
108. Mirsky, A. E., Pauling, L. 1936. *Proc. Natl. Acad. Sci. USA* 22: 439–47
109. Murphy, K. P., Privalov, P. L., Gill, S. J. 1990. *Science* 247: 559–61
110. Murray-Rust, P., Glusker, J. P. 1984. *J. Am. Chem. Soc.* 106: 1018–25
111. Nall, B. T., Dill, K. A., eds. 1991. *Conformations and Forces in Protein Folding*. Washington, DC: Am. Assoc. Adv. Sci.
112. Nemethy, G., Scherega, H. A. 1962. *J. Chem. Phys.* 36: 3382–3400
113. Nozaki, Y., Tanford, C. 1971. *J. Biol. Chem.* 246: 2211–17

114. Oas, T. G., Kim, P. S. 1988. *Nature* 336: 42–48
115. O'Neil, K. T., DeGrado, W. F. 1990. *Science* 250: 646–51
116. Osinga, M. 1979. *J. Am. Chem. Soc.* 101: 1621–22
117. Pace, C. N. 1990. *TIBS* 15: 14–17
118. Pace, C. N. 1992. *J. Mol. Biol.* 226: 29–35
119. Padmanabhan, S., Marqusee, S., Ridgeway, T., Laue, T. M., Baldwin, R. L. 1990. *Nature (London)* 344: 268–70
120. Pauling, L. 1939. *The Nature of the Chemical Bond.* Ithaca, NY: Cornell Univ. Press
121. Pauling, L., Corey, R. B. 1951. *Proc. Natl. Acad. Sci. USA* 37: 251–56
122. Pauling, L., Corey, R. B., Branson, H. R. 1951. *Proc. Natl. Acad. Sci. USA* 37: 205–11
123. Penrose, R. 1989. *Aperiodicity and Order* 2, ed. M. Jaric. New York: Academic
124. Perutz, M. F. 1951. *Nature* 167: 1053–54
125. Pimentel, G. C., McClellan, A. L. 1960. *The Hydrogen Bond.* San Francisco: Freeman
126. Pinker, R. J., Rose, G. D., Kallenbach, N. R. 1993. *J. Mol. Biol.* Submitted
127. Ponder, J. W., Richards, F. M. 1987. *J. Mol. Biol.* 193: 775–91
128. Presta, L. G., Rose, G. D. 1988. *Science* 240: 1632–41
129. Privalov, P. L. 1979. *Adv. Protein Chem.* 33: 167–241
130. Privalov, P. L., Gill, S. J. 1988. *Adv. Protein Chem.* 39: 191–243
131. Privalov, P. L., Gill, S. J., Murphy, K. P. 1990. *Science* 250: 297–98
132. Privalov, P. L., Khechinashvili, N. N. 1974. *J. Mol. Biol.* 86: 665–84
133. Ptitsyn, O. B., Rashin, A. A. 1975. *Biophys. Chem.* 3: 1–20
134. Quiocho, F. A. 1986. *Annu. Rev. Biochem.* 55: 287–315
135. Radzicka, A., Wolfenden, R. 1988. *Biochemistry* 27: 1664–70
136. Richards, F. M. 1974. *J. Mol. Biol.* 82: 1–14
137. Richards, F. M. 1977. *Annu. Rev. Biophys. Bioeng.* 6: 151–76
138. Richards, F. M. 1991. *Sci. Am.* 264: 54–63
139. Richards, F. M., Kundrot, C. E. 1988. *Proteins Struct. Funct. Genet.* 3: 71–84
140. Richardson, J. S. 1981. *Adv. Protein Chem.* 34: 167–339
141. Richardson, J. S., Richardson, D. C. 1988. *Science* 240: 1648–52
142. Richmond, T. J., Richards, F. M. 1978. *J. Mol. Biol.* 119: 537–55
143. Deleted in proof
144. Rose, G. D. 1978. *Nature* 272: 586–90
145. Rose, G. D. 1979. *J. Mol. Biol.* 134: 447–70
146. Rose, G. D., Geselowitz, A. R., Lesser, G. J., Lee, R. H., Zehfus, M. H. 1985. *Science* 229: 834–38
147. Rose, G. D., Gierasch, L. M., Smith, J. A. 1985. *Adv. Protein Chem.* 37: 1–109
148. Rose, G. D., Roy, S. 1980. *Proc. Natl. Acad. Sci. USA* 77: 4643–47
149. Salemme, F. R. 1983. *Prog. Biophys. Mol. Biol.* 42: 95–133
150. Sandberg, W. S., Terwilliger, T. C. 1990. *Science* 245: 54–57
151. Schellman, J. A. 1955. *C. R. Trav. Lab. Carlsburg Ser. Chim.* 29: 223–29
152. Scholtz, J. M., Baldwin, R. L. 1992. *Annu. Rev. Biophys. Biomol. Struct.* 21: 95–118
153. Schulz, G. E., Schirmer, R. H. 1979. In *Principles of Protein Structure*, pp. 166–205. New York: Springer-Verlag
154. Serrano, L., Fersht, A. R. 1989. *Nature* 342: 296–99
155. Serrano, L., Kellis, J. T. Jr., Cann, P., Matouschek, A., Fersht, A. R. 1992. *J. Mol. Biol.* 224: 783–804
156. Sharp, K. A., Nicholls, A., Fine, R. F., Honig, B. 1991. *Science* 252: 106–9
157. Sharp, K. A., Nicholls, A., Friedman, R., Honig, B. 1991. *Biochemistry* 30: 9686–97
158. Shirley, B. A., Stanssens, P., Hahn, U., Pace, C. N. 1992. *Biochemistry* 31: 725–32
159. Shortle, D., Stiles, W. E., Meeker, A. K. 1990. *Biochemistry* 29: 8033–41
160. Singh, J., Thornton, J. M., Snarey, M., Campbell, S. F. 1987. *FEBS Lett.* 224: 161–71
161. Small, D. M. 1986. *The Physical Chemistry of Lipids.* New York: Plenum
162. Stahl, N., Jencks, W. P. 1986. *J. Am. Chem. Soc.* 108: 4196–4205
163. Stickle, D. F., Presta, L. G., Dill, K. A., Rose, G. D. 1992. *J. Mol. Biol.* 226: 1143–59
164. Sueki, M., Lee, S., Powers, S. P., Denton, J. B., Konishi, Y., Scheraga, H. A. 1984. *Macromolecules* 17: 148–55
165. Sun, Y., Spellmeyer, D., Pearlman, D. A., Kollman, P. 1992. *J. Am. Chem. Soc.* 114: 6798–6801
166. Susi, H., Timasheff, S. N., Ard, J. S. 1964. *J. Biol. Chem.* 239: 3051–54
167. Tanford, C. 1978. *Science* 200: 1012–18
168. Tanford, C. 1980. *The Hydrophobic Effect.* New York: Wiley
169. Taylor, R., Kennard, O. 1984. *Acc. Chem. Res.* 17: 320–26

170. Taylor, R., Kennard, O., Versichel, W. 1983. *J. Am. Chem. Soc.* 105: 5761–66
171. Teller, D. C. 1976. *Nature* 260: 729–31
172. Traube, J. 1891. *Liebigs Ann. Chem.* 265: 27–55
173. Varadarajan, R., Connelly, P. R., Sturtevant, J. M., Richards, F. M. 1992. *Biochemistry* 31: 1421–26
174. Vedani, A., Dunitz, J. D. 1985. *J. Am. Chem. Soc.* 107: 7653–58
175. Voet, D., Voet, J. G. 1990. *Biochemistry*. New York: Wiley & Sons
176. Watson, J. D., Crick, F. H. C. 1953. *Nature* 171: 737–38
177. Weiner, S. J., Kollman, P. A., Case, D. A., Chandra Singh, U., Ghio, C., et al. 1984. *J. Am. Chem. Soc.* 106: 765–84
178. Wetlaufer, D. B. 1981. *Adv. Protein Chem.* 34: 61–92
179. Williams, D. H. 1992. *Aldrichimica Acta* 25: 9
180. Wolfenden, R. 1983. *Science* 222: 1087–93
181. Wolfenden, R., Andersson, L., Cullis, P. M., Southgate, C. B. 1981. *Biochemistry* 20: 849–55
182. Wolfenden, R., Lewis, C. A. Jr. 1976. *J. Theor. Biol.* 59: 231–35
182a. Wolfenden, R., Liang, Y. L., Matthews, M., Williams, R. 1987. *J. Am. Chem. Soc.* 109: 463–66
183. Wolfenden, R. V., Kati, W. M. 1991. *Acc. Chem. Res.* 24: 209–15
184. Woolfson, D. N., Cooper, A., Harding, M. M., Williams, D. H., Evans, P. A. 1993. *Science*. In press
185. Wüthrich, K., Wagner, G. 1978. *TIBS* 3: 227–30
186. Yun, R. H., Anderson, A., Hermans, J. 1991. *Proteins* 10: 219–28
187. Yun, R. H., Hermans, J. 1991. *Protein Eng.* 4: 761–66
188. Zehfus, M. H. 1987. *Proteins* 2: 90–110
189. Zehfus, M. H., Rose, G. D. 1986. *Biochemistry* 25: 5759–65

# GLYCOPROTEIN MOTILITY AND DYNAMIC DOMAINS IN FLUID PLASMA MEMBRANES

## Michael P. Sheetz

Department of Cell Biology, Duke University Medical Center, Durham, North Carolina 27710

KEY WORDS: membrane diffusion, membrane traffic, cell motility

CONTENTS

| | |
|---|---|
| INTRODUCTION | 418 |
| PASSIVE GLYCOPROTEIN MOVEMENTS | 419 |
|     *Experimental Measurements of Lateral Diffusion* | 419 |
|     *Percolation and Barrier Models* | 421 |
|     *Transient-Binding Models* | 422 |
|     *Protein-Density Models* | 422 |
|     *External, Bilayer, and Internal Membrane Layers as Independent Viscous Fluids* | 423 |
|     *Barriers to Lateral Diffusion: Location and Spacing* | 425 |
| ACTIVE GLYCOPROTEIN MOVEMENTS | 426 |
|     *Membrane Turnover and Flow* | 426 |
|     *Rearward Flow in Lamellipodia* | 428 |
|     *Forward Transport in Lamellipodia, Filopodia, and Flagella* | 429 |
| SUMMARY | 429 |

The concentration of membrane glycoproteins into domains for specific functions is important for efficient cell viability, but how concentration occurs in fluid membranes is not obvious. The predominant passive movements of plasma membrane glycoproteins measured by fluorescence-photobleaching-recovery and single particle–tracking methods have defined a fluid phase in which glycoproteins diffuse much more slowly than in a bilayer alone and in which cytoplasmic barriers restrict long-range movements. Apparently, three layers of the plasma membrane can independently contribute to the frictional drag on glycoprotein diffusion, the external, bilayer, and internal layers. For example, in glycoproteins with a single α-helical transmembrane segment, the major frictional drag is in the external layer, and neither percolation nor transient-binding models can explain their diffusive behavior. In those cases, protein concentration appears to be a major factor in determining the diffusion coefficient. Active movements of membrane glycoproteins and membrane turnover may augment other factors such as membrane curvature to produce dynamic domains of protein concentration. These new observations have important implications for the many important cell functions involving regional specializations of membranes.

## INTRODUCTION

Since the time of the diffusive mixing studies of Frye & Edidin (20) to the present, the analysis of protein diffusion in membranes has strongly influenced our understanding of membrane structure. Previous reviews have summarized much of the early data gathered using fluorescence methods to measure the diffusion coefficients and nondiffusing fractions of membrane glycoproteins and lipids (13, 17, 26). I compare and contrast those observations with more recent studies that used single-particle tracking (SPT) methods (21, 23, 49). All measurements of membrane glycoprotein diffusion have shown that diffusion is 5- to 100-fold slower in plasma membranes when compared with diffusion of the same glycoproteins in lipid bilayers. Several factors could contribute to the resistance encountered by glycoproteins in situ including the density of protein packing, percolation through cytoskeletal barriers at the membrane surface, and rapid binding to the cytoskeleton. Passive diffusion of glycoproteins probably does not concentrate glycoproteins, but active movements of glycoproteins attached to the cytoskeleton and active membrane turnover may create dynamic membrane domains.

A variety of cell functions of necessity involve the regional concentration of membrane glycoproteins such as rapid synaptic transmissions that require the acetylcholine receptor plaques on muscle cells at synaptic sites. The mechanisms of formation of similar concentrations of glycoproteins in other systems is not well understood. In a pure fluid mosaic membrane (51), lipid phase separations or passive diffusion and trapping by the cytoskeleton are the major mechanisms invoked to explain the regional concentrations of membrane glycoproteins [see review of membrane domains (16)]. Although those explanations fit with the behavior and dynamics of cells in tissues at steady state (a time scale of days to weeks), several very dynamic functions seen particularly in cells in vitro (a time scale of seconds to minutes) suggest that active motile processes are involved. For example, the directed migration of neuronal growth cones or chemotaxis of macrophages involves the directed sensing of environmental cues by a structure that is rapidly remodeling its cytoskeleton. Recent studies have defined the motile behavior of many different membrane glycoproteins in cells in culture, and those observations are inconsistent with some of the old models and suggest new concepts that have not been seriously considered. Primarily, by coupling glycoproteins to the dynamic processes of cytoskeleton and membrane turnover, glycoprotein domains can be formed routinely in plasma membranes. Domains created by cytoskeletal movements and membrane trafficking can be dynamic in that they could dissipate by diffusion when the active processes cease.

## PASSIVE GLYCOPROTEIN MOVEMENTS

To understand the radically lower diffusion coefficients of membrane glycoproteins in biological membranes, researchers have performed detailed theoretical treatments of several of the popular models including percolation (43, 44), transient binding (27), and protein density (36, 41, 42) (see Figure 1). Such treatments are useful in defining criteria for differentiating between the different models. For example, the dependence of diffusion on crosslinking is a potential discriminator because in the binding model, the diffusion coefficient should be strongly dependent on the extent of crosslinking. Clearly, biological diversity will mean that many different models will be represented somewhere. The ability to express chimeric proteins in a single cell type provides a powerful way to define which portions of a molecule are most important in determining the diffusion coefficient. In addition, the advent of optical tweezers has made it possible to apply probes to specific locations on the cell surface.

### Experimental Measurements of Lateral Diffusion

The measurement of diffusion of membrane glycoproteins and lipids has largely been through fluorescent techniques that typically involve the tagging of membrane proteins with antibodies or at minimum a Fab fragment (17). When possible, in the past the validity of the measurements made by this technology has been established, and photodamage is at worst a second-order factor in the measured diffusion coefficients (28). Thus, we assume that the diffusion coefficients measured using these techniques are generally valid unless there are extenuating circumstances. For example, it is often difficult to determine if microvilli or folds are present that could theoretically result in an underestimate of the true diffusion coefficient by a factor of 2.0–2.8, nearly independent of the length of villi (3). Com-

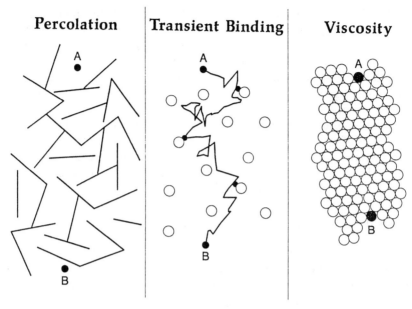

*Figure 1* This diagram illustrates the three basic models for the inhibition of diffusion of glycoproteins in biological membranes. In a percolation model, for the glycoprotein to traverse from point A to point B it would have to diffuse through the gaps in the barriers. Barriers could be on either the external or internal surface of the bilayer. In the transient-binding model, the glycoprotein would bind to immobilized structures (*open circles*) and would remain in one place for a significant fraction of the time. Reduction in the diffusion coefficient would be directly proportional to the fraction of time in the unbound state. In the viscosity model, the molecular crowding and immobility of other glycoproteins in the membrane would slow the diffusion of a glycoprotein from point A to B. The one important caveat in this model is that the highest viscosity need not be in the bilayer but could be in an external or internal membrane-protein layer.

parisons of the experimental diffusion coefficients from villated and nonvillated surfaces do not reveal significant differences (12, 59), which could be explained by a lack of flatness on the nonvillated membrane surfaces. An additional problem of the spot photobleaching method is that some of the cytoskeletally attached proteins could be moving actively, which would give a higher apparent diffusion coefficient.

Recently, the development of video analysis of two-dimensional movements of fluorescent (23), gold (21, 49), or latex (30) particle tags has provided a second way of following diffusion through single-particle tracking (SPT) (39). SPT has several weaknesses, including the fact that small numbers of glycoproteins are followed and the particle could be crosslinking membrane glycoproteins, thereby modifying diffusion coefficients (for review, see 47).

In general, there is good agreement on the mobile fraction of a given glycoprotein and the range of diffusion coefficients when measured using different techniques. But some details of the observations are problematic. For example, measurements of lipid-linked glycoprotein diffusion show immobile fractions that are difficult to explain (11, 37, 60). Antibody binding to surface ligands is generally thought to have an effect through crosslinking (35), but the extent of response varies dramatically between cells and depends on the size of the aggregates. In addition, the region of the cell is important because crosslinked glycoproteins are preferentially attached to the cytoskeleton at the leading edge of the lamellipodium (31). There are also potential problems with a monovalent Fab fragment, because in the red cell membrane even the binding of a monovalent Fab fragment can cause radical changes in membrane and presumably glycoprotein behavior (9).

## Percolation and Barrier Models

The early observations of diffusion in the erythrocyte membrane were consistent with the concept that the spectrin-actin membrane skeleton was the major resistance to lateral diffusion (22, 48). The finding of spectrin analogs in many other tissues and even the relationship of dystrophin to spectrin has reinforced the idea that spectrin-like proteins may have a significant role in resisting lateral movements of glycoproteins (4). If spectrin in all these systems lies on the membrane surface and forms a barrier to lateral diffusion, then the percolation models are most relevant.

In the percolation treatments (44, 45), a relatively high concentration of barriers is required to give the observed diffusion coefficients if one extrapolates from known membrane systems. At such a density of barriers ($>2/\mu^2$), the barrier-free pathlength (BFP), or distance between barriers to lateral movement, should be very short. When the distance between diffusion barriers was measured recently (14), the BFP was found to be very large ($\sim 3 \mu$) at physiological temperatures. Furthermore, the BFP depended greatly on temperature, whereas diffusion coefficients did not. The BFP and SPT measurements of glycoprotein diffusion do not detect tightly spaced barriers to diffusion. Even the fluorescence-photobleaching-recovery (FPR) measurements with different beam sizes define membrane domains that are on the size of microns (60). Thus, the diffusion coefficients measured in most biological membranes appear low irrespective of the presence of barriers, and evidence indicates that cytoskeletal barriers are the major factors in determining the diffusion coefficient only in the case of the erythrocyte membrane.

## Transient-Binding Models

Another mechanism that has been invoked is the transient binding of membrane glycoproteins to the cytoskeleton or extracellular matrix either directly or indirectly through interaction with a membrane protein bound to the cytoskeleton (27). In such a scenario, the real diffusion coefficient of the protein in the membrane is essentially the same in a pure lipid bilayer as in the membrane, but when the cytoskeleton is present, transient binding to the cytoskeleton immobilizes it for a fraction of the time. The strongest argument against this model is that there is only a weak dependence of the diffusion coefficient on glycoprotein crosslinking in SPT studies. The model would predict that, for a crosslinked aggregate, the fraction of the time free would be the product of the free fractions for the components of the aggregate (i.e. if two proteins with 10% time unbound were linked, then the aggregate should be unbound only 1% of the time). In SPT measurements, the particles are usually multivalent and could crosslink several glycoproteins, but the diffusion coefficients of the moving particles are the same as the measured values from FPR. Another concern is the dynamics in that the bound state should be observed as a pause in the position of the particle in SPT analysis, but even the shortest time periods (30 ms for video) show no evidence of regular attachments. When attachment is seen (49), it is for periods of seconds to minutes and often ends in endocytosis.

## Protein-Density Models

The third type of model extends the Saffman-Delbruck treatment (40) and assumes that the protein density is extremely high, with the added feature that some of the proteins are rigidly attached to the cytoskeleton (36, 41, 42). Experimental measurement of bovine serum albumin diffusion adsorbed on plastic surfaces gives a good description of the observations of membrane glycoproteins (57). The attractive feature of this model is that it explains the protein domain swapping results (see below). In contradiction to the model are the studies of the effect of electrophoretic concentration on diffusion coefficients (56) and of the rapid diffusion on membrane blebs (54). These studies highlight a concern about the FPR technique in general. RBL (rat basophilic leukemia) cells are spherical and when osmotically swollen form a sphere of twice the original surface area (56). Because new membrane was probably not added during osmotic shock, initially microvilli and membrane folds, which were lost during swelling, likely constituted over half of the cell's surface area. With conventional optics, we cannot know if the electrophoretic field caused a reorganization of the membrane folds; consequently, it is possible that extra surface membrane, rather than protein, was concentrated in the membrane. Furthermore, a distinct subset of membrane proteins could have been concentrated by displacing other proteins such that the total protein concentration was essentially constant. The measurements of diffusion on membrane blebs in cultured cells have shown that the removal of the cytoskeleton causes a dramatic increase in the diffusion coefficient (54). But also in blebs, the density of membrane glycoproteins may be altered in addition to the obvious loss of cytoskeletal attachments. The predictions of the protein-density model, which treats the membrane as a continuum, can explain most of the observations. What is lacking is an understanding of whether or not simply protein density, or the presence of proteins that are immobilized, provides the greatest frictional drag on the proteins. One approach to these and related questions is to simplify the description of the frictional factors contributing to the diffusion coefficient.

## External, Bilayer, and Internal Membrane Layers as Independent Viscous Fluids

Protein domain deletion and swapping among membrane glycoproteins has provided considerable information on the important determinants of diffusion coefficients (15, 34, 61). A method of codifying the observations is to assume that the frictional drag of each domain is additive such that a chimeric protein's diffusion coefficient can be determined from the reciprocal of the sum of the frictional coefficients of each separate domain (61):

$$D = kT/(f_e + f_b + f_i),$$

where $D$ is the diffusion coefficient, $k$ is Boltzmann's constant, and $f_e$, $f_b$, and $f_i$ are the frictional coefficients for the external, bilayer, and internal layers, respectively. External domains contribute the greatest resistivities in the case of the many proteins that have a single transmembrane $\alpha$-helix or are lipid-linked. On the other hand, erythrocyte band 3, which spans the membrane 11 times, has a large cytoplasmic domain (4) and its diffusion is primarily restricted by cytoplasmic proteins (22, 48). Note that this treatment assumes that the layers are independent, uniform fluids. The effects of stable matrices are considered below as secondary factors defining the boundaries of diffusion and not the diffusion coefficients.

The sequencing of many membrane proteins has shown that the majority of the protein mass does not lie in the plane of the lipid bilayer. Within the bilayer phase, almost all proteins have an $\alpha$-helical structure, and many have only a single $\alpha$-helix spanning the bilayer (50). $\alpha$-Helices are relatively compact structures when compared with carbohydrate side chains in the external face or the asymmetric cytoskeletal proteins on the internal face. If we then consider the membrane structure as an external protein layer, a lipid bilayer, and an internal protein layer, the protein concentration and presumably viscosity will be radically different for the different layers (Figure 2). For example, if we look at the human erythrocyte, which is seen as an 11-nm-thick structure in the electron microscope, we can assume that the three layers have approximate thicknesses of 3, 4, and 4 nm for the outer, bilayer, and inner layers, respectively. In a single erythrocyte, we estimate that there are $6.0 \times 10^{10}$ Daltons of protein and carbohydrate in the outer layer (corresponding to 333 mg/ml), $4.5 \times 10^{10}$ Daltons of protein in the lipid layer (188 mg/ml), and $22 \times 10^{10}$ Daltons of protein in the inner layer ($\sim 900$ mg/ml). In this case, the cytoplasmic layer is nearly fivefold higher in protein concentration and therefore could have a considerably higher viscosity than the lipid phase. A further complication is that the cytoplasmic layer in the erythrocyte has a spectrin-actin lattice structure that essentially blocks diffusion of spectrin and associated com-

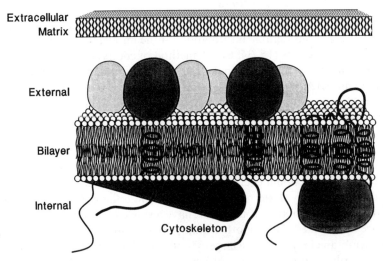

*Figure 2* Three different layers to the plasma membrane are noted in terms of their potentially separable behaviors. In the external layer, glycoproteins with a single α-helix spanning the membrane or lipid link are abundant. The viscosity of the external layer is determined by the external membrane proteins and not the extracellular matrix (32). The bilayer phase has primarily α-helical domains of glycoproteins, and there is little evidence of inhibition of lipid diffusion by glycoproteins. In the internal layer, the cytoskeleton and other proteins such as spectrin are very asymmetric and appear to play a significant role in inhibiting diffusion in the erythrocyte membrane. Long lateral movements are blocked primarily by the internal proteins, but outside of the erythrocyte the internal layer appears to only play a secondary role in inhibiting diffusion on the submicron scale.

ponents. Such detailed information is not available on many membranes, and the dynamics of membrane turnover and trafficking could alter it dramatically over a short period of time. The concept of separating the contributions of the external, bilayer, and intracellular layers has utility for the mechanical aspects of membrane function.

Of the three layers, we know the most about the external layer. It was characterized recently in several cultured cells by the use of gold attached to extracellular portions of proteins or lipids. Lipid diffusion is unhindered in biological membranes, and antifluorescein antibodies were used to attach gold to fluorescein phosphatidyl ethanolamine (32). The antibody and gold tag diffused fivefold slower than the lipid alone, from which the viscosity of the external phase was estimated to be 0.5–1.0 poise (32). In the same membranes, the diffusion coefficients of concanavalin A–coated latex particles (30) are less than a factor of two lower than the diffusion coefficients of the gold bound to lipid. This again indicates that the major frictional factor for those proteins was in the external membrane layer. Our understanding of the other layers as viscous fluids is not nearly as great. More research is needed on the proteins that have many $\alpha$-helices spanning the membrane and large intracellular domains.

## Barriers to Lateral Diffusion: Location and Spacing

Barriers to lateral diffusion have been measured by the change in the mobile fraction with bleaching-spot size in FPR (60), by the negative deviation from linearity of the mean-squared displacement with time in SPT measurements (11, 49), and by dragging proteins laterally with attached gold particles using optical tweezers until they are pulled from the trap by a barrier (14). In all of these cases, the barriers are predominantly cytoplasmic and are spaced on the scale of microns. The surprising finding was that the barrier-free pathlength measured by dragging glycoproteins greatly depended upon temperature [a fivefold increase in BFP with a ten-degree rise in temperature (14)]. In all of these cases the barriers did not contribute to the apparent diffusion coefficient of the glycoproteins, and it remains to be determined if stable boundaries influence the diffusion of band 3 in the erythrocyte.

One surprising finding that does not fit with any of the above theories is the observation that diffusion of a lipid-linked protein is bounded on the surface of sperm (38). This may represent a lipid-phase boundary or another type of barrier, and understanding in what other situations such a barrier might be used will be important.

Often in vivo, an extensive extracellular matrix of laminin, fibronectin, collagen, or other proteins is important for shaping the cell as well as in supporting mechanical processes. The presence of such a stable extracellular matrix can have a major effect on the diffusion of proteins; however, all the indications are that the role of this matrix is minimal in animal cells in culture. The strongest evidence comes from SPT measurements of lipid (32) and lipid-linked protein diffusion (14). These findings are somewhat selective in that measurements of diffusion are normally made on isolated cells in flattened regions of lamellipodia and not on cells with extensive extracellular matrix.

## ACTIVE GLYCOPROTEIN MOVEMENTS

The directed movement of cells and the focal assembly of cellular structures probably involves the active transport of membrane glycoproteins. Active movements could be driven by membrane flow (8) or by direct attachment to structures moving by motors (see below). Of the reported glycoprotein movements, the predominant mechanism appears to be attachment to moving cytoskeletal structures. In fact, in the plasma membrane there has not been a documented rearward lipid flow in any system (30, 33). Rather, we can speculate that attachment to the cytoskeleton and membrane protein turnover but not lipid flow are the major factors influencing membrane glycoprotein distribution and movement.

### Membrane Turnover and Flow

One of the major questions in dynamic cells is whether or not the membrane and lipid components are flowing toward the nucleus along with the cytoskeleton, which is clearly moving rearward (58). The observations of fluid-phase uptake indicate that the plasma membrane in many cells is rapidly being endocytosed and recycled [an area equivalent to the membrane surface area is endocytosed every 30–90 min (52)]. Although the concept of exocytosis at the front and endocytosis at the rear has not held up, sites of exocytosis and endocytosis can be separated physically, leading to dynamic gradients in components and some flows. For example, the measured rate of endocytosis of membrane glycoproteins corresponds to 100–800 $\mu^2$ in 10 min. The effective diffusion distance of a membrane glycoprotein with a diffusion coefficient of $10^{-9}$ cm$^2$ s$^{-1}$ is 5 $\mu$ in this time period. Thus, in addition to the classical concepts for domain formation involving barriers to glycoprotein diffusion (Figure 3a,b), exocytic or endocytic domains could concentrate or deplete membrane glycoproteins from specific regions (Figure 3c–e). These concentration gradients rely upon the continued traffic of membranes through the cell and would dissipate as soon as the endocytic rate decreased significantly.

The evidence on the location of the sites of endo- and exocytosis in

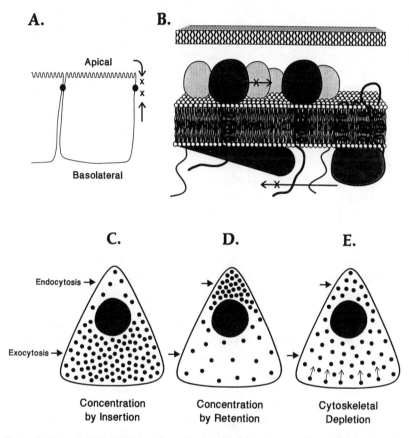

*Figure 3* Several models for the dynamic concentration of membrane glycoproteins. The common feature of these models is that they do not involve the stable attachment of the glycoproteins to cytoskeletal or external matrix structures. (*a*) The concentration of glycoproteins on the apical or basolateral surfaces of epithelial cells is the result of a barrier to lateral diffusion and the preferential transport of glycoproteins to either surface. (*b*) The barriers on the cytoplasmic surface block movement of selected glycoproteins into specific domains. (*c,d*) A separation of the site of exocytosis and endocytosis of a protein results in a depletion or concentration of the glycoprotein depending upon whether the protein is concentrated more (*c*) or less (*d*) in the endocytic vesicles than the membrane. (*e*) The glycoprotein preferentially attaches to the cytoskeleton over the lamellipodium and is depleted in that region. Freely diffusing glycoprotein could arise from reversal of cytoskeletal attachment (31) or from exocytosis.

most cells is fragmentary at present. Sites of insertion of viral membrane glycoproteins are near the front of the cell (5), but whether that means at the leading edge of the microtubules or the actual leading edge is not certain. Because of the absence of membranes in the lamellipodia, the primary sites of exocytosis are most likely near the ends of the microtubules or at the proximal boundary of the lamellipodium. Similarly, the major sites of endocytosis are proximal to the lamellipodium (10). A boundary has been described in the body of some cells toward which particles migrate [forward particles migrate rearward and rearward particles migrate forward toward this boundary (25)]. Much more effort is needed to give a better definition of the spatial patterning of endo- and exocytosis.

## *Rearward Flow in Lamellipodia*

The phenomenon of rearward movement of surface particles (1) is tied very clearly to the rearward migration of the cell cortex (7), which is composed of actin and associated proteins. Studies of particle movements clearly demonstrated that the rearward moving particles were attached to an actin-dependent cytoskeletal structure (18, 19). The exact identity of the moving structure, whether it is a subset of the actin or another substructure, has yet to be defined (55). Of note here is the fact that the leading edge of the lamellipodium is differentiated for rapid attachment of crosslinked glycoproteins to the cytoskeleton (31) and that this is the site for actin assembly as well (53). The fact that the attached surface proteins move rearward suggests that there should be a flow of material rearward and a concomitant depletion of proteins that bind to the cytoskeleton at the leading edge (Figure 3e).

Pgp-1, a major membrane glycoprotein, has an interesting distribution that occurs in migrating cells upon the binding of a monoclonal antibody (24). A dramatic gradient of Pgp-1 concentration forms; the lowest is at the leading edge and the highest is near the nucleus, but there is no decrease in the mobile fraction nor in the diffusion coefficient with antibody binding. One explanation of how the gradient could be maintained with a diffusing protein is that the glycoproteins are raked rearward by the rearward-moving cytoskeleton and attached membrane glycoproteins. An alternate explanation of this distribution is that the diffusing protein is exocytosed over the body of the cell, and as it diffuses over the lamellipodium, it binds to the cytoskeleton, which is moving rearward (Figure 3e). In such a model, the cytoskeletally attached protein could be endocytosed or released once over the body of the cell. An expected difference between the two models is that in the raking situation an antibody–gold particle complex would diffuse as it moved rearward whereas in the second case the particles would attach to the cytoskeleton before moving rearward.

## Forward Transport in Lamellipodia, Filopodia, and Flagella

The phenomenon of forward transport in cells is not nearly as well documented as rearward transport. Where it has been observed in actin-rich regions of the cell, it is most often rapid, in excess of 1 $\mu$ s$^{-1}$, and small objects are transported (29, 46). Another process of surface movement both forward and rearward is the microtubule-based bead movement on *Chlamydomonas* flagella (6). In both the actin and microtubule-based systems, movements are thought to involve the attachment of glycoproteins to already moving cytoskeletal structures. In fact, the behavior in the lamellipodia could best be described by suggesting that a barrier (rake) was physically pushing the glycoprotein forward but the glycoprotein was not specifically bound to the moving barrier. Small structures have been seen moving forward within thin filopodia at the rates of glycoprotein forward movements (46). This would be consistent with the known structure of the myosin 1 molecules since they have a lipid-binding site on their tails (2). Forward movement of the myosin through the lipid surface could push glycoproteins ahead of it.

## SUMMARY

The diffusion measurements of glycoproteins have further supported a fluid mosaic model of membrane structure, but the basis of the lower apparent diffusion coefficients in biological membranes remains incompletely understood. In the specific case of glycoproteins with a single $\alpha$-helix spanning the membrane, studies indicate that the major frictional drag is in the external protein layer and not the bilayer. Only in the erythrocyte membrane does the internal protein layer clearly control the lateral diffusion coefficient of a glycoprotein with a large cytoplasmic domain. In cultured cells, the barriers to lateral displacements over long distances are primarily on the cytoplasmic surface and not in the external matrix. Active movements of individual or small groups of glycoproteins both forward and rearward on cells appear to result from the interactions with moving cytoskeletal structures. Membrane turnover as well as transient attachment to the cytoskeleton can produce dynamic domains in the membrane that would depend on motile activity. Recent technological advances enable simultaneous monitoring of specific cell functions and glycoprotein motility, making it possible to correlate membrane fluidity and active glycoprotein movements with cell function.

ACKNOWLEDGMENTS

The writing of this review was supported by grants from the Muscular Dystrophy Association and the NIH.

## Literature Cited

1. Abercrombie, M., Heaysman, J., Pergrum, S. 1970. *Exp. Cell Res.* 60: 437–44
2. Adams, R. J., Pollard, T. D. 1989. *Nature* 340: 565–68
3. Aizenbud, B. M., Gershon, N. D. 1982. *Biophys. J.* 38: 287–93
4. Bennett, V. 1985. *Annu. Rev. Biochem.* 54: 273–304
5. Bergmann, J., Kupfer, A., Singer, S. J. 1983. *Proc. Natl. Acad. Sci. USA* 80: 1367–71
6. Bloodgood, R. A. 1990. In *Cilliary and Flagellar Membranes*, ed. R. A. Bloodgood, pp. 91–128. New York: Plenum
7. Bray, D., White, J. G. 1988. *Science* 239: 883–87
8. Bretscher, M. S. 1984. *Science* 224: 681–86
9. Chasis, J. A., Mohandas, N., Shohet, S. B. 1985. *J. Clin. Invest.* 75: 1919–26
10. DeBrabander, M., Nuydens, R., Geerts, H., Hopkins, C. R. 1988. *Cell Motil. Cytoskelet.* 9: 30–47
11. Debrabander, M., Nuydens, R., Ishihara, A., Holifield, G., Jacobson, K., Geerts, H. 1991. *J. Cell Biol.* 112: 111–24
12. Dragsten, P., Henkart, P., Blumenthal, R., Weinstein, J., Schlessinger, J. 1979. *Proc. Natl. Acad. Sci. USA* 76: 5163–67
13. Edidin, M. 1991. *The Structure of Cell Membranes*, ed. P. Yeagle. Boca Raton, FL: CRC Press
14. Edidin, M., Kuo, S., Sheetz, M. P. 1991. *Science* 254: 1379–82
15. Edidin, M., Zuniga, M. 1984. *J. Cell Biol.* 99: 2333–35
16. Edidin, M. E. 1992. *Trends Biochem. Sci.* In press
17. Elson, E. L. 1985. *Annu. Rev. Phys. Chem.* 36: 379–406
18. Fisher, G. W., Conrad, P. A., DeBiasio, R. L., Taylor, D. L. 1988. *Cell Motil. Cytoskelet.* 11: 235–47
19. Forscher, P., Smith, S. 1990. *Optical Microscopy in Biology*, ed. B. Herman, K. Jacobson, pp. 459–71. New York, NY: Liss
20. Frye, L. D., Edidin, M. 1970. *J. Cell Sci.* 7: 319–35
21. Geerts, H., De Brabander, M., Geuens, S., Nuyens, R. Moeremans, M., et al. 1987. *Biophys. J.* 52: 775–82
22. Golan, D. E., Veatch, W. 1980. *Proc. Natl. Acad. Sci. USA* 77: 2537–41
23. Gross, D. J., Webb, W. W. 1988. In *Spectroscopic Membrane Probes*, ed. L. M. Leow, pp. 19–48. Boca Raton, FL: CRC Press
24. Holifield, B. F., Ishihara, A., Jacobson, K. 1990. *J. Cell Biol.* 111: 2499–2512
25. Holifield, B. F., Jacobson, K. 1991. *J. Cell Sci.* 98: 191–203
26. Jacobson, K., Ishihara, A., Inman, R. 1987. *Annu. Rev. Physiol.* 49: 163–75
27. Koppel, D. E. 1981. *J. Supramol. Struct.* 17: 61–67
28. Koppel, D. E. 1986. *Biochem. Soc. Trans.* 14: 842–45
29. Kucik, D. F., Elson, E. L., Sheetz, M. P. 1989. *Nature* 340: 315–17
30. Kucik, D. F., Elson, E. L., Sheetz, M. P. 1990. *J. Cell Biol.* 111: 1617–22
31. Kucik, D. F., Kuo, S. C., Elson, E. L., Sheetz, M. P. 1991. *J. Cell Biol.* 115: 1029–36
32. Lee, G. M., Zhang, F., Ishihara, A., McNeil, C. L., Jacobson, K. A. 1992. *J. Cell Biol.* In press
33. Lee, J., Gustafsson, M., Magnusson, K., Jacobson, K. 1990. *Science* 247: 1229–33
34. Livneh, E., Benveniste, M., Prywes, R., Felder, S., Kam, Z., Schlessinger, J. 1986. *J. Cell Biol.* 103: 327–31
35. Menon, A. K., Holowka, D., Webb, W. W., Baird, B. 1986. *J. Cell Biol.* 102: 541–50
36. Minton, A. P. 1989. *Biophys. J.* 55: 805–8
37. Noda, M., Yoon, K., Rodan, G. A., Koppel, D. E. 1987. *J. Cell Biol.* 105: 1671–77
38. Phelps, B. M., Primakoff, P., Koppel, D. E., Low, M. G., Myles, D. G. 1988. *Science* 240: 1780–82
39. Qian, H., Sheetz, M. P., Elson, E. 1991. *Biophys. J.* 60: 910–21
40. Saffman, P. G., Delbruck, M. 1975. *Proc. Natl. Acad. Sci. USA* 72: 3111–13
41. Saxton, M. J. 1987. *Biophys. J.* 52: 989–97
42. Saxton, M. J. 1989. *Biophys. J.* 56: 615–22
43. Saxton, M. J. 1990. *Biophys. J.* 58: 1303–6
44. Saxton, M. J. 1990. *Biophys. J.* 57: 1167–77
45. Saxton, M. J. 1992. *Biophys. J.* 61: 119–28
46. Sheetz, M. P., Baumrind, N. L., Wayne, D. S., Pearlman, A. L. 1990. *Cell* 61: 231–41
47. Sheetz, M. P., Elson, E. E. 1993. In *Optical Microscopy: Emerging Methods and Applications*, ed. B. Herman, J. Lemasters, pp. 285–94. New York, NY: Academic
48. Sheetz, M. P., Schindler, M., Koppel, D. E. 1980. *Nature* 285: 510–12

49. Sheetz, M. P., Turney, S., Qian, H., Elson, E. L. 1989. *Nature* 340: 284–88
50. Singer, S. J. 1990. *Annu. Rev. Cell Biol.* 6: 247–96
51. Singer, S. J., Nicolson, G. L. 1972. *Science* 175: 720–31
52. Steinman, R. M., Mellman, I. S., Muller, W. A., Cohn, Z. A. 1983. *J. Cell Biol.* 96: 1–27
53. Symons, M. H., Mitchison, T. J. 1991. *J. Cell Biol.* 114: 503–13
54. Tank, D. W., Wu, E. S., Webb, W. W. 1982. *J. Cell Biol.* 92: 207–12
55. Theriot, J. A., Mitchison, T. J. 1991. *Nature* 352: 126–31
56. Thomas, J. L., Feder, T. J., Webb, W. W. 1992. *Biophys. J.* 61: 1402–12
57. Tilton, R. D., Gast, A. P., Robertson, C. R. 1990. *Biophys. J.* 58: 1321–26
58. Wang, J. L. 1986. *J. Cell Biol.* 101: 597–602
59. Wolf, D. E., Handyside, A. H., Edidin, M. 1982. *Biophys. J.* 38: 295–97
60. Yechiel, E., Edidin, M. 1987. *J. Cell Biol.* 105: 753–60
61. Zhang, F., Crise, B., Su, B., Hou, Y., Rose, J. K., et al. 1991. *J. Cell Biol.* 115: 75–84

# MECHANISMS OF MEMBRANE FUSION[1]

## Joshua Zimmerberg, Steven S. Vogel, and Leonid V. Chernomordik

Laboratory of Theoretical and Physical Biology, National Institute of Child Health and Human Development, National Institutes of Health, Bethesda, Maryland 20892

KEY WORDS:  exocytosis, viral infection, influenza, syncytia, lipids

CONTENTS

| | |
|---|---|
| PERSPECTIVES AND OVERVIEW | 434 |
| DOCKING AND INTIMATE CONTACT | 434 |
| FUSION PORES | 435 |
|    *Initial Pore* | 437 |
|    *Flicker* | 440 |
|    *Semistable States* | 440 |
|    *Final Irreversible Rise in Pore Conductance* | 442 |
| PROTEINS | 442 |
|    *Influenza Hemagglutinin* | 443 |
|    *The Fusion Peptide* | 445 |
|    *Nonviral Membrane Fusion Proteins* | 447 |
| LIPIDS | 448 |
|    *Membrane Tension* | 449 |
|    *Hydration Repulsion* | 449 |
|    *Curvature* | 450 |
|    *Pores in Lipid Bilayers* | 451 |
|    *Stalk Model of Lipid-Bilayer Fusion* | 451 |
|    *Lipids in Biological Fusion* | 454 |
| MODELS | 455 |
|    *Lipid-Protein Complexes* | 456 |
|    *The Cast and Retrieve Model for the Mechanism of HA-Mediated Fusion* | 458 |
| CONCLUSIONS | 461 |

[1]The US government has the right to retain a nonexclusive, royalty-free license in and to any copyright covering this paper.

## PERSPECTIVES AND OVERVIEW

Cells need to organize aqueous space in order to function, both to compartmentalize their interiors and to define inside from out. To accomplish this, cells employ oily, low dielectric lipid membranes that are diffusional barriers to hydrated, polar substances. Cells must therefore merge membranes in order to mix aqueous spaces. For example, sperm and egg pronuclei are safely sealed up, each within its own lipid coat. Quite a lot of biology goes into their meeting, but we do not understand how they locally fuse their lipid membrane barriers to allow the pronuclear merger to occur without spillage. The mechanism of this membrane fusion is obscure, as it is in other physiological membrane-fusion processes with diverse triggers (calcium, GTP, protons, etc) such as synaptic release of neurotransmitters; exocytotic release of digestive enzymes, immunoglobulins, and cytotoxic material; or intracellular fusion in the biosynthetic pathway. Even in influenza virus, for which the fusion protein for infection and syncytia formation has been unambiguously identified, sequenced, and crystallized (structure determined to 0.3-nm resolution), the fusion mechanism is not known.

At present, we think of at least seven distinct steps in membrane fusion: triggering, membrane docking, intimate membrane contact, protein/lipid mixing, fusion-pore opening, fusion-pore widening, and content swelling and/or mixing. The steps between binding and fusion-pore widening hold the most mystery for us at this time. How do fusion proteins act to cause lipid membrane rearrangement? Does biological membrane fusion proceed along the same reaction pathway as purely lipidic membrane fusion? Data must be integrated from different disciplines and fusion systems to build a coherent picture of these steps. Here, we review these diverse data, starting with ultrastructural and electrophysiological approaches. Next we review the structural data on known fusion proteins, and then we consider physical concepts relevant to lipid membrane interactions. Finally, models for the molecular mechanisms of biological fusion are discussed.

## DOCKING AND INTIMATE CONTACT

Contact between membranes is highly regulated in fusion, both prior to and after activation. Dimpling of the plasma membrane towards the secretory granule is seen in neutrophils, amoebocytes, and mast cells in which membranes are not docked (21, 35, 45) (Figure 1). In other systems, such as *Paramecium* trichocyst discharge and sea urchin egg cortical granule exocytosis, secretory granules are already in closer contact than can be measured with freeze-fracture microscopy (<5 nm) (22, 79, 125). In the sea urchin egg, these close contacts are maintained for months prior to activation with no obvious basal rate of secretion. In the frog neuromuscular junction, synaptic vesicles appear to align along the active zone of presynaptic nerve terminals. The steady-state number of docked vesicles and the rates of their replenishment appear to be limiting in synaptic exhaustion in model systems (89). Ultrastructural studies reveal that vesicles are tethered near intramembranous particles thought to be calcium channels. A synaptic vesicle protein, p65, that binds to calcium channels (67), α-latrotoxin receptors (124), and phospholipids but does not cause fusion, could act as the synaptic tether (15).

Although membrane docking is an essential prerequisite to fusion, it does not necessarily lead to fusion. Fibroblasts expressing the fusion protein of influenza virus, hemagglutinin (HA), can bind red blood cells with both precursor HA0 and the proteolytically activated HA, but fusion occurs only with HA (102). In sperm-egg fusion, binding occurs in sperm treated with two different monoclonal antibodies, but fusion is not detected (7). An inability to fuse also appears in mating mutants of yeast that have a normal binding phenotype (6). Of course, we do not know if these binding events are equivalent: is fusion prevented because of steric hindrance on the part of the antibody, or because the next step in the fusion mechanism is inhibited? In terms of energy, binding and fusion may be linked, with binding itself altering the free energy of the fusion system.

## FUSION PORES

In lipid membrane fusion, we cannot yet measure fusion pores because they expand so rapidly (Figure 2). Biological membrane fusion is characterized by the formation of a measurable fusion pore, as first noted in freeze fracture (21, 117, 119, 146). Later, a more sensitive technique was developed for measuring individual fusion events—whole-cell capacitance (107). Since the capacitance of a unit area of cell membrane is constant, cell capacitance increases when cell surface area increases during membrane fusion (such as what occurs when granular membrane incorporates into the plasma membrane). Measurements of the increase of capacitance accompanying exocytosis of giant secretory granules in the *beige* mouse mast cell revealed slow, fluctuating changes in capacitance (14, 188). Because the measurement of granule capacitance depends upon ionic accessibility of the granular membrane, these slow, fluctuating capacitance records suggests a small pore that impedes ionic currents. The equivalent circuit for exocytotic fusion was simplified after the resistive components for the granules and cell were found to be negligible (188). This gave an analytical solution for the measurement of cell admittance (188), from

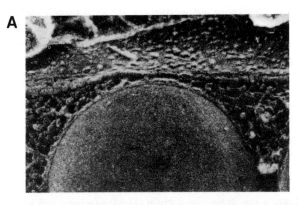

which pore conductances (permeability) could be calculated. A range of pore sizes was determined (188). Subsequently, another method for estimating fusion-pore conductance in *beige* mouse mast cells was derived by measuring the transient in current that results from the distribution of charge between the granule and the plasma membrane (13). In both cases, a small pore was seen to initiate membrane fusion. The following sections discuss in detail these properties of the fusion pore seen in electrical measurements: initial phase, flicker, semistable state, and final widening (Figure 3).

*Initial Pore*

A histogram of initial pore conductances is broad, and initial pores do not open abruptly. Although the first current-discharge measurements were consistent with instantaneous openings, subsequent experiments at low temperature showed a slow risetime for the initial pore in *beige* mouse mast cells (152) and HA-mediated syncytia formation (154, 153). This continuously growing initial pore is unlike the abrupt openings to defined levels seen in ionic channels (68), reconstituted gap junctions (61), and cell-cell gap junctions (136). The initial pore conductance, then, is an average value determined by the time response of the instrumentation. In mast cells, the initial pores ranged in conductance from 50 to 1500 pS [a gap junction is about 200 pS (152)]. In HA-mediated red blood cell/fibroblast fusion (154, 153), the initial pores ranged from 18 to 375 pS. Determining the dimensions of fusion pores from their conductances is difficult, because conductance is a function of width, length, and conductivity of the pore (35). Estimates of initial pore diameters range from 1 to 4 nm. In HA-mediated syncytia formation, pore diameter was estimated as 1–4 nm from cytoplasmic dye–diffusion studies (141).

For the first few milliseconds after opening, the fusion-pore conductance can increase or decrease, but it usually increases without obvious steps at rates spanning 50–200 pS/ms (35, 152). Although such continuous increases can be explained as rapid fluctuations between discrete conductance levels of a protein channel (such as those seen during the flickering block of ionic channels), they are also consistent with a pore whose dimensions vary continuously, such as one composed to some extent of lipids.

---

*Figure 1* Rapidly frozen, freeze-fracture replicas of rat peritoneal mast cells. (*a*) In unstimulated mast cells, a thin layer of cytoplasm separates granular and plasma membranes (25 nm here). (*b*) Fifteen seconds after stimulation with the secretagogue 48/80, the plasma membrane appears to dimple towards the granule membrane and contacts it in a small appositional zone. (*C*) A long hourglass-shaped pore joins a secretory granule membrane (*below*) with the extracellular space (*above*) (from Ref. 35).

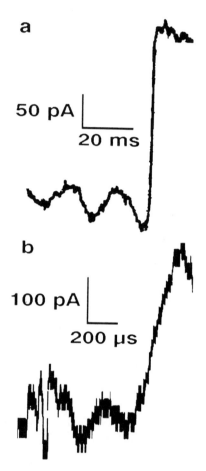

*Figure 2* The fusion of phospholipid vesicles to planar phospholipid membranes occurs instantaneously, limited only by the speed of the instrumentation. Vesicles containing multiple VDAC channels (189) fuse to an asolectin bilayer clamped to +10-mV transmembrane potential. In the presence of 40-mM calcium and an osmotic gradient, fusion is obtained (33). (*a*) The feedback resistor is $10^9$ ohms, and the time constant of the circuit is 2 ms. (*b*) The feedback resistor is $10^8$, and the time constant of the response is 200 $\mu$s. In both cases, the risetime of the fusion event was indistinguishable from the risetimes of channel opening and closing, and switch closure in model circuits.

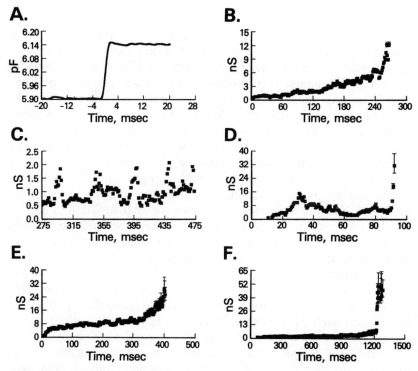

*Figure 3* The time course of the exocytotic pore widening is variable. The instrumentation can resolve millisecond events in whole-cell recordings of cell capacitance of beige mouse mast cells. Error bars in figures of individual pore conductance kinetics show the uncertainty in pore conductance calculated from one standard deviation of the noise in the admittance baseline. Because of the transformation from admittance data to fusion-pore conductances, uncertainties are not symmetric about mean pore conductances. (*A*) An instantaneous change in capacitance demonstrates the time resolution of the system. From slower changes in capacitance, the conductance of the exocytotic pore is calculated (*B–F*); (*B*) a gradual change in the conductance of a pore; (*C*) a portion of the time course of the conductance of a pore that shows rapid fluctuations; (*D*) an increase in pore conductance followed by a gradual decline before the final irreversible increase; (*E*) an initial increase in pore conductance followed by a long semistable pore conductance; (*F*) hovering of the pore conductance for 1 s between 1 and 5 nS. From the Curran collection (35).

Actual measurements of the transfer of lipid dyes during virus-cell (91), and cell-cell fusion (141) is slower than expected for simple lipidic junctions, which may reflect the nature of these fusion pores. The barrier to lipid flux could result from physical objects, i.e. proteins such as those found in axon hillocks (81), or from the curvature of the membrane in the fusion intermediate.

## Flicker

Fusion pore formation is reversible, that is, the fusion pore can close, a phenomenon referred to as *flicker* (49). Fusion pores can close in *beige* mouse mast cells even after reaching large conductances. No obvious morphological changes occur during flicker (Figure 4). Sperm-egg fusion is also characterized by a reversible pore formation, as demonstrated by capacitance measurements (96). Analysis of flicker in mast cells shows diminished cell capacitance after flickering, suggesting a loss of material from the plasma membrane during the flicker state (100). A tense secretory granule, which pulls membrane from the plasma membrane during the semistable state, is consistent with the fusion pore quickly attaining a bilayer configuration, as seen in freeze fracture (Figure 1). However, neither fusion-pore formation nor membrane transfer during flicker is affected by hyperosmotic treatment, so membrane tension probably does not play an important role in either process. Osmotically shrunken *beige* mouse mast cell granules require a 12% increase in diameter, or a 25% increase in area to lyse in hypotonic solution (188). Because taut lipid and biological membranes can only withstand a 3% stretching before lysing (83), the lipid bilayers of shrunken *beige* mast cell granules are presumably flaccid. Although the membrane involved in fusion may be contained by protein (see below), and local tension may be required for bilayer fusion to develop (187, 188), it is unlikely that local tension drives square microns of plasma membrane to the granule, as reported (100). Alternatively, endocytosis of small vesicles during flicker may also be an explanation, because endocytosis is often associated with exocytosis (66).

## Semistable States

In biological fusion, the fusion pore exhibits a continuum of conductances during its observable lifetime of 10 ms to 10 s (35, 101, 152, 153). However, the majority of pores exhibit a mode in their individual conductance histograms corresponding to a favored range of conductances, or plateau, in the time record that sometimes lasts seconds (Figure 3) (35, 153). Composite histograms of pore conductances from many experiments in *beige* mouse mast cell exocytosis show that this semistable conductance is between 0.6 and 20 nS, with a broad peak between 1 and 5 nS. As this corresponds to pore diameters large enough to visualize with electron microscopy (10–25 nm), and freezing times are on the order of 2 ms, the pores captured in fast-frozen secretory cells likely reflect these long-lived structures (21, 79, 117, 146). As in Figure 1, these pores seem to be of coplanar bilayer membranes with the normal complement of intramembrane particles. Thus, by the time the semistable pore is established,

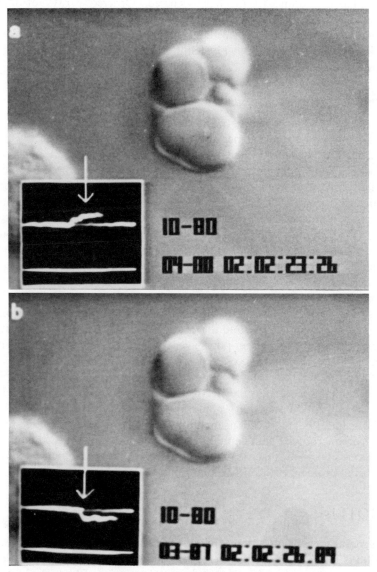

*Figure 4* Flicker fusion. A *beige* mouse mast cell is observed with simultaneous whole-cell capacitance and differential interference microscopy (186, 188). Degranulation is stimulated with GTP-$\gamma$-S. (*a*) An increase in cell capacitance is seen (*arrow*), corresponding to fusion-pore formation between a secretory granule and the extracellular space. (*b*) Three seconds later, the capacitance returns to baseline. No granular swelling is detected despite transient fusion-pore formation.

coplanar membrane fusion has occurred. Ordered arrays of intramembrane particles are absent within or adjacent to the hourglass figure in mast cells, but a rosette on the plasma membrane that expands after fusion is seen in *Paramecium* (79). Still, it is not clear if these phenomena represent fusion proteins.

## Final Irreversible Rise in Pore Conductance

After the semistable stage, fusion-pore conductance increases quickly, without steps, beyond the measurable range. This expansion occurs prior to detectable granule content swelling (35, 101), suggesting a membrane process for the rapid expansion of the hourglass shape of the fusion pore. Other processes involving content swelling probably participate in pore widening because hyperosmotic colloid treatment, which stabilizes granule contents, stabilizes such larger pores in mast cells (36) and in sea urchin eggs (173). In mast cells, ion exchange and swelling of secretory-granule contents are needed for final widening of the fusion pore to allow content expulsion (36, 101). In HA-mediated syncytia formation, hemoglobin and fluorescent dextran movement between the fused cells is seen much later than aqueous and lipid transfer—on the order of 30 minutes (141, 154). Fusion-pore widening needed for syncytia formation may be controlled by membrane tension (80) and the cytoskeleton (29). The measurement of the dynamics of these larger pores awaits the development of systems not limited by patch pipette access resistance (99).

Thus far, we find that fusion occurs in a specific, curved, dimpled appositional area seen in the electron microscope, but first detected electrically as a fusion pore. Fusion intermediates, such as those implicated in viral fusion (19), and initial fusion pores are too small to be visualized using electron microscopy with enough resolution to unambiguously determine structure. To better understand fusion-pore formation, we turn to studies of fusion proteins at the biochemical and structural level.

## PROTEINS

A general conclusion among investigators is that biological membrane fusion is mediated by proteins (2, 20, 69, 125, 156, 174, 177), but little is known about the proteins that mediate the majority of biological fusion reactions. Only for the enveloped virus influenza do we have primary, secondary, and tertiary structural data for an identified fusion protein, as well as functional and mutational studies that allow us to evaluate its mechanism. Because viral fusion proteins have been so thoroughly reviewed (69, 149, 156, 157, 174, 177), here we only describe studies of HA that are needed to evaluate models of biological membrane fusion. We hope the lessons from viral fusion may help illuminate other systems such as sperm-egg fusion and exocytosis.

## Influenza Hemagglutinin

After an influenza virus binds to the plasma membrane using viral sialic acid receptors, it is internalized into endosomes that become acidic, triggering the viral envelope to fuse with the endosomal membrane. Hemagglutinin (HA) is the viral spike protein responsible for this fusion (69, 174) as well as for the syncytial fusion described above. HA is a 225-kDa protein composed of three identical, disulfide-linked subunits coded by a single mRNA. For fusion activity and viral infectivity, the individual HA subunits must first be activated by proteolytic cleavage (HA0 → HA1-S-S-HA2). HA is extensively glycosylated (179) and acylated (84, 104, 145). The HA trimer forms approximately 500 rigid viral spikes on the surface of the influenza viral envelope. Although we do not know how many spikes must interact to promote fusion (30, 45), two or more HA trimers are believed to be required for fusion (30, 45). Molecular characterization of the gene encoding HA (see 85), in conjunction with X-ray crystallography of the ectodomain of the HA found in X-31 (HA liberated after cutting at the surface of the viral membrane with the protease bromelain is known as BHA), reveals five distinct protein domains (Figure 5) (179). The first domain consists of a short hydrophilic peptide inside the lumen enclosed by the viral envelope ($HA2_{212-221}$). The second domain is a hydrophobic membrane–anchoring segment ($HA2_{185-211}$) comprised of a trimer of membrane-spanning α-helices. The third domain consists of three small globular regions rising 3.5 nm above the surface of the viral membrane, each composed of a five-stranded β-sheet structure ($HA2_{22-35}$, $HA2_{131-184}$, $HA1_{1-14}$). This domain is topped by the N-terminal portion of HA2, often referred to as the fusion peptide ($HA2_{1-24}$), which lies parallel to and 3.5 nm above the viral membrane. This domain also contains the disulfide bond linking HA1 to HA2. The fourth domain is composed of a triple-stranded coiled coil of α-helices that extends 7.6 nm from the viral membrane and is formed primarily from $HA2_{36-130}$, but with a segment of HA1 running through its interior. It connects the globular domains adjacent to the viral membrane to a fifth large domain at the distal tip of the viral spike (13.5 nm from viral membrane to tip of viral spike). This most distal globular domain is comprised of antiparallel β-sheet ($HA1_{116-261}$) and contains a site for binding to sialic acid residues.

Upon exposure to low pH (5.0–6.5), HA undergoes conformational changes that first expose a hydrophobic fusion peptide normally sequestered within the protein (155, 158), and later conformational changes result

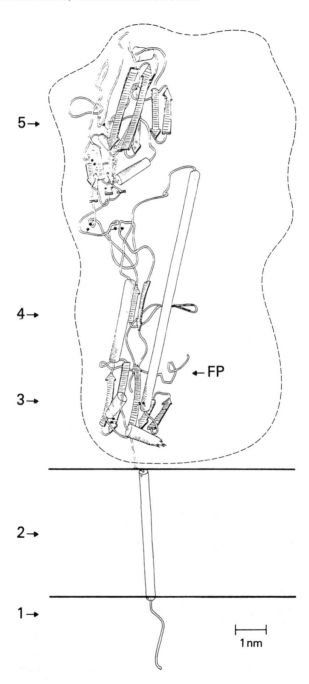

in the collapse of the spike-like structure (41). HA spikes from X31 become markedly more hydrophobic when acidified in the absence of target membranes (41) and aggregate in response to the exposure of their hydrophobic fusion peptides (in the case of bromelain-solubilized HA) (151). Avian influenza virus (FPV) loses its ability to fuse with subsequently added target membranes if acidified in the absence of target membranes (176). Other subtypes of influenza virus do not lose their ability to fuse after pretreatment at low pH (130). Although HA alone is responsible for the pH dependence of fusion to biological membranes (174), pH-independent, HA-mediated fusion to planar membranes in the absence of gangliosides occurs (111). Because the conformational changes of HA, which are related to fusion, can also be triggered by high temperatures at neutral pH (138), $H^+$ ions are thought to act only to enable the conformational changes in HA and not to directly participate in the fusion reaction per se. HA can also mediate the fusion of virus with purely lipidic liposomes (176) and planar phospholipid bilayer membranes (99, 110, 111), as well as fusion of plasma membrane to plasma membrane (30, 102) and plasma membrane to liposome (45). Cells expressing HA on their surface fuse to a variety of membranes, but the probability of fusion for red blood cells and liposomes differs significantly, even with optimal concentrations of gangliosides in the liposomes (174).

## The Fusion Peptide

Following proteolytic activation, influenza hemagglutinin, like the fusion proteins of Sendai, human immunodeficiency, and Rous sarcoma viruses, contains a stretch of apolar amino acids, typically ranging in length from 16–36 residues, on the same peptide fragment that contains the viral membrane-spanning hydrophobic anchor domain. These segments are hydrophobic, rich in glycine and alanine residues, and highly conserved within but not between viral families. Both the observed sequence conservation of the fusion peptide and analysis of isolated viral fusion mutants and site-directed mutants indicate that the HA fusion peptide is essential for fusion activity (37, 55, 174).

Topographical photolabeling using hydrophobic probes partitioned into liposomes has been used to determine which parts of HA can insert into

---

*Figure 5* The structure of HA. Schematic diagram of one of the three identical subunits of the influenza virus fusion protein HA and its five major domains: 1, the intraluminal domain; 2, membrane-spanning domain envisioned as an α-helix; 3, globular domain adjacent to surface of viral membrane containing the fusion peptide (FP); 4, coiled coil of α-helices; 5, distal globular domain. Domains 3, 4, and 5 are based on the crystal structure of BHA (179), whereas the structure of domains 1 and 2 are hypothetical. Outline represents the trimeric holoprotein (adapted from 179).

target membranes, how deeply these proteins penetrate the liposomal membrane, and how rapidly this occurs. The fusion peptides of both Sendai (113) and influenza (155) viruses can insert into liposomal target membranes, and for the X31 strain of HA, this labeling occurs prior to any detectable fusion at 0°C. It is not known if the fusion peptide can also imbed itself into the viral membrane, which is much closer to its tether site in the intact viral spike (3.5 nm) than to the target membrane (a distance of 10 nm). Photolabeling experiments using the bromelain-solubilized ectodomain of HA (BHA) show that the N-terminal 21 residues of BHA2 are all photolabeled. The relative abundance of labeling between these 21 residues suggests an amphipathic α-helical structure (17, 62) among other possible explanations (53). Synthetic peptides based on the sequences of the HA fusion peptide form α-helices when they contact membranes (86, 172), have an apparent membrane dissociation constant of 0.17 $\mu$M (86), and are membrane active (42). Like many amphipathic peptides, these synthetic peptides can promote the aggregation and fusion of liposomes to each other (9), probably through adhesion and tension (185). Although these forces undoubtedly play a role in the physiological mechanism, the fusion peptide is not sufficient for fusion when attached to BHA, which does not promote liposome-liposome fusion even at low pH, despite the fact that its fusion peptide clearly enters the target membrane.The protein containing the exposed fusion peptide should also be specifically anchored to the viral membrane (156), perhaps requiring transmembrane domains (190).

An experimental comparison of photolabeled BHA with different membrane probes suggests that the fusion peptide contacts only the outer monolayer of the liposomal target membrane (16). In HA on intact virus, this peptide is thought to reach beyond the first monolayer (155). It is not known if the fusion peptide traverses the whole target membrane and is exposed on the *trans* aqueous side, as would be expected if the fusion peptide formed a protein channel. Only the first 21 residues are photolabeled (albeit for BHA). Thus, with an α-helical axial rise of 0.15 nm/residue, the peptide length would be 3.15 nm. This is long enough to traverse the hydrocarbon region of most membranes, whose thickness, from capacitance, is on the order of 2–3 nm (112). Small molecules such as nystatin are thought to pinch membranes when active as pores (77, 95), and indeed, a 16–amino acid peptide thought to form an α-helix spans bacterial membranes (39). In addition, interaction of short signal peptides with liposomes can cause significant thinning of bilayer membranes (160). However, the fusion peptide may well be tilted in the membrane (12, 169), and thus only traverse a fraction of the hydrocarbon region, or it may even lie on the surface of the outer target membrane monolayer (16). Experiments with aqueous photolabels trapped in the lumen of liposomes could test for whether the fusion peptide traverses the entire liposome bilayer.

## Nonviral Membrane Fusion Proteins

The identities of fusion proteins mediating intracellular fusion events remain speculative (2, 20, 58, 69, 125, 137, 156, 159, 174). In part, this results both from the lack of good genetic models for membrane fusion and from the difficulty of distinguishing fusion proteins from proteins required for the assay of membrane fusion (membrane trafficking, vesicular content filling, docking, etc). Many intracellular proteins have been proposed to be fusion proteins in exocytosis (2, 20, 58, 125), intracellular trafficking (137), and synaptic transmission (159). A protein from guinea pig sperm, PH-30, was purified using an antibody that blocks sperm-egg fusion (7). The cloning and sequencing of this protein showed a putative fusion peptide in the $\alpha$-subunit—an intriguing similarity to the viral fusion proteins (8). Unfortunately, none of these putative fusion proteins have had their functions validated by the rigorous criteria used for viral fusion proteins: (*a*) reconstitution of nonleaky fusion (demonstrated by both membrane and aqueous continuity) between flaccid membranes composed of pure lipids and the purified candidate protein(s) specifically stimulated by the physiological trigger and (*b*) gene transfer of the candidate protein enabling a new and physiologically relevant fusion activity.

Identifying the locations of fusion proteins is important in understanding biological fusion (Figure 6). For example, the contents of exocytotic vesicles are not required for membrane fusion (142). For viral fusion, we know that fusion proteins that reside in only one membrane can mediate fusion (model E) (176). Several lines of evidence suggest the same arrangement for exocytotic membrane fusion as well. In calcium-triggered sea urchin egg secretion in vitro, fusion occurs in the absence of cytoplasmic factors (167, 168), ruling out model A. Because exocytotic granules can fuse with both the plasma membrane and with other granules both in vivo during compound exocytosis (21, 74, 90, 120) and in vitro (168, 167), models B and C are unlikely arrangements for fusion in these systems. Exocytotic granules, like viruses, can fuse with purely lipidic membranes (166). Thus, as in viral fusion, exocytotic fusion proteins can be arranged as in model E (rather than D), with fusion proteins initially in only one of the two fusing membranes. This arrangement seems to be an emerging motif for biological fusion.

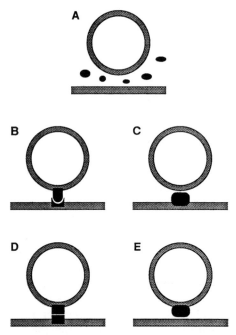

*Figure 6* Hypothetical distributions of proteins required for membrane fusion. (*A*) Cytoplasmic and (*B–E*) membrane-associated proteins. Distributions *B* and *D* require proteins in both membranes, either identical (*D*) or complementary (*B*), whereas distributions *C* and *E* require proteins in only one of the two membranes, either in the plasma membrane (*C*) or in the fusing virus/vesicle (*E*). Of these possibilities, only distribution *E* is consistent with the known distributions of both viral and exocytotic fusion proteins.

## LIPIDS

Fusion ultimately involves and requires the merging of the lipid bilayers of biological membranes. The fact that fusion proteins need only be present in one of the two membranes (69, 166) means that lipid itself can be the target for protein-mediated, biological membrane fusion. Lipids thus participate in protein-mediated fusion, and their energetics and conformations should be considered in any discussion of membrane fusion. Interactions of phospholipid bilayer membranes have been studied and reviewed in many systems: liposomes with liposomes (121, 178), liposomes with planar lipid bilayers (50), planar lipid bilayers with planar lipid bilayers (28), lipid bilayers formed on the surface of mica cylinders (64), and lipid bilayers next to each other in multilamellar structures (132).

Despite the relative simplicity of these systems, which have defined composition and geometry, we do not know how fusion occurs. Drawing from these systems, we summarize the physical concepts pertinent to biological membrane fusion and discuss in detail one molecular model for lipid bilayer fusion, the stalk hypothesis.

## Membrane Tension

The essential role of tension has been well documented for liposome–planar lipid bilayer membrane fusion (31–33, 123, 183, 186a). Fusion was not observed unless osmotic stress was applied to the liposome membrane. As noted, osmotic stress is not the only way to achieve high membrane tension. Any strong adhesion between membranes of fixed volume can result in flattening, internal pressure, and membrane tension (131). In addition, fusion of vesicles composed of acid lipids occurs at concentrations of divalent cations that increase the surface tension of lipid monolayers to a point corresponding to a 10% decrease in lipid surface area (115, 116). In liposome membranes, monolayers are connected crosswise and the condensation of the external monolayers exposed to $Ca^{2+}$ ions can provide membrane elastic stresses different from those of osmotic stress.

## Hydration Repulsion

Apposing lipid bilayer membranes are subject to long range electrostatic and van der Waals interaction forces (122). At distances between bilayers of about 1.0–2.5 nm, repulsive forces resulting from the work of dehydrating the lipid polar heads predominate over the other forces (132). The dependence of hydration forces on the distance between membranes is well approximated by an exponential dependence with a huge prefactor $[P_0 \approx 7 \times 10^9 \text{ dyn/cm}^2 \ (\sim 1000 \text{ atmospheres})]$ and a correlation length of $l_0 \approx 0.2–0.3$ nm. This tremendous repulsion must be overcome or reduced to obtain the intimate contact between bilayers required for fusion. Millimolar $Ca^{2+}$ can completely dehydrate the space between two phosphatidylserine bilayers by forming a *trans* complex involving lipid molecules on apposed surfaces (48, 129). However, the large energies of adhesion in this case (100 ergs/$cm^2$) can also rip membranes apart (73), resulting in content leakage. Obviously, given enough time and area of contact, thermal fluctuations of lipid bilayers should allow very small and short-lived points of intimate contact (87). Local dehydration of contacting surfaces due to lateral redistribution of lipids has also been hypothesized to lead to intimate contact (70). Lipids with lower hydration (phosphatidylethanolamine) promote fusion (28, 31, 178).

## Curvature

The promotion of lipid fusion by lipids that do not form bilayers and by agents ($Ca^{2+}$, $H^+$, heat) that disrupt bilayers suggests that nonbilayer lipid phases may be relevant to fusion mechanisms. Besides the bilayer, or lamellar phase, aqueous dispersions of different lipids can spontaneously form inverse hexagonal ($H_{II}$), cubic, and/or micellar phases (34, 60, 71). Analysis of these different phases, whose structures have been determined using X-ray diffraction and NMR, yield strikingly different molecular shapes of the component lipids. These shapes are separated into three groups: cones (e.g. phosphatidylethanolamines and cardiolipin in the presence of $Ca^{2+}$ ions), cylinders (phosphatidylcholine), and inverted cones (lysophosphatidylcholine) by comparing the surface area of the lipid polar head with the area of a cross-section of hydrocarbon tail (34, 71) in these different phases. For lipids with the same polar head, the more unsaturated their fatty acids, the more conical their shapes are (34). As a first approximation, this shape determines phase preferences: cones, cylinders, and inverted cones form inverse $H_{II}$ phases or inverted micelles, bilayers, and micelles, respectively. One must consider more than the conformation of an individual molecule, however. The interaction of molecules (electrostatic, van der Waals, etc) greatly affects the effective shape of lipid in a monolayer that determines its phase preference.

Alternatively, the shapes of lipids in monolayers can be described as the propensity of lipid monolayers to bend, i.e. the curvature of the monolayer in the unstrained state, or spontaneous curvature (27, 59, 63). For pure lipids with the molecular shapes of cones, cylinders, and inverted cones, the spontaneous curvature is negative, zero, and positive, respectively. Calculating the spontaneous curvature of monolayers composed of mixtures of lipids is more complicated. Some studies consider mixtures of lipids to be approximately additive in the spontaneous curvatures of the constituents (27), because bilayers are obtained from a mixture of two lipids, each of which separately forms only nonbilayer structures (93). Other studies show that spontaneous curvature may not be additive (134). A recent report showed that at certain temperatures and water concentrations dioleoylphosphatidylethanolamine/water dispersions demonstrate a hexagonal-lamellar-hexagonal transition sequence with little osmotic work required for the hexagonal-lamellar transition (54). This observation means that the spontaneous curvature of the lipid monolayer can sometimes depend on the curvature of the structure itself. Further, the lamellar-inversed $H_{II}$ phase transition energy depends not only on spontaneous curvature but also on packing constraints (163).

Lipid bilayer fusion is very sensitive to the spontaneous curvature of its constituents, with inverse $H_{II}$ phase–forming lipids generally promoting fusion (18). Diacylglycerol increases repulsion in phosphatidylcholine membranes, does not change repulsion for phosphatidylethanolamine membranes (38), but promotes fusion in phosphatidylethanolamine (150) and phosphatidylcholine/phosphatidylethanolamine/cholesterol membranes (108). Lysophosphatidylcholine inhibits bilayer fusion when added between bilayers (27) but does not significantly increase hydration repulsion at the mole fraction used (132). Although these lipids have different hydrations, their shapes, rather than their hydrations, are a better predictor of their effects on fusion.

## Pores in Lipid Bilayers

As discussed above, fusion pores during exocytosis are unlike proteinaceous ionic channels, which open instantaneously and have strictly defined levels of conductance. However, fusion pores do share some properties with pores in single protein-free lipid bilayers (98). Pores in lipid bilayers have a broad distribution of conductances (in the pS–nS range that corresponds to pore radii in the range of $\sim 0.5$–$7.0$ nm). The typical times of lipid pore development and expansion can be very fast (micro- to milliseconds) (1, 29, 31, 33, 56). However, lipid pores of $\sim 100$ pS can fluctuate, lasting up to minutes before abruptly closing (25, 26). Investigations on electroporation of planar lipid bilayers suggest that lipid pores have internal surfaces covered with polar lipid heads (1, 25, 27, 56). This implies a lipid monolayer that is bent at the edge of the pore. The elastic energy of such a pore will be small for membranes with positive curvature. In fact, the presence of inverted cone-shaped lipids like lysophosphatidylcholine in membranes promotes the development of such lipid pores (27, 56).

## Stalk Model of Lipid-Bilayer Fusion

A hypothesis for the mechanism of lipid bilayer fusion has been developed and tested (27, 28, 82, 87, 94) and is sketched here to consider its relevance to biological fusion. It is based, in part, on the concept of monolayer fusion. Monolayer (or semi-) fusion of membranes (the joining of contacting leaflets of membranes) was shown for planar lipid bilayers to occur before (or sometimes without) the second and subsequent step, complete fusion (28, 51, 88, 97, 98, 106). Monolayer fusion is also seen in some cases of liposome-liposome fusion (5, 11, 43, 46, 47). In contrast, a single bilayer does not comprise the prefusion state between vesicles and planar bilayers (109), but a single bilayer may transiently occur during fusion without detection.

In the stalk model, hydration repulsion between membranes is overcome transiently by thermal fluctuations, as described above. Rather than returning to the lamellar phase, transiently contacting lipids may relax their hydration repulsion by merging to form a stalk with a hydrophobic interior connecting the contacting membrane monolayers (Figure 7b). The net curvature of the stalk is negative, so the formation of a stalk is aided by lipids of negative spontaneous curvature in the contacting monolayers (27, 28, 87). The increase in stalk diameter after the initiation of monolayer fusion may lead to the formation of a single contact bilayer, resulting in the compression of external monolayers and extension of the internal ones (Figure 7c). Preexisting membrane tension in each membrane can affect

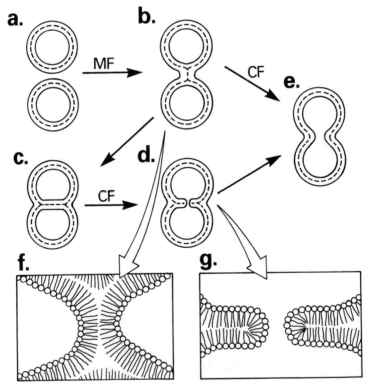

*Figure 7* Stalk hypothesis of lipid bilayer fusion. (*a*) Membranes are at equilibrium distance; (*b*) formation of a stalk between membranes leads to monolayer fusion (MF); (*c*) the contact bilayer, composed of the distal monolayers of the fusing membranes, expands; (*d*) formation of a hydrophilic pore in the contact bilayer results in complete fusion (CF); (*e*) hourglass configuration represents a completely fused state. Fine details of the stalk (*f*) and pore (*g*) structures.

the lifetime and size of this single adjoint bilayer. To complete fusion, a lipid pore of critical radius must be formed to expand in the contact bilayer, aided by lipids of positive spontaneous curvature in the internal monolayers (28, 82) (Figure 7d). The breaking of the inner monolayers of the fusing membranes, thereby forming the fusion pore, can also occur concomitantly with the development of the contact bilayer (Figure 7b–e). Thus, the stalk model is based on membrane hydration, curvature, and tension, and an initial stage of monolayer fusion. It predicts that membranes with contacting monolayers of negative spontaneous curvature and distal monolayers of positive curvature will be optimal for fusion, and that membrane tension promotes fusion-pore formation.

The main assumptions of the stalk hypothesis are consistent with experimental data obtained for different model systems (28, 82, 87). An experimental system using two interacting solvent-free planar lipid bilayers clearly showed monolayer fusion as the first stage after contact (28) and was the most useful system for testing the hypothesis. The inverted cone-shaped lysophosphatidylcholine inhibits monolayer fusion when added between interacting bilayers, but the same lipid promotes fusion-pore formation when added to distal monolayers. The lipid-dependence of the waiting time for monolayer fusion and the linear tension of hydrophilic pores (measured in parallel for phosphatidylethanolamine membranes in the presence of different concentrations of lysophosphatidylcholine) is in good quantitative agreement with the simplest theoretical description of the energies of pore and stalk formation (27) without any fitting parameters. Critical evaluation of this model depends upon correct estimation of the work of stalk formation in lipid monolayers of high negative curvature, additivity of spontaneous curvatures, independence of monolayer spontaneous curvature from the curvature of the structure (see above in section on curvature), and negligible effects of the solvent torus and lenses (28, 31, 123).

The above presents only a general qualitative description of the stalk hypothesis, not its theoretical development (27, 82, 87, 94). Further development of the stalk hypothesis includes the analysis of the role of hydrophobic voids in putative intermediate structures (162, 164) in stabilizing membranes against stalk formation (149). The changes in size of the fusion pore in single lipid bilayers as a function of membrane spontaneous curvature and tension were recently discussed (105).

An alternative mechanism is the formation of inverted micelles as fusion intermediates (148). Inverted micelle-type intermediates should only form under specific conditions close to the transition of bilayer phase into inverse $H_{II}$ phase and are significantly larger in diameter and so involve more lipid molecules than do stalks.

## Lipids in Biological Fusion

The physical ideas and the hypothetical mechanisms discussed above in relation to lipid bilayer fusion generally hold for continuous films in aqueous solution. Mechanical tension, hydration, and spontaneous curvature of the leaflets of biological membranes can be modulated by all components found in cell membranes including proteins and by the interactions of these components. Lipids themselves may play a role in forming the local, curved zone of apposition in which biological fusion-pore formation occurs.

The lipid composition of biological membranes is very asymmetric (118, 144, 147). Typically, sphyngomyelin and phosphatidylcholine (cylinders) are found in the outer leaflet of the plasma membrane of secretory cells (52), erythrocytes, and several other cells (118, 144, 147). The aminophospholipids, phosphatidylethanolamine (cone) and phosphatidylserine, which are known to support lipid bilayer fusion, mainly reside in the cytoplasmic leaflet of plasma membranes (118, 144, 147) and secretory (chromaffin) granules (181). The contacting monolayers in chromaffin granule exocytosis are thus composed of lipids with low hydration repulsion, a dehydration potential with calcium, and a curvature that promotes monolayer fusion. The inner monolayer of the chromaffin granules has a high concentration of lysolipids [inverted cones, up to 17% of the total phospholipid (40, 170, 171)], and so chromaffin granule membranes are optimal for fusion according to the stalk hypothesis.

Although the monolayer compositions of myoblast membranes are unknown, lipid changes occur just before fusion (139, 140). There is a decrease in phosphatidylcholine and phosphatidylinositol and an increase in phosphatidylethanolamine, cholesterol, and phosphatidic acid. Agents that delay fusion also delay these changes in lipid composition. Although the inositol lipid metabolite has been proposed to be physiologically fusagenic, its production via phospholipase C is not sufficient for fusion (44). Cholesterol is enriched in viral envelopes (78), and its presence in liposomes can facilitate or may be required for liposome/viral fusion (3, 57, 69, 165, 175). Cholesterol also facilitates secretory granule fusion with liposomes (166).

Recently, we found that lysolipids are potent inhibitors of $Ca^{2+}$-, GTP-$\gamma$-S-, GTP-, and $H^+$-triggered biological fusion reactions (23). Inhibition was reversible, did not correlate with lysis, and could not be attributed to any specific chemical moiety of lysolipids. Fusion was arrested at a stage preceding fusion-pore formation in *beige* mouse mast cells. Along with inhibition of myoblast fusion by lysophosphatidylcholine (133), these results suggest lysolipids may be universal inhibitors of biological fusion.

Inhibition can be readily interpreted as inhibition of highly curved stalk formation by incorporation of inverted cone-shaped molecules into contacting monolayers (27). Alternatively, lysolipids may inhibit fusion directly by binding the hydrophobic fusion peptide as it unfolds. Such lipid-binding regions of fusion proteins may be required for the protein's function and be structurally conserved in fusion proteins from different systems. Paradoxically, lysolipids were first suggested to be fusogenic (128). This observation was made for nonphysiological fusion, at lytic concentrations of lysolipids, and may relate to the enhancement of fusion-pore formation discussed above for lipid bilayer fusion. Alternatively, lysolipids could disrupt membrane lipid assymetry, and thus promote fusion (65, 143).

Proteins can interact with lipids to change membrane spontaneous curvature, hydration, and tension, either directly or indirectly through enzymatic activity. Small hydrophobic peptides such as carbobenzoxy-D-phenylalanyl-L- phenylalanylglycine inhibit viral fusion (75, 76) and stabilize bilayer structure (180). Peptides can also promote membrane fusion (92, 103, 161), and hydrophobic peptides such as cytochrome $c$, gramicidin, melittin, and cardiotoxin have been shown to promote $H_{II}$ formation or inverted micelle structures (34). Lipid-pore stability can also be increased by the presence of proteins, so that pores on the order of 1 nS can be stable for seconds and longer (24, 114). Proteins may also bring negatively charged lipids into contact and thereby increase their affinity for $Ca^{2+}$ by several orders of magnitude (121), bringing lipid-$Ca^{2+}$ binding into the physiological range.

## MODELS

Assimilating the body of knowledge on membrane fusion just presented, we reach the following conclusions: (*a*) Proteins in one membrane can promote a topological transformation of membrane components of two bound membranes into one. (*b*) Insertion of part of a fusion protein into the second membrane precedes membrane mixing. (*c*) A small pore of variable dimensions and lifetime links the aqueous space and rapidly widens to a plateau value. (*d*) An hourglass-shaped fusion pore of dimensions consistent with this plateau value is composed of a bilayer membrane coplanar with the two membranes that fuse. And (*e*) fusion pores can close. We do not know the pathway by which this takes place, nor whether all systems use the same pathway. Lipids can modulate fusion, and there are both similarities and differences between lipid and biological fusion. Energy, in the form of ATP or GTP, is clearly not required in viral systems and does not have to be added to exocytotic systems in vitro (72). The energy for fusion can be derived from membrane components themselves through binding or conformational changes.

Because the transition from the bound state to the hourglass state occurs below the limit of resolution of electron microscopy, hypothetical intermediates of membrane fusion are useful for conceptualization of future experiments. Many models for membrane fusion have been developed over the years (2, 10, 14, 105, 126, 127, 135, 182, 184, 187), each assuming different answers to the following questions: Is aqueous continuity established first, or does it follow formation of lipidic continuity? What is the role of proteins in bringing lipid bilayers into contact, and how do proteins help bilayers to overcome the hydration barrier and bend toward each other in order to merge? Do proteins act as catalysts to lower the activation energy of lipid-bilayer transformations? Is the initial fusion pore purely lipidic (105), proteinaceous (2, 14), or is it a lipid/protein complex (127, 135, 187)? Does the initial fusion pore span two bilayers or one? Does the lipid composition of the distal leaflets affect the fusion process? Unfortunately, clear answers have not been obtained for these questions. Some of them have never been addressed experimentally, while others are under intense scrutiny, such as the essential question—does membrane merger precede or follow fusion-pore formation? Answering this question—the order of lipid and aqueous continuity during fusion—has been difficult because the techniques used to measure these different continuities vary greatly in sensitivity and time response and because other factors, such as membrane microdomains, may prevent free lipid diffusion despite membrane continuity (4, 81).

Here we consider two specific fusion models, one based on pore-conductance data for the initial fusion pore and its development into a semistable structure in mast cells (lipid-protein complexes), and the other based on the characterization of HA-mediated fusion (cast and retrieve model).

## Lipid-Protein Complexes

To explain membrane fusion between liposomes and viruses or secretory granules, the proteinaceous pore model (2, 14) and the lipid-protein pore model (187) must rely upon protein insertion from one membrane into the other prior to fusion-pore formation. Both of these models predict the establishment of aqueous continuity prior to lipidic continuity. In models using protein pores, lipid would participate in pore widening only after the lumen of the granule was in contact with the extracellular space. Researchers once thought rapid conformational changes in fusion-inducing proteins led to the abrupt opening of the fusion pore. However, as discussed above, fusion pores do not open abruptly. In the lipid-protein complex model (187), lipids participate in the formation of the initial pore. Lipids can diffuse over molecularly significant distances in short times, so the short delays in synaptic transmission are not inconsistent with lipid involvement (187). With respect to HA, packing considerations are inconsistent with a purely proteinaceous pore consisting of the membrane-spanning domains of HA, because the extra-viral domains are too bulky to allow contact (187). If lipids form the fusion pore, there are no such packing constraints. If the lining of the pore were composed of polar headgroups of lipids, then a hydrophilic pore could result from the lipid rearrangement. One can imagine a protein helping to bring lipids into contact (see below), so that they spontaneously rearrange.

In the lipid-protein pore model, adhesion energy of proteins is used to bring lipids into intimate contact, but this model postulates that proteins act as a hydrophobic support to keep lipid head groups hydrated and separated (187). In this model, exposure of hypothetical hydrophobic surfaces would lead acyl chains of lipids to flip from the bilayer to a more energetically favorable position on the fusion proteins. As in Brownian dynamics, the lipid-protein model comprises a series of incremental steps with continuous movements of the constituent molecules. As a result of this spot welding between the membranes, bilayers are brought close together (Figure 8a). Because hydrated lipids repel at short distances, pressure must develop in this region of apposition, resulting in a blister (the pressure necessary to provide the force responsible for the subsequent molecular rearrangements derives from the protein conformational changes that restrict the pressure) (Figure 8b). Next, the acyl chains of the lipids in this strained region would reorient into a new configuration (Figure 8c). The lipids on the fusion proteins would then line a hydrated lumen, forming the pore. Following the initial formation of the pore, the hydration force between the lipid head groups that line the pore provides a repulsive force that progressively widens the pore. As the widening progresses, the neck stretches and the fusion-protein complexes dissociate (Figure 8d), allowing for rapid pore enlargement. Finally, as seen in Figure 8e, after fusion proteins dissociate, pore formation would become irreversible. There would be no leakage, because the rearranging lipids are constrained by the associated fusion protein complexes.

The steps shown in the cartoon of the lipid-protein pore complex have been described in kinetic terms to quantitatively account for the experimental observations [i.e. pore conductance histograms and time course (187)]. One prediction of the lipid-protein pore model (187) is that osmotic stress (189), decreasing the size of the water blister, should inhibit fusion. In contrast, it should facilitate intimate bilayer contact and stalk formation.

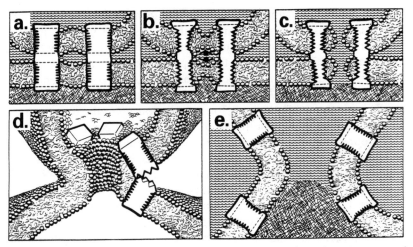

*Figure 8* Lipid-protein model. Hypothetical sequential steps of membrane contact and fusion (from 187). (*a*) After inserting hydrophobic domains from one membrane, fusion proteins bring membranes into close proximity. Lateral associations between more than one fusion protein trap a blister of water in which lipids try to stay maximally hydrated. (*b*) A conformation transition in the fusion proteins exposes more hydrophobic surface, and the acyl chains of the lipids move onto the newly created hydrophobic surface. (*c*) To maintain hydration and coat hydrophobic proteins at the same time, lipids move incrementally and a fusion pore develops—a hydrated lipid-lined pore. (*d*) Lipid eventually coats the entire surface of the protein, and (*e*) proteins dissociate to allow complete fusion.

Another test would be a measure of monolayer fusion, which should not precede fusion-pore formation in the lipid-protein model, but should in the stalk model.

With respect to influenza HA, we here propose a new mechanism for lipid contact wherein a fusion-peptide conformational change from random coil to α-helix provides the energy for contact and creates high membrane curvature as well.

## The Cast and Retrieve Model for the Mechanism of HA-Mediated Fusion

In this model, contact occurs because the HA proteins bend both the target membrane and viral membrane toward each other to a point where it is more energetically favorable for the membranes to fuse than to remain in a highly bent form. Thus, this model is an HA-specific variant of the stalk model. Initially, the viral membrane and target membrane are separated by 13.5 nm, the height of the viral spike protein, and the spacing between individual spikes is thought to be from 8 to 11 nm (see Figure 9). Upon

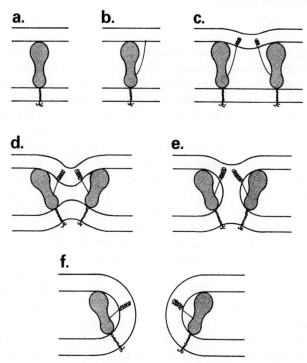

*Figure 9* Cast and retrieve model for HA-mediated fusion. (*a*) An HA spike in the viral membrane (*bottom*) makes contact with the target membrane (*top*). (*b*) Following acidification, the fusion peptide is released and makes contact with the target membrane 10 nm from where the peptide is covalently attached to the spike protein (note: the other two fusion peptides in this spike are not shown). The membrane-embedded fusion peptide undergoes a structural transformation into an α-helix, which results in an axial contraction. (*c*) The contraction of several fusion peptides from a few spike proteins cooperatively pulls the target membrane down toward the viral membrane. (*d*) Because of the geometry of the fusion-peptide attachment sites, the tension generated in response to peptide contraction causes the pivoting of the participating spike proteins. This pivoting then causes the viral membrane to bend toward the already bent target membrane. (*e*) Within the confined space of several viral spike proteins, the bent membranes form stalk structures. (*f*) Stalks resolve the unfavorable energetics of their extreme curvature by fusing the membranes.

acidification, individual viral spike proteins release their three fusion peptides ($HA2_{1-21}$) as well as a $\beta$-sheet structured tethering sequence ($HA2_{22-35}$), which we envision can pivot near residues Gly31 and Gly33 and Ala35 and 36. This 35–amino acid segment should extend to a maximum distance of 11.9 nm (assuming an axial length of 0.34 nm/residue) from its tethering point on the viral spike protein 3.5 nm above the surface of the viral membrane—ample length for contacting the target membrane 10 nm above (see Figure 9b). If the apolar 21-residue fusion peptide contacts the target membrane, it immediately embeds itself into the hydrophobic environment and begins to assume an α-helical conformation (Figure 9c), liberating 9.1 kcal/mol peptide bound to membrane (calculated from an apparent $K_d$ of 0.17 μM), roughly the same energy liberated by hydrolyzing one mole of ATP. Formation of an α-helical structure from an extended 21-residue fusion peptide would result in an axial contraction of 4 nm [(21 × 0.34)−(21 × 0.15)]; thus the 11.9-nm peptide extruded from the spike protein could contract to a 7.9-nm segment. Dividing energy by distance of contraction, we calculate that each peptide that binds to the target membrane will be able to apply a force of $1.6 \times 10^{-6}$ dynes on the target membrane and fusion protein (the two points where it is now attached). A simple estimate of the force required to bend a membrane through a ring of 10-nm diameter gives comparable values. If the fusion peptides of several adjacent spikes were embedded in the viral membrane, the combined force of their peptides (i.e. four spikes contributing two peptides each) could be sufficient to pull the target membrane down 3–4 nm between the upright spike proteins (Figure 9d). Because the peptides are attached to the viral spike protein at a point closer to the viral membrane, and because there would be tension on the peptide due to its contraction, the participating spike proteins could tilt outward, tending to bend the viral membrane up toward the target membrane (Figure 9c–d). Because the curvatures are so extreme within the confines of four spikes, the membranes may prefer to form a stalk structure (Figure 9e) and then fuse (Figure 9f) rather than remain bent.

Do any structural correlates support this model? Perhaps. Freeze-fracture analysis of influenza virus mixed with liposomes occasionally reveals small (9–14 nm) particles protruding from the liposomal surface at the virus-liposome interface that are thought to be trapped lipidic fusion intermediates (19). This size is consistent with a portion of the liposomal membrane bending toward the viral membrane between four spike proteins, though this could also reflect a protein or an inverted micelle. Acidification of the virus-liposome mixtures revealed a fusion-pore structure that is 11 nm in outer diameter [11−(4+4) = 3-nm inner diameter] and at least 22 nm long (19). A fusion pore on the order of 29 nm in length and 2–4 nm in inner diameter would yield a conductance of 140–550 pS, well within the range of the observed values (153). Thin-film cryoelectron microscopy also reveals structures that are consistent with the target membrane being drawn down between rigid spike proteins (Figure 10).

This model makes several specific predictions that can be experimentally tested. First, modifying HA2 by either the insertion or deletion of several

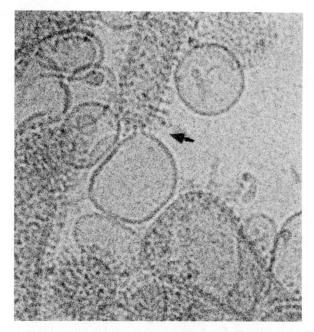

*Figure 10* Cryoelectron microscopy of influenza virus strain A-X31 and liposomes (egg phosphatidyl choline, egg phosphatidyl ethanolamine, gangliosides; 6:3:1). Virus and liposomes were preincubated for 3 h at pH 7.4 (0°C), then acidified at pH 5.1 for 25 min. Note liposomal membrane discontinuity (*arrow*), which might indicate a fusion site. (Photograph kindly provided by Dr. Frank Booy.)

amino acid residues anywhere in $HA2_{1-36}$ should inhibit fusion by making the fusion peptide either too slack to pull down the target membrane or conversely too short to reach the target membrane, and second, changing Gly31 and 33 and Ala35 and 36 to bulkier and/or less flexible amino acids (such as tryptophan or proline) should prevent the fusion peptide from extending out towards the target membrane, thus inhibiting membrane fusion.

## CONCLUSIONS

Biological membrane fusion, that is, the actual process wherein aqueous and lipidic continuity is established, has been studied with a vast array of experimental approaches in disparate physiological systems. These studies show that biological membrane fusion is a very complicated process. Because of the complexity and heterogeneity of the systems studied and the techniques used, which sometimes result in contradictory conclusions, it has often been easier to imagine that Nature has many different mechanisms for biological fusion. Still, recent developments have begun to reveal several unifying motifs. Biological membrane fusion emerges as a highly local phenomenon composed of short-lived and often reversible intermediates resulting from the actions of specific proteins. Fusion proteins residing in only one of the two fusing membranes can mediate membrane merger by a mechanism involving the insertion of hydrophobic peptides into a target membrane. Ultimately, the energy required to drive the fusion reaction, stored in the conformation of components of the membrane itself, is used to manipulate membrane lipids, the essential partners in fusion, in a manner that results in the establishment of both aqueous and lipidic continuity.

Similarities now seen between cell invasive membrane fusion and intracellular fusion, such as inhibition caused by the same lipid in different systems, encourage us to believe that the mechanism of fusion is similar in these various systems. If so, the modern molecular techniques being applied to study viral fusion proteins, in conjunction with complimentary functional assays for fusion using biophysical techniques such as capacitance measurements, light scattering, fluorescent spectroscopy, and electron and video enhanced microscopy, may ultimately lead to a better understanding of the molecular events involved in not only viral membrane fusion, but also perhaps all biological membrane fusion.

ACKNOWLEDGMENTS

We would like to acknowledge our many friends and colleagues with whom we have discussed membrane fusion. In particular we would like to thank: Adrian Parsegian, Fred Cohen, Klaus Gawrisch, Michael Kozlov, Michael Whitaker, Mitko Dimitrov, Peter Rand, Robert Blumenthal, Ron Holz, Sasha Sokoloff, Sergey Leikin, Teresa L. Z. Jones, Tim Whalley, Tom Reese, and Yuri Chizmadzhev. We are also very grateful to Don Wiley, Doug Chandler, and Frank Booy for kindly allowing us to use their data for figures. Finally we appreciate the valuable help of Lynn Kelly and Joan Glass in preparing this manuscript.

*Literature Cited*

1. Abidor, I. G., Arakelyan, V. B., Chernomordik, L. V., Chizmadzhev, Y. A., Pastushenko, V. F., Tarasevich, M. R. 1979. *J. Electroanal. Chem.* 104: 37
2. Almers, W. 1990. *Annu. Rev. Physiol.* 52: 607
3. Asano, K., Asano, A. 1988. *Biochemistry* 27: 1321
4. Bazzi, M. D., Nelsestuen, G. L. 1992. *Biochemistry* 31: 10406
4a. Bentz, J., ed. 1993. *Viral Fusion Mechanisms.* Boca Raton, FL: CRC Press
5. Bentz, J., Ellens, H., Lai, M.-Z., Szoka, F. C. 1985. *Proc. Natl. Acad. Sci. USA* 82: 5742
6. Berlin, V., Brill, J. A., Trueheart, J.,

Boeke, J. D., Fink, J. R. 1990. *Methods Enzymol.* 194: 774
7. Blobel, C. P., Myles, D. G., Primakoff, P., White, J. M. 1990. *J. Cell Biol.* 111: 69
8. Blobel, C. P., Wolfsberg, T. G., Turck, C. W., Myles, D. G., Primakoff, P., White, J. M. 1992. *Nature* 356: 248
9. Blumenthal, R. 1987. *Curr. Topics Membr. Transp.* 29: 203
10. Blumenthal, R., Puri, A., Walter, A., Eidelman, O. 1988. See Ref. 116a, p. 367
11. Bondeson, J., Sundler, R. 1985. *FEBS Lett.* 190: 283
12. Brasseur, R., Vandenbranden, M., Cornet, B., Burny, A., Ruysschaert, J.-M. 1990. *Biochim. Biophys. Acta* 1029: 267
13. Breckenridge, L. J., Almers, W. 1987. *Nature* 328: 814
14. Breckenridge, L. J., Almers, W. 1987. *Proc. Natl. Acad. Sci. USA* 84: 1945
15. Brose, N., Petrenko, A. G., Sudhof, T. C., Jahn, R. 1992. *Science* 256: 1021
16. Brunner J. 1989. *FEBS Lett.* 257: 369
17. Brunner, J., Zugliani, C., Mischler, R. 1991. *Biochemistry* 30: 2432
18. Burger, K. N., Verkleij, A. J. 1990. *Experientia* 46: 631
19. Burger, K. N. J., Knoll, G., Verkleij, A. J. 1988. *Biochim. Biophys. Acta* 939: 89
20. Burgoyne, R. D. 1990. *Annu. Rev. Physiol.* 52: 647
21. Chandler, D. E., Heuser, J. 1980. *J. Cell Biol.* 86: 666
22. Chandler, D. E., Heuser, J. E. 1979. *J. Cell Biol.* 83: 91
23. Chernomordik, L., Vogel, S. S., Sokoloff, A., Onaron, H. O., Leikina, E., Zimmerberg, J. 1993. *FEBS Lett.* 318: 71
24. Chernomordik, L. V. 1992. In *Guide to Electroporation and Electrofusion*, ed. D. Chang, B. Chassy, J. Saunders, A. Sowers, p. 63. San Diego: Academic
25. Chernomordik, L. V., Abidor, I. G. 1980. *J. Electroanal. Chem.* 116: 617
26. Chernomordik, L. V., Arakelyan, V. B., Abidor, I. G., Baskakov, V. A., Tarasevich, M. R. 1978. *Biophizika* 23: 806
27. Chernomordik, L. V., Kozlov, M. M., Melikyan, G. B., Abidor, I. G., Markin V. S., Chizmadzhev, Y. A. 1985. *Biochim. Biophys. Acta* 812: 643
28. Chernomordik, L. V., Melikyan, G. B., Chizmadzhev, Y. A. 1987. *Biochim. Biophys. Acta* 906: 309
29. Chernomordik, L. V., Sowers, A. E. 1991. *Biophys. J.* 60: 1026
30. Clague, M. J., Schoch, C., Blumenthal, R. 1991. *J. Virol.* 65: 2402
31. Cohen, F. S., Akabas, M. H., Zimmerberg, J., Finkelstein, A. 1984. *J. Cell Biol.* 98: 1054
32. Cohen, F. S., Niles, W. D., Curran, M., Zimmerberg, J. 1993. In *Membrane Fusion*, ed. P. Lelkes. New York: Springer. In press
33. Cohen, F. S., Zimmerberg, J. J., Finkelstein, A. 1980. *J. Gen. Physiol.* 75: 251
34. Cullis, P. R., Tilcock, C. P., Hope, M. J. 1991. See Ref. 178a, p.35
35. Curran, M., Cohen, F. S., Chandler, D. E., Munson, P. J., Zimmerberg, J. 1993. *J. Membr. Biol.* In press
36. Curran, M. J., Brodwick, M. S. 1991. *J. Gen. Phys.* 98: 771
37. Daniels, R. S., Downie, J. C., Hay, A. J., Knossow, M., Skehel, J. J., et al. 1985. *Cell* 40: 439
38. Das, S., Rand, R. P. 1986. *Biochemistry* 25: 2882
39. Davis, N. G., Model, P. 1985. *Cell* 41: 607
40. De Oliveira-Filqueiras, O. M., Van Den Besselaar, A. M. H. P., Van Den Gosch, H. 1979. *Biochim. Biophys. Acta* 558: 73
41. Doms, R. W., Helenius, A., White J. 1985. *J. Biol. Chem.* 260: 2973
42. Duzgunes, N., Gambale, F. 1988. *FEBS Lett.* 227: 110
43. Duzgunes, N., Straubinger, R. M., Baldwin, P. A., Friend, D. S., Papahadjopoulos, D. 1985. *Biochemistry* 24: 3091
44. Eberhard, D. A., Cooper, C. L., Low, M. G., Holz, R. W. 1990. *Biochem. J.* 268: 15
45. Ellens, H., Bentz, J., Mason, D., Zhang, F., White J. M. 1990. *Biochemistry* 29: 9697
46. Ellens, H., Bentz, J., Szoka, F. C. 1985. *Biochemistry* 24: 3099
47. Ellens, H., Bentz, J., Szoka, F. C. 1986. *Biochemistry* 25: 285
48. Feigenson, G. W. 1986. *Biochemistry* 25: 5819
49. Fernandez, J. M., Neher, E., Gomperts, B. D. 1984. *Nature* 312: 453
50. Finkelstein, A., Zimmerberg, J., Cohen, F. S. 1986. *Annu. Rev. Physiol.* 48: 163
51. Fisher, L. R., Parker, N. S. 1984. *Biophys. J.* 46: 253
52. Fontaine, R. N., Harris, R. A., Schroeder, F. 1979. *Life Sci.* 24: 395
53. Gallaher, W. R., Segrest, J. P., Hunter E. 1992. *Cell* 70: 531
54. Gawrisch, K., Parsegian, V. A., Hajduk, D. A., Tate, M. W., Gruner, S. M., et al. 1992. *Biochemistry* 31: 2856
55. Gething, M.-J., Doms, R. W., York, D., White, J. 1986. *J. Cell Biol.* 102: 11

56. Glaser, R. W., Leikin, S. L., Chernomordik, L. V., Pastushenko, V. F., Sokirko, A. 1988. *Biochim. Biophys. Acta* 940: 275
57. Gollins, S. W., Porterfield, J. S. 1986. *J. Gen. Virol.* 67: 157
58. Gomperts, B. D. 1990. *Annu. Rev. Physiol.* 52: 591
59. Gruner, S. M. 1985. *Proc. Natl. Acad. Sci. USA* 82: 3665
60. Gruner, S. M., Cullis, P. R., Hope, M. J., Tilcock, C. P. S. 1985. *Annu. Rev. Biophys. Biophys. Chem.* 14: 211
61. Harris, A. L., Walter, A., Pauls, D., Goodenough, D. A., Zimmerberg, J. 1992. *Mol. Brain Res.* 15: 269
62. Harter, C., James, P., Bachi, T., Seminza, G., Brunner, J. 1989. *J. Biol. Chem.* 264: 6459
63. Helfrich, W. 1973. *Z. Naturforsch.* 28: 693
64. Helm, C. A., Israelachivili, J. N., McGuiggan, P. M. 1992. *Biochemistry* 31: 1794
65. Herrmann, A., Clague, M. J., Puri, A., Morris, S. J., Blumenthal, R., Grimaldi, S., et al. 1990. *Biochemistry* 29: 4054
66. Heuser, J. E., Reese, T. S. 1973. *J. Cell Biol.* 57: 315
67. Heuser, J. E., Reese, T. S., Dennis, M. J., Jan, Y., Jan, L., Evans, L. 1979. *J. Cell Biol.* 81: 275
68. Hille, B. 1984. In *Ionic Channels of Excitable Membranes*, p. 186. Sunderland, MA: Sinauer
69. Hoekstra, D., Kok, J. W. 1989. *Biosci. Rep.* 9: 273
70. Hoekstra, D., Wilschut, J. 1989. In *Water Transport in Biological Membranes*, ed. G. Benga, p. 143. Boca Raton, FL: CRC Press
71. Israelachivili, J. N., Marcelja, S., Horn, R. G. 1980. *Q. Rev. Biophys.* 13: 121
72. Jackson, R. C., Crabb, J. H. 1988. *Curr. Topics Membr. Transp.* 32: 45
73. Kachar, B., Fuller, N., Rand, P. R. 1986. *Biophys. J.* 50: 779
74. Kagayama, M., Douglas, W. W. 1974. *J. Cell Biol.* 62: 519
75. Kelsey, D. R., Flanagan, T. D., Young, J., Yeagle, P. L. 1990. *J. Biol. Chem.* 265: 12178
76. Kelsey, D. R., Flanagan, T. D., Young, J. E., Yeagle, P. L. 1991. *Virology* 182: 690
77. Kleinberg, M. E., Finkelstein, A. 1984. *J. Membr. Biol.* 80: 257
78. Klenk, H. D. 1973. In *Biological Membranes*, ed. D. Chapman, D. F. H. Wallach, p. 145. London: Academic
79. Knoll, G., Braun, C., Plattner, H. 1991. *J. Cell Biol.* 113: 1295
80. Knutton, S. 1977. *J. Cell Sci.* 28: 189
81. Kobayashi, T., Storrie, B., Simons, K., Dotti, C. G. 1992. *Nature* 359: 647
82. Kozlov, M. M., Leikin, S. L., Chernomordik, L. V., Markin, V. S., Chizmadzhev, Y. A. 1989. *Eur. Biophys. J.* 17: 121
83. Kwok, R., Evans, E. 1981. *Biophys. J.* 35: 637
84. Lambrecht, B., Schmidt, M. F. G. 1986. *FEBS Lett.* 202: 127
85. Laver, G., Air, G. 1979. *Structure and Variation in Influenza Virus*, ed. G. Laver, G. Air. New York: Elsevier
86. Lear, J. D., DeGrado, W. F. 1987. *J. Biol. Chem.* 262: 6500
87. Leikin, S. L., Kozlov, M. M., Chernomordik, L. V., Markin, V. S., Chizmadzhev, Y. A. 1987. *J. Theor. Biol.* 129: 411
88. Liberman, E. A., Nenashev, V. A. 1972. *Biofizika* 17: 1017
89. Lim, N. F., Nowycky, M. C., Bookman, R. J. 1990. *Nature* 344: 449
90. Lindau, M., Scepek, S., Hartmann, J. 1992. *Biophys. J.* 61: A420
91. Lowy, R. J., Sarkar, D. P., Chen, Y., Blumenthal, R. 1990. *Proc. Natl. Acad. Sci. USA* 87: 1850
92. Lucy, J. A. 1984. *FEBS Lett.* 166: 223
93. Madden, T. D., Cullis, P. R. 1982. *Biochim. Biophys. Acta* 684: 149
94. Markin, V. S., Kozlov, M. M., Borovjagin, V. L. 1984. *Gen. Physiol. Biophys.* 3: 361
95. Marty, A., Finkelstein, A. 1975. *J. Gen. Physiol.* 65: 515
96. McCulloh, D. H., Chambers, E. L. 1992. *J. Gen. Physiol.* 99: 137
97. Melikyan, G. B., Abidor, I. G., Chernomordik, L. V., Chailakhyan, L. M. 1983. *Biochim. Biophys. Acta* 730: 395
98. Melikyan, G. B., Chernomordik, L. V., Abidor, I. G., Chailakhyan, L. M., Chizmadzhev, Y. A. 1983. *Dokl. Akad. Nauk. SSSR* 269: 1221
99. Melikyan, G. B., Niles, W. D., Peeples, M. E., Cohen, F. S. 1993. *Biophys. J.* (Abstr.) 64: A188
100. Monck, J. R., de Toledo, G. A., Fernandez, J. M. 1990. *Proc. Natl. Acad. Sci. USA* 87: 7804
101. Monck, J. R., Oberhauser, A., de Toledo, G. A., Fernandez, J. 1991. *Biophys. J.* 59: 39
102. Morris, S. J., Sarkar, D. P., White, J. M., Blumenthal, R. 1989. *J. Biol. Chem.* 264: 3972
103. Murata, M., Kagiwada, S., Takahashi, S., Ohnishi, S. 1991. *J. Biol. Chem.* 266: 14353
104. Naeve, C. W., Williams, D. 1990. *EMBO J.* 9: 3857

105. Nanavati, C., Markin, V. S., Oberhauser, A. F., Fernandez, J. M. 1992. *Biophys. J.* 63: 1118
106. Neher, E. 1974. *Biochim. Biophys. Acta* 373: 327
107. Neher, E., Marty, A. 1982. *Proc. Natl. Acad. Sci. USA* 79: 6712
108. Nieva, J.-L., Goni, F. N., Alonso, A. 1989. *Biochemistry* 28: 7364
109. Niles, W. D., Cohen, F. S. 1987. *J. Gen. Physiol.* 90: 703
110. Niles, W. D., Cohen, F. S. 1991. *J. Gen. Physiol.* 97: 1101
111. Niles, W. D., Cohen, F. S. 1991. *J. Gen. Physiol.* 97: 1121
112. Niles, W. D., Levis, R. A., Cohen, F. S. 1988. *Biophys. J.* 53: 327
113. Novick, S. L., Hoekstra, D. 1988. *Proc. Natl. Acad. Sci. USA* 85: 7433
114. Oberhauser, A. F., Fernandez, J. M. 1992. *Biophys. J.* 61: A421
115. Ohki, S. 1982. *Biochim. Biophys. Acta* 689: 1
116. Ohki, S. 1984. *J. Membr. Biol.* 77: 265
116a. Ohki, S., Doyle, D., Flanagan, T. D., Hui, S. W., Mayhew, E., eds. 1988. *Molecular Mechanisms of Membrane Fusion.* New York: Plenum
117. Olbricht, K., Plattner, H., Matt, H. 1984. *Exp. Cell Res.* 151: 14
118. Op den Kamp, J. A. F. 1979. *Annu. Rev. Biochem.* 48: 47
119. Ornberg, R. L., Reese, T. S. 1981. *J. Cell Biol.* 90: 40
120. Ornberg, R. L., Reese, T. S. 1981. *Methods Cell Biol.* 23: 301
121. Papahadjopoulos, D., Nir, S., Duzgunes, N. 1990. *J. Bioeng. Biomembr.* 22: 157
122. Parsegian, V. A., Rand, R. P. 1991. See Ref. 178a, p. 65
123. Perin, M. S., MacDonald, R. C. 1989. *J. Membr. Biol.* 109: 221
124. Petrenko, A. G. 1991. *Nature* 353: 65
125. Plattner, H. 1989. *Int. Rev. Cytol.* 119: 197
126. Plattner, H., Lampert, C. J., Gras, U., Vilmart-Seuwen, J., Stecher, B., et al. 1988. See Ref. 116a, p. 477
127. Pollard, H. B., Rojas, E., Burns, A. L. 1987. *Ann. N.Y. Acad. Sci.* 493: 524
128. Poole, A. R., Howell, J. I., Lucy, J. A. 1970. *Nature* 227: 810
129. Portis, A., Newton, C., Pangborn, W., Papahadjopoulos, D. 1979. *Biochemistry* 18: 483
130. Puri, A., Booy, F. P., Doms, R. W., White, J. M., Blumenthal, R. 1990. *J. Virol.* 64: 3834
131. Rand, R. P., Parsegian, V. A. 1988. See Ref. 116a, p. 73
132. Rand, R. P., Parsegian, V. A. 1989. *Biochim. Biophys. Acta* 988: 351
133. Reporter, M., Raveed, D. 1973. *Science* 181: 863
134. Rilfors, L., Lindblom, G. 1984. In *Membrane Fluidity*, ed. M. Kates, L. A. Manson, p. 205. New York: Plenum
135. Rojas, E., Pollard, H. B. 1987. *FEBS Lett.* 217: 25
136. Rook, M. B., Jongsma, H. J., van Ginneken, A. C. 1988. *Am. J. Physiol.* 255: H770
137. Rothman, J. E., Orci, L. 1992. *Nature* 355: 409
138. Ruigrok, R. W., Martin, S. R., Wharton, S. A., Skehel, J. J., Bayley, P. M., Wiley, D. C. 1986. *Virology* 155: 484
139. Santini, M. T., Indovina, P. L., Cantafora, A. 1991. *Biochim. Biophys. Acta* 1070: 27
140. Santini, M. T., Indovina, P. L., Cantafora, A., Blotta, I. 1990. *Biochim. Biophys. Acta* 1023: 298
141. Sarkar, D. P., Morris, S. J., Eidelman, O., Zimmerberg, J., Blumenthal, R. 1989. *J. Cell Biol.* 109: 113
142. Scheuner, D., Logsdon, C. D., Holz, R. W. 1992. *J. Cell Biol.* 116: 359
143. Schewe, M., Muller, P., Korte, T., Herrmann, A. 1992. *J. Biol. Chem.* 267: 5910
144. Schlegel, R. A., Williamson, P. 1988. See Ref. 116a, p. 289
145. Schmidt, M. F. G., Lambrecht, B. 1985. *J. Gen. Virol.* 66: 2635
146. Schmidt, W., Patzak, W., Lingg, G., Winkler, H. 1983. *Eur. J. Cell Biol.* 32: 31
147. Schroeder, F. 1985. In *Subcellular Biochemistry*, ed. D. B. Roodyn, p. 51. New York: Plenum
148. Siegel, D. P. 1986. *Biophys. J.* 49: 1171
149. Siegel, D. P. 1993. See Ref. 4a, p. 457
150. Siegel, D. P., Banschbach, J., Alford, D., Ellens, H., Lis, L. J., et al. 1989. *Biochemistry* 28: 3703
151. Skehel, J. J., Bayley, P. M., Brown, E. B., Martin, S. R., Waterfield, M. D., et al. 1982. *Proc. Natl. Acad. Sci. USA* 79: 968
152. Spruce, A. E., Breckenridge, L. J., Lee, A. K., Almers, W. 1990. *Neuron* 4: 643
153. Spruce, A. E., Iwata, A., Almers, W. 1991. *Proc. Natl. Acad. Sci. USA* 88: 3623
154. Spruce, A. E., Iwata, A., White, J. M., Almers, W. 1989. *Nature* 342: 555
155. Stegmann, T., Delfino, J. M., Richards, F. M., Helenius, A. 1991. *J. Biol. Chem.* 266: 18404
156. Stegmann, T., Doms, R. W., Helenius A. 1989. *Annu. Rev. Biophys. Biophys. Chem.* 18: 187
157. Stegmann, T., Helenius A. 1993. See Ref. 4a

158. Stegmann, T., White, J. M., Helenius, A. 1990. *EMBO J.* 9: 4231
159. Sudhof, T. C., Jahn, R. 1991. *Neuron* 6: 665
160. Tahara, Y., Murata, M., Ohnishi, S., Fujiyoshi, Y., Kikuchi, M., Yamamoto, Y. 1992. *Biochemistry* 31: 8747
161. Takahashi, S. 1990. *Biochemistry* 29: 6257
162. Tate, M. W., Eikenberry, E. F., Turner, D. C., Shyamsunder, E., Gruner, S. M. 1991. *Chem. Phys. Lipids* 57: 147
163. Tate, M. W., Gruner, S. M. 1987. *Biochemistry* 13: 231
164. Tate, M. W., Gruner, S. M. 1987. *Biochemistry* 26: 231
165. Umeda M., Nojima, S., Inoue, K. 1985. *J. Biochem.* 97: 1301
166. Vogel, S. S., Chernomordik, L. V., Zimmerberg, J. 1992. *J. Biol. Chem.* 267: 25640
167. Vogel, S. S., Delaney, K., Zimmerberg, J. 1991. *Ann. N.Y. Acad. Sci.* 635: 35
168. Vogel, S. S., Zimmerberg, J. 1992. *Proc. Natl. Acad. Sci. USA* 89: 4749
169. Voneche, V., Portetelle, D., Kettmann, R., Willems, L., Limbach, K., et al. 1992. *Proc. Natl. Acad. Sci. USA* 89: 3810
170. Voyta, J. C., Slakey, L. L., Westhead, E. W. 1978. *Biochem. Biophys. Res. Commun.* 80: 413
171. Westhead, E. W. 1987. *Ann. N.Y. Acad. Sci.* 493: 92
172. Wharton, S. A., Martin, S. R., Ruigrok, R. W. H., Skehel, J. J., Wiley, D. C. 1988. *J. Gen. Virol.* 69: 1847
173. Whitaker, M., Zimmerberg, J. 1987. *J. Physiol.* 389: 527
174. White, J. M. 1990. *Annu. Rev. Physiol.* 52: 675
175. White, J., Helenius, A. 1980. *Proc. Natl. Acad. Sci. USA* 77: 3273
176. White, J., Kartenbeck, J., Helenius, A. 1982. *EMBO J.* 1: 217
177. White, J., Kielian, M., Helenius, A. 1983. *Q. Rev. Biophys.* 16: 151
178. Wilschut, J. 1991. See Ref. 178a, p. 89
178a. Wilschut, J., Hoekstra, D., eds. 1991. *Membrane Fusion.* New York: Marcel Dekker
179. Wilson, I. A., Skehel, J. J., Wiley, D. C. 1981. *Nature* 289: 366
180. Yeagle, P. L., Young, J., Hui, S. W., Epand, R. M. 1992. *Biochemistry* 31: 3177
181. Zachowski, A., Henry, J.-P., Devaux, P. F. 1989. *Nature* 340: 75
182. Zieseniss, E., Plattner, H. 1985. *J. Cell Biol.* 101: 2028
183. Zimmerberg, J. 1987. *Biosci. Rep.* 7: 251
184. Zimmerberg, J. 1988. See Ref. 116a, p. 181
185. Zimmerberg, J. 1991. See Ref. 178a, p. 183
186. Zimmerberg, J. 1993. *Methods Enzymol.* 221: In press
186a. Zimmerberg, J., Cohen, F. S., Finkelstein, A. 1980. *J. Gen. Physiol.* 75: 241
187. Zimmerberg, J., Curran, M., Cohen, F. 1991. *Ann. N.Y. Acad. Sci.* 635: 307
188. Zimmerberg, J., Curran, M., Cohen, F. S., Brodwick, M. 1987. *Proc. Natl. Acad. Sci. USA* 84: 1585
189. Zimmerberg, J., Parsegian, A. 1986. *Nature* 323: 36
190. Gilbert, J. M., Kemble, G. W., Weiss, C. D., Hernandez, L., White, J. M. 1992. *Mol. Biol. Cell* 3: 1226

# FAST CRYSTALLOGRAPHY AND TIME-RESOLVED STRUCTURES

## Janos Hajdu

Laboratory of Molecular Biophysics and Oxford Centre for Molecular Sciences, Oxford University, The Rex Richards Building, South Parks Road, Oxford OX1 3QU, United Kingdom

## Inger Andersson

Department of Molecular Biology, Swedish University of Agricultural Sciences, Uppsala Biomedical Center, S-751 24 Uppsala, Sweden

KEY WORDS: X-ray crystallography, four-dimensional X-ray studies, Weissenberg photography, Laue diffraction sets, catalysis in enzyme crystals, structural clusters, refinement of crystal structures

CONTENTS

| | |
|---|---|
| PERSPECTIVES AND OVERVIEW | 468 |
|    Structural Dynamics and X-Ray Crystallography | 469 |
|    Methods in Fast and Kinetic Crystallography | 470 |
|    Towards Femtosecond Time Resolution in X-Ray Diffraction Experiments | 470 |
| ACTIVITY IN CRYSTALLINE ENZYMES | 471 |
|    Initiation of Reactions in Enzyme Crystals | 471 |
| FAST DATA-COLLECTION TECHNIQUES—A CRITICAL ASSESSMENT | 474 |
|    Monochromatic Techniques | 475 |
|    White Radiation Techniques: the Laue Method | 475 |
|    Limitations of the Laue Method | 476 |
| FAST DIFFRACTION EXPERIMENTS AND TIME-RESOLVED STRUCTURES | 481 |
|    Glycogen Phosphorylase | 484 |
|    Catalytic Domain of H-ras p21 Protein | 484 |
|    Trypsin | 485 |
|    γ-Chymotrypsin | 486 |
| IMPROVING RESULTS FROM PARTIAL DATA SETS | 486 |
|    Modeling Missing Data | 487 |
|    Improving Difference Maps from Partial Data Sets | 487 |

MIXED STRUCTURAL STATES AND PARTIALLY OCCUPIED SITES..................................... 488
   *Difference Fourier Maps with Partial Occupancies* ................................................. 488
   *Refinement of the Structure* ....................................................................................... 489
FURTHER THOUGHTS ON FOUR-DIMENSIONAL STRUCTURES ........................... 492
   *Structural Information on Intermediates from a Single Image?* ........................... 493

# PERSPECTIVES AND OVERVIEW

In the structure of biological macromolecules, rigidity is combined with a certain degree of flexibility. Superimposed on continuous thermal fluctuations (85), biological macromolecules undergo distinct structural changes during their function or as a control of this function. A fundamental understanding of macromolecular structure must therefore include a description of structural changes involved in their function. Rigidity is important for maintaining specificity, and flexibility is required for adjusting to changing conditions in function. The correct mixture of the two is of evolutionary significance. Active sites of enzymes, for instance, take up different structural states in different phases of the catalytic cycle. We estimate that at the present rate of growth, about 30,000–50,000 X-ray data sets will be collected by the year 2000 alone. We suggest (3, 26) that one of the major goals of structural biology should be to achieve four-dimensional structure determination, with time being the fourth dimension, and four-dimensional structures and coordinate files being the norm, in which conventional three-dimensional structures are stills from the movie. How much of this goal will eventually be achieved with X-ray crystallography is a matter of speculation. Diffraction methods require crystalline order in the sample, and the constraints of the lattice may affect mobility within the crystal. A combination of various experimental techniques with diffraction methods and computational procedures offers the most obvious route for four-dimensional structural studies. Note that 23% of macromolecular structures published last year were NMR structures. The technique complements X-ray methods, offering a different window on structural dynamics (see e.g. 99, 104). X-ray diffraction techniques can be used to detect motion over the complete time scale of atomic, molecular, and lattice vibrations as well as disorder. This method was the first to produce a four-dimensional structure for a protein molecule in the form of a short three-dimensional movie of a catalytic reaction in an enzyme crystal (54, 55). This paper gives an overview of what has been achieved since and outlines possible future directions.

For reasons discussed below, information on the time scale of motion cannot be extracted easily from a single X-ray data set. First we give a brief introduction to the problems, then we describe fast diffraction methods to acquaint the nonspecialized reader with the possibilities and restrictions of the techniques, and finally we summarize the results obtained so far. A number of reviews (53, 56, 57, 93) deal with time-resolved diffraction studies, including a conference volume (28a). Time-resolved X-ray scattering experiments on noncrystalline biological materials have also been described (50, 71, 77, 102; see also relevant papers in the conference volume of *J. Appl. Cryst.* 1991, 24(5): 413–974).

## Structural Dynamics and X-Ray Crystallography

The fastest chemical processes involving displacements of atoms (thermal vibrations or the breakage of chemical bonds) happen on the femtosecond-picosecond time scale. These processes are still very slow compared with the frequency of X-ray photons ($10^{18}$–$10^{19}$ s$^{-1}$). Consequently, X-ray photons detect atoms as static entities although somewhat displaced from their mean position. Structures obtained from diffraction experiments are therefore averaged over the volume of the crystal irradiated during data collection and also over the time needed to obtain the data set. There is no simple way of distinguishing moving atoms from static disorder in the crystal from a single X-ray data set. To achieve this, additional information is needed.

The first experimental evidence from X-ray data for atomic movements in proteins came from the analysis of the variation of isotropic temperature factors in two different crystal forms of lysozyme (5) and from the temperature-dependent variations of individual temperature factors in myoglobin (44). Papers from Phillips (101), Willis (143), Stuart & Phillips (124), Frauenfelder (43), and Tilton et al (130) have bearing on this approach. One can also look at diffuse scattering around and between Bragg peaks (22, 23, 25, 33, 112). Diffuse scattering is essentially an out-of-register diffraction originating from a disorder (or motion) of larger scale than simple atomic vibrations. Variations in diffuse scattering carry information on breathing motions and on the dynamics of the crystal lattice.

The time-dependent analysis of externally induced structural changes in crystals provides another opportunity to obtain information on structural changes in crystals and is the subject of this review. The kinetic approach requires the rapid collection of diffraction data sets (just like shooting a movie) following the triggering event (see e.g. 54, 55, 67, 72, 113, 115, 117). Apart from enabling the investigator to look at changes in the average structure, the technique also gives information on changes in temperature factors and diffuse scattering. The raw result of a kinetic X-ray experiment is a moving average of all structural states, weighted by their concentrations, as the reaction proceeds in the crystal. The deconvolution of the observed mixed structural states into their major individual components is, therefore, an essential requirement. This may not always be possible and it has not yet been attempted. Instead, the experiments published so far were performed on systems in which a transient build-up of an intermediate could be achieved.

## Methods in Fast and Kinetic Crystallography

Kinetic diffraction studies require the fast measurement of a sequence of three-dimensional sets of structure-factor amplitudes with an accuracy and completeness that is sufficient to describe the structures correctly during the structural change or reaction in the crystal. Data-collection times of minutes to seconds are currently available at synchrotron sources with monochromatic techniques, and of seconds to milliseconds with white radiation techniques. Attempts have already been made with both monochromatic and Laue techniques to achieve nanosecond-picosecond exposures using flash X-ray tubes or plasma sources (see e.g. 42, 67, 69, 70, 86, 97, 138, 145) and lately also with synchrotron sources (82, 127). Perhaps the most exciting achievement so far is a sequence of Laue photographs taken at 200-ns intervals on an exploding aluminium single crystal by Jamet (67). There seems to be little communication between synchrotron users attempting to reach this time domain and those who are already there. Fast experiments with laser plasmas and flash X-ray tubes may not be widely known in the synchrotron community as these studies are not quoted in recent papers trying to achieve similarly short Laue exposures at synchrotron sources. Such barriers need to be broken down. No structure has yet been obtained in any of these studies, but results are bound to emerge soon.

## Towards Femtosecond Time Resolution in X-Ray Diffraction Experiments

Although picosecond Laue exposures are of considerable importance, the really exciting chemical steps are usually over within a few hundred femtoseconds. In addition, mechanical perturbations are essentially localized on a time scale of 100 fs within a molecule (for an excellent review see 90). Consequently, the perturbation of the system can be much better characterized in the femtosecond regime than in picoseconds. A femtosecond laser excitation may, in principle, synchronize molecules in a crystal for a short time. On this time scale, real synchrony could be achieved before the structures Boltzman out. Ideas have recently been outlined on how to achieve femtosecond time resolution in Laue diffraction experiments (J. Hajdu, submitted). This time resolution may perhaps allow the observation of transition states in three dimensions.

## ACTIVITY IN CRYSTALLINE ENZYMES

A surprisingly small proportion of enzymes have had their activity tested in the crystalline state (from interviews with colleagues, we estimate it to be less than 5%). Most of those analyzed were active. This is not entirely surprising because 35–95% of the volume of a protein crystal is fluid of crystallization (average 50%). The average distance between neighboring macromolecules in dense protein solutions, such as the cytosol (water content: 70–80%), is only a few Ångströms longer than the average distance between neighboring molecules in the three-dimensional lattice of a protein crystal. NMR structures from protein solutions show close similarities with the structures determined using diffraction techniques. Most differences can be traced back to lattice contacts (135). Nevertheless, in going from the solution to the crystalline state, it is reasonable to anticipate certain changes in the structure. The mobility of the molecule is likely to be influenced, and the changes may affect activity in the crystal. The activity of crystalline enzymes is generally lower than the activity in solution (reviewed in 89). For example, Chance et al (24) determined that the reaction of ferrimyoglobin with azide was 21-fold slower in the crystalline state. Similarly, Theorell et al (128) recorded a 1000-fold reduction in the activity of crystalline alcohol dehydrogenase with NADH and isobutyramide, and an even more dramatic reduction of activity can be anticipated where the active sites are blocked to some extent by lattice contacts [e.g. hen egg-white lysozyme (11)]. A decrease in activity can often be attributed to restricted conformational flexibility within the lattice, resulting in a smaller probability of the essential "transitional conformation" in the crystalline state (10, 24, 146). In addition, a decrease in the measured activity can be expected if the dimensions of the crystal exceed the critical dimensions at which diffusion becomes a limiting factor (32, 73, 74, 81, 89, 118). Similar considerations for nonbiological catalysts have been described (129).

### Initiation of Reactions in Enzyme Crystals

DIFFUSION OF REAGENTS INTO CRYSTALS Currently, fast diffraction methods permit the simple analysis of short-lived structures that accumulate transiently in the crystal during a reaction. This requires a relatively fast binding followed by a relatively slow reaction. Because of the generally lower activity of crystalline enzymes, uniform catalysis can often be triggered by diffusing reagents (e.g. substrates) into crystals. This is the most straightforward way of initiating a reaction in a protein crystal. However, the speed of diffusion and ligand binding sets an upper limit to the speed of reactions that can be analyzed. In favorable cases, half-saturation binding with small ligands can be reached within about a minute (68). Results with this technique are described elsewhere (8, 54, 55, 107, 117).

PHOTOACTIVATION OF CAGED SUBSTRATES  If the rate of diffusion relative to the catalytic reaction is slow, then the concentrations of intermediates will not be high in the crystal. In such cases, photochemical methods can be used to bypass much of the diffusion barrier and to initiate the reaction more rapidly. Photolabile protecting groups are used to make one of the reactants biologically inert. The group can be removed from the substrate by a light pulse to liberate the active component. The physiological and physico-chemical aspects of the use of such caged compounds have been extensively reviewed (1, 7, 28, 51, 91). Caged substrates and effectors will probably play an important role in triggering reactions in protein crystals. However, their availability and successful application is unlikely to be universal. Limitations may arise because of the high concentrations required to saturate active sites in crystals, a low quantum yield, a slow photolysis rate, or the formation of reactive by-products that modify the protein.

In dilute solutions, the photochemical release of molecules from photolabile caged compounds can be relatively fast ($>100$ s$^{-1}$), but within protein crystals, it has been frustratingly slow. In the experiments published so far, the time required to achieve nearly complete photolysis has been between 1 and 30 min (35, 113, 115, 123), for several reasons. The concentration of the caged substrate in crystals has to be at least as high as the concentration of the active sites (5–15 mM). To achieve saturation, a much higher concentration of the caged substrate is usually required (see e.g. 28, 35). This makes these crystals practically black at the wavelength of photolysis, so that very often a thin skin of product can only be created by the light pulse (J. Hajdu & A. Hadfield, submitted), similar to initiating the reaction by diffusion in a flow cell. Higher intensity light pulses could produce deeper photolysis, but they also heat up the crystal more, leading to side reactions and disorder (see e.g. 95, 113). The emergence of disorder has implications for the method of data collection (see below). In order to overcome the absorption problem, longer wavelength radiation could be used, albeit with a reduced quantum efficiency. One could try the so-called two-photon excitation technique at double the wavelength of the single photon excitation. Two-photon excitations are based on a quantum effect (47) in which two red or infrared photons are simultaneously absorbed by the chromophore and simulate a single UV photon at half the wavelength. This technique could achieve deeper photolysis, but it requires a much higher photon intensity in a much shorter time. Note the difference between having the same number of photons spread out in time or compressed into a very short pulse here. A larger temperature rise during the shorter light pulse may be a problem. A major challenge will be to achieve rapid photochemical conversions that approach 100% throughout the whole body of the crystal without substantial heating.

A good photoreactive caged substrate should meet the following requirements: solubility in water (>10 mM); excitation at more than 300 nm; high quantum efficiency; breakdown products should not react with the protein and should remain soluble after photolysis; and if possible, the primary breakdown products should not absorb at the wavelength of the excitation to permit a steadily deepening penetration of the photolysing light pulse in the crystal. In addition, it is useful to cocrystallize the caged substrate with the enzyme in order to keep the concentration of the caged compound low (e.g. 113, 115).

PHOTOACTIVATION OF CAGED ENZYMES A much wider application of photochemical triggering in crystallography can be expected if one cages the enzyme instead of its substrate. With caged enzymes, the concentration of the photolabel is minimal (one per active site) whereas the concentration of the substrates (uncaged in this case) can be as high as necessary. The natural substrates can be prediffused into the crystal under inactive conditions at the most suitable temperatures for diffusion before the caged enzyme is liberated by a light pulse, e.g. at cryogenic temperatures. Two routes are possible for caging the enzyme. One of them requires active-site specific reagents that can be removed photochemically later (122, 132). The other and more exciting possibility is the introduction of unnatural caged residues at specific sites in the protein during the synthesis of the protein. This latter approach has immense potential and would be the technique of our choice. With a handful of basic building blocks (not more than eight caged amino acids: lysine, arginine, aspartate, glutamate, serine, threonine, tyrosine, cysteine), almost all known active sites could be blocked. The chemistry of the caged residues could be fine-tuned for efficient photolysis, low toxicity, small size, etc, bypassing a whole range of problems in trying to synthesize hundreds of chemically different caged substrates.

In a pioneering work, Mendel et al (92) describe the synthesis of a photoactivatable caged lysozyme, in which a 2-nitrobenzyl-caged aspartate was incorporated in the active site. The caged enzyme was expressed in an in vitro transcription-translation system after mutating the gene sequence to the nonsense amber codon TAG at the required position. A complementary tRNA was charged with aspartyl-$\beta$-nitrophenyl ester and was used to produce the caged enzyme. The enzyme was inactive in the dark but gained catalytic activity during photolysis. The expression of such proteins in sufficient quantity is a challenge. For effective use in X-ray crystallography, milligram quantities of synthetic caged enzymes would be required.

OTHER TECHNIQUES Attempts with temperature jump have not yet produced interpretable three-dimensional results with Laue diffraction (94). Problems may include uneven heating, movement of the crystal, and loss of diffraction due to increased mosaicity (see below).

In a recent study, Bartunik et al (8) described cryoenzymology studies on elastase, in which the enzyme was soaked with a peptide substrate at around 200–220 K in the presence of 70% methanol. The crystal was then heated, and the formation of the acyl-enzyme intermediate was followed using fast monochromatic data-collection methods. The appropriate temperature range for these studies was determined by taking so-called scanning-Laue photographs (9). The temperature at which the reaction began in earnest was marked by a streaking out of the Laue reflections.

Some reactions in the crystal will need to be slowed down to the time scale of the reaction initiation (or data collection) in order to allow one to study them in any detail. Apart from working at lower temperatures, one can use slower substrate analogues (54, 55) and mutant proteins (37, 125) with reduced activity to help with the accumulation of a reaction intermediate. Lowering the temperature is an obvious way to slow down reactions or to freeze out intermediates (see e.g. 2, 8, 34, 40, 89, 136). Unfortunately, suitable cryoprotectants are not always available, and cooling may change the mechanism of the reaction. Arrhenius behavior can almost never be assumed with biological macromolecules in a wide enough temperature range because the shape of the potential well for the substrate and intermediates changes with cooling. The mobility of the enzyme also undergoes abrupt changes at certain temperatures (8, 30, 105). Even so, cooling will likely be used extensively in diffraction studies (46).

# FAST DATA-COLLECTION TECHNIQUES—A CRITICAL ASSESSMENT

The introduction of synchrotron radiation for structural studies in 1971 by Rosenbaum and his coworkers (109) brought the time resolution of macromolecular X-ray techniques into the domain where biological action takes place. Small molecules have been studied at similar or faster time scales before with flash X-ray tubes and laser-induced plasma sources. Laser plasmas produce more intense bursts of radiation than synchrotron sources but the radiation is more divergent and needs focusing. Current technology with laser plasmas can give picosecond X-ray bursts with a similarly smooth spectrum as synchrotron radiation and with energies right into the MeV range (76).

## Monochromatic Techniques

With parallel monochromatic X-rays, a small proportion of lattice planes diffract at any particular orientation of the crystal. The crystal has to be rotated in the beam in order to record the full reflections and to bring the next set of lattice planes to diffraction. Reflections are integrated through a small angular range. Data rates of 50–1000 reflections $s^{-1}$ can be achieved at existing synchrotron sources, and an increase of two to three orders of magnitude could be expected with the next generation of storage rings. This increase could bring monochromatic data-collection techniques into the millisecond time range. Monochromatic techniques tolerate crystals with high mosaicity (even as much as a few degrees) and give full data sets that are relatively easy to process. There are no inherent problems with the technique. One of the recent advances in detector development was the introduction of imaging plates with very low background and wide dynamic range. Sakabe (110) exploited the excellent properties of this detector with his giant Weissenberg camera. Data collection times of 100–200 s could be achieved with this camera at the Photon Factory in Japan. To bring monochromatic techniques into the millisecond time range, higher-intensity X-ray sources are needed and the crystal must be rotated fast. This is not as difficult as earlier thought (96). An average sized protein crystal (0.5 mm in diameter) centered on a goniometer axis and rotated at 1000 RPM experiences a maximal centrifugal acceleration of 0.27 g. This is less than the acceleration that occurs when taking the crystal off the shelf.

## White Radiation Techniques: the Laue Method

The Laue method employs a beam of polychromatic X-rays to illuminate a stationary crystal. Reflections are integrated through a small wavelength range instead of a small angular range. Numerous lattice planes diffract simultaneously under these circumstances as the Bragg condition is satisfied for each of these planes by at least one wavelength of the spectrum. With crystals of high symmetry, a high proportion of the unique data set may be recorded on a single photograph (27). Exposure times as short as picoseconds are possible (42, 67, 69, 70, 86, 97, 127, 137, 138, 145). The tremendous speed advantage that the Laue method has over monochromatic techniques is largely lost as soon as more than one exposure needs to be taken to complete the data set. Speed is the greatest asset of this technique, and perhaps the only one. The processing of Laue diffraction data is generally more complicated than the processing of monochromatic data sets. The technique is sensitive to small errors in the application of the wavelength-normalization curve to compensate for wavelength-dependent factors in the intensity measurements (21, 38, 61,

103, 144). The quality of data obtained this way is currently lagging behind the quality of monochromatic data sets.

The other technique for obtaining structure-factor amplitudes from Laue photographs is based on a difference method and can be used for the analysis of structural changes relative to a known starting structure (58). The limitations of the latter approach are obvious, but it gives more reliable structure-factor amplitudes than the currently available wavelength-normalization methods with crystals that tolerate more than one exposure.

Apart from difficulties in data processing, the Laue method suffers from two fundamental limitations inherent in the physics of Laue diffraction (56). These restrict its use in routine data collection and have implications for the use of the method in time-resolved experiments. As it stands, the Laue method is not the method of choice for time-resolved structural studies. The identification of these problems is an important step for the next stage of development.

## Limitations of the Laue Method

EXTREME SENSITIVITY TO DISORDER  In practice, reciprocal lattice points have finite dimensions. The size and shape of reflections on Laue photographs depend on the dimensions of the reciprocal lattice points and on the divergence of the incident X-ray beam. Large reciprocal lattice points (i.e. crystals with slightly imperfect real-space lattices or moderate mosaicity of about 0.1–0.2°) produce radially elongated reflections (Figures 1 and 2b). Higher degrees of mosaicity or a larger beam divergence will completely blur the pattern. A transient degradation of the diffraction pattern often accompanies successful reactions in crystals. To avoid problems with streaking, only crystals of the highest quality can be used in Laue diffraction studies. Unfortunately, the size of reciprocal lattice points usually increases during a structural transition or reaction, and therefore, useful structural data may not always be attainable from reactive crystals with this technique. The first application of the Laue method in 1984 to follow an enzymatic reaction in a protein crystal (glycogen phosphorylase $b$) ended in disaster (54). Substrate binding and the enzymatic reaction were accompanied by a marked increase in the mosaicity of the crystal. This produced reflections with long, radially streaked profiles (Figure 2b). Nine out of the ten ligands tested induced transient disorder in the crystal; the tenth killed it. Similar phenomena were observed in subsequent time-resolved Laue experiments on other proteins (8, 64, 95, 107, 113, 115, 117). Many enzymes undergo volume changes during their catalytic cycle (19), and the pulsation of the protein molecule may inflate the size of reciprocal lattice points, leading to streaking. With phosphorylase, reflections were streaked in a particular direction, suggesting that the most pronounced

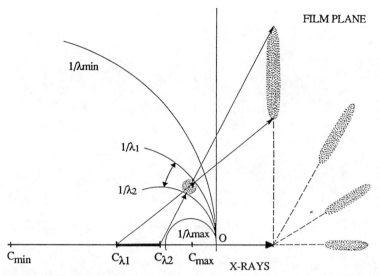

*Figure 1* Radially elongated reflections on Laue photographs (adapted from a drawing by N. Watanabe, Photon Factory, Tsukuba). Mosaicity and disorder inflate the size of reciprocal lattice points. A reciprocal lattice point from an isotropically mosaic crystal can be represented by a sphere with a finite radius. The reflection (corresponding to the image of a reciprocal lattice point) is integrated over the wavelength range of $1/\lambda_1$ to $1/\lambda_2$ in the figure. Values of $1/\lambda_1$ and $1/\lambda_2$ depend on the size and position of the reciprocal lattice point. With reciprocal lattice points of uniform size, differences between $1/\lambda_1$ and $1/\lambda_2$ are large at low resolution and small at high resolution, so that streaks are longest with the lattice point close to the origin (O) of the reciprocal lattice. The shape of the streaks carries information on the shape of the reciprocal lattice points. The projection axis giving rise to reflections on the film is the X-ray axis. As a consequence, reflections on the photograph are radially streaked from the direct beam mark. $C_{min}$, $C\lambda_1$, $C\lambda_2$, and $C_{max}$ are centers of Ewald spheres with radii $1/\lambda_{min}$, $1/\lambda_1$, $1/\lambda_2$, and $1/\lambda_{max}$, respectively. Adapted from Ref. 56.

movements occurred in the $a^*,b^*$ plane of the crystal. This is the plane in which the molecular two-fold axis lies. Attempts to use the Laue method on this system were eventually abandoned and the studies were completed with monochromatic radiation (55).

Monochromatic techniques are much less sensitive to mosaicity and give nearly complete data sets. Figure 2d shows a monochromatic still photograph taken under conditions identical to those in Figure 2b. Fast monochromatic techniques are better suited to kinetic structural studies than the Laue method. On a more positive tone, without the extreme sensitivity of the Laue method to crystal quality (Figures 2a,b), the phenomenon of transient disorder in reactive crystals could not have been discovered (54). The technique is eminently suitable to study such disorders, and this is why white synchrotron radiation is used in X-ray topography (131).

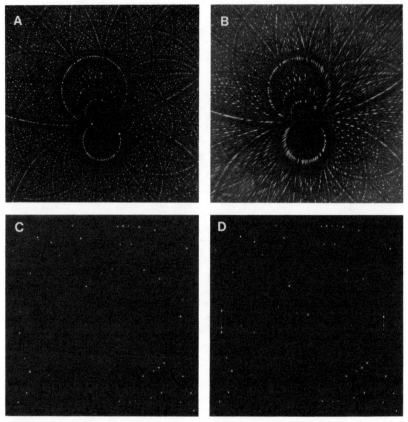

*Figure 2* Effect of moderate lattice disorder on a Laue photograph (*A,B*) and on a monochromatic still photograph (*C,D*). A crystal of glycogen phosphorylase *b* (P4$_3$2$_1$2, $a = b = 128.8$ Å, $c = 116.2$ Å) was placed in a flow cell surrounded by the crystallization buffer. The buffer was exchanged with a solution containing phosphate. Binding of phosphate triggers a rapid transient disorder in the crystal (54). Laue photographs were taken before (*A*) and 40 seconds after the addition of phosphate (*B*). The experiment was repeated in an identical fashion using monochromatic X rays (*C*, no phosphate; and *D*, 40 seconds after phosphate addition). Mosaicity in both *B* and *D* increased to about 0.2°. Adapted from Ref. 56.

THE LOW-RESOLUTION HOLE  Laue data sets from crystals of macromolecules often contain no reliable data in the range from about 4- to 5-Å resolution to infinity. The two limiting Ewald spheres ($1/\lambda_{max}$ and $1/\lambda_{min}$ on Figure 1) touch each other at the origin of the reciprocal lattice, and so the efficiency of the Laue method in this area is similar to the efficiency of monochromatic still photographs. In addition, most of the low resolution reflections in this narrow region are harmonic overlaps (29), and those that are singlets are at the short-wavelength limit of the spectrum (56) where the scattering power is the lowest and the drop in spectral intensity is the steepest. These reflections give more unreliable data than others in the pattern. Factors listed above produce a low-resolution hole in the data set (27, 56). The consequence of the low-resolution hole is that structural elements with high B factors are difficult to identify from Laue data sets. The increase in motion in a structure causes a weakening of reflections at higher resolutions and an increase in the background, so that the errors increase faster. Little information can be expected for structural motifs with high B factors from weak reflections at high resolutions (141). Given this fact, one must conclude that Laue studies on reacting systems give more biased results on multiple structural states than monochromatic techniques. This is a feature of Laue diffraction and we will have to live with it. Note that multiple structural states are deliberately produced in kinetic experiments.

The seriousness of the these problems is not generally recognized. The only advantage of the Laue method seems to be its speed in collecting medium- to high-resolution data from perfect crystals. Unfortunately, reactive crystals are seldom perfect, and partial Laue data sets pose problems with mixed structures. Not all reactions that take place in crystals are thus suitable for straight kinetic crystallographic analysis with the Laue technique. It is therefore important to concentrate efforts to circumvent these limitations by developing computational and experimental techniques in order to get the most out of Laue exposures and also by investigating alternative monochromatic techniques for the fast collection of more complete diffraction-data sets. Currently, the only possible way for the fast collection of monochromatic-data sets is the Weissenberg method (4, 110). We believe that the Weissenberg method using monochromatic X rays could eventually turn into a credible alternative to the Laue technique in studies in which data-collection times on the second-millisecond range are required. Figure 3 shows a section of a predicted Weissenberg photograph containing a practically full monochromatic data set on a single exposure from a crystal of glycogen phosphorylase $b$. Such data sets could be obtained in about 100–200 s at most currently available synchrotrons and about two to three orders of magnitude faster with the next generation of synchrotrons. In the nanosecond-picosecond range however, there seems to be no substitute for the Laue method. Thus, continued development of this technique is well justified.

ON THE RANGE OF USEFUL EXPOSURE TIMES IN LAUE CRYSTALLOGRAPHY
Heating by the intense white synchrotron radiation may produce a time

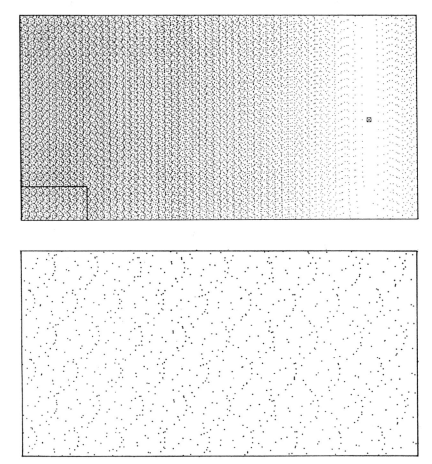

*Figure 3* Simulated Weissenberg photograph containing a data set to 2.3-Å resolution on glycogen phosphorylase $b$ (P4$_3$2$_1$2, $a = b = 128.8$ Å, $c = 116.2$ Å). Rotation axis: $c$ (vertical in figure). Rotation range: 0–45°; Weissenberg coupling constant: 2° mm$^{-1}$; wavelength: 1.040 Å. Camera radius: 860 mm, image plate size: 20 × 40 cm. A long crystal-to-detector distance is necessary to improve the signal-to-noise ratio. In a real experiment, four image plates of this size would be used simultaneously to provide a larger detector area (40 × 80 cm). A cross on the top figure indicates the position of the X-ray beam. The bottom picture is an enlargement of the lower left corner of the top figure. Preliminary experiments suggest that monochromatic data sets could be collected within 100–200 s at current synchrotron sources with the Weissenberg method, and data-collection times of about 1.0–0.1 s can be expected with the next generation of synchrotrons. Courtesy of A. Nakagawa (Photon Factory, Tsukuba).

domain within which the Laue method is effectively blind (3, 52, 117). It has been estimated that approximately $10^{10}$–$10^{13}$ X-ray photons are needed to produce an interpretable Laue pattern from a protein crystal (52). The time span within which these photons are delivered to the sample has an effect on the resulting diffraction image. The rate of heat deposition in the crystal depends on the dose rate. At high dose rates, data collection turns into a combined diffraction/T-jump exercise. Even moderate temperature rises can inflate the size of reciprocal lattice points, producing streaked reflections on Laue photographs (95, 117). The resulting electron density maps are influenced by higher atomic temperature factors. This effect is accentuated by the lack of low resolution reflections in Laue data sets. Extensive heating may eventually prevent the collection of interpretable Laue data from protein crystals. At extremely high dose rates (requiring subnanosecond exposures), the damage may not have enough time to develop and distort the lattice. Structural relaxations by various acoustic modes require many picoseconds within a protein molecule. Mechanical perturbations in the protein crystal are therefore essentially localized at the lattice point within a picosecond. Data collection within this time domain could therefore give structural results that are hardly influenced by radiation damage. A major advantage of Laue data collection with such high dose rates is the possibility of breaking the radiation-damage barrier (46, 52, 62).

## FAST DIFFRACTION EXPERIMENTS AND TIME-RESOLVED STRUCTURES

Due to a lack of space, we cannot list fast diffraction experiments performed on small molecules or metals using flash X-ray sources or laser plasmas here. Over a hundred publications deal with picosecond-nanosecond experiments in this field, many of them using Laue diffraction. Representative entries and reviews are given in the section Perspectives and Overview. Table 1 sums up results on fast diffraction experiments and time-resolved structures with macromolecules. For reasons of speed, in many of these experiments Laue diffraction was used for data collection. In the studies on the H-ras p21 protein (113, 115), 70–95% of the crystals had to be rejected because they gave streaky Laue patterns. The rejected crystals were still good enough for monochromatic data collection to about 1.0- to 1.3-Å resolution on the mutant protein. Fast monochromatic methods will certainly substitute for Laue diffraction as long as the time scale of data collection can be made to match the time scale of the dynamic event. Moreover, because of difficulties with fast reaction initiation in protein crystals, reactions will probably need to be slowed down (e.g. by

**Table 1** Fast diffraction experiments in macromolecular crystallography

| Protein (reference) | Exposure time for data set[a] | Experiment | 3D result |
|---|---|---|---|
| Collagen (12) | 200 s | First use of white synchrotron radiation in structural biology. Structure factor amplitudes obtained from an energy dispersive fiber pattern. | None. |
| Calcium binding protein (96) | 1 s | Test Laue photographs. | None. |
| Pea lectin (60) | 4 × 45 s | Test Laue photographs. | None. |
| Gramicidin A (59) | 50 s | Laue diffraction from small crystals (20 μm³); investigation of radiation damage. | None. |
| Glycogen phosphorylase $b$[b] (54, 55) | 30–480 min[c] | Catalytic reaction followed with a sequence of monochromatic data sets. | 3D movie of a catalytic reaction. |
| Glycogen phosphorylase $b$[b] (54) | 250 ms | Kinetic Laue experiment. Demonstration of transient disorder during ligand binding and catalysis. | None. |
| Hen egg white lysozyme[b] (94) | 64 ms | Test Laue photograph during crystal heating. | None. |
| Glycogen phosphorylase $b$ (58) | 3 × 1 s | Laue data set collected (52% complete to 2.4 Å) on native and maltoheptose-soaked enzyme. | First electron density map from protein Laue data. |
| Xylose isomerase (39) | 3 × 1 s | Laue data set collected (23% complete to 3 Å) on native and Eu³⁺-soaked enzyme. | Difference electron density map showed bound $Eu^{3+}$. |
| Zinc insulin[b] (107) | 3 s | Laue exposures taken during the transformation of 4Zn-insulin to 2Zn-insulin. Transient disorder observed. | None. |
| Pea lectin (61) | 4 × 45 s | Data processed (54% complete to 2.6 Å) from earlier experiment (60). | None. |
| Tomato bushy stunt virus (20) | 24 s | Single Laue shots taken from native, Ca²⁺-free, and Gd³⁺-soaked virus (30% complete to 3.5 Å). | $Ca^{2+}$ and $Gd^{3+}$ sites visible in difference Fourier maps. |
| ω-amino acid: pyruvate amino transferase[b] (140) | 5 × 1 ms in 10-ms intervals | Dynamic Laue experiment using the dot-streak method to follow laser excitation in the protein. | None. |
| ω-amino acid: pyruvate amino transferase (139) | 10 × 100 ms | Laue data set collected on enzyme inhibitor complex. | Difference Fourier maps show bound inhibitor at the active site. |
| Rubisco (4) | 20 min[c] | Fast monochromatic data collection on a large protein in Weissenberg geometry. A million reflections measured to 1.4 Å. | 277,448 unique reflections obtained, conformation of bound sugar determined. |
| H-ras p21 protein guanidine-binding domain[b] (115) | 3 × 10–15 s | Laue data collected on p21 crystals complexed with caged GTP before and after photolysis. | 3D movie on catalysis shows GTP hydrolysis. Caged GTP binds differently from GTP. |
| H-ras p21 protein guanidine-binding domain[b] (113) | 4 × 0.5–0.1 s | Laue data collected on cellular p21 plus caged GTP before and after photolysis. | 3D movie on catalysis shows GTP hydrolysis. Caged GTP binds like GTP, GDP, or GppNHp. |

CRYSTALLOGRAPHY IN FOUR DIMENSIONS    483

| | | | |
|---|---|---|---|
| Turkey egg white lysozyme (64) | 2 × 2 s | Laue data collected on 2 photographs to yield 67% of data between 5.0- and 2.5-Å resolution (I > 2σ). | First protein structure solved[d] from a Laue data set (R = 0.19). |
| GAPDH from *Trypanosoma brucei* (133) | 4 × 2 s, 3 × 3 s | Laue data set collected from 3 crystals (37% complete to 3.2 Å). Iterative density averaging used. Missing data modeled. | Protein structure solved[d] from Laue data. |
| Hen egg white lysozyme (127) | 120 ps (nominal) | 45–135 reflections collected and processed from a single X-ray bunch emitted by an undulator. | None. |
| Photoactive yellow protein[b] (95) | Not given | Dot-streak Laue study on photobleaching. Heating by laser made pattern disordered. | None. |
| Trypsin[b] (117, 148) | 25 ms, 800 ms | Kinetic Laue experiment. Deacylation of guanobenzoyl Ser195 followed after pH-jump. | Electron density maps show attacking water. |
| Elastase (8) | 10 s (Laue); Not given (mono) | Scanning Laue experiment at low temperature to follow the hydrolysis of a peptide. Map calculated from monochromatic data. | Monochromatic electron density map shows acyl-elastase intermediate. |
| Human carbonic anhydrase II (83, 84) | 3 s, 20 s | Laue data sets collected from native and derivative enzyme (55–67% complete to 2.2 Å). | Electron density maps confirm tetra-coordination of $Zn^{2+}$. |
| γ-Chymotrypsin[b] (108, 122) | 5 s | Laue data collected on cinnamate-soaked enzyme following UV irradiation in the presence of substrates. | Reaction is believed to take place, but "contaminant" tetrapeptide makes interpretations difficult. |
| Glycogen phosphorylase $b^{b}$ (35) | 3 × 0.8 s | Laue data collected to monitor photochemically induced catalysis using caged $P_i$ in the crystal. | Missing low-resolution data made interpretation difficult. New software improved maps. |
| Glycogen phosphorylase $b^{b}$ (In preparation[e]) | 120 s[c] | Fast Weissenberg data set collected on transiently disordered crystal during phosphate binding. | $P_i$ not visible. Electron density maps show small structural changes. |
| Human thymidylate synthetase (Submitted[f]) | 3 × 1.5 s + 50 min | Combined Laue-monochromatic data set collected. | Structure solved.[d] |
| Cytochrome *c* peroxidase[b] (In preparation[g]) | 4 × 300 ms | Free radical formation followed in the enzyme (compound I observed). | Oxyferryl group and structural changes visible. |
| Cytochrome P450[b] (In preparation[h]) | | Formation of the oxyferric form of the enzyme followed in the crystal. | Electron density maps show intermediate. |

[a] Note that the actual data collection times are longer than the exposure times listed here.
[b] Kinetic experiment.
[c] Monochromatic data collection.
[d] Structure solved by molecular replacement.
[e] J. Hajdu & V. Fülöp, in preparation.
[f] C. A. Schiffer, I. J. Clifton, V. J. Davidson, D. V. Santi, J. Hajdu & R. M. Stroud, submitted.
[g] V. Fülöp, P. Phizackerley, M. Soltis, I. J. Clifton, J. Hajdu & S. L. Edwards, in preparation.
[h] I. Schlichting, in preparation.

cooling), and data-collection times of $10–10^{-2}$ s will be sufficiently fast in the overwhelming majority of kinetic experiments. Monochromatic techniques will provide this speed within the next year or so. A few points from results in Table 1 are highlighted below:

## Glycogen Phosphorylase

Attempts to reproduce the original monochromatic experiments on the reaction of heptenitol with phosphate (54, 55) by Laue diffraction have been hampered by transient streaking of Laue reflections during the introduction of substrates into the crystal (Figure 2b). To overcome this problem, Duke et al (35) used photochemical techniques with 3,5-dinitrophenyl-phosphate as a caged substrate to initiate the reaction. Formation of the tightly binding end product (heptulose-2-phosphate) was demonstrated after the photolytic release of phosphate in the crystal. Reflection streaking was not observed. Unfortunately, the photochemical reaction did not produce as much phosphate as that used in the monochromatic work (approximately four to five times less). It is possible that the lack of streaking was due to this lower phosphate release. The lower phosphate concentration also meant lower occupancies for intermediates at the active sites, making the interpretation of the results more difficult. This was further complicated by the systematic incompleteness of the Laue data sets (35). This effect is more pronounced with a large protein like phosphorylase than with smaller proteins, e.g. trypsin (117, 148) or the H-ras p21 protein (113, 115) in which the bound ligand represents a higher proportion of scattering material in the structure than in a larger protein. Hence, Laue maps from these smaller proteins were of better quality than the initial maps from phosphorylase. In the meantime, a powerful new program package (LEAP) has been developed to process Laue data (S. Wakatsuki, in preparation) with improved algorithms for spot alignment, background estimation, profile fitting, spatial deconvolution, harmonic deconvolution, and spectral correction to give results with comparable precision (if not completeness) to monochromatic measurements. Data sets processed with this program retained somewhat more low-resolution terms and produced improved electron-density maps.

## Catalytic Domain of H-ras p21 Protein

H-ras p21 belongs to a family of guanine nucleotide-binding proteins thought to be involved in signal transduction pathways. H-ras p21 exists in two states: an active GTP-binding state that passes a signal onto the next molecule in the transduction pathway and an inactive GDP-binding state. Mutant ras proteins with a decreased rate of GTP hydrolysis have been implicated in the development of tumors. While the structure of the H-ras p21–GDP complex of the catalytic domain (residues 1–166) could be solved easily, no structure for the short-lived H-ras p21-GTP complex could be obtained. In the impressive first paper on the structure of this complex, Schlichting et al (115) used mixed diastereomers of a photosensitive caged GTP. This could be cocrystallized in a 1:1 complex with the protein. Photolysis of this complex gave the H-ras p21-GTP complex. The structure of the H-ras p21–GTP complex was determined about 4 min after the photolytic removal of the protecting group. A data set collected about 14 min after photolysis (during GTP-hydrolysis) led to an electron-density map that lacked most of the density of the $\gamma$-phosphate group. Structural changes observed on GTP hydrolysis were directly or indirectly related to the loss of the $\gamma$-phosphate. A change in coordination of the active-site magnesium ion was seen. After the monochromatic movie of a catalytic reaction in glycogen phosphorylase, this was the first successful color movie using the Laue method. Molecular-dynamics simulations corroborated the results (41). In a careful and thorough follow-up study, Scheidig (113) used a single diastereomer of caged GTP (the R isomer about the chiral center of the protecting group) both with the native and with a mutant variant of the H-ras p21 protein. The results of the new experiments differ in certain details from the earlier results. The protecting group of the caged GTP (R-isomer) is in a different position in the complex. This position is maintained when using the S-isomer or mixing the two diastereomers. All of the interactions between the protein and the phosphate groups seen in the monochromatic H-ras p21–GppNHp structure are also present in this H-ras p21-caged GTP structure. Caged GTP binds to H-ras p21 in a GTP-like manner and with identical magnesium coordination. Studies are underway to resolve the apparent contradiction between the two sets of experiments.

## *Trypsin*

Singer et al (117, 148) recently published a detailed time-resolved Laue study on the catalytic reaction in trypsin. They showed with great clarity the attacking water in the acyl enzyme intermediate of the enzyme. The reaction studied was the deacylation of the transiently stable $p$-guanidinobenzoyl trypsin triggered by a pH jump. The course of the reaction was monitored using Laue diffraction. Data sets obtained in these experiments were nearly complete from 3.6- to 1.8-Å resolution; they were in the 40% range from 3.6 to 5.0 Å; and near 0% from 5 Å to infinity. The attacking water molecule was not visible in the original difference Fourier maps ($F_{obs(3\ min)} - F_{obs(0\ min)}$). The detection of this water molecule depended on $F_{obs}$-$F_{calc}$ maps calculated after some cycles of anhydrous refinement. This shows that each kinetic study needs to be considered individually and calls for care in data processing.

## γ-Chymotrypsin

Results with γ-chymotrypsin crystals are more difficult to interpret. Dixon & Matthews (31) have shown that the γ-crystal form of chymotrypsin (referred to as γ-chymotrypsin) contains a tightly bound tetrapeptide inhibitor at the active site of the enzyme. Replacing this inhibitor with any other ligand is extremely difficult in the crystal. No activity measurements have been reported on the crystals. Solution studies on chymotrypsin (without the inhibitor peptide) show that the active-site serine can be modified with *trans-p*-diethylamino-*o*-hydroxy-α-methylcinnamate (referred to as cinnamate) to form a stable acyl-enzyme adduct. The reagent can be removed from the active site by a UV flash to regenerate activity in solution. Stoddard et al (122) soaked crystals of γ-chymotrypsin with this reagent for six weeks in an attempt to replace the tetrapeptide inhibitor with cinnamate. To what extent the tightly bound peptide could be displaced is not clear. In the experiment, only 25% of the unique data set between 2.5- to 5.0-Å resolution could be extracted from Laue photographs, and practically 0% from 5.0 Å to infinity. The binding of cinnamate was probably weak with low occupancy, giving a mixture of cinnamate and the inhibitor tetrapeptide at the active sites. Shortly before the experiment, another inhibitor (3-benzyl-6-chloro-2-pyrone) was diffused into the crystal to react with the liberated enzyme after photolysis. No activity measurements were performed on this type of crystal before or after UV irradiation. Assuming activity in the crystal and also that the tetrapeptide inhibitor was replaced by cinnamate, the interpretation of the maps was that they were "suggestive of a bicyclic coumarin species produced by photolysis and deacylation; however, the electron density is difficult to model unambiguously by one unique chemical state" (122, p. 5503).

# IMPROVING RESULTS FROM PARTIAL DATA SETS

Improving results from partial data sets has implications in all fields of crystallography where, for one reason or another, less than a full data set could be collected. In most cases, crystallographers work with data sets that are not quite complete, but Laue crystallographers work with data sets that are frighteningly incomplete.

## Modeling Missing Data

The three-dimensional structure of glycosomal glyceraldehyde-3-phosphate dehydrogenase (gGAPDH) from *Trypanosoma brucei* was solved from a partial Laue data set that was 37% complete (133). The crystals contain six subunits per asymmetric unit, enabling the investigator to overcome the absence of more than 60% of reflections by sixfold density averaging. Crystals of glycosomal glyceraldehyde-3-phosphate dehydrogenase are difficult to grow, and very few could be grown because of the minute amounts of enzyme that can be purified from *T. brucei*. Three crystals were available at the time of data collection, but data collection by conventional methods would have required many more (around 10 crystals). The structure was solved from a partial Laue data set obtained from these crystals. Following the initial molecular replacement solution, rigid body refinement was applied and the missing data were filled in using calculated structure factors in an iterative density-modification procedure that was part of the overall refinement process. The coefficients used during map calculations were $m*F_{obs}*\exp(i\varphi_{calc})$ for the measured reflections, and $0.9*\langle m \rangle *F_{calc}*\exp(i\varphi_{calc})$ for the missing reflections, where $m$ is the Sim weight for the observed reflections in a resolution range. Subsequent map inversion amplitudes and phases were gradually introduced to minimize the effect of missing measurements on the electron-density map. Data from map inversion were added gradually in shells of resolution in steps of 1 reciprocal lattice point from the point where the completeness was the highest, and proceeded step by step towards the extremes. After each addition, two cycles of sixfold density averaging were carried out. The final map was calculated using 22,022 measured reflections and an additional 38,215 "imagined" reflections calculated between 3.2- and 7.0-Å resolution.

## Improving Difference Maps from Partial Data Sets

ISOMORPHOUS DIFFERENCE PATTERSON MAPS  Isomorphous difference Patterson maps from Laue data sets are often noisy and uninterpretable. A substantial part of the noise in these maps can be attributed to the incompleteness of the data sets. This noise can be separated mathematically from the noise emerging from errors in the measurements and can be described in terms of a point-spread function. The point-spread function can then be used to clean up the maps (56, 111).

DIFFERENCE FOURIER MAPS  Noise originating from the incompleteness of a data set is essentially removed in simple difference Fourier maps calculated with coefficients $(F_{obs} - F_{calc})$, $\alpha_{calc}$. The point-spread function produces identical features in the native and the derivative structures, and these features cancel each other in these difference maps. However, difference Fourier maps in which the mean electron density is not zero (i.e. maps with coefficients of $2F_{obs} - F_{calc}$, $3F_{obs} - 2F_{calc}$, etc) will retain a certain contribution from the point-spread function making them substantially more fragmented than simple $F_{obs} - F_{calc}$ maps. These maps can be cleaned up using a procedure similar to the one described above for isomorphous difference Patterson maps.

A difference Fourier synthesis is a rough approximation of the new structure in which the derivative structure factor is assumed to be parallel with the protein structure factor so that the protein phase is used in Fourier imaging of the somewhat unrelated difference structure. A study by P. Nordlund (Oxford University, in preparation) assesses the possibility of improving the difference Fourier by refeeding approximate information on the difference structure back to the derivative to obtain a derivative-structure factor that is nonparallel to the protein-structure factor. This can be done by filtering an initial difference Fourier in a procedure similar to the solvent flattening of protein electron-density maps, but in this case the "solvent content" is 90–98% of the unit cell. The filtered difference information can be backtransformed to generate a nonparallel and improved derivative structure factor that can be fed back (using the experimental derivative amplitude) to further cycles of phase improvement. The method can give good estimates of heavy atom difference structure, as shown for protein B2 of ribonucleotide reductase (98). The method has the potential to improve difference information obtained with the Laue method. Laue data suffer from the so-called low-resolution hole, and filtered difference structure factors as described here can be used to estimate fractions for the components of harmonically overlapped reflections or to sample reciprocal space where no measurement is available.

## MIXED STRUCTURAL STATES AND PARTIALLY OCCUPIED SITES

Mixed structural states and partially occupied sites are commonplace in protein crystallography. Mixed structures are deliberately produced in kinetic experiments, so methods to analyze and refine them are worth considering.

### Difference Fourier Maps with Partial Occupancies

As shown already by Luzzatti (87), for an almost complete structure, maps calculated with the commonly used coefficients $(2|F_N| - |F_P|) \exp(i\alpha_P)$, where N denotes the complete structure and P the model structure, are biased towards the partial structure. Luzzatti's work has since been extended by others (88, 100, 106, 134) to include calculations of more appropriate coefficients. Smith et al (121) derived empirically the coefficient $k$ for the computation of difference maps for antiviral drugs using the formula $[F_{drug} - (1-k)F_{nat}] \exp(i\alpha_{nat})$. The phases in this case were derived from molecular replacement real-space averaging.

## Refinement of the Structure

It was recognized at an early stage that individual side chains may take up different conformations in different molecules so that separate images of them may be seen in the electron-density map (75, 141). The result of a refinement of the structure most often consists of a list of four parameters per atom: $x$, $y$, and $z$ coordinates and an isotropic thermal motion factor, $B$. The refined structure can be obtained by least-squares (63), simulated annealing (13), or molecular-dynamics (45) refinement. The model is a smeared-out static model that does not describe the dynamical properties of the protein properly. The following points need to be considered:

VIBRATIONAL PARAMETERS  In reality, atomic vibrations are anisotropic and anharmonic, and a full structural description would require nine or more parameters per vibrating atom. However, for the refinement to make sense, the information available in the experimental data must be sufficient compared with the number of parameters refined. With increasing protein size, the problem becomes increasingly underdetermined and/or computationally demanding. Only a few examples of anisotropic refinement in proteins are available (65, 116, 142). In these cases, anisotropy of thermal vibrations is treated in a restricted way, and knowledge of the structure is used to reduce the number of necessary parameters. An example of how to overcome the consequences of the lack of data for anisotropic refinement comes from Kuriyan & Weis (79), who used a rigid-body 10-parameter model. Stuart & Phillips (124) presented another idea, but this has not yet been put into practice. The combinations of an anisotropic refinement with anharmonic corrections of vibrational parameters would require even more observations to substantiate the model. Unfortunately, protein crystals do not diffract to the required resolution, and therefore such corrections have not been attempted with macromolecules. With small molecules, anharmonic refinement has been carried out in a few cases (see e.g. 147).

MULTIPLE CONFORMATIONAL STATES  A protein crystal is a collection of many structures. Macromolecular refinement schemes use Bragg data to arrive at a single "best" model for the average macromolecular structure. To describe a structure in terms of a single model introduces severe constraints on the model, even with the inclusion of anisotropic and anharmonic thermal vibrations around mean atomic positions. The structure of proteins suggests that an ensemble of different conformations should be taken into account (preferably each with its own vibrational parameters). However, the number of observations relative to the number of parameters is small with protein crystals, requiring the use of external information in the refinement.

High resolution data and improved refinement methods have resulted in a thorough investigation of disorder and multiple conformations in some proteins (23, 80, 120, 126). Around 10% of the protein side chains were found to exist in more than one unique conformation, as can be readily seen from the maps. These presumably represent the cases in which the distribution between the two forms/occupancies is nearly 50:50. Other cases are more likely to go undetected. A more recent example is the structure refinement of the scorpion toxin AaHII to 1.3-Å resolution. In this work, three protein side chains were modeled with a double conformation, and eight waters had double sites (D. Housset & J.-C. Fontecilla-Camps, personal communication). In the reported cases, multiple conformers were detected by visual inspection of electron-density maps. For this to be possible, atomic positions must be separated by enough distance and have similar occupancies to result in distinct peaks in the electron-density maps. This may not always be the case.

Molecular-dynamics calculations show multiple minima in the potential energy surface of proteins (36) and suggests that the best strategy for including anisotropic and anharmonic effects for crystallographic refinement is to include multiple occupancies (66). Kuriyan et al (78) used a "twin-refinement" search procedure to locate and characterize different structures. These studies use two noninteracting structures that contribute equally to the calculated structure-factor amplitude in a restrained molecular-dynamics refinement. Because time and space in diffraction experiments are equivalent as long as the frequency of the diffracted beam is much higher than the frequency/speed of motion, averaging in time and averaging between unit cells in refinement are equivalent. Gros (48, 49) proposed a refinement method in which the restraints are enforced over an ensemble of structures averaged over time. This gives a better agreement between the model and the data, especially at high resolution (1.7 to 3.0 Å), produces difference maps that are much cleaner than the classical $F_{obs}$-$F_{calc}$ maps, and may account for the anisotropy and anharmonicity of the structural fluctuations. In a recent application of this procedure, J. B. Clarage & G. N. Phillips (in preparation) obtained a crystallographic residual of $R = 0.09$ for an ensemble of structures in myoglobin compared to $R = 0.16$ for the unique single-conformation model. Free $R$-value analysis (14) suggests that the decrease of $R$ was statistically meaningful in the refinement.

SOLVENT IN THE CRYSTAL   Further improvement of the model can be achieved by including a better description of the solvent density distribution throughout the whole unit cell. Badger & Caspar (6) addressed this problem with the cubic form of insulin, using data to 1.7-Å resolution and an iterative difference Fourier method. Their procedure can be summed up as follows: The starting point is phases from the refined protein structure and a uniform electron density outside the molecular envelope. A difference map is calculated with these phases and the diffraction amplitudes and is added outside the envelope. This modified map is then inverted to recalculate modified phases for the iterative refinement. At convergence, a model is obtained that provides a better description of the density throughout the cell and that demonstrates that the nonrandom arrangement of the water molecules extends several layers outside the well-ordered hydration shell around the protein. The resulting $R$-factor is very low, especially at low resolution, and improvement is observed at higher resolution.

CLUSTERING OF STRUCTURES   The molecular interface between lattice points in protein crystals is similar to but usually less extensive than the interface between subunits in oligomeric proteins. With mixed structural states, the arrangement of various species in the crystal can be random, but in some truly lucky cases, the mixture is regular and periodic as in the allosteric L-lactate dehydrogenase from *Bifidobacterium longum* (Figure 4). This crystal form of the enzyme (obtained by S. Iwata & T. Ohta from Tokyo University) contains a regular and periodic 1:1 mixture of the tetrameric R- and T-state form of this protein. There is practically a full story in a single crystal, and the story is clearly visible because of the regular arrangement of the two components within the lattice. We propose that between the two extremes of the distribution, i.e. the completely random and the completely regular, there might be intermediate states in which certain structures occur more frequently around each other's vicinity than in other places in the crystal. The likely driving force behind such a clustering of structures can be a cross-talk (cooperativity) between neighboring lattice points, just like allosteric interactions between subunits in certain oligomeric enzymes. The effect can be dynamic, with clusters of certain sizes regrouping constantly. The possibility that clustered structures might exist within protein crystals has not yet been considered in the literature. A more complete description of macromolecular crystals may be possible by considering structural clusters (static or dynamic) within the crystal. The approach could lead to a better explanation of diffuse scattering, the shape and size of reciprocal lattice points, weak and variable superlattices, and the effect of radiation damage on the diffraction pattern within protein crystals.

*Figure 4* The crystal lattice of the allosteric L-lactate dehydrogenase from *Bifidobacterium longum* viewed down the $c$ axis (F222, $a = 148.4$ Å, $b = 295.9$ Å, $c = 71.0$ Å). T-state tetramers are drawn in dark lines, R-state tetramers in light lines. Courtesy of S. Iwata & T. Ohta (Tokyo University).

# FURTHER THOUGHTS ON FOUR-DIMENSIONAL STRUCTURES

The mechanistic interpretation of data on chemical reactions implies a sequence of structural transitions of the components along the reaction coordinate (see e.g. 15, 119). The inferred transitions are usually not very well defined and have never been observed in three dimensions. To relate three-dimensional structures to structural transitions, and thus to reaction mechanisms, Bürgi & Dunitz studied several closely related compounds

CRYSTALLOGRAPHY IN FOUR DIMENSIONS    493

and structural elements in different chemical environments and crystal forms (see e.g. 16–18). They could derive a scheme for the mechanism of a nucleophilic substitution reaction from these observations. Placing the individual crystal structures in order gave a recognizable pattern of molecular events for the reaction in accordance with evidence from other sources and provided a fairly good estimate of the relative magnitudes of potential constants in the potential-energy surface. With macromolecules, a similar synthesis needs to be made—probably by using computational techniques in molecular mechanics—to produce likely four-dimensional structures and coordinate files from similar sets of data.

## *Structural Information on Intermediates from a Single Image?*

When a reaction is looked at in equilibrium, the diffraction picture contains information on all components (weighted by their concentrations) on the reaction pathway. Certain developments in this field suggest that the approach of looking at a single data set for multiple structural states and dynamic information is not completely hopeless (6, 14, 23, 48, 49, 112). Reactions at equilibrium could be studied in which intermediates are not too numerous and their concentrations not too different. The equilibrium may be altered by changing temperature, pH, etc. Additional measurements and information on the system (e.g. spectral signals from the crystal, molecular dynamics calculations with lattice restraints, etc) are likely to play a role here. The equilibrium could be altered to bring other species into the foreground if they were suppressed before.

Although routine structure refinement from Bragg peaks has been highly successful in protein crystallography, macromolecular crystal structures are still far behind the quality of small-molecule crystal structures. Inadequacies of the refinement model are probably responsible for the failure to reduce the crystallographic $R$-factors much below 15% with current refinement procedures. The actual error in Bragg data sets is much less than that. Routine structure determination and refinement uses only part of the scattered radiation from the crystal—the half that goes into Bragg reflections. The other half, producing background and diffuse scattering around and between Bragg peaks, is generally disregarded. The term *data set* certainly needs to be redefined in current protein crystallography; information from the noncrystalline diffraction should be included. A three-dimensional data set should, in principle, contain entries (*a*) for the integrated intensities of Bragg reflections, (*b*) for the three-dimensional shapes of Bragg reflections, and (*c*) for the three-dimensional distribution of intensities within the background. Elements of four-dimensional structures could in principle be obtained from individual three-dimensional data sets with improved refinement procedures. Putting these into order requires added knowledge and information on the system.

The structures that can be studied today in a dynamic experiment are set primarily by the system and not by the experimenter, and the favorable case of a single conformation being predominant in the crystal (subject to thermal motion) may not always be obtainable. Besides enhancing hidden structures in kinetic studies by computational, chemical, physical, and genetic means, we must also invest in fast data-collection methods. A detailed analysis of mixed structures would actually require better-quality and higher-resolution data than data used in routine monochromatic studies. One of the questions is this: How good is the Laue technique—as it stands today—as a structural method? The streaking of reflections—especially in kinetic experiments—is a serious restriction to the use of this technique. The incompleteness of Laue data sets is another one. I. J. Clifton (Oxford University) sees the problem in the following way (26, p. 147): "It is important not to judge Laue diffraction too harshly because of the artifacts introduced by missing data. For the Laue crystallographer, missing data, particularly missing low resolution data, is a fact of life that one has to live with just as all crystallographers live with the phase problem. It may be that our expectations of what can come out of a kinetic experiment will have to change in the future. Crystallographers are used to seeing a structure as the final result of an experiment. Perhaps in a difficult time-resolved experiment including mixtures of structural states one will not be able to relate the maps obtained to any single meaningful structure—rather the results would be probable structures with probable concentrations of various species through time."

Hopefully, the interest in studies of motion and reactions will lead to a renewed interest in improving experimental and computational methods in this field.

Acknowledgments

We wish to thank Drs. S. L. Edwards, J. Fontecilla-Camps, D. Housset, S. Iwata, A. Nakagawa, T. Ohta, E. Pai, G. A. Petsko, G. N. Phillips, P. Phizackerley, D. Ringe, A. Scheidig, I. Schlichting, R. M. Sweet, and N. Watanabe for sending us material prior to publication. Thanks are also due to I. J. Clifton, V. Fülöp, A. Hadfield, L. N. Johnson, P. Nordlund, J. Snaith, and S. Wakatsuki for discussions. Financial support from the U. K. Medical Research Council and the Swedish Research Council for Agriculture and Forestry is gratefully acknowledged. I. A. wishes to thank the Royal Society for a Guest Research Fellowship.

Literature Cited

1. Adams S. R., Tsien, R. Y. 1993. *Annu. Rev. Physiol.* 55: 755–84
2. Alber, T., Petsko, G. A., Tsernoglou, D. 1976. *Nature* 263: 297–300
3. Andersson, I., Clifton, I. J., Edwards, S. L., Fülöp, V., Hadfield, A. T., et al. 1992. See Ref. 28a, pp. 97–104
4. Andersson, I., Clifton, I. J., Fülöp, V., Hajdu, J. 1991. See Ref. 96a, pp. 20–28
5. Artymiuk, P. J., Blake, C. C. F., Grace, D. E. P., Oatley, S. J., Phillips, D. C., Sternberg, M. J. E. 1979. *Nature* 280: 563–68
6. Badger, J., Caspar, D. L. D. 1991. *Proc. Natl. Acad. Sci. USA* 88: 622–26
7. Baldwin, J. E., McConnaughie, A. W., Moloney, M. G., Pratt, A. J., Bo Shim, S. 1990. *Tetrahedron* 46: 6879–84
8. Bartunik, H. D., Bartunik, L. J., Viehmann, H. 1992. *Philos. Trans. R. Soc. London Ser. A* 340: 209–20
9. Bartunik, H. D., Borchert, T. 1989. *Acta Crystallogr. Sect. A* 45: 718–26
10. Bialek, W., Onuchic, J. N. 1988. *Proc. Natl. Acad. Sci. USA* 85: 5908–12
11. Blake, C. C. F., Koenig, D. F., Mair, G. A., North, A. C. T., Phillips, D. C., Sarma, V. R. 1965. *Nature* 206: 757–63
12. Bordas, J., Munro, I. H., Glazer, A. M. 1976. *Nature* 262: 541–45
13. Brünger, A. T., Kuriyan J., Karplus M. 1987. *Science* 235: 458–60
14. Brünger, A. 1992. *Nature* 355: 472–75
15. Bürgi, H. B. 1973. *Inorg. Chem.* 12: 2321–25
16. Bürgi, H. B. 1982. See Ref. 111a, pp. 430–39
17. Bürgi, H. B., Dunitz, J. D., Shefter, E. 1973. *J. Am. Chem. Soc.* 95: 5065–67
18. Bürgi, H. B., Lehn, J. M., Wipff, G. 1974. *J. Am. Chem. Soc.* 96: 1956–56
19. Butz, P., Greulich, K. O., Ludwig, H. 1988. *Biochemistry* 27: 1556–63
20. Campbell, J. W., Clifton, I. J., Greenhough, T. J., Hajdu, J., Harrison, S. C., et al. 1990. *J. Mol. Biol.* 214: 627–32
21. Campbell, J. W., Habash, J., Helliwell, J. R., Moffat, K. 1986. *Inf. Q. Protein Crystallogr. Daresbury Lab.* 18: 23–31
22. Caspar, D. L. D., Clarage, J., Salunke, D. M., Clarage, M. 1988. *Nature* 332: 659–62
23. Chacko, S., Phillips, G. N. Jr. 1992. *Biophys. J.* 61: 1256–66
24. Chance, B., Ravilly, A., Rumen, N. 1966. *J. Mol. Biol.* 17: 525–34
25. Clarage, J. B., Clarage, M. S., Phillips, W. C., Sweet, R. M., Caspar, D. L. D. 1992. *Proteins Struct. Funct. Genet.* 12: 145–57
26. Clifton, I. J. 1992. *Laue crystallography of proteins and viruses.* D. Philos. thesis. Oxford Univ.
27. Clifton, I. J., Elder, M., Hajdu, J. 1991. *J. Appl. Crystallogr.* 24: 267–77
28. Corrie, J. E. T., Katayama, Y., Reid, G. P., Anson, M., Trentham, D. R. 1992. *Philos. Trans. R. Soc. London Ser. A* 340: 233–44
28a. Cruickshank, D. W. J., Helliwell, J. R., Johnson, L. N., eds. 1992. *Time-Resolved Macromolecular Crystallography.* Oxford: Oxford Univ. Press for the Royal Society
29. Cruickshank, D. W. J., Helliwell, J. R., Moffat, K. 1987. *Acta Crystallogr. Sect. A* 43: 656–74
30. Cusack, S. 1992. *Curr. Biol.* 2: 411–13
31. Dixon, M. M., Matthews, B. W. 1989. *Biochemistry* 28: 7033–38
32. Doscher, M. S., Richards, F. M. 1963. *J. Biol. Chem.* 238: 2399–6.
33. Doucet, J., Benoit., J.-P. 1987. *Nature* 325: 643–46
34. Douzou, P., Petsko, G. A. 1984. *Adv. Protein Chem.* 36: 245–61
35. Duke, E. M. H., Hadfield, A., Walters, S., Wakatsuki, S., Bryan, R. K., Johnson, L. N. 1992. *Philos. Trans. R. Soc. London Ser. A* 340: 245–61
36. Elber, R., Karplus, M. 1987. *Science* 235: 318–21
37. Escobar, W. A., Tan, A. T., Fink A. T. 1991. *Biochemistry* 30: 10783–87
38. Ewald, P. P. 1914. *Ann. Phys.* 44: 257–82
39. Farber, G. K., Machin, P. A., Almo, S. C., Petsko, G. A., Hajdu, J. 1988. *Proc. Natl. Acad. Sci. USA* 85: 112–15
40. Fink, A. L., Ahmed, I. A. 1976. *Nature* 263: 294–97
41. Foley, C. K., Pedersen, L. G., Charifson, P. S., Darden, T. A., Wittinghofer, A., et al. 1992. *Biochemistry* 31: 4951–59
42. Frankel, R. D., Forsyth, J. M. 1979. *Science* 204: 622–24
43. Frauenfelder, H. 1988. *Annu. Rev. Biophys. Biophys. Chem.* 17: 451–79
44. Frauenfelder, H., Petsko, G. A., Tsernoglou, D. 1979. *Nature* 280: 558–63
45. Fujinaga M., Gros P., van Gunsteren W. F. 1989. *J. Appl. Crystallogr.* 22: 1–8
46. Gonzalez, A., Thompson, A. W., Nave, C. 1992. *Rev. Sci. Instrum.* 63: 1177–80
47. Göppert-Mayer, M. 1931. *Ann. Phys.* 9: 273–94
48. Gros, P. 1991. See Ref. 96a, pp. 409–18
49. Gros, P., van Gunsteren, W. F., Hol, W. G. 1990. *Science* 249: 1149–52

50. Gruner, S. M. 1987. *Science* 238: 305–12
51. Gurney, A. M., Lester, H. A. 1987. *Physiol. Rev.* 67: 583–617
52. Hajdu, J. 1990. In *Frontiers in Drug Research, Alfred Benzon Symposium 28*, ed. B. Jensen, F. S. Jørgensen, H. Kofod, pp. 375–95. Copenhagen: Munksgaard
53. Hajdu, J., Acharya, K. R., Stuart, D. I., Johnson, L. N. 1988. *Trends Biochem. Sci.* 13: 104–9
54. Hajdu, J., Acharya, K. R., Stuart, D. I., McLaughlin, P. J., Barford, D., et al. 1986. *Biochem. Soc. Trans.* 14: 538–41
55. Hajdu, J., Acharya, K. R., Stuart, D. I., McLaughlin, P. J., Barford, D., et al. 1987. *EMBO J.* 6: 539–46
56. Hajdu, J., Almo, S. C., Farber, G. K., Prater, J. K., Petsko, G. A., et al. 1991. See Ref. 96a, pp. 29–49
57. Hajdu, J., Johnson, L. N. 1990. *Biochemistry* 29: 1669–78
58. Hajdu, J., Machin, P., Campbell, J. W., Greenhough, T. J., Clifton, I., et al. 1987. *Nature* 329: 178–81
59. Hedman, B., Hodgson, K., Helliwell, J. R., Liddington, R. C., Papiz, M. Z. 1985. *Proc. Natl. Acad. Sci. USA* 82: 7604–6
60. Helliwell, J. R. 1984. *Rep. Prog. Phys.* 47: 1403–97
61. Helliwell, J. R., Habash, J., Cruickshank, D. W. J., Harding, M. M., Greenhough, T. J., et al. 1989. *J. Appl. Crystallogr.* 22: 483–97
62. Henderson, R. 1990. *Philos. Trans. R. Soc. London Ser. B* 241: 6
63. Hendrickson W. A., Konnert J. H. 1980. In *Computing in Crystallography*, ed. R. Diamond, S. Ramasheshan, K. Venkatesan, pp. 13.01–13.23. Bangalore: Indian Institute of Science
64. Howell, P. L., Almo, S. C., Parsons, M. R., Hajdu, J., Petsko, G. A. 1992. *Acta Crystallogr. Sect. B* 48: 200–7
65. Howlin, B., Moss, D. S., Harris, G. W. 1989. *Acta Crystallogr. Sect. A* 45: 851–61
66. Ichiye, T., Karplus, M. 1988. *Biochemistry* 27: 3487–97
67. Jamet, M. F. 1970. *C. R. Acad. Sci. Paris Ser. B* 271: 714–17 (In French)
68. Johnson, L. N., Hajdu, J. 1990. In *Biophysics and Synchrotron Radiation*, ed. S. Hasnian, pp. 142–55. Chichester: Ellis Horwood
69. Johnson, Q., Keeler, R. N., Lyle, J. W. 1967. *Nature* 213: 1114–15
70. Johnson, Q., Mitchell, A. C., Keeler, R. N., Evans, L. 1970. *Phys. Rev. Lett.* 25: 1099–1101
71. Jones, G. R., Bordas, J., Clarke, D., Diakun, G. P., Mant, G. R. 1989. In *Synchrotron Radiation in Structural Biology*, ed. R. M. Sweet, A. D. Woodhead, pp. 67–75. New York: Plenum
72. Joshi, N. R., Green, R. E. 1980. *J. Mater. Sci.* 15: 729–38
73. Kasvinsky, P. J., Madsen, N. B. 1976. *J. Biol. Chem.* 251: 6852–59
74. Kelly, J. A., Waley, S. G., Adam, M., Frere, J.-M. 1992. *Biochim. Biophys. Acta* 1119: 256–60
75. Kendrew, J. C., Watson, H. C., Strandberg, B. E., Dickerson, R. E., Phillips, D. C., Shore, V. C. 1961. *Nature* 190: 666–72
76. Kmetec, J. D., Gordon, C. L. III, Macklin, J. J., Lemoff, B. E., Brown, G. S., Harris, S. E. 1992. *Phys. Rev. Lett.* 68: 1527–30
77. Kress, M., Huxley, H. E., Faruqi, A. R., Hendrix, J. 1986. *J. Mol. Biol.* 188: 325–42
78. Kuriyan, J., Ösapay, K., Burley, S. K., Brünger, A. T., Hendrickson, W. A., Karplus, M. 1991. *Proteins Struct. Funct. Genet.* 10: 340–58
79. Kuriyan, J., Weis, W. I. 1991. *Proc. Natl. Acad. Sci. USA* 88: 2773–77
80. Kuriyan, J., Wilz, S., Karplus, M., Petsko, G. A. 1986. *J. Mol. Biol.* 192: 133–54
81. Laidler, K. J., Bunting, P. S. 1980. *Methods Enzymol.* 64: 227–48
82. Larson, B. C., White, C. W., Noggle, T. S., Barhorst, J. F., Mills, D. 1983. *Appl. Phys. Lett.* 42: 282–83
83. Liljas, A., Carlsson, M., Håkansson, K., Lindahl, M., Svensson, S. A., Wehnert, A. 1992. *Philos. Trans. R. Soc. London Ser. A* 340: 301–9
84. Lindahl, M., Liljas, A., Habash, J., Harrop, S., Helliwell, J. R. 1992. *Acta Crystallogr. Sect. B* 48: 281–85
85. Linderstrøm-Lang, K. U., Schellman, J. A. 1959. *The Enzymes* 1: 443–510
86. Lunney, J. G., Dobson, P. J., Hares, J. D., Tabatabaei, S. D., Eason, R. W. 1986. *Opt. Commun.* 58: 269
87. Luzzatti, V. 1953. *Acta Crystallogr.* 6: 142–52 (In French)
88. Main, P. 1979. *Acta Crystallogr. Sect. A* 35: 779–85
89. Makinen, M. W., Fink, A. L. 1977. *Annu. Rev. Biophys. Bioeng.* 6: 301–43
90. Martin, J.-P., Vos, M. H. 1992. *Annu. Rev. Biophys. Biomol. Struct.* 21: 199–222
91. McCray, J. A., Trentham, D. R. 1989. *Annu. Rev. Biophys. Biophys. Chem.* 18: 239–70
92. Mendel, D., Elman, J. A., Schultz, P. G. 1991. *J. Am. Chem. Soc.* 113: 2758–60

93. Moffat, K., 1989. *Annu. Rev. Biophys. Biophys. Chem.* 18: 309–32
94. Moffat, K., Bilderback, D., Schildkamp, W., Volz, K. 1986. *Nucl. Instrum. Methods Sect. A* 246: 627–35
95. Moffat, K., Chen, Y., Ng, K., McRee, D., Getzoff, E. D. 1992. *Philos. Trans. R. Soc. London Ser. A* 340: 175–90
96. Moffat, K., Szebenyi, D. M. E., Bilderback, D. H. 1984. *Science* 223: 1423–25
96a. Moras, D., Podjarny, A. D., Thierry, J. C., eds. 1991. *Crystallographic Computing 5: From Chemistry to Biology.* Oxford: Oxford Univ. Press
97. Murnane, M. M., Kapteyn, H. C., Rosen, M. D., Falcone, R. W. 1991. *Science* 251: 531–36
98. Nordlund, P., Sjöberg, B.-M., Eklund, H. 1990. *Nature* 345: 593–98
99. Otting, G., Liepinsh, E., Wüthrich, K. 1991. *Science* 254: 974–80
100. Phillips, D. C. 1968. *Use of difference syntheses in studies of substrate binding conformational changes and refinement.* Presented at the Workshop for Protein Crystallography, Hirschegg, Kleinwalzertal, 17–22 March 1968; See also Blundell, T., Johnson, L. N. 1976. *Protein Crystallography,* p. 417. New York: Academic
101. Phillips, D. C. 1981. *Biochem. Soc. Symp.* 46: 1–15
102. Potschka, M., Koch, M. H. J., Adams, M. L., Schuster, T. M. 1988. *Biochemistry* 27: 8481–91
103. Rabinovich, D., Lourie, B. 1987. *Acta Crystallogr. Sect. A* 43: 774–80
104. Radford, S. E., Dobson, C. M., Evans, P. A. 1992. *Nature* 358: 302–7
105. Rasmussen, B. F., Stock, A. M., Ringe, D., Petsko, G. A. 1992. *Nature* 357: 423–24
106. Read, R. J. 1986. *Acta Crystallogr. Sect. A* 42: 140–49
107. Reynolds, C. D., Stowell, B., Joshi, K. K., Harding, M. M., Maginn, S. J., Dodson, G. G. 1988. *Acta Crystallogr. Sect. B* 44: 512–15
108. Ringe, D., Stoddard, B. L., Bruhnke, J., Koenigs, P., Porter, N. 1992. *Philos. Trans. R. Soc. London Ser. A* 340: 273–84
109. Rosenbaum, G., Holmes, K. C., Witz, J. 1971. *Nature* 230: 434–37
110. Sakabe, N. 1983. *J. Appl. Crystallogr.* 16: 542–47
111. Samudzi, C. T., Rosenberg, J. M. 1992. *J. Appl. Crystallogr.* 25: 65–68
111a. Sayre, D., ed. 1982. *Computational Crystallography.* Oxford: Clarendon
112. Scaringe, R. P., Comès, R. 1990. In *Physical Methods of Chemistry,* ed. B. W. Rossiter, J. F. Hamilton, 5: 517–601. New York: Wiley. 2nd ed.
113. Scheidig, A., Pai, E. F., Schlichting, I., Corrie, J. E. T., Reid, G. P., et al. 1992. *Philos. Trans. R. Soc. London Ser. A* 340: 263–72
114. Deleted in proof
115. Schlichting, I., Almo, S. C., Rapp., G., Wilson, K., Petratos, K., et al. 1990. *Nature* 345: 309–15
116. Sheriff, S., Hendrickson, W. A. 1987. *Acta Crystallogr. Sect. A* 43: 118–21
117. Singer, P. T., Carty, R. P., Berman, L. E., Schlichting, I., Stock, A., et al. 1992. *Philos. Trans. R. Soc. London Ser. A* 340: 285–300
118. Sluyterman, L. A. A., De Graff, M. J. M. 1969. *Biochim. Biophys. Acta* 171: 277–87
119. Smith, I. W. M. 1992. *Nature* 358: 279–80
120. Smith, J. L., Hendrickson, W. A., Honzatko, R. B., Sheriff, S. 1986. *Biochemistry* 25: 5018–27
121. Smith, T. J., Kremer, M. J., Luo, M., Vriend, G., Arnold, E., et al. 1986. *Science* 233: 1286–93
122. Stoddard, B. L., Koenigs, P., Porter, N., Petratos, K., Petsko, G. A., Ringe, D. 1991. *Proc. Natl. Acad. Sci. USA* 88: 5503–7
123. Stoddard, B. L., Koenigs, P., Porter, N., Ringe, D., Petsko, G. A. 1990. *Biochemistry* 29: 8042–51
124. Stuart, D. I., Phillips, D. C. 1985. *Methods Enzymol.* 115: 117–42
125. Strynadka, N. C. J., Adachi, H., Jensen, S. E., Johns, K., Sielecki, A., et al. 1992. *Nature* 359: 700–5
126. Svensson, L. A., Sjölin, L., Gilliland, G. L., Finzel, B. C., Wlodawer, A. 1986. *Proteins Struct. Funct. Genet.* 1: 370–75
127. Szebenyi, D. M. E., Bilderback, D. H., LeGrand, A., Moffat, K., Schieldkamp, W., et al. 1992. *J. Appl. Crystallogr.* 25: 414–23
128. Theorell, H., Chance, B., Yonetani, T. 1966. *J. Mol. Biol.* 17: 513–24
129. Thomas, J. M. 1988. *Angew. Chem. Int. Ed. Engl.* 27: 1673–91
130. Tilton, R. F., Dewan, J. C., Petsko, G. A. 1992. *Biochemistry* 31: 2469–81
131. Tuomi, T., Naukkarinen, K., Rabe, P. 1974. *Phys. Status Solidi A* 25: 93–106
132. Turner, A. D., Pizzo, S. V., Rozakis, G., Porter, N. A. 1988. *J. Am. Chem. Soc.* 110: 244–50
133. Vellieux, F. M. D., Hajdu, J., Verlinde, C. M. L. J., Groendijk, H., Read, R. J., et al. 1992. *Proc. Natl. Acad. Sci. USA* In press
134. Vijayan, M. 1980. *Acta Crystallogr. Sect. A* 36: 295–98

135. Wagner, G., Hyberts, S. G., Havel, T. F. 1992. *Annu. Rev. Biophys. Biomol. Struct.* 21: 167–98
136. Walter, J., Steigemann, W., Singh, T. P., Bartunik, H. D., Bode, W., Huber, R. 1982. *Acta Crystallogr. Sect. B* 38: 1462–72
137. Wark, J. S., Whitlock, R. R., Hauer, A. A., Swain, J. E., Solone, P. J. 1989. *Phys. Rev. B* 40: 5705–14
138. Wark, J. S., Woolsey, N. C., Whitlock, R. R. 1992. *Appl. Phys. Lett.* 61: 651–53
139. Watanabe, N., Higashi, T., Yonaha, K., Sakabe, N. 1991. *Photon Factory Activity Rep. Tsukuba, Japan.* p. 248
140. Watanabe, N., Sakabe, N. 1990. *Photon Factory Activity Rep. Tsukuba, Japan.* p. 89
141. Watenpaugh, K. D., Sieker, L. C., Herriott, J. R., Jensen, L. H. 1973. *Acta Crystallogr. Sect. B* 29: 943–56
142. Watenpaugh, K. D., Sieker, L. C., Jensen, L. H. 1980. *J. Mol. Biol.* 138: 615–33
143. Willis, B. T. M. 1982. See Ref. 111a, pp. 479–87
144. Wood, I. G., Thompson, P., Matthewman, J. C. 1983. *Acta Crystallogr. Sect. B* 39: 543–47
145. Woolsey, N. C., Wark, J. S., Riley, D. 1990. *J. Appl. Crystallogr.* 23: 441–43
146. Zhang, K., Chance, B., Reddy, K. S., Ayene, I., Stern, E. A., Bunker, G. 1991. *Biochemistry* 30: 9116–20
147. Zucker, U. H., Schulz, H. 1982. *Acta Crystallogr. Sect. A* 38: 563–68
148. Singer, P. T., Smalås, A., Carty, R. P., Mangel, W. F., Sweet, R. M. 1993. *Science* 259: 669–73

# SUBJECT INDEX

## A

Acetylcholine
　M current and, 178
Acetylcholine receptor
　membrane thickness and, 168
Acetyltetraglycine ethyl ester
　solubility of, 90
Acids
　prolyl isomerization and, 139
Acrosomal bundle
　electron cryomicroscopy and, 240
g-Actin
　X-ray crystallography and, 246
Alanine scanning, 338–39
Albumin
　Cu(II)- binding site of, 259–60
Alcohol dehydrogenase
　crystalline
　　activity of, 471
　zinc binding sites in, 258
Aldolase
　diffusion in solution, 54
Alkanes
　lipid bilayer hydrophobic thickness and, 168
　transfer from solution to vapor, 396
α-helical coiled-coil protein
　X-ray diffraction and, 244
Amino acids
　potassium channel P region and, 189–92
　protein stabilization due to, 84–85
　sequences
　　potassium channels and, 185–88
　surface tension of water and, 87
　transfer from solution
　　free energy of, 398–400
Aminophospholipids
　in biologic membranes, 454
Ammonium sulfate
　enzyme isolation and, 68
　protein crystallization and, 202
Amoebocytes
　secretory granules in
　　dimpling of plasma membrane towards, 434
Amphiphilic polypeptides
　synthetic, 164–66
Amyl alcohol
　transfer from vapor to water
　　enthalpy of, 392
Andersson, I., 467–94
Antibiotics
　amide-amide hydrogen bonding and, 385
Apamin
　potassium channels and, 178

*Aplysia* neurons
　S channel in, 178
Aqueous solution
　self-diffusion of proteins in
　　macromolecular crowding and, 54
Artificial neural networks, 285
　applications of, 291–96
　knowledge-based, 293
Atomic vibrations
　crystallography and, 489
Atrium
　muscarinic cholinergic potassium channel in, 178
Available volume theory (AVT), 33–34, 40, 45
AVT
　See Available volume theory
Azurin
　copper binding site in, 273

## B

*Bacillus subtilis*
　glucose permease of, 222–23
　HPr of, 219–22
Bacteria
　phosphoenolpyruvate sugar phosphotransferase system of, 200–1, 218–23
Bacterial chemotaxis, 201, 223–26
Bacteriophage T4
　endonuclease VII of, 322–24
Bacteriophage T4 replication
　branched DNA and, 300
Bacteriophage T7
　dsDNA in
　　visualization of, 249
Bacteriorhodopsin
　electron cryomicroscopy and, 240
　in fluid lipid bilayers
　　aggregation of, 159
　　asymmetry of, 167
　　periodic array of structure of, 234
　　polypeptide chain trace of, 241
　　structural resolution of, 238
Barford, D., 199–229
Barnase
　intramolecular hydrogen bonding in, 395
　prolyl isomerization and, 127
Barrier model
　diffusion in fluid plasma membranes and, 421–22
Bee venom toxin
　potassium channels and, 178
Betaine
　as protein structure protectant, 92
*Bifidobacterium longum*
　L-lactate dehydrogenase of
　　crystal lattice of, 491–92

Biologic medium
　macromolecular crowding in, 28–30
Biologic membrane fusion
　fusion pores and, 435
　lipids in, 454–55
　semistable states in, 440–42
Bloom, M., 145–69
Bone
　growth and differentiation of
　　pituitary hormones and, 330
Bovine pancreatic trypsin inhibitor (BPTI)
　α-carbons of, 367
　high-temperature dynamics
　　simulation of, 376–77
　main-chain atoms of, 368
　in water
　　simulation of, 369–73
Bovine serum albumin (BSA)
　diffusion in solution, 54
BPTI
　See Bovine pancreatic trypsin inhibitor
Brain
　potassium channels in
　　inactivation of, 186
　muscarinic cholinergic, 178
Brown, A. M., 173–95
BSA
　See Bovine serum albumin
Bulk solvents
　protein dynamics simulations and, 359–60
Butane
　transfer from solution to vapor, 396
Butanol
　transfer from vapor to water
　　enthalpy of, 392

## C

*Caenorhabditis elegans*
　*unc*-13 gene product of
　　binding phorbol esters, 5
Calcineurin
　immunosuppressants and, 129
Calcium channels
　peptides associated with, 174
　synaptic vesicle protein binding to, 435
Calmodulin
　calcium binding sites in, 258
Calorimetry
　erythrocyte band 3 protein receptors and, 158
Campbell, A. P., 99–120
Carbobenzoxy-D-phenylalanyl-L-phenylalanylglycine
　viral fusion and, 455
Carbohydrate metabolism
　isocitrate dehydrogenase and, 210

499

# SUBJECT INDEX

Carbonic anhydrase
  electrophilic catalysis in, 258, 263–65
  prolyl isomerization and, 127
Carboxymyoglobin
  in water
    simulation of, 372
Carboxypeptidase
  active site of
    aspartate-histidine-zinc triad of, 279
  electrophilic catalysis in, 258, 263–65
Cardiotoxin
  membrane fusion and, 455
Cartilage
  growth and differentiation of
    pituitary hormones and, 330
Cast and retrieve model
  hemagglutinin-mediated membrane fusion and, 458–61
CD4
  extracellular domain of
    structure of, 334
Cell division
  protein phosphorylation and, 200
Cell division-control protein kinase, 217
Cell growth
  protein phosphorylation and, 200
  regulation of
    protein kinase C and, 5
Cellular folding
  prolyl isomerase in, 137–38
Chaperone
  molecular, 124
Chaperone protein PapD
  second domain of
    structure of, 334
Charybdotoxin
  potassium channels and, 178, 189–90
Chemical equilibrium
  macromolecular crowding and, 31–36
  thermodynamic control of
    cosolvents and, 70–76
Chemical exchange
  limits of
    relaxation matrix in, 104–5
  multiple spin system in
    relaxation matrix analysis for, 102–4
Chemical potential perturbation, 70
Chemical probes
  DNA junctions and, 305–7
Chemotaxis
  bacterial protein of, 201, 223–26
Chernomordik, L. V., 433–62
n-Chimaerin
  binding phorbol esters, 5–7
Chiu, W., 233–52
2-Chloroethanol
  as denaturant, 74–76
Cholera toxin
  transferred nuclear Overhauser effect difference spectroscopy and, 107

Cholesterol
  membrane-thickening effect of, 166
  proteins sensitive to, 2
  secretory granule fusion with liposomes and, 454
Chromaffin granule exocytosis
  monolayers in, 454
Chromatography
  gel filtration
    metal-binding sites and, 272
  high-performance liquid
    receptor:ligand interactions and, 340
Chymotrypsin
  conformational specificity of, 128
  crystallography and, 486
Chymotrypsin inhibitor CI2
  prolyl isomerization and, 127
Chymotrypsinogen
  denaturation of, 72, 82
Ciliary neurotrophic factor receptor, 334
Citric acid cycle
  isocitrate dehydrogenase and, 209–10
Clegg, R. M., 299–326
Cohen, F. E., 283–97
Collagen
  denaturation by glycerol, 82
  in vivo maturation of
    cyclosporin A and, 137–38
Collagen fibrils
  self-assembly of
    glycerol and, 73, 85
Concanavalin A
  prolyl isomerization and, 127–28
Concanavalin A receptor
  on erythrocyte band 3 proteins, 157–58
Conformational states
  crystallography and, 489–90
Coronary artery
  potassium channels of
    inactivation of, 186
Cosolvents
  salting-out, 85–89
  thermodynamic control of equilibria by, 70–76
Crotoxin complex crystal
  glucose-embedded
    electron diffraction pattern of, 244
Cryoprotectants, 69
Cryospecimens
  preparation of, 237–38
Crystalline enzymes
  activity in, 471–74
  diffusion of reagents into, 471–72
  photoactivation of, 472–74
Crystallography, 467–94
  chymotrypsin and, 486
  data-collection techniques in, 474–81
  enzyme crystals in, 471–74
  fast
    methods in, 470
  fast diffraction experiments in, 481–86

four-dimensional, 492–94
glycogen phosphorylase and, 484
H-ras p21 protein and, 484–85
kinetic
  methods in, 470
mixed structural sites and, 488–91
monochromatic techniques in, 475
partial data sets in, 486–88
partially occupied sites and, 488–91
photochemical triggering in, 472–74
protein electron, 234
receptor:ligand interactions and, 341–42
trypsin and, 485–86
white radiation techniques in, 475–76
See also X-ray crystallography
Crystals
  electron cryomicroscopy and, 239–46
Cyanines
  four-way DNA junctions and, 307
Cyclic AMP-dependent protein kinase, 213–18
  autophosphorylation site on, 216–17
  fold of, 215
Cyclins
  cell division-control mutants and, 217
Cycloalkanes
  activity coefficients in water, 398
Cyclohexane
  interactions with solutes, 398
Cyclophilin, 124, 128
  structure of, 129
Cyclophilin A
  three-dimensional structure of, 130
Cyclosporin A
  immunosuppressive effects of
    mediation of, 129
  isomerase binding to, 124, 128
  protein maturation and endoplasmic reticulum and, 124
  in vivo, 137–38
Cytochrome b5
  in water
    simulation of, 370
Cytochrome c
  membrane fusion and, 455
  prolyl isomerization and, 127
  in water
    simulation of, 370
Cytochrome c-rubredoxin complex
  in water
    simulation of, 371–72
Cytochrome P-450
  affinity for lipid, 3–4
Cytomatrix, 60
Cytoplasm
  diffusion of probe molecules in, 57–58

## SUBJECT INDEX     501

macromolecular crowding in, 56–58
Cytoskeleton
  transient binding of membrane glycoproteins to, 422

### D

Daggett, V., 353–79
Denaturants, 82–83
Denaturation, 68
  chymotrypsinogen, 72, 82
  collagen, 82
    metal binding and, 261
  protein, 68, 82–83
    metal binding and, 261
de Vos, A. M., 329–50
Dextran
  viral particle/protein precipitation due to, 43–45
Diacylglycerol
  generation of, 4
  insulin receptor and, 2
  interaction with protein kinase C, 9–10
  protein kinase C:phosphatidylserine interactions and, 16–17
Dialysis equilibrium, 76–81
Difference Fourier maps, 487–88
  with partial occupancies, 488–89
Digestive enzymes
  exocytic release of membrane fusion and, 434
Dimyristoyl phosphatidylcholine
  lipid-concentration profiles of, 162–63
Dioxan
  solubilities of amino acids in, 398
Dipalmitoyl phosphatidylcholine
  peptide incorporated into nuclear magnetic resonance spectra for, 164–65
Distearoyl phosphatidylcholine
  lipid-concentration profiles of, 162–63
Disulfide bonds
  protein folding and, 134
DNA
  binding sites in
    affinity of *lac* repressor for, 57
  complementary base-pairing in hydrogen bonding and, 390
  condensed
    macromolecular crowding and, 61
  cruciform structures of, 301
  duplex
    stacking free energy in, 317–18
  parallel/antiparallel conformations of
    stability of, 320–21
DNA double helix
  simulation of, 361
DNA-helix destabilizing protein
  X-ray diffraction and, 244

DNA junctions
  fluorescence resonance energy transfer and, 305
  gel electrophoresis and, 301–5
  helical arms of
    coaxial stacking of, 317
    interactions with proteins, 322–24
  probing experiments and, 305–7
  stability of, 319–20
  stereochemistry of, 310–13
  strand exchange at
    electrostatic repulsions and, 318–19
  structure of, 299–326
    dynamics of, 321–22
    folded, 300–10
    metal ions and, 313–16
    stacked X, 300–1, 308–10
    thermodynamics of, 316–21
DNA ligases
  macromolecular crowding and, 47
DNA polymerases
  macromolecular crowding and, 47
DNA promoter recognition
  artificial neural networks and, 291–93
DNA replication systems
  macromolecular crowding and, 47
DNase I
  four-way DNA junctions and, 307
DNA sequences
  eukaryotic
    artificial neural networks and, 293–94
*Drosophila melanogaster*
  Shaker phenotype in
    potassium channels responsible for, 177
  Slowpoke phenotype in
    gene responsible for, 181–83

### E

Electron cryomicroscopy, 233–52
  asymmetric particles and, 249–51
  biologic specimens amenable to, 239–51
  crystals in, 239–46
  helical arrays and, 246
  icosahedral particles and, 247–49
  image reconstruction in, 236–37
  structural resolution attainable with, 238–39
Electron microscopy
  biologic specimens in, 238
  biophysics of, 234–51
  electron energy used in, 235
Electron paramagnetic resonance (EPR) spectroscopy
  metal-binding sites and, 258

protein-Cu(II) complexes and, 274–75
Electrostatic interactions
  proteins in solution and, 359, 361–62
Endocytosis
  location of sites of, 426–28
Endonuclease I, 322
Endonuclease VII
  DNA junctions and, 322–24
Endoplasmic reticulum
  prolyl isomerases in, 135
  protein maturation in
    cyclosporin A and, 124
Enzyme(s)
  digestive
    membrane fusion and, 434
  DNA junction-resolving, 322–324
Enzyme activity
  protein phosphorylation and, 200
  two-histidine metal-binding site and, 262–63
Enzyme-catalyzed reactions
  macromolecular crowding and, 36–37
Enzyme crystals
  activity in, 471–74
  diffusion of reagents into, 471–72
  photoactivation of, 472–74
Enzyme probes
  DNA junctions and, 305–7
EPO
  *See* Erythropoietin
Equilibrium-exchange model, 81
Erythrocyte(s)
  lipids in, 454
Erythrocyte band 3 proteins
  concanavalin A receptor of, 157–58
Erythrocyte membrane
  diffusion in, 421
Erythropoietin
  hematopoietic-cell growth/differentiation and, 330
  signal transduction and, 346
Erythropoietin receptor
  WSXWS box mutations
    ligand binding and, 337
*Escherichia coli*
  chemotactic response in, 223–26
  cytoplasmic prolyl isomerase of, 134
  DNA polymerase I of
    macromolecular crowding and, 47
  HPr of, 219–22
  isocitrate dehydrogenase of, 209–13
  promoter sequences in
    knowledge-based artificial neural networks and, 293
  recombinant human growth hormone produced in, 337
Ethane
  transfer from solution to vapor, 396

Ethanol
 solubilities of amino acids in, 398
 transfer from vapor to water enthalpy of, 392
Eukaryotes
 cellular volume changes in compensatory mechanisms for, 58
 plasma membrane of, 146
Eukaryotic cells
 protein phosphorylation in, 200
Eukaryotic DNA sequences
 artificial neural networks and, 293–94
Eukaryotic transcription factors
 zinc fingers and clusters of, 258
Excluded volume theory, 36
Exocytosis
 chromaffin granule monolayers in, 454
 fusion proteins in, 447
 location of sites of, 426–28
Extracellular matrix
 protein diffusion and, 425–26
 transient binding of membrane glycoproteins to, 422

F

Ferrimyoglobin
 reaction with azide, 471
Fibroblasts
 expressing hemagglutinin, 435
 folding of procollagen I in, 137
Fibronectin
 extracellular domain of structure of, 334
Filopodia
 forward transport in, 429
FK 506
 binding protein for, 124, 128
 immunosuppressive effects of mediation of, 129
 isomerase binding to, 124
FKBP12
 structure of, 129
Flagella
 forward transport in, 429
 X-ray fiber diffraction and, 246
Flicker, 440
Fluid-mosaic model, 147
Fluid plasma membranes, 417–29
 glycoprotein motility in
  active, 426–29
  passive, 419–26
Fluorescence assay
 receptor:ligand interactions and, 341
Fluorescence recovery after photobleaching
 tracer diffusion of probe molecules in polymers and, 55
Fluorescence resonance energy transfer
 DNA junctions and, 305

Four-helix bundle protein
 tetrahedral binding site in, 270–73
Fourier maps
 difference, 487–88
 with partical occupancies, 488–89
Frog neuromuscular junction synaptic vesicles in
 alignment along nerve terminals, 435
Fusion pores, 435–42
 in lipid bilayers, 451

G

Gel electrophoresis
 nucleic acid structures and, 301–5
Gel filtration chromatography
 metal-binding sites and, 272
Generalization
 vs. memorization, 288–90
Genetic recombination
 homologous
  four-way DNA junction in, 300
Gene transcription
 protein phosphorylation and, 200
GH
 See Growth hormone
Gibbs adsorption isotherm, 87
Globins
 iron and, 258
Globular proteins
 hydrophilic/hydrophobic groups in, 402
 packing densities of, 405–7
Glucose permease, 222–23
Glyceraldehyde-3-phosphate dehydrogenase
 three-dimensional structure of, 487
Glycerol
 collagen denaturation due to, 82
 organelle dissection from cell and, 67–68
 preferential hydration in, 88
 protein stabilization due to, 68, 84
 self-assembly of collagen fibrils and, 73, 85
Glycogen phosphorylase, 202–209
 crystallography and, 484
Glycosomal glyceraldehyde-3-phosphate dehydrogenase
 three-dimensional structure of, 487
Glyoxylate bypass pathway
 isocitrate dehydrogenase and, 210
GM-CSF
 See Granulocyte-macrophage colony stimulating factor
Gramicidin
 membrane fusion and, 455
Granulocyte-macrophage colony stimulating factor (GM-CSF)

crystal structure of, 333
hematopoietic-cell growth/differentiation and, 330
receptor-binding sites on, 344
Growth hormone (GH)
 porcine
  crystal structure of, 331
 structural model of, 337
 signal transduction and, 346
 See also Human growth hormone
Growth hormone receptor
 activation of mechanisms for, 346–48
Guanidine hydrochloride
 preferential binding for unfolded proteins in, 80
 preferential interactions of, 82–83
 protein denaturation and, 68
Guanidine sulfate
 protein stabilization due to, 85

H

Hajdu, J., 467–94
*Halobacterium halobium*
 periodic array of bacteriorhodopsin in, 234
Heat capacity
 folded/unfolded proteins and, 396
Helical arrays
 electron cryomicroscopy and, 246
Hemagglutinin
 fibroblasts expressing, 435
 membrane fusion mediated by, 442–45
 cast and retrieve model for, 458–61
Hematopoietic cells
 growth and differentiation of regulation of, 330
Hematopoietic hormones, 344–45
Hematopoietic receptors, 330
 extracellular domains of, 334–37
 ligands for, 331–34
Hematopoietic superfamily, 330–31
 crystallographic analysis of, 341–42
 mutational analysis of, 337–41
Hematopoietic superfamily receptors, 334
Hematopoietin(s), 344–45
 applications of, 348–49
 signal transduction and, 346
Hematopoietin receptor
 extracellular domains of, 345
Hemoglobin
 deoxygenated sickle solubility of, 46
 diffusion in solution, 54
 viscosity of, 51
Heparin
 viral particle/protein precipitation due to, 43

## SUBJECT INDEX    503

Heptane
  transfer from solution to vapor, 396
Hexane
  transfer from solution to vapor, 396
hGH
  See Human growth hormone
High-performance liquid chromatography (HPLC)
  receptor:ligand interactions and, 340
HIV
  See Human immunodeficiency virus
Homeostasis
  macromolecular crowding and, 60–61
Homologue-scanning mutagenesis, 338
Hormonal response
  protein phosphorylation and, 200
Hormone(s)
  hematopoietic, 344–45
  pituitary
    physiologic processes regulated by, 330
Hormone-induced receptor oligomerization, 340
HPLC
  See High-performance liquid chromatography
H-ras p21 protein
  crystallography and, 484–85
Human growth hormone (hGH), 329–50
  epitopes on
    functional/structural, 342–44
  hydrophobic side chains of, 331–33
  pharmacologic effects of, 348–49
Human growth hormone binding protein, 338
  cross-linking of, 340
  monoclonal antibodies to, 340–41
Human growth hormone receptor
  activation of
    mechanisms for, 346–48
  cloning of, 337–38
  crystallographic analysis of, 341–42
  extracellular domain of, 334, 336
  mutational analysis of, 337–41
  N-terminal domain of, 334–35
Human immunodeficiency virus (HIV)
  fusion proteins of, 445
  transactivating protein of, 272
Human immunodeficiency virus protease
  in water
    simulation of, 371
Hyaluronic acid
  viral particle/protein precipitation due to, 43
Hydration
  preferential, 85–89

Hydration repulsion
  liposome-planar lipid bilayer membrane fusion and, 449
Hydrocarbons
  saturated
    transfer from water, 396–97
Hydrogen atoms
  protein dynamics simulations and, 358–59
Hydrogen bonding, 384–96
  polar groups in proteins and, 385–90
  protein folding and, 390–96
Hydrophobic effect, 382–84, 396–403
Hydrophobicity
  protein molecule self-assembly and, 383
  scales of, 398–400
Hydrophobicity profile, 401
Hydrophobic matching, 146–47, 150–56
  membrane function and, 167–68
-Hydroxybutyrate dehydrogenase
  phosphatidylcholine and, 2

I

Icosahedral particles
  electron cryomicroscopy and, 246–49
Image reconstruction
  three-dimensional, 236–37
    biologic specimens amenable to, 239–51
    structural resolution attainable with, 238–39
Immunoglobulin(s)
  constant domains of
    topology of, 334
  exocytic release of
    membrane fusion and, 434
Immunoglobulin light chain
  prolyl isomerization and, 127
Immunoprecipitation assay
  receptor:ligand interactions and, 340–41
Indole
  solvent interactions with, 398–99
Inelastic laser light scattering
  tracer diffusion of probe molecules in polymers and, 55–56
Influenza hemagglutinin
  membrane fusion and, 442–45
Influenza neuraminidase-antibody fragment complex
  X-ray diffraction and, 244
Influenza virus
  fusion protein of, 434
  fibroblasts expressing, 435
Inositol trisphosphate
  generation of, 4
Insulin
  metal-induced dimerization in, 272
  in water
    simulation of, 371

Insulin receptor
  tyrosine kinase activity of, 2
Intercalating agents
  four-way DNA junctions and, 307
Interleukin(s)
  hematopoietic-cell growth/differentiation and, 330
Interleukin-1
  in water
    simulation of, 372–73
Interleukin-2
  growth hormone-like topology of, 333
Interleukin-2 receptor
  WSXWS box mutations
    ligand binding and, 337
Interleukin-4
  crystal structure of, 333
  receptor-binding sites on, 344–45
  signal transduction and, 346
Intracellular trafficking
  fusion proteins in, 447
Iron
  oxygen transport and, 258
Isobutyramide
  crystalline alcohol dehydrogenase and, 471
Isocitrate dehydrogenase, 209–13
  catalytic site of, 212
Isomerization
  cis/trans, 124
  of peptide bonds, 125–26
  prolyl, 125–28
    catalysis in unfolding proteins, 134–35
    mechanism of, 138–39
    in protein folding, 126–28
Isomorphous difference Patterson maps, 487
Isopiestic equilibrium, 78

J

Johnson, L. N., 199–229

K

Knowledge-based artificial neural networks, 293

L

L-Lactate dehydrogenase
  crystal lattice of, 491–92
Lamellipodia
  forward transport in, 429
  rearward flow in, 428
α-Latrotoxin receptor
  synaptic vesicle protein binding to, 435
Laue method, 475–76
  limitations of, 476–81
Learning
  machine, 284–85
  protein phosphorylation and, 200
Learning surface
  complexities of, 288
Levitt, M., 353–79

# SUBJECT INDEX

Ligand:receptor interactions, 330
  crystallographic analysis of, 341–42
  mutational analysis of, 337–41
Ligand:macromolecular systems
  nuclear magnetic resonance spectroscopy and, 100
Light adaptation
  rhodopsin phosphorylation and, 5
Light harvesting
  protein phosphorylation and, 200
Light harvesting complex
  X-ray crystallography and, 241–42
Light scattering
  solution composition perturbation and, 78
Lilley, D. M. J., 299–326
*Limulus* sperm
  acrosomal bundle of
    electron cryomicroscopy and, 240
Lipid(s)
  biologic membrane fusion and, 454–55
  hydrolysis of
    transduction of signals promoting, 5
  lateral diffusion of
    measurement of, 419–21
  membrane fusion and, 448–55
  as regulators of protein function, 2–4
  sensory transduction and, 4
Lipid:protein complexes
  models of, 456–57
Lipid:protein interactions
  membrane function and, 167–68
  in membranes, 145–69
  microscopic interaction model for, 155–56
  models of, 150–56
  model system for, 164–67
  phenomenological thermodynamic models for, 150–55
  selectivity in, 2
  specificity in, 2
Lipid:protein kinase C interactions, 1–23
  model for, 19–21
Lipid bilayer(s)
  bacteriorhodopsin in
    asymmetry of, 167
  fusion pores in, 451
  protein kinase C:phosphatidylserine interactions in, 13
  short and long polypeptides in, 164–66
Lipid-bilayer fusion
  stalk model of, 451–53
Lipid matrix
  protein structure/function and, 1–2
Lipid membrane
  lateral distribution of proteins in, 159–60
  structural transitions in, 148–49

Lipid membrane fusion
  fusion pores and, 435
Liposomes
  secretory granule fusion with cholesterol and, 454
Liver
  phosphorylase activation in, 203
Lysolipids
  biologic membrane fusion and, 454–55
  in chromaffin granule monolayers, 454
Lysozyme
  high-temperature dynamics simulation of, 377
  internal packing in, 404
  photoactivatable caged synthesis of, 473–74
  temperature factors in, 469
  in water
    simulation of, 372–73
α-Lytic protease
  high-temperature dynamics simulation of, 376–77

## M

Machine learning, 284–85
Macromolecular complexes
  formation of
    crowding and, 36
Macromolecular crowding, 27–61
  cellular evolution and, 59–60
  cellular volume regulation and, 58
  condensed DNA and, 61
  in cytoplasm, 56–58
  cytoplasmic structure and, 61
  diffusive transport of solutes and, 37–40
  enzyme-catalyzed reactions and, 36–37
  equilibria and reaction rates in model systems and, 46–51
  homeostasis and, 60–61
  reaction rates and, 36–37
  solution equilibria and, 31–36
  thermodynamic activity of tracer proteins and, 43–47
  transport properties in model systems and, 51–56
Macromolecular drugs
  efficacy of, 58–59
Macromolecular electron microscopy, 234
Macromolecular equilibrium reaction
  effect of cosolvent on, 70–71
Macromolecules
  effective specific volume of, 40–42
Macrophages
  human growth hormone and, 348
Magnetic birefringence
  four-way DNA junctions and, 308
Mast cells
  flicker in, 440

secretory granules in
  dimpling of plasma membrane towards, 434
M current, 178
Melittin
  membrane fusion and, 455
Membrane(s)
  erythrocyte
    diffusion in, 421
  fluid plasma, 417–29
    glycoprotein motility in, 419–29
  lipid
    lateral distribution of proteins in, 159–60
    structural transitions in, 148–49
  as many-particle systems, 147–50
  organization of
    protein-induced, 157–60
  protein diffusion in, 419–29
  protein:lipid hydrophobic matching and, 167–68
  protein:lipid interactions and, 145–69
  self-diffusion of proteins in
    macromolecular crowding and, 51–54
Membrane fusion, 433–62
  docking and, 434–35
  fusion pores and, 435–42
  hemagglutinin-mediated, 442–45
  cast and retrieve model for, 458–61
  lipids and, 448–55
  models of, 455–61
  proteins and, 442–47
Membrane glycoproteins
  motility of
    active, 426–29
    passive, 419–26
Membrane proteins
  electron crystallography and, 240–42
  integral, 146–47
    lipid acyl-chain length profile near, 161
    lipid structure/composition near, 160–64
  lipid-binding sites on, 4
Membrane transport
  protein phosphorylation and, 200
Memorization
  vs. generalization, 288–90
Memory
  protein phosphorylation and, 200
-Mercaptoethanol
  protein-Co(II) complex and, 268
Metabolic buffering
  macromolecular crowding and, 60–61
Metal-binding sites, 257–80
  as discrete units, 259–60
  formed by histidine residues, 260–63
  with four protein ligands, 270–77

with three protein ligands, 263–70
Metal ions
  DNA junctions and, 313–16
Metalloantibodies, 265–66
Metalloproteins
  metal-binding sites in, 258
Methane
  transfer from solution to vapor, 396
Methanol
  transfer from vapor to water enthalpy of, 392
2-Methyl-2,4-pentanediol
  protein stabilization due to, 84–85
Methylamines
  protein stabilization due to, 84–85
Microscopic interaction model, 155–56
Microtubules
  self-assembly of, 85
  X-ray solution scattering and, 246
Minton, A. P., 27–61
Mitogen-activated protein kinase, 217
  activation of, 218
Molecular chaperones, 124
Molecular dynamics
  protein dynamics simulations and, 357–58
Molecular-occupancy model, 81
Molecular recognition
  mechanisms of, 329–30
Monoclonal antibodies
  human growth hormone binding protein and, 340–41
Monte Carlo simulation
  lateral distribution of proteins and, 159
Mouritsen, O. G., 145–69
mRNA splicing donor/acceptor-site recognition
  artificial neural networks and, 293
Multivariate statistical analysis
  electron cryomicroscopy and, 250
Muscle
  glucose-1-P and, 202
  growth and differentiation of pituitary hormones and, 330
  phosphorylase activation in, 203
Muscle contraction
  protein phosphorylation and, 200
Mutational analysis
  potassium channel pore and, 191
  receptor:ligand interactions and, 337–41
Myelin basic protein
  affinity for lipid headgroups, 4
Myoglobin
  diffusion in solution, 54
  structures in
    crystallographic residual for, 490
  temperature factors in, 469

Myohemerythrin
  iron and, 258

N

NADH
  crystalline alcohol dehydrogenase and, 471
*Necturus* smooth muscle
  potassium channels of inactivation of, 186
Network(s)
  multilayer, 287–88
  single layer, feed-forward, 286
  See also Neural networks
Network-training analysis, 290–91
Neural networks, 283–97
  applications of, 291–96
  artificial, 285
  knowledge-based, 293
  memorization vs. generalization in, 288–90
  topologies for, 285–88
Neuromuscular junction
  synaptic vesicles in alignment along nerve terminals, 435
Neurons
  M current in, 178
Neurotransmitters
  synaptic release of membrane fusion and, 434
Neutrophils
  secretory granules in dimpling of plasma membrane towards, 434
Newton, A. C., 1–23
NMR
  See Nuclear magnetic resonance
NOE
  See Nuclear Overhauser effect
Nuclear magnetic resonance (NMR)
  proteins in solution and, 354
  solution equilibria of model amides and, 390
Nuclear magnetic resonance (NMR) spectroscopy
  atomic structures of biomolecules and, 233–34
  biophysics of, 234–51
  erythrocyte band 3 protein receptors and, 158
  ligand-macromolecular systems and, 100
  macromolecules and, 239
  metal-binding sites and, 258
  molecular conformation and, 235
Nuclear Overhauser effect (NOE), 99–120
  proteins in solution and, 354
Nucleic acids
  electron cryomicroscopy and, 243–44
  macromolecular crowding and, 46–51
  structures of
    gel electrophoresis and, 301–5

O

1-Octanol
  polar interactions with solutes, 398–99
Oligomerization
  receptor
    hormone-induced, 340
Oligopeptides
  prolyl isomerization in, 138–39
Optical absorption spectroscopy
  metal-binding sites and, 258, 271–72
  protein-Cu(II) complexes and, 274
Organelles
  self-assembly of, 85
Osmolytes, 69, 92–94
Ovalbumin
  phosphorylation of, 226
Ovomucoid
  in water
    simulation of, 370–71

P

1-Palmitoyl-2-oleoyl phosphatidylcholine bilayers
  hydrophobic thickness of cholesterol and, 166
Pancreatic ribonuclease A
  folding of
    disulfide-bond formation and, 136
Pancreatic RNase
  prolyl isomerization and, 127
Pancreatic trypsin inhibitor
  prolyl isomerization and, 127
  See also Bovine pancreatic trypsin inhibitor
*Paramecium* trichocyst discharge
  secretory granules in, 434
Parvalbumin
  in water
    simulation of, 370
Patterson maps
  isomorphous difference, 487
Pentane
  transfer from solution to vapor, 396
Pepsin
  phosphorylation of, 226
Pepsinogen
  phosphorylation of, 226
Peptide:antibody complexes
  transferred nuclear Overhauser effect difference spectroscopy and, 107
Peptide(s)
  membrane fusion and, 455
Peptide bonds
  isomerization of, 125–26
Perceptrons, 285–87
Percolation model
  diffusion in fluid plasma membranes and, 421–22
pGH
  See Porcine growth hormone
Phorbol esters
  protein kinase C activation and, 4–5

Phosphate-recognition sites, 201–2
Phosphatidic acid
 insulin receptor and, 2
 protein kinase C:phosphatidylserine interactions and, 16
Phosphatidylcholine
 in biologic membranes, 454
 -hydroxybutyrate dehydrogenase and, 2
Phosphatidylethanolamine
 in biologic membranes, 454
 protein kinase C and, 13
Phosphatidylinositol bisphosphate
 hydrolysis of
  receptor-mediated, 4
Phosphatidylserine
 in biologic membranes, 454
 protein kinase C and, 2, 4, 10–17
Phosphoenolpyruvate sugar phosphotransferase system, 200–1, 218–23
Phospholipid(s)
 synaptic vesicle protein binding to, 435
Phospholipid turnover
 signal transduction provoking, 4
Phosphorylase kinase
 phosphorylation and, 202
Phosphorylation, 199–229
 cyclic AMP-dependent protein kinase and, 213–18
 glycogen phosphorylase and, 202–9
 isocitrate dehydrogenase and, 209–13
 M channel-receptor coupling and, 178
 phosphoenolpyruvate sugar phosphotransferase system and, 218–23
 silent, 226
Photoactivation
 crystalline enzymes and, 472–74
Photosynthesis
 protein phosphorylation and, 200
Photosynthetic reaction center
 membrane-protein complex in structure of, 240
Photosynthetic reaction center proteins
 phase behavior of, 157
Pituitary hormones
 physiologic processes regulated by, 330
Plasma membrane
 fluid, 417–29
  glycoprotein motility in, 419–29
 lipids in, 454
Plastacyanin
 copper binding site in, 258, 273
Polyethylene glycol
 denaturation of
  chymotrypsinogen by, 72, 82

protein stabilization due to, 84–85
viral particle/protein precipitation due to, 43–45
Polymers
 tracer diffusion of probe molecules in, 55–56
 water-soluble
  viral particle/protein precipitation due to, 43
Polyols
 preferential hydration in, 88
 protein stabilization due to, 84
Polypeptides
 in lipid bilayers, 164–66
 synthetic amphiphilic, 164–66
Porcine growth hormone (pGH)
 crystal structure of, 331
 structural model of, 337
Porins
 X-ray crystallography and, 241–42
Porphyrins
 four-way DNA junctions and, 307
Potassium channels, 173–95
 amino acid sequences in function of, 185–86
 calcium-activated, 177–78
 G protein-activated, 178
 inactivation of
  C-terminus, 187
  N-terminus, 186–87
 pore/P-type, 187–88
 ligand-gated, 181–83
 P region of
  amino acids forming, 189–92
 topography of, 183–85
 voltage-dependent
  classification of, 175–83
  electromechanical model of, 175
 voltage sensor and, 192–94
Potassium currents
 voltage-dependent
  functional classification of, 176–78
Potassium pore
 localization of, 189–92
Potential energy function
 protein dynamics and, 355–57
Preferential binding parameter, 70
Preferential hydration, 85–89
Preferential interaction(s)
 molecular contacts and, 76–81
Preferential interaction parameter, 70
Presnell, S. R., 283–97
Pressure
 high
  nonnative protein-water simulations and, 378
PRL
 See Prolactin
Procollagen
 folding of, 137

Prokaryotes
 auxotrophic/temperature-sensitive mutants of
  high osmolarity media and, 59
Prokaryotic cells
 protein phosphorylation in, 200
Prolactin (PRL)
 signal transduction and, 346
Prolactin receptor
 dimerization mechanism for, 348
 extracellular domains of, 345
 human growth hormone receptor and, 338
 WSXWS box mutations
  ligand binding and, 337
Proline
 hydrophilic nature of, 398, 400
Prolyl isomerase, 128–29
 catalysis of slow steps in folding by, 129–37
 cellular folding and, 137–38
 efficiency of, 136–37
 reaction catalyzed by, 138–39
 ribonuclease T1 folding by catalysis of, 133–35
 structure of, 129
 substrate specificity of, 128–29
Prolyl isomerization, 125–28
 catalysis in unfolding proteins, 134–35
 mechanism of, 138–39
 in protein folding, 126–28
Propane
 transfer from solution to vapor, 396
Propanol
 transfer from vapor to water
  enthalpy of, 392
Protein(s)
 atomic movements in
  X-ray evidence of, 469
 conformational states of, 489–90
 denaturation of, 68, 82–83
  metal binding and, 261
 denaturation equilibrium of
  effect of cosolvents on, 72
 folded
  prolyl isomerization in, 127–28
 functions of
  lipids as regulators of, 2–4
 globular
  hydrophilic/hydrophobic groups in, 402
  packing densities of, 405–7
 hydrophobic effect and, 396–403
 interactions with DNA junctions, 322–24
 internal packing in, 403–7
 lateral distribution in lipid membranes, 159–60
 membrane
  electron crystallography and, 240–42

integral, 146–47, 160–64
lipid-binding sites on, 4
membrane fusion and, 442–47
metal-binding sites on
  design of, 257–80
phosphate-recognition sites
  on, 201–2
phosphorylation sites on, 228
polar groups in
  hydrogen bonds between,
    385–90
self-diffusion in aqueous solution
  macromolecular crowding
    and, 54
self-diffusion in membranes
  macromolecular crowding
    and, 51–54
stability of
  solvent effects on, 67–95
  stabilization of, 84–85
unfolding
  catalysis of prolyl isomerization in, 134–35
Protein coding-region recognition
  artificial neural networks and,
    293–94
Protein Data Bank, 201
Protein-density model
  diffusion in fluid plasma membranes and, 422–23
Protein disulfide isomerase
  protein folding and, 135–36
Protein dynamics
  simulations of, 353–79
  bulk solvents and, 359–60
  energy minimization and,
    357
  hydrogen atoms and, 358–59
  molecular dynamics and,
    357–58
  potential energy function
    and, 355–57
  procedures and problems
    in, 358–62
  truncating interactions and,
    360–62
  simulations of
    water models in, 362–64
Protein electron crystallography,
  234
Protein folding, 381–411
  catalysts of
    prolyl isomerase/protein disulfide isomerase and,
      135–36
  hydrogen bonds and, 390–96
  hydrophobic effect and, 400–3
  internal packing and, 404–7
  model for, 407–11
  prolyl isomerization in, 126–28
  slow
    enzymatic catalysis of, 123–41
    stereochemical code for, 409–11
Protein kinase
  cell division-control, 217
  cyclic AMP-dependent, 213–18

autophosphorylation site
  on, 216–17
fold of, 215
mitogen-activated, 217
  activation of, 218
Protein kinase C, 4–7
  diacylglycerol and, 9–10
  function of, 4–5
  interaction with lipid headgroups, 7–21
    measurement of, 7–9
  lipid-binding domain of, 7
  phosphatidylserine and, 2, 10–17
  structure of, 5–7
Protein kinase C isozymes
  diacylglycerol and, 9–10
  pseudosubstrate sequence of, 6
Protein kinase C:lipid interactions, 1–23
  model for, 19–21
Protein:lipid complexes
  models of, 456–57
Protein:lipid interactions
  membrane function and, 167–68
  in membranes, 145–69
  microscopic interaction model
    for, 155–56
  models of, 150–56
  model system for, 164–67
  phenomenological thermodynamic models for, 150–55
  selectivity in, 2
  specificity in, 2
Protein phosphatase 1, 203
Protein phosphorylation, 199–229
  cyclic AMP-dependent protein kinase and, 213–18
  glycogen phosphorylase and, 202–209
  isocitrate dehydrogenase and, 209–13
  phosphoenolpyruvate sugar
    phosphotransferase system and, 218–23
  silent, 226
Protein ribonuclease T1
  stability of
    hydrogen bonding and,
      385
Protein secondary-structure recognition
  artificial neural networks and,
    294–96
Protein tertiary-structure recognition
  artificial neural networks and,
    296
Protein:water interactions, 373–75
  simulations of
    native, 364–75
    nonnative, 375–78
Prothrombin
  prolyl isomerization in, 127–28
Prothrombin fragment 1
  in water
    simulation of, 373

R

Reaction rates
  macromolecular crowding
    and, 36–37
Receptor:ligand interactions, 330
  crystallographic analysis of,
    341–42
  mutational analysis of, 337–41
Regan, L., 257–80
Relaxation matrix analysis
  transferred nuclear Overhauser effect and, 102–7
Rhinoviruses
  X-ray crystallographic structure of, 249
*Rhodopseudomonas sphaeroides*
  photosynthetic apparatus of,
    157
Rhodopsin
  protein kinase C and, 5
Ribonuclease
  stabilization by sorbitol, 72,
    82
Ribonuclease A
  preferential binding for, 82–83
  refolding rates of, 126
Ribonuclease T1
  folding by prolyl isomerase
    catalysis of, 133–35
  folding mechanism of, 130–33
  prolyl isomerization and, 127
  slow refolding of
    kinetic model for, 132
Ribosomal protein L7/L12
  in water
    simulation of, 371
Ribosome(s)
  X-ray crystal structure of, 245
Ribosome 70S particle
  ice-embedded
    stereo-pair image of, 251
Ristocetin
  binding affinities of small peptides and, 387
Rose, G. D., 381–411
Rotavirus-Fab complex
  three-dimensional surface density map of, 248
Rous sarcoma virus
  fusion proteins of, 445
Rubredoxin
  iron binding site in, 258

S

*Salmonella typhimurium*
  chemotactic response in, 223–26
  flagellar filament of, 246
Salt(s)
  enzyme isolation and, 68
  protein stabilization due to,
    84–85
  surface tension of water and,
    87
Salting-in, 68
Salting-out, 68–69, 84–85
Salting-out cosolvents, 85–89
Sarcosine
  as protein structure protectant,
    92

# SUBJECT INDEX

Scaled particle theory (SPT), 33–35, 40, 45
*Schizosaccharomyces pombe*
  Thr167 phosphorylation in, 217–18
Schmid, F. X., 123–41
Scorpion toxin
  potassium channels and, 178, 189–90
  protein side chains of, 490
Sea urchin cortical granule exocytosis
  secretory granules in, 434–35
Second messengers
  generation of, 4
Secretory cells
  lipids in, 454
Sedimentation equilibrium, 78
Selectivity filter
  potassium channel and, 192
Sendai virus
  fusion proteins of, 445
Sensory transduction
  lipids in, 4
SHAKE algorithm, 359
Sheetz, M. P., 417–29
Sickle hemoglobin
  deoxygenated
    solubility of, 46
Signal transduction
  hematopoietins and, 346
  lipids in, 4
  prolyl isomerases and, 124–25
  protein phosphorylation and, 200, 207–9
Single-particle tracking, 420–21
Skeletal muscle
  potassium channels of
    inactivation of, 186
Small-angle X-ray scattering
  solution composition perturbation and, 78
Smooth muscle
  potassium channels of
    inactivation of, 186
Sodium channels
  peptides associated with, 174
Solubility constant
  thermodynamic definition of, 71
Solutes
  diffusive transport of
    macromolecular crowding and, 37–40
Solution
  self-diffusion of proteins in
    macromolecular crowding and, 54
Solution equilibria
  macromolecular crowding and, 31–36
Solvents
  bulk
    protein dynamics simulations and, 359–60
  nonpolar
    prolyl isomerization and, 139
Solvophobic effect, 88

Sorbitol
  ribonuclease stabilization due to, 72, 82
Spectroscopy
  See specific type
Sphingolipids
  proteins sensitive to, 2
Sphingomyelin
  in biologic membranes, 454
SPT
  See Scaled particle theory
Squid giant axon
  delayed rectifiers in, 176
Stalk model
  lipid-bilayer fusion and, 451–53
Staphylococcal nuclease
  prolyl isomerization and, 127
*Staphylococcus aureus*
  HPr of, 219–22
Steric exclusion, 85–87
Steroids
  activity coefficients in water, 398
Streptavidin
  two-dimensional structure of, 243
Sucrose
  organelle dissection from cell and, 67–68
  protein stabilization by, 68
  as stabilizer, 74–76
Sugars
  protein stabilization due to, 84–85
  surface tension of water and, 87
Superosmolytes, 94
Surface tension
  protein stabilization and, 87–88
SV40
  X-ray diffraction and, 247
Sykes, B. D., 99–120
Synaptic transmission
  fusion proteins in, 447
Synaptic vesicles
  alignment along nerve terminals, 435
Synchrotron radiation
  structural studies and, 474

# T

Temperature
  high
    nonnative protein-water simulations and, 375–78
Tension
  liposome-planar lipid bilayer membrane fusion and, 449
Tetraethyl ammonium
  potassium currents and, 180–81
Tetrodotoxin
  sodium channels and, 189
Thermal diffuse scattering, 469
Thermodynamics
  cosolvent control of equilibria and, 70–76

Thermolysin
  Ca(II)-binding loop of, 260
  zinc binding sites in, 258
*Thermus thermophilus*
  DNA ligase of
    macromolecular crowding and, 47
Thioredoxin
  copper binding site in, 274
  prolyl isomerization and, 127
Timasheff, S. N., 67–95
TMV
  See Tobacco mosaic virus
Tobacco mosaic virus (TMV)
  electron microscopy of, 239
  three-dimensional density map of, 247
  X-ray fiber diffraction and, 246
Transfer free energy, 70
Transferred nuclear Overhauser effect, 99–120
  average relaxation matrix and, 108–9
  chemically equivalent protons and, 109–10
  experiments with, 108–10
  fraction bound/mixing time and, 111–15
  intermolecular vs. intramolecular, 107–8
  internal motions and, 110, 115–18
  relaxation matrix and, 102–7
  theory of, 102–8
Transferrin
  in vivo maturation of
    cyclosporin A and, 137–38
Transient-binding model
  diffusion in fluid plasma membranes and, 422
Transient electric birefringence
  four-way DNA junctions and, 307–8, 313
Trimethylamine-N-oxide
  as protein structure protectant, 92
Triose-phosphate isomerase
  sulfate- and phosphate-binding sites of, 202
Triton X-100 mixed micelles
  lipid:protein kinase C interactions and, 7–8, 11
Tropomyosin
  X-ray diffraction and, 244
Troponin C
  calcium binding sites in, 258
  transferred nuclear Overhauser effect and, 111
Troponin I
  transferred nuclear Overhauser effect and, 111
Truncating interactions
  protein dynamics simulations and, 360–62
*Trypanosoma brucei*
  glyceraldehyde-3-phosphate dehydrogenase of
    three-dimensional structure of, 487

## SUBJECT INDEX   509

Trypsin
  crystallography and, 485–86
  metal binding and, 262
Tyrosyl-tRNA synthetase
  mutations in
    hydrogen bonds in, 387

## U

Urea
  cellular osmotic pressure and, 92
  preferential binding for unfolded proteins in, 80
  preferential interactions of, 82–83
  protein denaturation and, 68

## V

Vancomycin
  binding affinities of small peptides and, 387
van der Waals interactions
  cyclohexane with solutes, 398
  proteins in solution and, 356, 359
Vibrational parameters
  crystallography and, 489
Viral oncogene response
  protein phosphorylation and, 200
Viruses
  electron cryomicroscopy and, 248–49
  helical, 246
  X-ray fiber diffraction and, 246
Vogel, S. S., 433–62
Voltage sensor
  potassium channels and, 192–94

## W

Water
  hydrogen bonds in
    strengths of, 387
  self-cohesive properties of, 397
  solubilities of amino acids in, 398
  surface tension of
    protein stabilization and, 87–88
Water:protein interactions, 373–75
  simulations of
    native, 364–75
    nonnative, 375–78
Water models
  protein dynamics simulations and, 362–64
Weisenberg photography, 479
Wells, J. A., 329–50
Wolfenden, R., 381–411

## X

*Xenopus* oocytes
  potassium current expression in, 185
X-ray crystallography
  atomic positions of
    biomolecules and, 233
  biophysics of, 234–51
  electron density in, 235
  femtosecond time resolution in, 470
  macromolecular crystals in, 239
  phosphate-recognition sites and, 201
  protein structure and, 403
  structural dynamics and, 469–70
X-ray diffraction
  glycogen phosphorylase and, 203
X-ray fiber diffraction
  helical arrays and, 246
X-ray solution scattering
  helical arrays and, 246

## Z

Zimmerberg, J., 433–62
Zimmerman, S. B., 27–61
Zinc
  n-chimaerin binding, 6–7
Zinc finger(s), 258
Zinc-finger peptides, 268–70, 275–77

# CUMULATIVE INDEXES

## CONTRIBUTING AUTHORS, VOLUMES 18–22

### A

Åhqvist, J., 20:267–98
Allewell, N. M., 18:71–92
Altschul, S. F., 20:175–203
Ames, J. B., 20:491–518
Anderson, C. F., 19:423–65
Andersson, I., 22:467–98
Arndt-Jovin, D. J., 18:271–308

### B

Baldwin, R. L., 21:95–118
Banaszak, L., 20:221–46
Barford, D., 22:199–232
Beck, W. F., 18:25–46
Beratan, D. N., 21:349–77
Berg, J. M., 19:405–21
Berzofsky, J. A., 19:69–82
Beveridge, D. L., 18:431–92
Bloom, M., 22:145–71
Bormann, B. J., 21:223–42
Boxer, S. G., 19:267–99
Bråndén, C.-I., 21:119–43
Brendel, V., 20:175–203
Brown, A. M., 22:173–98
Brudvig, G. W., 18:25–46
Bucher, P., 20:175–203
Bustamante, C., 20:415–46

### C

Caffrey, M., 18:159–86
Campbell, A. P., 22:99–122
Cerdan, S., 19:43–67
Chan, H. S., 20:447–90
Chance, B., 20:1–28
Chernomordik, L. V., 22:433–66
Chiu, W., 22:233–55
Clegg, R. M., 22:299–328
Clore, G. M., 20:29–63
Cohen, F. E., 22:283–98
Cohn, M., 21:1–24
Coleman, J. E., 21:441–83
Coleman, W. J., 19:333–67
Cornette, J. L., 19:69–82

### D

Daggett, V., 22:353–80
Davies, D. R., 19:189–215
de Paula, J. C., 18:25–46
de Vitry, C., 19:369–403
de Vos, A. M., 22:329–51
Deisenhofer, J., 20:247–66
DeLisi, C., 19:69–82
Devaux, P. F., 21:417–39
DiCapua, F. M., 18:431–92
Dill, K. A., 20:447–90
Doms, R. W., 18:187–211
Dyson, H. J., 20:519–38

### E

Endo, S., 18:1–24
Engel, A., 20:79–108
Engel, J., 20:137–52
Engelman, D. M., 21:223–42
Englander, S. W., 21:243–65
Erickson, H. P., 21:145–66
Erie, D. A., 21:379–415

### F

Fenselau, C., 20:205–20
Fitzgerald, P. M. D., 20:299–320
Fleischer, S., 18:333–64
Freire, E., 19:159–88

### G

Gray, H. B., 21:349–77
Gronenborn, A. M., 20:29–63

### H

Hajdu, J., 22:467–98
Hartl, F. U., 21:293–322
Havel, T. F., 21:167–98
Helenius, A., 18:187–211
Honig, B., 19:301–32
Hope, H., 19:107–26
Hyberts, S. G., 21:167–98

### I

Inui, M., 18:333–64

### J

Jennings, M. L., 18:397–430
Johnson, L. N., 22:199–232

Jovin, T. M., 18:271–308

### K

Kabsch, W., 21:49–76
Karlin, S., 20:175–203
Kataoka, R., 19:69–82
Kidokoro, S., 18:1–24
Kowalczykowski, S. C., 20:539–75

### L

Lauffenburger, D. A., 20:387–414
LeMaster, D. M., 19:243–66
Leslie, A. G. W., 20:363–86
Lester, H. A., 21:267–92
Levitt, M., 22:353–80
Lilley, D. M. J., 22:299–328
Lin, S. W., 20:491–518
Lindqvist, Y., 21:119–43
Lukat, G. S., 20:109–36

### M

Margalit, H., 19:69–82
Marr, K., 20:343–62
Martin, J., 21:293–322
Martin, J.-L., 21:199–222
Mathies, R. A., 20:491–518
Mattice, W. L., 18:93–111
Mayne, L., 21:243–65
Mayorga, O. L., 19:159–88
McCray, J. A., 18:239–70
McLaughlin, S., 18:113–36
Meyer, T., 20:153–74
Michel, H., 20:247–66
Minton, A. P., 22:27–65
Moffat, K., 18:309–32
Mouritsen, O. G., 22:145–71

### N

Neupert, W., 21:293–322
Newton, A. C., 22:1–25

### O

O'Brien, E. T., 21:145–66
Onuchic, J. N., 21:349–77

## P

Peersen, O. B., 21:25–47
Peters, K. S., 20:343–62
Pollard, W. T., 20:491–518
Popot, J.-L., 19:369–403
Presnell, S. R., 22:283–98
Privalov, P. L., 18:47–69
Prockop, D. J., 20:137–52

## R

Record, M. T. Jr., 19:423–65
Regan, L., 22:257–81
Rose, G. D., 22:381–415

## S

Sanchez-Ruiz, J. M., 19:159–88
Sarvazyan, A. P., 20:321–42
Schmid, F. X., 22:123–43
Schneider, G., 21:119–43
Scholtz, J. M., 21:95–118
Scott, R. A., 18:137–58
Seelig, J., 19:43–67
Senior, A. E., 19:7–41
Sharp, K. A., 19:301–32
Sharrock, W., 20:221–46

Shaw, W. V., 20:363–86
Sheetz, M. P., 22:417–31
Silvius, J. R., 21:323–48
Smith, S. O., 21:25–47
Springer, J. P., 20:299–320
Stegmann, T., 18:187–211
Stock, A. M., 20:109–36
Stock, J. B., 20:109–36
Stryer, L., 20:153–74
Stühmer, W., 20:65–78
Sykes, B. D., 22:99–122

## T

Teeter, M. M., 20:577–600
Timasheff, S. N., 22:67–97
Timmins, P., 20:221–46
Trentham, D. R., 18:239–70
Tsong, T. Y., 19:83–106
Tullius, T. D., 18:213–37

## V

Vajda, S., 19:69–82
van Osdol, W. W., 19:159–88
Vandekerckhove, J., 21:49–76
Vogel, S. S., 22:433–66
von Hippel, P. H., 21:379–415

Vos, M. H., 21:199–222

## W

Wada, A., 18:1–24
Wagner, G., 21:167–98
Wallace, B. A., 19:127–57
Warshel, A., 20:267–98
Watson, T., 20:343–62
Weber, G., 19:1–6
Wells, J. A., 22:329–51
Widom, J., 18:365–95
Winkler, J. R., 21:349–77
Wittenberg, B. A., 19:217–41
Wittenberg, J. B., 19:217–41
Wolfenden, R., 22:381–415
Wright, P. E., 20:519–38

## Y

Yager, T. D., 21:379–415
Yonath, A., 21:77–93
Youvan, D. C., 19:333–67

## Z

Zimmerberg, J., 22:433–66
Zimmerman, S. B., 22:27–65

# CHAPTER TITLES, VOLUMES 18–22

**Actin**
    Structure and Function of **Actin**        W. Kabsch, J. Vandekerckhove        21:49–76

**Adhesion**
    Analysis of **Receptor**-Mediated Cell
        Phenomena: **Adhesion** and Migration        D. A. Lauffenburger        20:387–414

**Alkaline Phosphatase**
    Structure and Mechanism of **Alkaline**
        **Phosphatase**        J. E. Coleman        21:441–83

**Aspartate Transcarbamoylase**
    *Escherichia coli* **Aspartate**
        **Transcarbamoylase**: Structure, Energetics,
        and Catalytic and Regulatory Mechanisms        N. M. Allewell        18:71–92

**Aspartic Proteinases**
    The Structure and Function of **Aspartic**
        **Proteinases**        D. R. Davies        19:189–215

**Bacteriorhodopsin**
    From Femtoseconds to Biology: Mechanism
        of **Bacteriorhodopsin's** Light-Driven
        **Proton Pump**        R. A. Mathies, S. W. Lin, J. B. Ames, W. T. Pollard        20:491–518

**Biomembrane**
    Solubilization and Functional **Reconstitution**
        of **Biomembrane** Components        J. R. Silvius        21:323–48

**Caged Compounds**
    Properties and Uses of Photoreactive **Caged**
        **Compounds**        J. A. McCray, D. R. Trentham        18:239–70

**Calcium**
    **Calcium** Spiking        T. Meyer, L. Stryer        20:153–74

**Calorimetry**
    **Calorimetrically** Determined **Dynamics** of
        Complex **Unfolding** Transitions in **Proteins**        E. Freire, W. W. van Osdol, O. L. Mayorga, J. M. Sanchez-Ruiz        19:159–88

**Chaperones**
    Protein **Folding** in the Cell: The Role of
        Molecular **Chaperones** Hsp70 and Hsp60        F. U. Hartl, J. Martin, W. Neupert        21:293–322

**Chemotaxis**
    Bacterial **Chemotaxis** and the Molecular
        Logic of Intracellular **Signal Transduction**
        Networks        J. B. Stock, G. S. Lukat, A. M. Stock        20:109–36

**Chloramphenicol Acetyltransferase**
   Chloramphenicol Acetyltransferase    W. V. Shaw, A. G. W. Leslie    20:363–86

**Chromatin**
   Toward a Unified Model of **Chromatin**
      **Folding**    J. Widom    18:365–95

**Coil**
   The β-**Sheet** to **Coil** Transition    W. L. Mattice    18:93–111

**Collagen Triple Helices**
   The Zipper-Like Folding of **Collagen Triple Helices** and the Effects of **Mutations** that Disrupt the Zipper    J. Engel, D. J. Prockop    20:137–52

**Computer**
   Expanding Roles of **Computers** and **Robotics** in Biological Macromolecular Research    A. Wada, S. Kidokoro, S. Endo    18:1–24

**Conformations**
   Defining Solution **Conformations** of Small Linear **Peptides**    H. J. Dyson, P. E. Wright    20:519–38

**Crystallography**
   **Time-Resolved** Macromolecular **Crystallography**    K. Moffat    18:309–32
   **Crystallography** of Biological Macromolecules at **Ultra-Low Temperature**    H. Hope    19:107–26
   What Does **Electron Cryomicroscopy** Provide that **X-Ray Crystallography** and **NMR Spectroscopy** Cannot?    W. Chiu    22:233–55
   Fast **Crystallography** and **Time-Resolved Structures**    J. Hajdu, I. Andersson    22:467–98

**Cytochrome c Oxidase**
   **X-Ray Absorption Spectroscopic** Investigations of **Cytochrome c Oxidase** Structure and Function    R. A. Scott    18:137–58

**DNA**
   Physical Studies of **Protein-DNA Complexes** by Footprinting    T. D. Tullius    18:213–37
   **Ion Distributions** around **DNA** and Other Cylindrical **Polyions**: Theoretical Descriptions and Physical Implications    C. F. Anderson, M. T. Record, Jr.    19:423–65
   Statistical Methods and Insights for **Protein** and **DNA Sequences**    S. Karlin, P. Bucher, V. Brendel, S. F. Altschul    20:175–203
   Direct Observation and Manipulation of Single **DNA** Molecules by **Fluorescence Microscopy**    C. Bustamante    20:415–46
   Biochemistry of Genetic **Recombination**: **Energetics** and Mechanism of **DNA** Strand Exchange    S. C. Kowalczykowski    20:539–75
   The Structure of the **Four-Way Junction** in **DNA**    D. M. J. Lilley, R. M. Clegg    22:299–328

**Dynamics**
   **Calorimetrically** Determined **Dynamics** of Complex **Unfolding** Transitions in **Proteins**    E. Freire, W. W. van Osdol, O. L. Mayorga, J. M. Sanchez-Ruiz    19:159–88

## CHAPTER TITLES 515

**Electrical Modulation**
Electrical Modulation of **Membrane Proteins**: Enforced Conformational Oscillations and Biological **Energy** and Signal Transductions — T. Y. Tsong — 19:83–106

**Electron Cryomicroscopy**
What Does **Electron Cryomicroscopy** Provide that **X-Ray Crystallography** and **NMR Spectroscopy** Cannot? — W. Chiu — 22:233–55

**Electron Transfer**
Mechanisms of Long-Distance **Electron Transfer** in **Proteins**: Lessons from **Photosynthetic Reaction Centers** — S. G. Boxer — 19:267–99
Pathway Analysis of Protein **Electron-Transfer** Reactions — J. N. Onuchic, D. N. Beratan, J. R. Winkler, H. B. Gray — 21:349–77

**Electrostatic**
The **Electrostatic** Properties of **Membranes** — S. McLaughlin — 18:113–36
**Electrostatic** Interactions in Macromolecules: Theory and Applications — K. A. Sharp, B. Honig — 19:301–32
**Electrostatic Energy** and **Macromolecular Function** — A. Warshel, J. \:hqvist — 20:267–98

**Energy**
Electrical Modulation of **Membrane Proteins**: Enforced Conformational Oscillations and Biological **Energy** and Signal Transductions — T. Y. Tsong — 19:83–106
**Electrostatic Energy** and **Macromolecular Function** — A. Warshel, J. Aqvist — 20:267–98

**Energetics**
Biochemistry of Genetic **Recombination**: **Energetics** and Mechanism of **DNA** Strand Exchange — S. C. Kowalczykowski — 20:539–75

**Enzyme**
Atomic and Nuclear Probes of **Enzyme** Systems — M. Cohn — 21:1–24

*Escherichia coli*
The **Proton-Translocating ATPase** of *Escherichia coli* — A. E. Senior — 19:7–41

**Excitation-Contraction**
Biochemistry and Biophysics of **Excitation-Contraction** Coupling — S. Fleischer, M. Inui — 18:333–64

**Femtosecond Biology**
Femtosecond Biology — J.-L. Martin, M. H. Vos — 21:199–222

**Fluorescence Microscopy**
Direct Observation and Manipulation of Single **DNA** Molecules by **Fluorescence Microscopy** — C. Bustamante — 20:415–46

**Folding**
Toward a Unified Model of **Chromatin Folding** — J. Widom — 18:365–95

## CHAPTER TITLES

**Four-Way Junction**
 The Structure of the **Four-Way Junction** in
  **DNA** — D. M. J. Lilley, R. M. Clegg — 22:299–328

**Glycoprotein**
 **Glycoprotein** Motility and Dynamic Domains
  in Fluid **Plasma Membranes** — M. P. Sheetz — 22:417–31

**Gramicidin**
 **Gramicidin** Channels and Pores — B. A. Wallace — 19:127–57

**GTP**
 **Microtubule** Dynamic Instability and **GTP**
  **Hydrolysis** — H. P. Erickson, E. T. O'Brien — 21:145–66

**α-Helix**
 The Mechanism of **α-Helix** Formation by
  **Peptides** — J. M. Scholtz, R. L. Baldwin — 21:95–118

**Hematopoietins**
 Structure and Function of **Human Growth**
  **Hormone**: Implications for the
  **Hematopoietins** — J. A. Wells, A. M. de Vos — 22:329–51

**Hemoglobin**
 Mechanisms of Cytoplasmic **Hemoglobin** and
  **Myoglobin** Function — J. B. Wittenberg, B. A. Wittenberg — 19:217–41

**Human Growth Hormone**
 Structure and Function of **Human Growth**
  **Hormone**: Implications for the
  **Hematopoietins** — J. A. Wells, A. M. de Vos — 22:329–51

**Hydrogen Bonding**
 **Hydrogen Bonding, Hydrophobicity,**
  **Packing**, and **Protein Folding** — G. D. Rose, R. Wolfenden — 22:381–415

**Hydrogen Exchange**
 **Protein Folding** Studied Using
  **Hydrogen-Exchange** Labeling and
  **Two-Dimensional NMR** — S. W. Englander, L. Mayne — 21:243–65

**Hydrophobicity**
 **Hydrogen Bonding, Hydrophobicity,**
  **Packing**, and **Protein Folding** — G. D. Rose, R. Wolfenden — 22:381–415

**Intramembrane Helix**
 **Intramembrane Helix-Helix** Association in
  **Oligomerization** and Transmembrane
  **Signaling** — B. J. Bormann, D. M. Engelman — 21:223–42

**Ion Channels**
 Structure-Function Studies of **Voltage-Gated**
  **Ion Channels** — W. Stuhmer — 20:65–78
 The Permeation Pathway of
  **Neurotransmitter**-Gated **Ion Channels** — H. A. Lester — 21:267–92

**Ion Distributions**
 **Ion Distributions** around **DNA** and Other
  Cylindrical **Polyions**: Theoretical
  Descriptions and Physical Implications — C. F. Anderson, M. T. Record, Jr. — 19:423–65
 Structure-Function Studies of **Voltage-Gated**
  **Ion Channels** — W. Stuhmer — 20:65–78

## CHAPTER TITLES 517

**K⁺ Channels**
    Functional Bases for Interpreting Amino Acid
    Sequences of Voltage-Dependent **K⁺**
    **Channels**      A. M. Brown      22:173–98

**Lipid Phase Transition**
    The Study of **Lipid Phase Transition**
    Kinetics by Time-Resolved **X-Ray**
    **Diffraction**      M. Caffrey      18:159–86

**Lipids**
    Protein Involvement in Transmembrane **Lipid**
    **Asymmetry**      P. F. Devaux      21:417–39
    Interaction of Proteins with **Lipid**
    **Headgroups**: Lessons from **Protein Kinase**
    **C**      A. C. Newton      22:1–25
    Models of **Lipid-Protein Interactions** in
    **Membranes**      O. G. Mouritsen, M. Bloom      22:145–71

**Lipoproteins**
    Structure and Function of a **Lipoprotein**:
    **Lipovitellin**      L. Banaszak, W. Sharrock, P.      20:221–46
         Timmins

**Lipovitellin**
    Structure and Function of a **Lipoprotein**:
    **Lipovitellin**      L. Banaszak, W. Sharrock, P.      20:221–46
         Timmins

**Luminescence**
    **Luminescence** Digital Imaging **Microscopy**      T. M. Jovin, D. J. Arndt-Jovin      18:271–308

**Macromolecular Crowding**
    **Macromolecular Crowding**: Biochemical,
    Biophysical, and Physiological
    Consequences      S. B. Zimmerman, A. P. Minton      22:27–65

**Macromolecular Function**
    **Electrostatic Energy** and **Macromolecular**
    **Function**      A. Warshel, J. Aqvist      20:267–98

**Mass Spectrometry**
    Beyond Gene Sequencing: Analysis of
    **Protein** Structure with **Mass Spectrometry**      C. Fenselau      20:205–20

**Membrane Fusion**
    Protein-Mediated **Membrane Fusion**      T. Stegmann, R. W. Doms, A.      18:187–211
         Helenius
    Mechanisms of **Membrane Fusion**      J. Zimmerberg, S. S. Vogel, L. V.      22:433–66
         Chernomordik

**Membrane Proteins**
    **Electrical Modulation** of **Membrane**
    **Proteins**: Enforced Conformational
    Oscillations and Biological **Energy** and
    Signal Transductions      T. Y. Tsong      19:83–106
    On the Microassembly of Integral **Membrane**
    **Proteins**      J.-L. Popot, C. de Vitry      19:369–403
    **Solid-State NMR** Approaches for Studying
    **Membrane Protein** Structure      S. O. Smith, O. B. Peersen      21:25–47

**Membranes**
    The **Electrostatic** Properties of **Membranes**      S. McLaughlin      18:113–36

| | | |
|---|---|---|
| Models of **Lipid-Protein Interactions** in **Membranes** | O. G. Mouritsen, M. Bloom | 22:145–71 |

**Metabolism**
| | | |
|---|---|---|
| **NMR** Studies of **Metabolism** | S. Cerdan, J. Seelig | 19:43–67 |

**Metal-Binding Sites**
| | | |
|---|---|---|
| The Design of **Metal-Binding Sites** in **Proteins** | L. Regan | 22:257–81 |

**Microscopes**
| | | |
|---|---|---|
| Biological Applications of **Scanning Sensor Microscopes** | A. Engel | 20:79–108 |

**Microscopy**
| | | |
|---|---|---|
| **Luminescence** Digital Imaging **Microscopy** | T. M. Jovin, D. J. Arndt-Jovin | 18:271–308 |

**Microtubules**
| | | |
|---|---|---|
| **Microtubule** Dynamic Instability and **GTP** Hydrolysis | H. P. Erickson, E. T. O'Brien | 21:145–66 |

**Molecular Simulation**
| | | |
|---|---|---|
| Free Energy Via **Molecular Simulation**: Applications to Chemical and Biomolecular Systems | D. L. Beveridge, F. M. DiCapua | 18:431–92 |

**Mutations**
| | | |
|---|---|---|
| The Zipper-Like Folding of **Collagen Triple Helices** and the Effects of **Mutations** that Disrupt the Zipper | J. Engel, D. J. Prockop | 20:137–52 |

**Myoglobin**
| | | |
|---|---|---|
| Mechanisms of Cytoplasmic **Hemoglobin** and **Myoglobin** Function | J. B. Wittenberg, B. A. Wittenberg | 19:217–41 |
| Time-Resolved **Photoacoustic Calorimetry**: a Study of **Myoglobin** and **Rhodopsin** | K. S. Peters, T. Watson, K. Marr | 20:343–62 |

**Neural Networks**
| | | |
|---|---|---|
| Artificial **Neural Networks** for **Pattern Recognition** in Biochemical Sequences | S. R. Presnell, F. E. Cohen | 22:283–98 |

**Neurotransmitter**
| | | |
|---|---|---|
| The Permeation Pathway of **Neurotransmitter**-Gated **Ion Channels** | H. A. Lester | 21:267–92 |

**NMR**
| | | |
|---|---|---|
| **NMR** Studies of **Metabolism** | S. Cerdan, J. Seelig | 19:43–67 |
| Uniform and **Selective Deuteration** in **Two-Dimensional NMR** of Proteins | D. M. LeMaster | 19:243–66 |
| **Two-, Three-, and Four-Dimensional NMR** Methods for Obtaining Larger and More Precise **Three-Dimensional Structures** of **Proteins** in Solution | G. M. Clore, A. M. Gronenborn | 20:29–63 |
| **Solid-State NMR** Approaches for Studying **Membrane Protein** Structure | S. O. Smith, O. B. Peersen | 21:25–47 |
| **NMR Structure Determination** in Solution: A Critique and Comparison with X-Ray Crystallography | G. Wagner, S. G. Hyberts, T. F. Havel | 21:167–98 |
| **Protein Folding** Studied Using **Hydrogen-Exchange** Labeling and **Two-Dimensional NMR** | S. W. Englander, L. Mayne | 21:243–65 |

## CHAPTER TITLES 519

What Does **Electron Cryomicroscopy**
Provide that **X-Ray Crystallography** and
**NMR Spectroscopy** Cannot?     W. Chiu     22:233–55

**Nuclear Overhauser Effect**
The Two-Dimensional Transferred **Nuclear
Overhauser Effect**: Theory and Practice     A. P. Campbell, B. D. Sykes     22:99–122

**Oligomerization**
Intramembrane **Helix-Helix** Association in
**Oligomerization** and Transmembrane
**Signaling**     B. J. Bormann, D. M. Engelman     21:223–42

**Optical**
**Optical** Method     B. Chance     20:1–28

**Packing**
**Hydrogen Bonding, Hydrophobicity,
Packing**, and **Protein Folding**     G. D. Rose, R. Wolfenden     22:381–415

**Pattern Recognition**
Artificial **Neural Networks** for **Pattern
Recognition** in Biochemical Sequences     S. R. Presnell, F. E. Cohen     22:283–98

**Peptides**
Defining Solution **Conformations** of Small
Linear **Peptides**     H. J. Dyson, P. E. Wright     20:519–38
The Mechanism of $\alpha$-**Helix** Formation by
**Peptides**     J. M. Scholtz, R. L. Baldwin     21:95–118

**Phosphorylation**
The Effects of **Phosphorylation** on the
Structure and Function of **Proteins**     L. N. Johnson, D. Barford     22:199–232

**Photoacoustic Calorimetry**
Time-Resolved **Photoacoustic Calorimetry**: a
Study of **Myoglobin and Rhodopsin**     K. S. Peters, T. Watson, K. Marr     20:343–62

**Photosynthetic**
Mechanism of **Photosynthetic Water**
Oxidation     G. W. Brudvig, W. F. Beck, J. C. de Paula     18:25–46

Mechanisms of Long-Distance **Electron
Transfer** in **Proteins**: Lessons from
**Photosynthetic Reaction Centers**     S. G. Boxer     19:267–99
Spectroscopic Analysis of Genetically
Modified **Photosynthetic Reaction Centers**     W. J. Coleman, D. C. Youvan     19:333–67
High-Resolution Structures of **Photosynthetic
Reaction Centers**     J. Deisenhofer, H. Michel     20:247–66

**Plasma Membranes**
**Glycoprotein** Motility and Dynamic Domains
in Fluid **Plasma Membranes**     M. P. Sheetz     22:417–31

**Polyions**
**Ion Distributions** around **DNA** and Other
Cylindrical **Polyions**: Theoretical
Descriptions and Physical Implications     C. F. Anderson, M. T. Record, Jr.     19:423–65

**Polymer**
**Polymer** Principles in **Protein Structure** and
**Stability**     H. S. Chan, K. A. Dill     20:447–90

## Prolyl Isomerase
**Prolyl Isomerase**: Enzymatic Catalysis of
Slow **Protein-Folding** Reactions    F. X. Schmid    22:123–43

## Protein
**Thermodynamic** Problems of **Protein** Structure    P. L. Privalov    18:47–69
**Calorimetrically** Determined **Dynamics** of Complex **Unfolding** Transitions in **Proteins**    E. Freire, W. W. van Osdol, O. L. Mayorga, J. M. Sanchez-Ruiz    19:159–88
Mechanisms of Long-Distance **Electron Transfer** in **Proteins**: Lessons from **Photosynthetic Reaction Centers**    S. G. Boxer    19:267–99
Two-, Three-, and Four-Dimensional **NMR** Methods for Obtaining Larger and More Precise **Three-Dimensional Structures** of **Proteins** in Solution    G. M. Clore, A. M. Gronenborn    20:29–63
Statistical Methods and Insights for **Protein** and **DNA** Sequences    S. Karlin, P. Bucher, V. Brendel, S. F. Altschul    20:175–203
Beyond Gene Sequencing: Analysis of **Protein** Structure with **Mass Spectrometry**    C. Fenselau    20:205–20
**Polymer** Principles in **Protein Structure** and **Stability**    H. S. Chan, K. A. Dill    20:447–90
**Water-Protein Interactions**: Theory and Experiment    M. M. Teeter    20:577–600
The Effects of **Phosphorylation** on the Structure and Function of **Proteins**    L. N. Johnson, D. Barford    22:199–232
The Design of **Metal-Binding Sites** in **Proteins**    L. Regan    22:257–81

## Protein Dynamics
Realistic Simulations of **Native-Protein Dynamics** in Solution and Beyond    V. Daggett, M. Levitt    22:353–80

## Protein Folding
**Protein Folding** Studied Using **Hydrogen-Exchange** Labeling and **Two-Dimensional NMR**    S. W. Englander, L. Mayne    21:243–65
**Protein Folding** in the Cell: The Role of Molecular **Chaperones** Hsp70 and Hsp60    F. U. Hartl, J. Martin, W. Neupert    21:293–322
**Prolyl Isomerase**: Enzymatic Catalysis of Slow **Protein-Folding** Reactions    F. X. Schmid    22:123–43
**Hydrogen Bonding, Hydrophobicity, Packing,** and **Protein Folding**    G. D. Rose, R. Wolfenden    22:381–415

## Protein-DNA Complexes
Physical Studies of **Protein-DNA Complexes** by Footprinting    T. D. Tullius    18:213–37

## Protein Kinase C
Interaction of Proteins with **Lipid Headgroups**: Lessons from **Protein Kinase C**    A. C. Newton    22:1–25

## Protein Stability
The Control of **Protein Stability** and Association by Weak Interactions with **Water**: How Do Solvents Affect These Processes?    S. N. Timasheff    22:67–97

## Proton Pump
From Femtoseconds to Biology: Mechanism of **Bacteriorhodopsin's** Light-Driven **Proton Pump** — R. A. Mathies, S. W. Lin, J. B. Ames, W. T. Pollard — 20:491–518

## Proton-Translocating ATP
The **Proton-Translocating ATPase** of *Escherichia coli* — A. E. Senior — 19:7–41

## Receptor
Analysis of **Receptor**-Mediated Cell Phenomena: **Adhesion** and Migration — D. A. Lauffenburger — 20:387–414

## Recombination
Biochemistry of Genetic **Recombination**: **Energetics** and Mechanisms of **DNA** Strand Exchange — S. C. Kowalczykowski — 20:539–75

## Reconstitution
Solubilization and Functional **Reconstitution** of **Biomembrane Components** — J. R. Silvius — 21:323–48

## Red Blood Cell Anion Transport Protein
Structure and Function of the **Red Blood Cell Anion Transport Protein** — M. L. Jennings — 18:397–430

## Retroviral Proteases
Structure and Function of **Retroviral Proteases** — P. M. D. Fitzgerald, J. P. Springer — 20:299–320

## Rhodopsin
Time-Resolved **Photoacoustic Calorimetry**: a Study of **Myoglobin and Rhodopsin** — K. S. Peters, T. Watson, K. Marr — 20:343–62

## Ribosomes
Approaching Atomic Resolution in Crystallography of **Ribosomes** — A. Yonath — 21:77–93

## Robotics
Expanding Roles of **Computers** and **Robotics** in Biological Macromolecular Research — A. Wada, S. Kidokoro, S. Endo — 18:1–24

## Rubisco
**Rubisco**: Structure and Mechanism — G. Schneider, Y. Lindqvist, C.-I. Branden — 21:119–43

## Scanning Microscopy
Biological Applications of **Scanning Sensor Microscopes** — A. Engel — 20:79–108

## Selective Deuteration
Uniform and **Selective Deuteration** in **Two-Dimensional NMR** of Proteins — D. M. LeMaster — 19:243–66

## β-Sheet
The **β-Sheet** to **Coil** Transition — W. L. Mattice — 18:93–111

## Signal Transduction
Bacterial **Chemotaxis** and the Molecular Logic of Intracellular **Signal Transduction** Networks    J. B. Stock, G. S. Lukat, A. M. Stock    20:109–36

## Signaling
Intramembrane **Helix-Helix** Association in **Oligomerization** and Transmembrane **Signaling**    B. J. Bormann, D. M. Engelman    21:223–42

## Thermodynamic
**Thermodynamic** Problems of **Protein** Structure    P. L. Privalov    18:47–69

## Three-Dimensional Structures
Two-, Three-, and Four-Dimensional NMR Methods for Obtaining Larger and More Precise **Three-Dimensional Structures** of **Proteins** in Solution    G. M. Clore, A. M. Gronenborn    20:29–63

## Time-Resolved
**Time-Resolved** Macromolecular **Crystallography**    K. Moffat    18:309–32

Fast **Crystallography** and **Time-Resolved** Structures    J. Hajdu, I. Andersson    22:467–98

## Transcription
The Single-Nucleotide Addition Cycle in **Transcription:** A Biophysical and Biochemical Perspective    D. A. Erie, T. D. Yager, P. H. von Hippel    21:379–415

## Ultra-Low Temperature
**Crystallography** of Biological Macromolecules at **Ultra- Low Temperature**    H. Hope    19:107–26

## Ultrasonic Velocimetry
**Ultrasonic Velocimetry** of Biological Compounds    A. P. Sarvazyan    20:321–42

## Unfolding
**Calorimetrically** Determined **Dynamics** of Complex **Unfolding** Transitions in **Proteins**    E. Freire, W. W. van Osdol, O. L. Mayorga, J. M. Sanchez-Ruiz    19:159–88

## Vaccine Design
Molecular Structure and **Vaccine Design**    S. Vajda, R. Kataoka, C. DeLisi, H. Margalit, J. A. Berzofsky, J. L. Cornette    19:69–82

## Voltage-Gated Ion Channels
Structure-Function Studies of **Voltage-Gated Ion Channels**    W. Stuhmer    20:65–78

## Water
Mechanism of **Photosynthetic Water** Oxidation    G. W. Brudvig, W. F. Beck, J. C. de Paula    18:25–46

**Water-Protein Interactions**: Theory and
    Experiment      M. M. Teeter      20:577–600
The Control of **Protein Stability** and
    Association by Weak Interactions with
    **Water**: How Do Solvents Affect These
    Processes?      S. N. Timasheff      22:67–97

**X-Ray Absorption Spectroscopy**
    **X-Ray Absorption Spectroscopic**
    Investigations of **Cytochrome** $c$ **Oxidase**
    Structure and Function      R. A. Scott      18:137–58

**X-Ray Diffraction**
    The Study of **Lipid Phase Transition**
    Kinetics by Time-Resolved **X-Ray**
    **Diffraction**      M. Caffrey      18:159–86

**Zinc Finger**
    **Zinc Finger Domains**: Hypotheses and
    Current Knowledge      J. M. Berg      19:405–21